Design of Digital Computers
An Introduction

By

Hans W. Gschwind
Chief, Engineering Development Division
Computation Directorate
Air Force Missile Development Center

Adjunct Professor of Electrical Engineering
University of New Mexico

Sixth Printing

With 447 Figures

Springer-Verlag
Wien GmbH 1967

© 1967 by Springer-Verlag Wien

Originally published by Springer-Verlag/Wien in 1967

Softcover reprint of the hardcover 1st edition 1971

Sixth Printing, March 1971

Library of Congress Catalog Card Number 66-28998

Title No. 9169

ISBN 978-3-662-37358-3   ISBN 978-3-662-38101-4 (eBook)
DOI 10.1007/978-3-662-38101-4

Preface

This book is intended as an introductory text concerned with the *design* of digital computers; computer programming and operation are mentioned only when they have a direct bearing on the equipment itself.

The need for such an engineering text has been felt by the author while teaching computer courses for electronic engineers: Adequate texts on the programming of digital computers exist but, as far as the engineering is concerned, presently available treatments are either not in the form of textbooks or are concerned only with specific individual aspects.

The organization of this book is such that a well rounded concept of digital computer systems is gradually constructed from building blocks in an engineering sense. Even though this approach may be disadvantageous to the casual reader who would like to begin with the overall concept of the computer and then, perhaps, select some topics of interest, the arrangement is preferable in a didactic respect. Of course, nothing prohibits the reader from glancing through chapters in any sequence he likes.

The material contained within a chapter, generally increases in complexity beyond the requirements of a one-semester introductory course. On the other hand, an attempt has been made to provide a ,,fresh start" with the beginning of each chapter. Thus, with the proper selection of material, the book may serve as text for a one-, two- or three-semester curriculum. Especially for longer curricula, it is recommended that the classroom work be supplemented by electronic lab periods and by the programming of an available machine.

Several subsequent versions of the manuscript have been used in the extension program of the University of New Mexico conducted at the Air Force Missile Development Center. The author is indebted to numerous students pointing out errors and inadequacies, and to his colleagues at the Air Force Missile Development Center for many helpful suggestions. The author is particularly grateful to Mr. JAMES L. KRONE for his detailed suggestions on the final version of the manuscript, to Mr. FRANK A. KEIPERT for his comments on the galley proofs, and to Mrs. NORMA J. KENEALLY of the Holloman Graduate Center for the repeated typing of the manuscript.

Bowie, Maryland, July 1965   HANS W. GSCHWIND

Table of Contents

Page

1. **Introduction** 1
   Bibliography 3

2. **Number Systems and Number Representations** 5
   2.1. Counting in Unconventional Number Systems 5
   2.2. Arithmetic Operations in Unconventional Number Systems ... 6
   2.3. Conversions 8
       2.3.1. Conversions of Integers 9
       2.3.2. Conversions of Fractions 12
   2.4. Number Representations 14
       2.4.1. Binary Coded Decimal Numbers 14
       2.4.2. The Inverted Binary Code 16
       2.4.3. Other Representations 18
   2.5. The Residue Number System 19
   Bibliography 22

3. **Boolean Algebra** 23
   3.1. Binary Variables 24
   3.2. Functions of One Variable 24
   3.3. Functions of Two Variables 26
       3.3.1. The Logic Product 27
       3.3.2. The Logic Sum 27
       3.3.3. Combinations of Logical AND's, OR's, NOT's 28
       3.3.4. Theorems Concerning Functions of One or Two Variables 29
   3.4. Functions of Three or More Variables...................... 31
       3.4.1. Diagrammatic Representations 31
       3.4.2. Theorems Involving Three or More Variables 33
       3.4.3. The Standard Sum 35
       3.4.4. The Standard Product 36
   3.5. Minimization 37
       3.5.1. Minimization by Algebraic Operations 38
       3.5.2. Minimization by Graphical Methods 39
       3.5.3. Special Cases...................................... 41
   Bibliography 43

4. **Logic Circuits** 45
   4.1. Diode Circuits 46
       4.1.1. Positive OR-Circuits 46
       4.1.2. Positive AND-Circuits.............................. 49

  Page

      4.1.3. Combinations of Positive OR- and AND-Circuits 50
      4.1.4. Diode Matrices 53
      4.1.5. Negative AND- and OR-Circuits 55
   4.2. Transistor Circuits 56
      4.2.1. The Inverter or NOT-Circuit 56
      4.2.2. Combinations of NOT-Circuits with AND- and OR-Circuits 59
      4.2.3. The Transistor Gate 62
      4.2.4. The Implementation of Logic Functions by Transistors
             Exclusively 63
      4.2.5. Pulse-Type Switching Circuits 66
      4.2.6. The Emitter Follower 68
      4.2.7. Circuits with *npn* Transistors 69
   4.3. Vacuum Tube Circuits 70
      4.3.1. Triode Circuits 70
      4.3.2. Pentode Circuits 73
   4.4. Circuits with Inverted Levels 75

   Bibliography 76

5. Storage Elements 78
   5.1. Static Storage Elements 80
      5.1.1. The Static Flip-Flop 80
      5.1.2. Magnetic Cores 89
      5.1.3. Magnetic Recording 92
   5.2. Dynamic Storage Elements 96
      5.2.1. One-Shots 96
      5.2.2. Delay Lines 99
      5.2.3. Dynamic Flip-Flops............................... 101

   Bibliography 103

6. Computer Circuits... 104
   6.1. Registers 104
      6.1.1. Flip-Flop Registers 104
      6.1.2. Recirculating Registers 107
      6.1.3. Shift Registers 109
   6.2. Counters 115
      6.2.1. Binary Counters 115
      6.2.2. Counters for Other Number Systems 123
      6.2.3. Counters with Special Features 129
   6.3. Adders 134
      6.3.1. Binary Adders 134
      6.3.2. Decimal Adders 142
      6.3.3. Subtracters 155

   Bibliography 159

7. The Basic Organization of Digital Computers 160
   7.1. Design Philosophy...................................... 160
   7.2. The Basic Internal Functions of an Automatic Digital Computer 164
   7.3. The Layout of Early Computers.......................... 166

Page

7.4. The Concept of the Stored Program Computer 170
7.5. A Model Layout of a Digital Computer 174

Bibliography 178

8. **The Functional Units of a Digital Computer** 179

8.1. The Arithmetic Unit 179
    8.1.1. Addition and Subtraction 179
    8.1.2. Multiplication 200
    8.1.3. Division 220
    8.1.4. Extraction of the Square Root 235
    8.1.5. Logic Operations 243
    8.1.6. Floating-Point Arithmetic Operations 252
    8.1.7. The Layout of the Main Arithmetic Registers 264

8.2. The Control Unit 273
    8.2.1. The Sequencing of Operations 274
    8.2.2. Function Translation 282
    8.2.3. Addressing 285
    8.2.4. The Instruction Format and Repertoire 295
    8.2.5. The Registers of the Control Unit 308

8.3. The Memory 310
    8.3.1. Core Memories with Linear Selection 312
    8.3.2. Core Memories with Coincident Selection 315
    8.3.3. Delay Line Memories 320
    8.3.4. Drum or Disk Memories 321
    8.3.5. Associative Memories 326

8.4. The Input/Output Unit 331
    8.4.1. A Model Input/Output Unit 331
    8.4.2. The Monitoring of External Operations 335
    8.4.3. Variations in the Layout of Input/Output Units 341

8.5. Communications in Digital Computer Systems 347

Bibliography 365

9. **Unorthodox Concepts** 367

9.1. Polymorphic Computer Systems 367

9.2. Arithmetic Units with Problem-Dependent Interconnections ... 369

9.3. Hybrid Computations 370
    9.3.1. Concurrent Operations of Analog and Digital Computers 370
    9.3.2. Combinations of Analog and Digital Representations.... 373
    9.3.3. Computers Using Hybrid Elements 374

9.4. Digital Differential Analyzers 375

9.5. Micro-Programmed Computers 379

9.6. Machines with Cellular Organization 387
    9.6.1. Cellular Logic 388
    9.6.2. Highly Parallel Machines 392

9.7. List Processors and List Processing Features 400

9.8. Associative Computers 411

Page

9.9. Learning Structures, Adaptive and Self-Organizing Systems ... 416
   9.9.1. A Sample Perceptron 417
   9.9.2. Other Learning Structures.......................... 422

Bibliography 428

10. Miscellaneous Engineering and Design Considerations 431
  10.1. Capability versus Cost 432
  10.2. Speed versus Cost 434
  10.3. Reliability versus Cost 438
  10.4. Error Detection and Correction Techniques 441
    10.4.1. Error Detection 441
    10.4.2. Error Correction 448
  10.5. Computer Evaluations 451
  10.6. Engineering Check List 455

Bibliography 457

11. Unusual Computer Components................................... 459
  11.1. Magnetic Components.................................... 459
    11.1.1. Magnetic Cores as Logic Elements 459
    11.1.2. The Transfluxor 463
    11.1.3. The Parametron 469
    11.1.4. Magnetic Film Memory Devices 474
    11.1.5. Non-Destructive Read for Magnetic Devices 483
    11.1.6. Miscellaneous Magnetic Devices or Techniques....... 494
  11.2. Superconductive Components 498
    11.2.1. The Cryotron 499
    11.2.2. Superconductive Storage Cells 504
  11.3. The Tunnel Diode....................................... 507
  11.4. Miscellaneous Devices 514

Bibliography 515

Index.. 521

1. Introduction

"Computers" have attracted general interest only rather recently although computing devices have been known for a long time. The Antikythera mechanism, supposedly used by ancient Greeks to determine the motions of the stars and planets [1], the astrolabes of the middle ages [2], and Pascal's calculator [3], are only a few examples of early computational devices. However, the present usage of the term "computer" includes neither those relatively primitive (though certainly effective) aids for computation, nor later developments like the slide rule, the planimeter, or the desk calculator. What we mean nowadays by a computer is a machine which performs a computation automatically and without human intervention, once it is set up for a specific problem. If we want to emphasize this distinction, we speak of automatic computers as opposed to calculators or computing devices.

The present use of the term "computer" has a second connotation. It usually refers to an electronic device, although there have been (and still are) automatic computers which operate mechanically or electromechanically. There are mainly two reasons for this close association between electronics and modern computers: no known principle other than electronics allows a machine to attain the speeds of which modern computers are capable; and no other principle permits a design of comparable convenience.

Even though practically all modern computers operate electronically, there are several distinct types of machines. Here, we do not mean differences concerning circuit elements such as tubes or transistors, but basic differences in the design philosophy. The most characteristic distinction is probably the analog or digital nature of a computer.

An analog computer represents the values which it uses in its calculation by physical quantities. The slide rule, which is an analog device (although, of course, no computer according to our definition), uses the physical quantity "length" to represent computational values. An electronic analog computer uses voltages as convenient analog quantities (higher voltages for larger values, lower voltages for smaller values, etc.). In contrast, a digital computer employs numbers, as we usually do in paper and pencil calculations. Numerical values are represented by the presence or absence of electric potentials or pulses on certain lines. The magnitude of these potentials or pulses is of no particular significance, as long as it is adequate for the fault-free operation of the computer. Of course, both basic representations have their merits and their disadvantages.

Analog Computers are of relatively uncomplicated design. It is quite feasible to build small and inexpensive machines. Moreover, problems put on an analog computer usually are simulated by an electronic network of resistors, capacitors, amplifiers, etc., which has an intelligible relationship to the problem to be solved [4]. On the other hand, the electronic model usually comes only to within about 1% or .1% of a true representation of the actual problem. Even though this inherent error is of no importance for many problems, there are calculations in which it cannot be tolerated. Furthermore, there are several types of problems which, due to the nature of its design, cannot be solved by an analog computer[1].

Digital Computers are relatively complex machines. In many instances, it is difficult for an expert to recognize from the program alone even the type of problem to be solved. However, digital computers have the great advantage that they can solve practically all problems which can be stated in mathematical language. Their accuracy is not limited by the operating principle, but only by practical considerations. Furthermore, they can be employed to translate their own internal language into very concise and intelligible statements or, conversely, interpret instructions given in almost everyday language for their own use.

In addition to analog and digital computers, there are a few computer types which attempt to combine the advantages of both principles. The *Digital Differential Analyzer*, similar to an analog computer, represents problems by a network of units (the integrators) but, like a digital computer, uses numbers to represent computational values [5]. In *Hybrid Computers*, analog computations are combined with digital computations [6].

Of these four types of computers, only the digital computer will be considered here[2]. The topics may be roughly divided into four categories. Chapters 2 and 3 contain fundamental information on number systems and Boolean algebra. The detail included provides more than a prerequisite for the understanding of the then following material. Chapters 3, 4, 5, and 6 are concerned with individual components and circuits which constitute the computer hardware. Chapters 7, 8, and 9 are devoted to the organization of

---

[1] For instance, inventory control, bookkeeping, playing of mathematical games, or, perhaps, finding all prime numbers between 0 and 10^6.

[2] It is, however, worthwhile to note that some of the indicated basic design techniques may be applied to any digital equipment, including that of digital differential analyzers and hybrid computers, and that the organization of these latter two types of computers is at least sketched in order to provide a reference against which the organization of digital computers may be viewed.

computers. Chapters 10 and 11 contain miscellaneous topics not included elsewhere.

Problem 1 (Voluntary): What do you think is the reason that some computer centers use both, analog and digital computers ?

Problem 2 (Voluntary): What type of computer should in your opinion be acquired:
a) by a banking institution,
b) by a manufacturer of servo mechanisms for airplane control ?
Try to justify your choice.

Problem 3 (Voluntary): You are provided a desk calculator and an operator for it. The operator can execute only simple instructions such as "add the value in column 7 to the value in column 5" or "copy down result in column 9". Try to devise a set of instructions and a worksheet for the operator so that he can calculate the value of sin x for any given x by the approximation:

$$\sin x \approx x - \frac{x^3}{6} + \frac{x^5}{120}$$

References

1. DEREK, and DE SOLLA PRICE: An Ancient Greek Computer, Scientific American, vol. 200, No. 6, pp. 60–67. June 1959.
2. SCHROEDER W.: Practical Astronomy. London: T. Werner Laurie Ltd. 1956.
3. WILLERS A.: Mathematische Instrumente. Munich: R. Oldenbourg. 1943.
4. KORN, and KORN: Electronic Analog Computers. New York: McGraw-Hill. 1952.
   JOHNSON C. L.: Analog Computer Techniques. New York: McGraw-Hill. 1956.
5. FORBES G. F.: Digital Differential Analyzers. Private Print: 13745 Eldridge Ave, Sylmar, California. 1957.
   HANDEL P. VON: Electronic Computers. Vienna: Springer. Englewood Cliffs: Prentice Hall. 1961.
6. BURNS M. C.: High Speed Hybrid Computer, National Symposium on Telemetering, San Francisco. 1959.
   BIRKEL G., JR.: Mathematical Approach to Hybrid Computing, National Symposium on Telemetering, San Francisco. 1959.
   TRUITT T. D.: Hybrid Computation, AFIPS Conference Proceedings, vol. 25, pp. 249–269, Spring Joint Computer Conference. 1964.

Selected Bibliography

RICHARDS R. K.: Arithmetic Operations in Digital Computers. New York: D. Van Nostrand. 1955.
RICHARDS R. K.: Digital Computer Components and Circuits. New York: D. Van Nostrand. 1957.

GRABBE, RAMO, and WOOLDRIDGE: Handbook of Automation, Computation and Control. New York: John Wiley. 1959.

LEDLEY R. S.: Digital Computer Control Engineering. New York: McGraw-Hill. 1960.

HUSKEY, and KORN: Computer Handbook. New York: McGraw-Hill. 1962.

A history of early digital computers is contained in:

Staff of the Computation Laboratory Harvard, A Manual of Operation for the Automatic Sequence Controlled Calculator. Cambridge. 1946.

2. Number Systems and Number Representations

The familiar decimal system is by no means the only possible number system. Considered impartially, it merely constitutes one among possible and practical systems which became propagated, probably for the sole reason that human beings happen to have ten fingers. The Mayas used the vigesimal number system (based upon 20, i.e. fingers and toes) [1] and even in our days, there are some endeavors to introduce the duodecimal system (based on 12) for general use [2]. Since computers are not bound by tradition and since the decimal system has no unique merits, the designer of a computer is free to select that number system which suits his purpose best.

2.1. Counting in Unconventional Number Systems

Before we set out on unfamiliar grounds, let us shortly review the decimal number system. Any decimal number is made up of the ten symbols: 0, 1, 2, ... 9. When we count, we use these symbols consecutively: 0, 1, 2, ... 9. Then, if we have exhausted all available symbols, we place the symbol 1 in a new position and repeat the cycle in the old position: 10, 11, ... 19. If we run out of symbols again, we increase the digit in the second position and repeat the cycle in the first position: 20, 21, ... 29, etc. If we have no more symbols for the second position, we create a third position: 100, 101, and so on.

Counting in a different number system follows the same procedure. Let us count in the ternary system (base 3). We have only three symbols: 0, 1, and 2. We proceed as follows: 0, 1, 2. Then having no other symbols for this position, we continue: 10, 11, 12. Running out of symbols again, we write: 20, 21, 22. Having no more symbols for the second position, we create a third position: 100, 101, 102, 110, 111, 112, 120, 121, 122, 200, 201, and so on.

Problem 1: Count in the binary system (base 2) and in the duodecimal system (base 12) up to the equivalent of the decimal number 25. Use the letters T and E as symbols for ten and eleven in the duodecimal system.

Problem 2: Try to state some advantages and disadvantages which the duodecimal system might have over the decimal system for computers and for everyday calculations.

2.2. Arithmetic Operations in Unconventional Number Systems

We can perform calculations in other number systems equally well as in the decimal system, once we are familiar with a few simple rules. For arithmetic operations in the decimal system, we (mentally) use the addition and multiplication tables reproduced below.

Table 2.1. *Decimal Addition Table*

+	0	1	2	3	4	5	6	7	8	9
0	0	1	2	3	4	5	6	7	8	9
1	1	2	3	4	5	6	7	8	9	10
2	2	3	4	5	6	7	8	9	10	11
3	3	4	5	6	7	8	9	10	11	12
4	4	5	6	7	8	9	10	11	12	13
5	5	6	7	8	9	10	11	12	13	14
6	6	7	8	9	10	11	12	13	14	15
7	7	8	9	10	11	12	13	14	15	16
8	8	9	10	11	12	13	14	15	16	17
9	9	10	11	12	13	14	15	16	17	18

Table 2.2. *Decimal Multiplication Table*

×	0	1	2	3	4	5	6	7	8	9
0	0	0	0	0	0	0	0	0	0	0
1	0	1	2	3	4	5	6	7	8	9
2	0	2	4	6	8	10	12	14	16	18
3	0	3	6	9	12	15	18	21	24	27
4	0	4	8	12	16	20	24	28	32	36
5	0	5	10	15	20	25	30	35	40	45
6	0	6	12	18	24	30	36	42	48	54
7	0	7	14	21	28	35	42	49	56	63
8	0	8	16	24	32	40	48	56	64	72
9	0	9	18	27	36	45	54	63	72	81

Table 2.3. *Ternary Addition and Multiplication Tables*

+	0	1	2
0	0	1	2
1	1	2	10
2	2	10	11

×	0	1	2
0	0	0	0
1	0	1	2
2	0	2	11

Let us construct corresponding tables for, let us say, the ternary system. Having only three symbols, we will obtain nine entries. Instead of the decimal symbols for three and four, we will show their ternary equivalents 10 and 11. (See Table 2.3.)

We can use these tables for calculations in the ternary system in the same manner as we use the decimal tables for computations in the decimal system. Suppose we want to add the two ternary numbers 1021220 and 210121. The computation is given below:

```
        1   1       1   1
        1   0   2   1   2   2   0
    +           2   1   0   1   2   1
    ─────────────────────────────────
        2   0   0   2   1   1   1
```

The carries to be brought forward are indicated in the top line. Similarly, for the product of the two ternary numbers 1120 and 12, we obtain:

```
                    1   1   2   0
        ×                       1   2
    ─────────────────────────────────
                1   0   0   1   0
                1   1   2   0
    ─────────────────────────────────
            2   1   2   1   0
```

The simplest addition and multiplication tables are obtained for the binary system:

Table 2.4. *Binary Addition and Multiplication Tables*

+	0	1		×	0	1
0	0	1		0	0	0
1	1	10		1	0	1

The simplicity of these tables is perhaps one of the reasons why the binary number system is so attractive to computer designers.

From now on, we will indicate the base of a number by an appropriate index if the base is not apparent from the context. For instance, a number like 453_8 shall indicate an octal number (base 8).

Problem 3: Construct the addition and multiplication tables for the quinary (base 5) and octal (base 8) number systems. Be sure to make all entries in the appropriate number system.

Problem 4: Construct the addition tables for the duodecimal (base 12) and the hexadecimal (base 16) systems. Use the letters T and E as symbols for ten and eleven in the duodecimal system and the letters A, B, C, D, E, F as symbols for numbers from ten to fifteen for the hexadecimal system.

Problem 5: Perform the following arithmetic operations:
a) $10111_2 + 1101_2$
b) $11010_2 - 10110_2$
c) $101101_2 \times 1011_2$
d) $11011_2 \div 11_2$
e) $2431_5 + 132_5$
f) $324_5 \times 14_5$
g) $6327_8 + 4530_8$
h) $124_8 - 76_8$
i) $1256_8 \times 27_8$

Check your computations by converting these problems and their results to the decimal system after you have studied paragraph 2.3.

2.3. Conversions

As long as there is more than one number system in use, it will be necessary to convert numbers from one system to another. Such a conversion is required if we want to insert decimal numbers into a binary computer, or vice versa, if we want to interpret results computed by such a machine. If we are to do this conversion ourselves, we prefer to perform the required arithmetic in the familiar decimal system. If the computer performs the conversion, an algorithm in its number system is preferable.

Each position in a decimal number like 2536 has a certain weight associated with it. The digit 2 in the above number represents, for example, two thousand or its position has the weight 10^3. Writing the number 2536 in longhand, we have:

$$2536_{10} = 2 \times 10^3 + 5 \times 10^2 + 3 \times 10^1 + 6 \times 10^0$$

An arbitrary decimal number has the form:

$$N_{10} = \cdots d_3 \times 10^3 + d_2 \times 10^2 + d_1 \times 10^1 + d_0 \times 10^0 + d_{-1} \times 10^{-1} + \cdots$$

$$(2.1)$$

A number written in a system other than decimal has the same general structure; only the weights will be different. For an arbitrary number written in the octal system, we obtain for instance:

$$N_8 = \cdots C_3 \times 8^3 + C_2 \times 8^2 + C_1 \times 8^1 + C_0 \times 8^0 + C_{-1} \times 8^{-1} + C_{-2} \times 8^{-2} + \cdots$$

$$(2.2)$$

The coefficients C_n are octal integers ranging from 0_8 to 7_8.

Conversion formulae derive the coefficients in one number system (e.g. Equation 2.1) from the coefficients of another number system (e.g. Equation 2.2). Since the procedures are different for different conversions, we will consider one case at a time.

2.3.1. Conversion of Integers

Let us start with a specific example. Suppose we want to convert the number 3964_{10} to the octal system. This number is an integer in the decimal system and consequently also an integer in the octal system. (We can derive it by counting "units".) According to Equation (2.2) we can write in general terms:

$$3964_{10} = \cdots C_3 \times 8^3 + C_2 \times 8^2 + C_1 \times 8^1 + C_0 \times 8^0 \qquad (2.3)$$

All C's are positive integers smaller than 8, but not yet determined.

Suppose we split the right-hand side of Equation (2.3) into two parts:

$$3964_{10} = (\cdots C_3 \times 8^2 + C_2 \times 8^1 + C_1) \times 8 + C_0 \qquad (2.4)$$

The first term, apparently, is that part of our original number which is dividable by 8 (the integral part of the quotient $3964_{10} \div 8$), whereas the term C_0 is that part of the original number which is not dividable by 8 (the remainder of the quotient $3964_{10} \div 8$).

If we divide 3964_{10} by 8, we obtain:

$$3964_{10} \div 8 = 495 + 4/8$$

We can therefore write:

$$3964_{10} = 495 \times 8 + 4 \qquad (2.5)$$

Comparing (2.4) and (2.5), we find $C_0 = 4$, or we can write:

$$3964_{10} = 495_{10} \times 8^1 + 4_8 \times 8^0 \qquad (2.6)$$

Again, we can split the decimal coefficient 495 in an integral multiple of 8 and a remainder. The new integral part may then be split again. If we continue this procedure, we are able to find all the octal coefficients of the number. The consecutive steps are given below:

$$
\begin{aligned}
3964_{10} &= 3964_{10} \times 8^0 \\
&= 495_{10} \times 8^1 + 4 \times 8^0 \\
&= 61_{10} \times 8^2 + 7 \times 8^1 + 4 \times 8^0 \\
&= 7 \times 8^3 + 5 \times 8^2 + 7 \times 8^1 + 4 \times 8^0 \\
&= 7574_8
\end{aligned}
\qquad
\begin{aligned}
3964 \div 8 &= 495 + 4/8 \\
495 \div 8 &= 61 + 7/8 \\
61 \div 8 &= 7 + 5/8
\end{aligned}
$$

If we write down only the essential items of our computation, we derive a very simple conversion scheme:

decimal	0	7	61	495	3964
octal		7	5	7	4

Starting on the right-hand side with the original number, we divide by 8, write the integral part of the quotient above, and the remainder below the line. Then we divide the integral part again by 8, noting the new integral part and the remainder. We continue in this manner until the number above the line is reduced to zero. The bottom line contains then the desired octal number.

This scheme can be applied to conversions from the decimal number system to any other number system. The divisor used in this repeated division must always be the base of the new number system.

Example: Convert 3964_{10} to base 5.

decimal	0	1	6	31	158	792	3964
quinary	1	1	1	3	2	4	

Problem 6: Convert 3964_{10} to the binary, ternary, and duodecimal system.

If we want to convert numbers *to* the decimal system, we can follow the same procedure except, now, we have to perform the subsequent division in the number system from which we want to convert.

Example: Convert 7574_8 to the decimal system.

octal	0	3_8	47_8	614_8	7574_8
decimal	3_8	11_8	6_8	4_8	

We divide by 12_8 which is equivalent to 10_{10}, the new base we want to convert to. Being not too versed in the use of octal arithmetic, we might have to resort to a scratch pad or an octal desk calculator to derive above figures. Inspecting the results, we find the number 3 (11) 6 4. Here, we have to remember that the octal system in which we performed our arithmetic, contains no symbol for the equivalent of the decimal numbers eight and nine. We interpret, therefore, 11_8 as 9_{10} and the result correctly as 3964_{10}.

Problem 7: Convert the binary number 101101 to base 10.

If we do not like to do arithmetic in unfamiliar number systems, we can devise another method which avoids this inconvenience. We can evaluate a number directly from Equation (2.2).

Example: $7574_8 = 7 \times 8^3 + 5 \times 8^2 + 7 \times 8^1 + 4 \times 8^0$

$$= 7 \times 512 + 5 \times 64 + 7 \times 8 + 4$$
$$= 3584 + 320 + 56 + 4$$
$$= 3964_{10} \tag{2.7}$$

Or, using a somewhat different notation, we obtain:

$$7574_8 = [(7 \times 8 + 5) \times 8 + 7] \times 8 + 4$$

$$\underbrace{\hspace{1cm}56\hspace{1cm}}$$
$$\underbrace{\hspace{2cm}61\hspace{2cm}}$$
$$\underbrace{\hspace{3cm}488\hspace{2cm}}$$
$$\underbrace{\hspace{4cm}495\hspace{2cm}}$$
$$\underbrace{\hspace{5cm}3960\hspace{1cm}}$$
$$\hspace{6cm}3964 \tag{2.8}$$

This computation can again be written in form of a simple scheme which, together with the previous notation, should be self-explanatory:

octal	7	5	7	4
decimal		56	488	3960
		61	495	3964

Problem 8: Convert 1101101_2 to base 10.

Problem 9: We have now seen a method to convert decimal numbers to octal numbers, doing the arithmetic in the decimal system, and methods to convert octal numbers to decimal numbers, doing the arithmetic either in the decimal or in the octal system. Following the just previously outlined procedure, devise a method to convert decimal integers to octal integers, doing the arithmetic in the octal number system.

Problem 10: Use this method to convert 3964_{10} to base 8.

Conversions between number systems become extremely simple when the base of one number system is a power of the base of the other. Let us illustrate this with the conversion of binary to octal numbers. A binary number can be written as:

$$\cdots + a_8 \times 2^8 + a_7 \times 2^7 + a_6 \times 2^6 + a_5 \times 2^5 + a_4 \times 2^4 + a_3 \times 2^3 + a_2 \times 2^2 +$$
$$+ a_1 \times 2^1 + a_0 \times 2^0$$
$$= \cdots + (a_8 \times 2^2 + a_7 \times 2^1 + a_6 \times 2^0) \times 8^2 + (a_5 \times 2^2 + a_4 \times 2^1 + a_3 \times 2^0) \times 8^1 +$$
$$+ (a_2 \times 2^2 + a_1 \times 2^1 + a_0 \times 2^0) \times 8^0 \tag{2.9}$$

The right-hand side of Equation (2.9) has the form of an octal number with its coefficients given in the binary system. The conversion of the coefficients (each an integer which is smaller than 8) is given in the following table:

Table 2.5. Binary and Octal Equivalents

binary	octal		binary	octal
000	0		100	4
001	1		101	5
010	2		110	6
011	3		111	7

If we memorize this table, we are able to convert any binary number to base 8 and vice versa by inspection[1].

Example:

$$010 \quad 110 \quad 101 \quad 001_1$$
$$= 2 \quad\quad 6 \quad\quad 5 \quad\quad 1_8$$

2.3.2. Conversion of Fractions

A fraction has a numerical value smaller than 1. Since "1" has the same meaning in any number system, a fraction will always convert to a fraction[2].

Suppose we want to convert a given decimal fraction to an octal fraction. The result, in general terms, will be:

$$N = C_{-1} \times 8^{-1} + C_{-2} \times 8^{-2} + C_{-3} \times 8^{-3} + \cdots \tag{2.10}$$

If we multiply this number by 8, we obtain:

$$8N = C_{-1} + C_{-2} \times 8^{-1} + C_{-3} \times 8^{-2} + \cdots \tag{2.11}$$

This is the sum of an integer (C_{-1}) and a new fraction ($C_{-2} \times 8^{-1} + C_{-3} \times 8^{-2} + \cdots$)

Multiplying the original (decimal) fraction by 8, we likewise obtain the sum of an integer and a fraction. We, therefore, are able to determine the coefficient C_{-1}. Applying the same method to the new fraction, we can determine C_{-2}; then, using the remaining fraction, we get C_{-3}, and so on.

Example: Convert $.359375_{10}$ to base 8.

decimal	octal
$N_{10} = \mathbf{0}.359375$	$N_8 = 0. \ldots$
$N_{10} \times 8 = \mathbf{2}.875$	$N_8 = 0.2 \ldots$
$.875 \times 8 = \mathbf{7}.000$	$N_8 = 0.27$

$$.359375_{10} = .27_8$$

---

[1] The octal notation is frequently used as "shorthand" for binary numbers in computer programming and operation.

[2] However, a fraction which terminates in one number system may be non-terminating in another. For example: $.1_3 = .3333\ldots_{10}$

Multiplying by the base of the number system we want to convert to, we are able to convert decimal fractions to any other number system.

Problem 11: Convert $.359375_{10}$ to base 2, 5 and 12.

The outlined method can also be adapted for conversions *to* decimal fractions. Here, however, the consecutive multiplications have to be performed in the number system we want to convert from.

Example: Convert $.27_8$ to base 10.

octal	decimal	
0.27_8	$0.\ldots\ldots$	
$.27_8 \times 12_8 = \mathbf{3}.46_8$	$0.3\ldots..$	
$.46_8 \times 12_8 = \mathbf{5}.74_8$	$0.35\ldots.$	
$.74_8 \times 12_8 = \mathbf{11}.30_8$	$0.359\ldots$	$(11_8 = 9_{10})$
$.3_8 \ \times 12_8 = \mathbf{3}.6_8$	$0.3593..$	
$.6_8 \ \times 12_8 = \mathbf{7}.4_8$	$0.35937.$	
$.4_8 \ \times 12_8 = \mathbf{5}.0_8$	0.359375	
$.27_8 = .359375_{10}$		

Problem 12: Convert $.01011_2$ to base 10.

If we prefer arithmetic operations in the number system we are converting to, we can use a straightforward multiplication, similar to the one we used for integers.

Example: Convert $.27_8$ to base 10.

$$
\begin{aligned}
.27_8 &= 2 \times 8^{-1} + 7 \times 8^{-2} \\
&= 2 \times .125 + 7 \times .015625 \\
&= .25 + .109375 \\
&= .359375_{10}
\end{aligned}
\tag{2.12}
$$

This calculation becomes somewhat shorter if we use the following notation:

$$
\begin{aligned}
C_{-1} \times 8^{-1} &+ C_{-2} \times 8^{-2} + C_{-3} \times 8^{-3} + C_{-4} \times 8^{-4} = \\
&= \{[(C_{-4} \div 8 + C_{-3}) \div 8 + C_{-2}] \div 8 + C_{-1}\} \div 8
\end{aligned}
$$

$$
.27_8 = \underbrace{(7 \div 8 + 2) \div 8}
$$

$$
\underbrace{.875}
$$

$$
\underbrace{2.875}
$$

$$
.359375_{10}
\tag{2.13}
$$

We can show this calculation in a scheme similar to the one on page 11:

octal	.		2	7
decimal	.359375		.875	
			2.875	

As for integers, we have now methods to convert fractions from one number system to another performing the arithmetic either in the number system to be converted to, or the number system to be converted from.

Problem 13: Convert $.359375_{10}$ to base 8, doing all arithmetic in base 8.

When the base of one number system is a power of the base of the other number system, conversions of fractions, like those of integers, become very simple. For instance, conversions of fractions from the octal to the binary system and vice versa can be done by inspection.

It is appropriate to conclude this paragraph on number conversions with some general observations: Most conversion methods are based on a repeated shifting of the point (decimal, octal, binary point, etc.) and a subsequent comparison of integral or fractional parts of the result in both, the new and the old number system. The shifting of the point is accomplished by multiplication or division. For instance, a decimal number is "shifted by one octal place" when we multiply or divide it by eight. When the original numbers to be converted consist of both, an integral and a fractional part, they may be either "scaled" so that they have only an integral or only a fractional part, or the integral and the fractional part may be converted separately and the results added.

2.4. Number Representations

2.4.1. Binary Coded Decimal Numbers

Practically all electronic digital computers represent numbers by binary configurations. Binary machines use the number itself. Octal or hexadecimal computers group three, respectively four, binary digits to make up one octal or hexadecimal digit. Octal or hexadecimal numbers are thereby used mainly for input or output purposes while computations are performed in the binary number system. Practically the only computers which perform their computations not in the binary system, are decimal machines. Every digit of a decimal number is represented by a group of binary digits ($=$ bits) in such a machine. We frequently speak of binary coded decimal numbers or of BCD numbers, codes and characters for short. Table 2.6 below lists several commonly employed codes.

Table 2.6. *Common Binary Codes for Decimal Digits*

Decimal Digit	8421 Code	2421 Code	Excess-3 Code	2 Out of 5 Code	Biquinary Code
0	0000	0000	0011	11000	01 00001
1	0001	0001	0100	00011	01 00010
2	0010	0010	0101	00101	01 00100
3	0011	0011	0110	00110	01 01000
4	0100	0100	0111	01001	01 10000
5	0101	0101	1000	01010	10 00001
6	0110	0110	1001	01100	10 00010
7	0111	0111	1010	10001	10 00100
8	1000	1110	1011	10010	10 01000
9	1001	1111	1100	10100	10 10000

The above is a small sample from the theoretically unlimited number of possible codes [3], but each of them has some property which makes the code particularly useful in one respect or another. Weighted codes such as the 8421, 2421 and biquinary code make it relatively easy to determine the numeric value of a given binary configuration. Take the binary configuration 1001 in the 8421 code as an example. The respective weights in the 8421 code are, as the name implies: 8, 4, 2 and 1. The numeric value of the configuration 1001 can, therefore, be expressed as: $1 \times 8 + 0 \times 4 + 0 \times 2 + 1 \times 1 = 8 + 1 = 9$. Similarly, the configuration 1110 in the 2421 code has the numeric value: $1 \times 2 + 1 \times 4 + 1 \times 2 + 0 \times 1 = 8$. The weights in the biquinary code are 50 43210 so that the binary configuration 01 10000 has a numeric value of $1 \times 0 + 1 \times 4 = 4$.

The *8421 code* gives the straight binary equivalent of a decimal digit. Most of the rules for the addition of numbers in the 8421 code follow, therefore, from the simple rules for the addition of binary numbers. However, precautions have to be taken that "illegal" codes such as 1011, which may appear as a result of a straight binary addition, are eliminated.

The *2421 code* makes it simple to derive the 9's complement of a decimal digit, which is advantageous if a subtraction is performed by the addition of the complement[1]. For instance, the 9's complement of 8 is 1. Reversing 0's and 1's in the code for 8, we obtain: 0001, which is the code for 1. Reversing 0's and 1's in the code for 7, we obtain 1000 which, although being an "illegal" code, according to the weights, is a code for the digit 2.

The *excess-3 code* derives its name from the property that its representation of a decimal digit corresponds to a straight binary representation except that all representations are by 3 too large. Like in the 2421 code,

---

[1] See also paragraph 8.1.1.1.

the 9's complement of a decimal digit can be derived by complementing binary zeros and ones. It has the advantage over the 2421 code that only legal codes result from such a complementation. The excess-3 code, like the 2 out of 5 code is a non-weighted code.

The *2 out of 5 code*, as the name implies, uses two binary ones in five available binary positions to represent a decimal digit. This allows a relatively simple detection of illegal codes if they ever should appear due to a malfunction of the computer. All single errors (a binary one instead of a binary zero or vice versa) and a large number of multiple errors can be detected.

The *biquinary code* permits an even simpler detection of errors but also wastes more digital positions.

2.4.2. The Inverted Binary Code

The inverted binary code may be considered as an unusual notation for binary counts. Its structure can be seen in the following table:

Table 2.7. *Inverted versus True Binary Numbers*

Inverted Binary Code	Binary Equivalent
0000	0000
0001	0001
001$\overline{1}$	0010
0010	0011
01$\overline{10}$	0100
0111	0101
0101	0110
0100	0111
$\overline{1100}$	1000
etc.	etc.

Counting in the inverted binary code, we start in the same manner as in the true binary number system: 0, 1. Then forced to introduce a second position, we reverse the count in the first position: 11, 10. Installing the third position, we repeat all previous counts in the first two positions in inverted sequence: 110, 111, 101, 100. Following the same principle, we obtain the first four-digit number 1100 and count up to 1000, the highest four-digit number. Points at which we install a new position and reverse the count are marked by a horizontal line in Table 2.7.

Problem 14: Count up to the equivalent of 16_{10} using the inverted binary code.

The inverted binary code has a very useful property: Counting from one number to the next, only one digit is changed[1]. This property makes the inverted binary code a natural choice for applications where a momentary count has to be read, while a counter is counting (asynchronous read out or reading "on the fly"). If we happen to read while a counter is in transition from one state to the next, we might or might not read the changing digits correctly. Using the inverted binary code, the resulting error is at most one unit whereas using the true binary system (or any other true number system, for that matter), the resulting error might be disastrous, as the following example will show. Reading while a binary counter passes from 0111 to 1000, we might obtain any binary number ranging from 0000 to 1111, numbers which have not the least resemblance to the one we are supposed to read.

Despite the apparent awkwardness of the inverted binary code, conversions between true and inverted binary numbers are not too difficult. The following rules are given without proof:

Conversion from True Binary to Inverted Binary Numbers: Write down the true binary number and underneath the true binary number shifted by one digit. Add both numbers according to the binary addition table, disregarding any carries[2]. Disregarding the digit on the utmost right, the result is the original number in the inverted binary code.

Example:

IGNORE CARRIES ⤳

$$
\begin{array}{r}
1\ 1\ 0 \\
\oplus\ 1\ 1\ 0 \\
\hline
1\ 0\ 1\ (0
\end{array}
\qquad 110_2 = 101_{IB}
$$

Conversion from Inverted to True Binary Numbers: If the inverted binary number has n digits, write down the number n times, each time shifted by 1 digit. Add these n numbers disregarding any carries. Crossing off the last $(n—1)$ digits, the result is the binary number equivalent to the original inverted binary code.

Example:

$$
\begin{array}{r}
1\ 1\ 1 \\
1\ 1\ 1 \\
\oplus\ 1\ 1\ 1 \\
\hline
1\ 0\ 1\ (0\ 1
\end{array}
\qquad 111_{IB} = 101_2
$$

---

[1] Such codes are referred to as *Gray Codes*. The inverted binary code is a Gray code which makes the conversion to the true binary system fairly simple.

[2] This is also called "addition modulo 2", and is frequently denoted by the symbol $\oplus$.

An equivalent conversion can be performed in which the original inverted binary number is written only once: Copy all binary digits as they are, if they have an even number of ones or no ones to the left. Complement all binary digits which have an odd number of ones to the left.

Problem 15: Convert 10111 from inverted binary to true binary.

Problem 16: Convert 10111 from true to inverted binary.

2.4.3. Other Representations

Computers can interpret binary configurations not only as numbers, but in various other ways. For instance, the bit in the most significant position of a number is usually not given any numeric value but is used to signify the sign of the number (e.g. "0" for positive numbers and "1" for negative numbers). Computer instructions (like add or multiply) have various binary codes. Alphanumeric codes are used when a computer has to handle letters in addition to numbers. Table 2.8 shows such an alphanumeric code which is used by many computers for communication with an electric typewriter.

Table 2.8. *Example of a Typewriter Code*

Character	Code	Character	Code	Character	Code	Character	Code
0	37	A	30	K	36	U	34
1	52	B	23	L	11	V	17
2	74	C	16	M	07	W	31
3	70	D	22	N	06	X	27
4	64	E	20	O	03	Y	25
5	62	F	26	P	15	Z	21
6	66	G	13	Q	35	Space	04
7	72	H	05	R	12	Car. Return	45
8	60	I	14	S	24	Shift up	47
9	37	J	32	T	01	Shift down	57

The code for each character is a 6-digit binary number. Table 2.8 shows the equivalent 2-digit octal number. For the communication with other devices like card punches or readers, or high-speed printers, other codes may be used.

Floating-Point Numbers: A computer usually considers the decimal or binary point at a fixed position either to the left, or to the right of a number. We speak then of fixed-point numbers. If numbers of a wide range or of unpredictable magnitude are expected for a computation, a floating-point format for numbers may be used. A floating-point number consists of a coefficient and an exponent[1] (like $.63 \times 10^7$, or its binary equivalent), both of which occupy predetermined positions within a binary configuration:

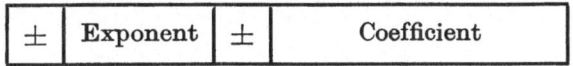

Fig. 2.1. Floating-Point Representation of a Number

Some machines have built-in instructions to operate with floating-point numbers; others require an interpretation program. We shall have more to say about floating-point numbers in paragraph 8.1.6.

The few indicated examples, by no means, exhaust the number of possible number representations or even those which are in actual use [3, 6]. Computer designers do not hesitate to invent their own representations when they see some advantage in doing so. For instance, some computers represent negative numbers by their arithmetic complement[2]. Others may represent decimal digits by a string of pulses, the number of which represents the numerical value. In some cases computational variables are not represented by their numerical value, but rather by their increments[3]. Finally, there is the vast field of error detection and correction codes [4]. Their intent is to represent numbers (or information in general) in such a manner that it becomes possible to detect and correct errors produced by faulty transmission and/or faulty arithmetic operations[4].

2.5. The Residue Number System

If we divide a number by another number (which we shall call here the base) we obtain a quotient and a remainder, or residue. For instance, if we divide the number 8 by the base 5, we obtain a quotient of 1 and a residue of 3. Table 2.9 arbitrarily lists the numbers from 0 to 29 and their residues to base 2, 3 and 5.

The residues shown in Table 2.9 uniquely identify a number. In other words the listed residue combinations can be used as codes to represent numbers. The configuration 104 represents the decimal number 9 just as uniquely as a binary 1001 or an octal 11.

We might suspect that the proposed number representation requires rather awkward manipulations in actual use, and certainly this is true to some extent. For instance, there is no straightforward procedure to deter-

---

[1] Frequently the terms mantissa and characteristic are used.
[2] This is advantageous when subtractions are performed by the addition of the complement. See paragraph 8.1.1.1.
[3] Frequently, less digits are required to express a change in a variable than to express the total value of the variable.
[4] Compare also paragraph 10.4.

Table 2.9. *Table of Residues to Base 2, 3 and 5*

Number N	Residue to Base 2	Residue to Base 3	Residue to Base 5	Number N	Residue to Base 2	Residue to Base 3	Residue to Base 5	Number N	Residue to Base 2	Residue to Base 3	Residue to Base 5
0	0	0	0	10	0	1	0	20	0	2	0
1	1	1	1	11	1	2	1	21	1	0	1
2	0	2	2	12	0	0	2	22	0	1	2
3	1	0	3	13	1	1	3	23	1	2	3
4	0	1	4	14	0	2	4	24	0	0	4
5	1	2	0	15	1	0	0	25	1	1	0
6	0	0	1	16	0	1	1	26	0	2	1
7	1	1	2	17	1	2	2	27	1	0	2
8	0	2	3	18	0	0	3	28	0	1	3
9	1	0	4	19	1	1	4	29	1	2	4

mine the larger of two numbers. On the other hand, some operations are susprisingly simple. An addition requires only the addition of corresponding residues in their appropriate number system, disregarding any carries.

Example:
$$\begin{array}{r} 1\ 1\ 3 \\ +\ 0\ 1\ 4 \\ \hline 1\ 2\ 2 \end{array} \qquad (13_{10} + 4_{10} = 17_{10})$$

The addition of individual residues is as follows:

$$1_2 + 0_2 = 1_2;\ 1_3 + 1_3 = 2_3;\ 3_5 + 4_5 = 1)2_5$$

Since there are no carries from one position to another, a computer can perform an addition (or subtraction) in all positions simultaneously.

A multiplication is performed by the multiplication of corresponding residues, disregarding any carries.

$$\begin{array}{r} 1\ 1\ 2 \\ \times\ 0\ 1\ 4 \\ \hline 0\ 1\ 3 \end{array} \qquad (7_{10} \times 4_{10} = 28_{10})$$

The individual residues are multiplied as follows:

$$1_2 \times 0_2 = 0_2;\ 1_3 \times 1_3 = 1_3;\ 2_5 \times 4_5 = 1)3_5$$

Again, a computer may operate upon all positions of a number simultaneously and derive the product of two numbers in the residue system much more rapidly than in any of the more conventional number systems.

Problem 17: Perform the following operations using the base 2, 3 and 5 residue system:

   a) $7_{10} + 13_{10}$
   b) $15_{10} - 9_{10}$
   c) $2_{10} \times 8_{10}$

The representation of numbers by their residues to base 2, 3 and 5 as shown in Table 2.9 is adequate for integers between 0 and 29. Beyond that, the representation becomes ambiguous. We note, for instance, that the representation of 30_{10} would be 000, the same as for 0_{10}. This problem is solved rather easily by introducing more bases. Suppose we use 7 as a fourth base. The representation of 30_{10} would then be 0002 whereas the representation of 0_{10} is 0000. As a matter of fact, the use of the bases 2, 3, 5 and 7 allows us to represent $2 \times 3 \times 5 \times 7 = 210$ different numbers (or integers from 0 to 209) without ambiguity.

Problem 18: Perform the following operations using the base 2, 3, 5 and 7 residue number system:

   a) $69_{10} + 43_{10}$
   b) $8_{10} \times 18_{10}$

Although it makes no basic difference what bases we select in the residue number system, it is advantageous to use only numbers which are relatively prime (numbers which have no common divisor). This gives the maximum number of unique codes without redundancy.

Problem 19: Using the bases 2 and 4, we might expect to represent $2 \times 4 = 8$ numbers without ambiguity. How many can you actually represent? List the "codes" for all integers, starting with zero, until the codes repeat.

The clumsiness in comparing two numbers, determining the sign of a result, and particularly the complicated division process, make the residue number system not a very practical system. However, due to the extremely fast multiplication it is seriously considered for special purpose computers in applications where the more awkward operations are not, or not very frequently required. In addition, it has a potential application in checking the operation of arithmetic units using more conventional number systems.

References

1. CERAM C. W.: Gods, Graves, and Scholars. New York: Alfred A. Knopf. 1952.
2. TERRY G. S.: The Dozen-System. London, New York, Toronto: Longmans, Green & Co. 1941.
   Duodecimal Bulletin of the Duodecimal Society of America. 20, Carlton Place. Staten Island, N.Y.
3. WHITE G. S.: Coded Decimal Number Systems for Digital Computers, Proceedings IRE, vol. 41, No. 10, pp. 1450–1452. Oct. 1953.

4. HAMMING R. W.: Error Detecting and Error Correcting Codes, Bell System Technical Journal, vol. 29, pp. 147–160. Apr. 1950.

REED I. S.: A Class of Multiple-Error-Correcting Codes and the Decoding Scheme, 1954 Symposium on Information Theory, Transactions IRE, vol. IT-4, pp. 38–49. Sept. 1954.

ULRICH W.: Non-Binary Error Correction Codes, Bell System Technical Journal, vol. 36, No. 6, pp. 1341–1382. Nov. 1957.

GARNER H. L.: Generalized Parity Checking, Transactions IRE, vol. EC-7, No. 3, pp. 207–213. Sept. 1958.

KILMER W. L.: An Idealized Over-All Error Correcting Digital Computer, etc., Transactions IRE, vol. EC-8, No. 3, pp. 321–325. Sept. 1959.

BROWN D. T.: Error Detecting and Correcting Binary Codes for Arithmetic Operations, Transactions IRE, vol. EC-9, No. 3, pp. 333–337. Sept. 1960.

PETERSON, and BROWN: Cyclic Codes for Error Detection, Proceedings IRE, vol. 49, No. 1, pp. 228–235. Jan. 1961.

MARCUS M. P.: Minimum Polarized Distance Codes, IBM Journal of Research and Development, vol. 5, No. 3, pp. 241–248. July 1961.

GOROG E.: Some New Classes of Cyclic Codes Used for Burst-Error Correction, IBM Journal of Research and Development, vol. 7, No. 2, pp. 102–111. Apr. 1963.

5. GARNER H. L.: The Residue Number System, Transactions IRE, vol. EC-8. No. 2, pp. 140–147. June 1959.

GUFFIN R. M.: A Computer ... Using the Residue Number System, Transactions IRE, vol. EC-11, No. 2, pp. 164–173. Apr. 1962.

SZABO N.: Sign Detection in Nonredundant Residue Systems, Transactions IRE, vol. EC-11, No. 4, pp. 494–500. Aug. 1962.

KEIR, CHENEY, and TANNENBAUM: Division and Overflow Detection in Residue Number Systems, Transactions IRE, vol. EC-11, No. 4, pp. 501–507. Aug. 1962.

MERRILL R. D.: Improving Digital Computer Performance Using Residue Number Theory, Transactions IRE, vol. EC-13, No. 2, pp. 93–101. Apr. 1964.

6, SONGSTER G. F.: Negative-Base Number-Representation Systems, Transactions IEEE, vol. EC-12, No. 3, pp. 274–277. June 1963.

LIPPEL B.: Negative-Bit Weight Codes and Excess Codes, Transactions IEEE, vol, EC-13, No. 3, pp. 304–306. June 1964.

Selected Bibliography

DICKSON L. E.: Modern Elementary Theory of Numbers. Chicago: University of Chicago Press. 1939.

Staff of Engineering Research Associates, High-Speed Computing Devices. New York: McGraw-Hill. 1950.

RICHARDS R. K.: Arithmetic Operations in Digital Computers. Princeton: D. Van Nostrand. 1955.

HARDY, and WRIGHT: An Introduction to the Theory of Numbers. London: Oxford University Press. 1956.

McCRACKEN D. D.: Digital Computer Programming. New York: John Wiley and Sons. 1957.

CROWDER N. A.: The Arithmetic of Computers. New York: Doubleday & Co. 1960. (A "Tutor Text".)

3. Boolean Algebra

Mathematical or symbolic logic[1] is frequently referred to as Boolean algebra in honor of George Boole, one of its early contributors. Even though we will not use Boolean algebra for its originally anticipated purpose, the subject still can be introduced with a quotation from the preface to George Boole's fundamental work [1]: "Whenever the nature of the subject permits the reasoning process to be without danger carried on mechanically, the language should be constructed on as mechanical principles as possible...". In the spirit of this statement, Boolean algebra uses a symbolism as short-form notation for problems in logic. The reasoning process itself may then be reduced to mathematical operations of a particular kind.

One or two examples might help to explain the types of problems which were anticipated by the originators of symbolic logic. Suppose we find the following statement to be true: "If it rains, the street is wet". Are we now allowed to conclude: "It rains, if the street is wet"[2]? The original statement has two essential parts: "It rains"; and "the street is wet". These parts are called propositions. In Boolean algebra, propositions are represented by letters. For instance, the letter A may represent the proposition "it rains", and B the proposition "the street is wet". The logic dependence or relation of propositions is expressed by symbols like $=$, $+$, $>$, etc. and can be investigated by methods similar to the ones used in ordinary algebra.

Boolean algebra provides answers not only to problems of the above type but it can be applied to questions like: How many conditions do we have to consider? Let us take one of Boole's original illustrations. Suppose we consider the three propositions: "It rains"; "it hails"; and "it freezes". One possible condition is: "It rains and it hails, but it does not freeze". But how many possible different conditions are there altogether?

Even from these short introductory remarks, it may be conceivable that this type of symbolic logic can be applied to certain types of electric or electronic problems. Suppose that we have some complex circuitry employing electromagnetic relays. Boolean algebra might provide us with answers to the following problem: What will happen if certain conditions

---

[1] Also: Set Theory.
[2] For an answer to this question see footnote on page 30.

are present, i.e. this and that relay is energized and those relays are not energized ? Or, alternately: Have we considered all possible cases, or are there some conditions under which the circuit might fail ? We may even hope that Boolean algebra is able to tell us the minimum number of relays or other elements required to perform a certain function. If this should be the case, we have a most powerful tool which enables us to investigate electric and electronic circuits unbiased and independent from irrelevant considerations[1].

The following paragraphs deal with fundamentals of Boolean Algebra applicable to switching circuits. A more complete coverage of mathematical logic, and many specialized topics can be found in the literature[2].

3.1. Binary Variables

Many, if not all, problems in logic can be reduced to simple yes-no decisions, e.g.: is $A = B$?; is $x > 0$?; can I afford a new car ?

Accordingly, Boolean Algebra considers only binary variables. This is to say, propositions can assume only one of two possible states: true or false; existing or non-existing; yes or no, etc. Electric and electronic switching elements lend themselves very readily to this treatment: a switch is either on or off; a tube or a transistor is conducting or non-conducting; a relay is energized or not energized; contacts are closed or open; pulses or potentials are present or absent, etc.

The two possible states of a binary variable may be indicated by the two symbols 0 and 1. It is customary to use the symbol 1 to represent affirmative conditions like: yes, true, existing, on, present, and the symbol 0 for their opposites: no, false, non-existing, off, absent, etc.

Boolean algebra may state: $A = 1$, if switch A is turned on; and $A = 0$, if switch A is turned off.

Since there are only two possible states considered, the mathematical definition of a binary variable can be given as:

$$x = 0, \text{ if, and only if } x \neq 1; \tag{3.1}$$

$$x = 1, \text{ if, and only if } x \neq 0$$

3.2. Functions of One Variable

Like an algebraic variable, a binary variable may be a function of one or more other variables. The simplest function of a binary variable is the *identity* and denoted as follows:

$$L = A \tag{3.2}$$

---

[1] It is entirely possible to describe or even design digital computers without the formal use of Boolean algebra, but equivalent logic reasoning must be carried out in any case.

[2] See "Selected Bibliography" at the end of this chapter.

Let us illustrate this function (and following examples) by a contact network. A is a normally open contact of switch A. The light L will be on, if and only if the switch A is "on". The equality sign in Equation (3.2) expresses the logical identity of the two propositions A, and L.

Fig. 3.1. Illustration of the Function $L=A$

We can either say: light L is on, if switch A is on, or: switch A is on, if the light L is on. There is no condition under which one proposition is true, while the other is false.

$$\text{Definition of Identity: if } X=Y, \text{ then } Y=X \qquad (3.3)$$

Logic functions are frequently presented in diagrammatic form. Fig. 3.2 shows the function $L=A$ in two of the most commonly used diagrams:

Fig. 3.2a gives the Venn diagram. The square encloses all elements (all propositions) under consideration. It is sometimes called the universal set or universe for short. The circle (or

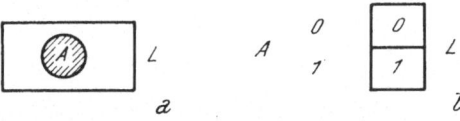

Fig. 3.2. Venn Diagram and Karnaugh Map for Equation (3.2)

circles in more complicated diagrams) divides the area of the square into parts corresponding to the different existing conditions. In this simple diagram, we consider only two different cases: those for which the switch A is turned on (inside of circle); and those for which the switch A is turned off (outside of circle). The shaded area indicates that the proposition under investigation (L) is present if we are within the circle.

Fig. 3.2b shows the Karnaugh map[1]. It also indicates that there are only two conditions under consideration ($A=0$, and $A=1$). The entry "1" in the table represents the existence of the output function (L) for $A=1$, and "0", the non-existence of an output for $A=0$.

The second of the two possible functions of one variable involves the logic "complementation" and is given by:

$$L=A' \qquad (3.4)$$

Equation (3.4) is to be interpreted as: proposition L is present if, and only if proposition A is *not* present. The condition "not A" is usually indicated by a prime or a bar, like: A', or $\overline{A}$.

We can give the following definition of complementation or logic "NOT":

$$1'=0, \text{ and } 0'=1 \qquad (3.5)$$

---

[1] The term "Veitch diagram" is used synonymously with "Karnaugh map".

The circuit corresponding to Equation (3.4) uses a normally closed contact of switch A, so that the light L is turned on, if the switch A is turned "off". (Corresponding to a closed contact.)

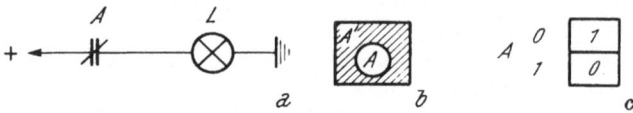

Fig. 3.3. Illustrations for Equation (3.4)

The Venn diagram and the truth table should be self-explanatory.

The equality sign of Equation (3.4) permits us to state: the Light L is on if, and only if switch A is turned off. Consequently, we can make the statement: the light L is off, if the switch A is on. In other words, we can deduce:

$$\text{if } L = A', \text{ then } L' = A \tag{3.6}$$

Using the obvious relation[1]:

$$(A')' = A \tag{3.7}$$

we see that Equation (3.6) is obtained by performing a complementation on both sides of the Equation (3.4). (This operation is permissible for all logic equations and sometimes helpful in reducing the complexity of an expression.)

3.3. Functions of Two Variables

Considering two binary variables, say A and B, we can have any one of the following four distinct conditions:

$$\text{I. } A = 0, \ B = 0;$$
$$\text{II. } A = 0, \ B = 1;$$
$$\text{III. } A = 1, \ B = 0;$$
$$\text{IV. } A = 1, \ B = 1.$$

Accordingly, the Venn diagram and the Karnaugh map will show the universal set divided into four distinct areas:

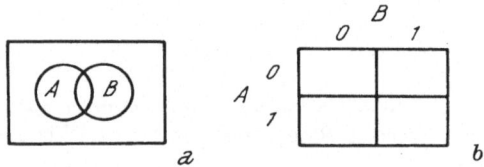

Fig. 3.4. Venn Diagram and Karnaugh Map for Two Variables

Dependent upon which areas we shade (Fig. 3.4a) or where we enter 0's or 1's (Fig. 3.4b), we will have a variety of different output functions.

---

[1] See e.g. Fig. 3.3b.

3.3.1. The Logic Product

Let us suppose we want an output only for condition IV above. This requires that both proposition A *and* proposition B are present. Boolean algebra uses the multiplication sign (or sometimes the sign: $\wedge$ or $\cap$, read "cap") to express such a requirement for the simultaneous existence of two propositions. Accordingly, we would state[1]:

$$L = A \cdot B \qquad (3.8)$$

The corresponding contact network and logic diagrams are given in Fig. 3.5.

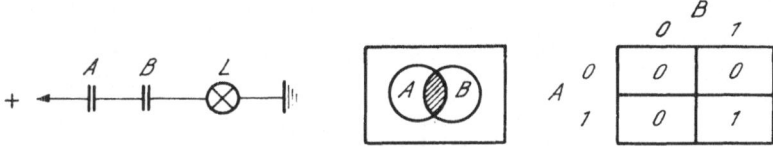

Fig. 3.5. Illustrations of the Logic Product

The light L is on if, and only if both switches A and B are turned on. The mathematical definition of the logic product is given below:

$$0 \cdot 0 = 0;$$
$$0 \cdot 1 = 1 \cdot 0 = 0; \qquad (3.9)$$
$$1 \cdot 1 = 1.$$

The terms: logic AND, conjunction, and intersection are frequently used in place of logic product. By priming (complementing) one or both input variables we can also represent any one of the remaining conditions II through IV by a logic product.

Problem 1: Show the logic equation, the Venn diagram, the Karnaugh map, and the contact network for an output if:

a) $A = 0$ and $B = 0$;

b) $A = 0$ and $B = 1$;

c) $A = 1$ and $B = 0$.

3.3.2. The Logic Sum

While a logic product requires the simultaneous presence of two propositions (A and B), the logic sum pertains to the presence of one *or* the other proposition. The "or" is inclusive, i.e. it includes also the case where both

---

[1] As in ordinary algebra, the multiplication sign is frequently omitted. Equation (3.8) would then read $L = AB$.

propositions are present (A or B, or both). Equation (3.10) and Fig. 3.6 may serve as illustrations.

$$L = A + B \qquad (3.10)$$

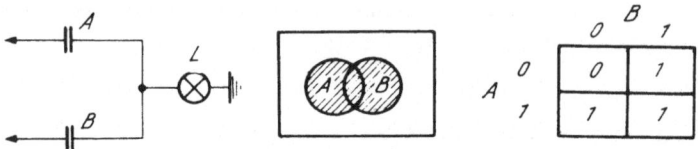

Fig. 3.6. Illustration of the Logic Sum

The mathematical definition of a logic sum can be given as:

$$0+0=0;$$
$$0+1=1+0=1;$$
$$1+1=1. \qquad (3.11)$$

The terms: logic OR, join, joint set, union, and disjunction are used synonymously with logic sum. Sometimes the signs $\lor$ or $\cup$ (read cup) are used instead of $+$.

Problem 2: Show the logic equation, the corresponding contact network, the Venn diagram, and the truth table for the remaining three cases where A or A' is combined with B or B' in a logic sum.

3.3.3. Combinations of Logical AND's, OR's, NOT's

It is possible to define and use a variety of logical operations other than AND, OR, NOT, and logical expressions are sometimes given in terms of NOR, exclusive OR, and others. Rather than introducing new symbols for these operations, we shall represent more complicated functions by combinations of several AND, OR, NOT operations. These three functions lend themselves most readily to algebraic treatment and even in cases where circuits would be most simply described in other terms, the translation to AND, OR, NOT, presents no problem.

Let us illustrate such a combination of logical operations by a particular function of two variables. Suppose we have the two switches A and B, and we want an indication if the positions of the two switches disagree (A is on and B is off; or A is off and B is on). Apparently, we want an output for conditions II and III listed on page 26[1]. The design of the Venn diagram or Karnaugh map is straightforward.

---

[1] This is the logic function of EXCLUSIVE OR: One or the other; but not both.

The Karnaugh map reveals the conditions for an output most easily: $A'B$ or AB'. We, therefore, state:

$$L = A'B + AB' \qquad (3.12)$$

In logical expressions of this type, it should be understood that logic multiplication has the priority over logic addition, just as algebraic multiplication has the priority over algebraic addition. In other words, the logical product should always be considered as contained in parentheses like:

$$A'B + AB' = (A'B) + (AB') \neq A'(B+A)B' \qquad (3.13)$$

The latter inequality can easily be shown by shading a Venn diagram for the terms on both sides of the inequality sign.

The contact network corresponding to Equation (3.12) is given in Fig. 3.8:

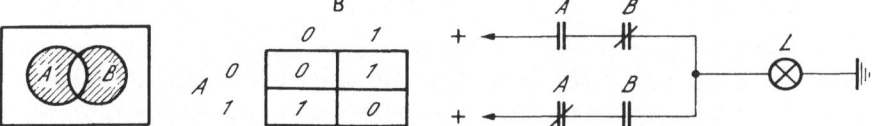

Fig. 3.7. Venn Diagram and Karnaugh Map for an Output if $A \neq B$ Fig. 3.8. Contact Network Equivalent to Equation (3.12)

Later we shall see a different logic equation and a different network also equivalent to Fig. 3.8. However, before deriving these, we shall show a few theorems which are quite helpful in rearranging or simplifying logical expressions.

Problem 3: Design a circuit which gives an output if and only if the position of switch A is the same as the position of switch B.

Problem 4: There are sixteen different ways to shade the Venn diagram given in Fig. 3.4. Show for these sixteen distinct cases the Venn diagram, the Karnaugh map, the logic equation, and the contact network. Leave space for alternate circuits and equivalent electronic implementations to be discussed later. (This problem should not be considered compulsory, although the resulting collection may be quite helpful later.)

3.3.4. Theorems Concerning Functions of One or Two Variables

Most of the following theorems are obvious or self-explanatory. Where there is any doubt, the reader can easily verify the correctness (see problem 5 below).

$X \cdot X = X$	$X + X = X$	(Idempotent)	(3.14)
$X \cdot 0 = 0$	$X + 0 = X$	(Union)	(3.15)
$X \cdot 1 = X$	$X + 1 = 1$	(Intersection)	(3.16)
$X \cdot Y = Y \cdot X$	$X + Y = Y + X$	(Commutative)	(3.17)
$X \cdot X' = 0$	$X + X' = 1$	(Complementary)	(3.18)
$(X \cdot Y)' = X' + Y'$	$(X + Y)' = X' Y'$	(Dualization)	(3.19)

The 0's in Equations (3.15) and (3.18) should be interpreted as propositions which are never true, whereas the 1's in Equations (3.16) and (3.18) represent propositions which are always true.

Problem 5: Prove the correctness of Equations (3.18) and (3.19) in one of the following ways:

a) Draw Venn diagrams for both sides of an equation and show that *all* shaded areas in both diagrams are identical.

b) Show that *all* entries in the two Karnaugh maps corresponding to the two sides of an equation are identical.

The two methods indicated in problem 5 are equivalent and commonly known as proof by *Perfect Induction*. Although not too highly esteemed by theorists, the proof is valid and quite practical for engineering approaches (If two functions behave the same for *each unique* condition then they behave the same for *all* conditions or, in other words, they are logically equivalent).

Sometimes the relation between variables is expressed by signs like: $<$ or $>$[1]. In ordinary algebra we would state: One value is smaller (or larger) than the other one. In Boolean algebra we say: One proposition is contained in the other one, or is a subset of the other one. A simple illustration for this relation is given in Fig. 3.9.

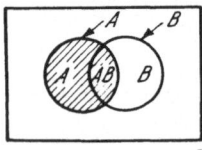
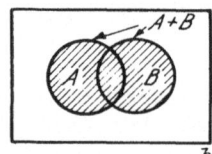

Fig. 3.9. Illustration of Subsets

From Fig. 3.9 a we see that the area AB is contained in the area A; therefore, we can state $AB < A$. (If AB is true, then certainly A is true.) From Fig. 3.9b we derive: $A < A + B$[2]. From these two relations, we can deduce: $AB < A < A + B$. In general, we have the following transitive properties:

$$\text{if } X \leq Y \text{ and } Y \leq Z, \text{ then } X \leq Z \qquad (3.20)$$

---

[1] Alternately, the symbols $\subset$ or $\supset$ are used.

[2] In the statement "If it rains, the street is wet", the proposition "it rains" is a subset of "the street is wet". There are conditions under which the street may be wet without raining, but no condition under which it may rain without the street being wet.

Problem 6: Establish the validity of Equations (3.21) and (3.22):

$$\text{if } X \le Y, \text{ then } XY = X \text{ and } X + Y = Y \quad \text{(Consistency)} \qquad (3.21)$$

$$X + XY = X, \quad X(X + Y) = X \qquad (3.22)$$

The following theorems can be derived from previous theorems, but their correctness is most easily established by perfect induction.

$$X + X'Y = X + Y \qquad\qquad X(X' + Y) = XY \qquad (3.23)$$

$$XY + X'Y' = (X + Y')(X' + Y) \qquad (X + Y)(X' + Y') = XY' + X'Y \quad (3.24)$$

It is to be noted that in all these theorems, as in ordinary algebra, a single letter may represent a more complex expression.

3.4. Functions of Three or More Variables

3.4.1. Diagrammatic Representations

A Venn diagram for three variables is given in Fig. 3.10a. It contains eight areas corresponding to the eight distinct states which a combination of three variables can assume. Fig. 3.10b shows the equivalent three-variable Karnaugh map.

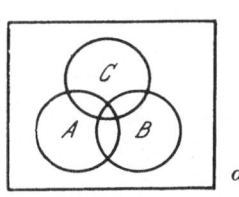

C $A\ B$	0	1
0 0		
0 1		
1 1		
1 0		

a   *b*

Fig. 3.10. Venn Diagram and Karnaugh Map for Three Variables

The states of variables in the Karnaugh map are listed in Gray code. We note that in both diagrams only one variable changes its state when we cross a border between two adjacent areas. Also only one variable changes for a transition from the top row to the bottom row and from the leftmost to the rightmost column. This feature can be conveniently used for the simplification of logical expressions, as we shall see in paragraph 3.5.2.

Unfortunately, it is not possible to construct two-dimensional Venn diagrams for more than three variables while retaining this feature, although a Karnaugh map for four variables can be designed.

Problem 7: Show the border of the four propositions A, B, C', D' in the Karnaugh map of Table 3.1 in four different colors. (If necessary, color first all individual fields in which a specific proposition is true and then draw the border line around the colored area.)

Table 3.1. *Four-Variable Karnaugh Map*

A B	C D	0 0	1 0	1 1	0 1
0 0					
0 1					
1 1					
1 0					

More than four variables can be represented by several diagrams. For instance, we may have one Karnaugh map (Table 3.1) for the state $E=0$, and another one for $E=1$. For a high number of variables, the search for "adjacent" areas becomes cumbersome and one may resort to a listing of output values in a *Truth Table*.

Table 3.2. *Five-Variable Truth Table*

A	B	C	D	E	Output
0	0	0	0	0	
0	0	0	0	1	
0	0	0	1	0	
·	·	·	·	·	
·	·	·	·	·	
·	·	·	·	·	
1	1	1	1	1	

A truth table simply tabulates all possible conditions of input variables in a binary order. One or more ouput columns list the desired ouputs for each of these specific conditions. The truth table does not have the property that only one variable changes its state for a transition from one entry to an adjacent one, but it can be expanded to an arbitrary number of variables.

One may also use a *Vertex Frame* representation equivalent to a Venn diagram or a Karnaugh map for more than four variables. Fig. 3.11 a is such a representation for one variable. The two nodes correspond to the two states which a binary variable can assume. The solid node represents an output, the empty node no output. Fig. 3.11 a corresponds therefore to a logical "NOT". Fig. 3.11 b represents two variables. The corresponding four

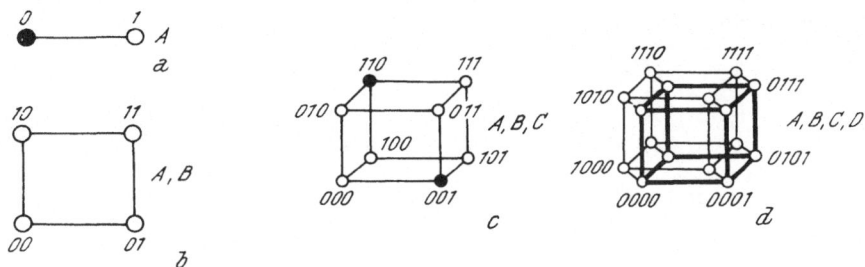

Fig. 3.11. Geometrical Representation of Binary Variables

distinct states are indicated by four nodes. It is to be noted that for a transition from one node to an adjacent one only one variable changes its state. The cube in Fig. 3.11 c is a three-variable diagram. The logic function represented by the solid nodes is: $ABC' + A'B'C$. The "hypercube" in Fig. 3.11 d is a four-variable diagram. The 16 nodes correspond to the 16 different conditions we have to consider for four variables. Again for a transition from one node to an adjacent one, only one variable changes its state.

The representations in Figs. 3.11 b, c, d are obtained consecutively by a doubling of the previous diagram. (The consideration of one additional input-variable doubles the number of possible distinct states). By doubling Fig. 3.11 d one may obtain two sets of "hypercubes": one for $E = 0$ and the other one for $E = 1$, and so on [3]. For more complicated diagrams it may be preferable to construct three dimensional wire models.

Problem 8: Represent the function $L = AB + C'$
a) in a Venn diagram,
b) in a Karnaugh map,
c) in a truth table,
d) in a vertex frame.

3.4.2. Theorems Involving Three or More Variables

The following theorems have equivalents in regular algebra:

$$X + Y + Z = (X + Y) + Z = X + (Y + Z) \quad \text{(Associative)} \tag{3.25}$$

$$XYZ = (XY)Z = X(YZ) \tag{3.26}$$

$$XY + XZ = X(Y+Z) \qquad\qquad \text{(Distributive)} \qquad (3.27)$$

$$(X+Y)(X+Z) = X + YZ \qquad\qquad\qquad\qquad (3.28)$$

Problem 9: Establish the correctness of Equations (3.27) and (3.28) by perfect induction.

De Morgan's Theorem is an extension of Equation (3.19) for an arbitrary number of variables:

$$(U+V+W+\cdots+Z)' = U'V'W'\cdots Z' \qquad (3.29)$$

$$\text{(De Morgan)}$$

$$(UVW\cdots Z)' = U'+V'+W'+\cdots+Z' \qquad (3.30)$$

The proof is not too complicated. Suppose we use the following truth table.

Table 3.3. *Truth Table for n Variables*

U	V	W	. . .	Z
0	0	0	. . .	0
.	.	.		.
.	.	.		.
.	.	.		.
1	1	1	. . .	1

The left-hand member of Equation (3.29) is true only if all the propositions $U, V, W \ldots Z$ are in the zero state (which corresponds to the first line of the truth table). The right-hand member of Equation (3.29) is obviously true only for this line. Therefore, the truth tables for the left- and for the right-hand members are identical and Equation (3.29) is valid. By similar reasoning, the correctness of Equation (3.30) may be established.

Problem 10: Apply De Morgan's theorem to the following expressions:

a) $(B+C)'$
b) $(AB)'$
c) $(AB+CD)'$

The *General Theorem* states that it is possible to represent any logical function by a sum of products or a product of sums. Such forms, when fully expanded, are frequently called the standard sum (or disjunctive normal), respectively the standard product (or conjunctive normal).

3.4.3. The Standard Sum

Let us suppose we have the following truth table for a function, X, of three variables A, B, C:

Table 3.4. *Truth Table for a Function of Three Variables*

A	B	C	X	X'
0	0	0	0	1
0	0	1	1	0
0	1	0	1	0
0	1	1	0	1
1	0	0	1	0
1	0	1	1	0
1	1	0	1	0
1	1	1	0	1

The function X gives an output for the following five out of eight possible conditions: $A'B'C$, $A'BC'$, $AB'C'$, $AB'C$, ABC'. We, therefore, can state:

$$X = A'B'C + A'BC' + AB'C' + AB'C + ABC' \qquad (3.31)$$

Each of the three-letter products corresponds to a unique input condition. The standard sum simply lists all unique input conditions under which we have an output. For a function of n variables we obtain then a sum of n-letter products but, no matter how 0's and 1's are distributed in the truth table, we obviously can always write a function in this manner. The contact network equivalent to Equation (3.31) is given in Fig. 3.12 below:

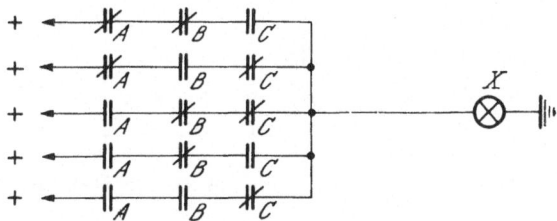

Fig. 3.12. Implementation of a Standard Sum

Each of the three-letter products in Equation (3.31) is implemented here by a serial arrangement of three contacts. Each of these combinations provides current to the light X for only one unique condition of input variables. This is by no means the simplest possible circuit for this function and, as a matter of fact, the general theorem states only that it is always possible to represent a function by a standard sum but it does not say that this is the shortest form.

Problem 11: Show the contact network corresponding to the standard sum of the following functions:

a) $AB + ABC + BC$
b) $(A + B')\ (B + C')\ (C + A')\ (ABC + A'B'C')$

If necessary write down the truth table.

3.4.4. The Standard Product

While each term in the standard sum provides an output for a certain unique input combination, the terms of a standard product make sure that there is no output for certain unique input conditions.

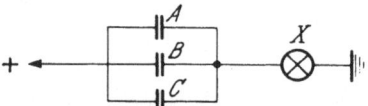

Fig. 3.13. Partial Implementation of Table 3.4

Let us suppose we wanted to inhibit an output for the condition $A'B'C'$ (the first entry in Table 3.4). We could use three contacts arranged in parallel as shown in Fig. 3.13:

This set of contacts provides a current path for all conditions except for the one unique combination: $A'B'C'$. If we disregard, for the moment, any further contact sets which might be required, we would state:

$$X = (A + B + C) \tag{3.32}$$

Inspecting Table 3.4, we find two more input conditions for which we have to inhibit an output: $A'BC$ and ABC. Two more sets of three contacts will suppress an output also for these conditions:

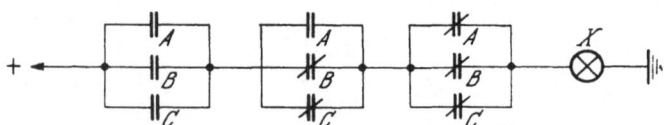

Fig. 3.14. Implementation of Table 3.4 by the Standard Product

Expressing parallel connections by logical OR's and series connections by logical AND's, we derive the standard product for the function X from Fig. 3.14:

$$X = (A + B + C)\ (A + B' + C')\ (A' + B' + C') \tag{3.33}$$

For a function of n variables, we would find a product of n-letter sums. We would have one sum for each zero in the output column of the truth table.

Again, no matter how the 0's are distributed in a truth table, it will always be possible to represent a function by a standard product.

Problem 12: Implement the function of Fig. 3.7 by a standard product.

The standard product can be derived from the truth table in a rather mechanical manner. Suppose we first implement the function X' by a standard sum (see X' column in Table 3.4). Following the methods outlined in paragraph 3.4.3 we obtain:

$$X' = A'B'C' + A'BC + ABC \qquad (3.34)$$

Applying De Morgan's theorem we find:

$$X = (A'B'C')' \; (A'BC)' \; (ABC)'$$
$$= (A+B+C) \; (A+B'+C') \; (A'+B'+C') \qquad (3.35)$$

which is the standard product.

Problem 13: Implement the functions given in problem 11 by standard products.

Problem 14 (Voluntary): Add a new column to the collection started in Problem 4 which shows the standard product. If the solution is already in form of a standard product give the standard sum.

Both, the standard sum and the standard product are logically equivalent. Their electric or electronic implementations may use a different number of elements but neither constitutes the minimum implementation in general.

3.5. Minimization

The term minimization in connection with Boolean algebra may have several different meanings. One might seek the shortest logical expression for a certain function, or the circuit with the minimum number of components to perform a logic function. One might also try to find the least expensive circuit, or perhaps the most obvious or simplest implementation. A minimization in a particular respect might not be a minimization in some other repect and the balance of factors, as minimization in general, is to a large extent a matter of intuition. The following paragraphs will provide several tools with which problems of minimization can be attacked. Together with the knowledge of electronic circuits (to be dealt with in later chapters), a combination of these methods should assure a reasonable engineering solution.

3.5.1. Minimization by Algebraic Operations

The theorems of Boolean algebra can be used to reduce the complexity of logic expressions. Let us show this in a specific example. Suppose we want to simplify the expression $(A+B)(A+C)$. Equation (3.28) provides the answer immediately. But in general, we will not be so fortunate, so let us try to simplify this expression by other theorems. Using Equations (3.27) and (3.17) we can write:

$$(A+B)(A+C)=A(A+C)+B(A+C)=AA+AC+AB+BC \qquad (3.36)$$

With Equation (3.14) we obtain

$$(A+B)(A+C)=A+AC+AB+BC \qquad (3.37)$$

and with Equation (3.22):

$$(A+B)(A+C)=A+BC \qquad (3.38)$$

From Equation (3.36), we could also have proceeded like:

$$AA+AC+AB+BC=A(1+C+B)+BC=A\cdot 1+BC=A+BC \qquad (3.39)$$

Problem 15: Reduce in a similar manner the expressions:

a) $AB+AB'+A'B+A'B'$
b) $(A+B)(A+B')$
c) $AB+AC+(A'+B')'$
d) $(A+B')(A+C)(B+C')$

Surprisingly, by addition of redundant terms one may sometimes quickly simplify an expression.

Example: $(AB'+A'B+A'B')'$

$$=(AB'+A'B'+A'B+A'B')'$$
$$=(\underbrace{\quad B'\quad}+\underbrace{\quad A'\quad})'$$
$$= AB \qquad (3.40)$$

The redundant term $A'B'$ has been added in the second line. This can be justified since there was already a term $A'B'$ and $A'B'+A'B'=A'B'$ (Equation 3.14).

Example: $B'+AB=B'+AB+AB'$

$$=B'+A(B+B')$$
$$=B'+A$$

The redundant term AB' in the first line may be added since it is contained in, or is a subset of the already present B' (Equation 3.22).

Problem 16: Simplify:

a) $A'BC + AB'C + ABC' + ABC$

b) $A'B + C(A + B')$

The simplification of logic expressions by algebraic operations may seem here purely intuitional. However, with a little practice, many possible simplifications become apparent. Furthermore, there exist certain rules or "recipes" which allow to find simplifications in a more or less mechanical manner [2].

3.5.2. Minimization by Graphical Methods

Graphic representations of logical expressions tend to make possible simplifications apparent [3]. Suppose we have the expression: $X = AB + AB'$. From the Venn diagram, we see immediately that we can simplify to $X = A$.

In general, the standard representations of a function (the standard sum or the standard product) can be simplified if there are "neighboring" areas in a Venn diagram or a Karnaugh map in which the output function is true (respectively false). In order to obtain the simplest representation, one has to look for as large "patches" of areas and as few patches as possible. The patches as

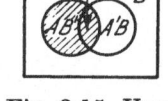

Fig. 3.15. Venn Diagram for $AB + AB'$

represented in a Karnaugh map must thereby be rectangular and must comprise 2, 4, 8, 16... individual areas.

Let us take the following Karnaugh map as an example:

Table 3.5. *Sample Function*

		C	
A	B	0	1
0	0	0	1 — B'C
0	1	1	0 — BC'
1	1	1	0
1	0	1	1

AC'   AB'

We find four different patches, of two entries each, as indicated in Table 3.5. For the lowest horizontal patch, the standard sum would list: $AB'C' + AB'C$. The diagram tells us that we don't have to care about the state of C as long as we have: AB'. Therefore, we can "cover" this patch by AB'.

In the same manner, the remaining three patches can be labelled: AC', BC', $B'C$. Furthermore, the diagram tells us we can omit either the patch AB' or the patch AC' (but not both) because the 1's in these patches are covered by the remaining three patches. We, therefore, can write:

$$X = AB' + BC' + B'C \qquad\qquad (3.41)$$

or alternately:

$$X = AC' + BC' + B'C \qquad\qquad (3.42)$$

Compared to the standard sum (a sum of five three-letter terms), these sums are much shorter. They are called the "minimum sum" when no further simplification is possible. The larger the patches are, the more pronounced is the simplification. In the following Karnaugh map, for instance, two patches are sufficient to cover all 1's.

Table 3.6. *Sample Function*

		C	0	1	1	0
A	B	D	0	0	1	1
0	0		1	0	0	1
0	1		1	0	0	1
1	1		0	1	1	0
1	0		0	1	1	0

The standard sum can, therefore, be simplified as follows:

$$ABCD + A'B'C'D' + A'BC'D' + A'B'C'D + A'BC'D + ABCD' +$$
$$+ AB'CD' + AB'CD = A'C' + AC \qquad\qquad (3.43)$$

In a similar manner, the standard product can be simplified to the minimum product by covering large patches of 0's. For example, the standard product as given in Equation (3.35) may be simplified as follows:

Table 3.7. *Minimum Product*

		C	
A	B	0	1
0	0	0	1
0	1	1	0
1	1	1	0
1	0	1	1

$$X' = A'B'C' + BC \qquad \text{or} \qquad X = (A + B + C)(B' + C') \qquad (3.44)$$

Again, the savings are the more pronounced, the larger the patches are.

Problem 17: Show the minimum sum and the minimum product for the functions given in problems 15 and 16.

Problem 18: Write the following expression in form of a minimum product. The result is a frequently used theorem.

$$XY + X'Z$$

Problem 19: Write the following expression in form of a minimum sum. The result is a frequently used theorem.

$$(X + Y) (X' + Z)$$

The implementation of the minimum sum or the minimum product frequently results in a circuit with the least number of circuit elements. This is notably true for two-level diode circuits[1]. When other circuit elements are used, other "minimum" forms may be preferable.

3.5.3. Special Cases

In some instances, simplifications can be made when there are conditions where we "don't care" whether there is an output or not. Let us take a practical example. Suppose we want to "decode" the digit 9 in the 8421 code[2]. The proposition A shall represent the bit with the weight 8,

Table 3.8. *Karnaugh Map for the Decoding of the Digit 9 in the 8421 Code*

		C	0	1	1	0
A	B	D	0	0	1	1
0	0		0	0	0	0
0	1		0	0	0	0
1	1		×	×	×	×
1	0		0	×	×	1

proposition B has the weight 4, C and D have the weights 2 and 1 respectively. The digit 9 is then given by the configuration $AB'C'D$. The circuit shall give an output for this condition but no output for the other nine conditions which correspond to the digits 0 through 8. The Karnaugh map is given in Table 3.8.

The "1" represents the output corresponding to the digit 9. Zeros are entered for input combinations corresponding to the nine other digits. The crosses represent conditions which are not defined by the 8421 code. In these cases, we do not care whether or not our decoding circuit would provide a true or false output, since we never will have these conditions, as long

---

[1] See paragraph 4.1.   [2] See Table 2.6.

as the machine works properly. We are therefore free to enter 0's or 1's into the Karnaugh map as we please, or better, as it may be practical to reduce the complexity of the decoding circuit.

If we enter 1's instead of the crosses inside the patch indicated in Table 3.8, and replace all other crosses by 0's, then we can simplify the original condition for the digit 9: $(AB'C'D)$ to the simpler term: AD.

Problem 20: Give the shortest logic expression for the decoding of the digit 2 and of the digit 8 in the 8421 code.

Problem 21: Show the shortest logic expression for the function Z defined by the following Karnaugh map:

		Y	
		0	1
X	0	0	0
	1	1	×

Frequently, situations are encountered where already derived logic expressions can be used as part of newly to be implemented functions. In such cases, it usually does not pay to reduce the expression for the new function to its minimum form.

Suppose we have already implemented the function:

$$F_1 = A'B'CD' + A'BC'D' + AB'C'D \qquad (3.45)$$

In addition, we have to derive:

$$F_2 = A'D' + AB'C'D \qquad (3.46)$$

F_2 is here given as a minimum sum.

In order to make eventual simplifications apparent, we draw Karnaugh maps for both functions:

Table 3.9. *Karnaugh Map for Equations (3.45) and (3.46)*

A	B	C:0 D:0	C:1 D:0	C:1 D:1	C:0 D:1		A	B	C:0 D:0	C:1 D:0	C:1 D:1	C:0 D:1
0	0	0	1	0	0		0	0	1	1	0	0
0	1	1	0	0	0		0	1	1	1	0	0
1	1	0	0	0	0		1	1	0	0	0	0
1	0	0	0	0	1		1	0	0	0	0	1

$$F_1 \qquad\qquad\qquad F_2$$

All 1's of the function F_1 are represented in F_2. Those 1's of the function F_2 which are not covered by F_1 can be covered by the patch $A'D'$. Therefore, we can write:

$$F_2 = F_1 + A'D' \tag{3.47}$$

Note that in this equation, some of the 1's of the function F_2 are unnecessarily covered twice, which normally would correspond to a lengthier expression or a waste of circuitry. But since F_1 is already implemented here, a shorter expression (and a saving in circuitry) is realized.

Problem 22: Suppose you have to implement the two functions: $Y = A' + B'C'$ and $X = A'B'C + A'BC' + AB'C'$. How can you achieve some saving by using one function as part of the other?

Problem 23: In many cases not only an established function but also its complement can be used to simplify other functions. Suppose you have to implement the two functions:

$$X = A'C'D' + ABCD' + AB'CD + ABC'D$$
$$Y = AC'D' + AB'D' + ABCD + AB'C'$$

How can you achieve some saving by using the complement of one function as part of the other?

References

1. BOOLE G.: The Mathematical Analysis of Logic, 1847. Oxford: Basil Blackwell. 1951.
   BOOLE G.: The Laws of Thought, 1853. New York: Dover Publications, Inc.
2. QUINE W. V.: The Problem of Simplifying Truth Functions, American Mathematical Monthly, vol. 59, pp. 521–531. Oct. 1952.
   QUINE W. V.: A Way to Simplify Truth Functions, American Mathematical Monthly, vol. 61, pp. 627–631. Nov. 1955.
   McCLUSKEY E. J.: Minimization of Boolean Functions, Bell System Technical Journal, vol. 35, pp. 1417–1444. Nov. 1956.
   GIMPEL J. F.: A Reduction Technique for Prime Implicant Tables, Transactions IEEE, vol. EC-14, No. 4, pp. 535–541. Aug. 1965.
3. VEITCH E. W.: A Chart Method for Simplifying Truth Functions. Proceedings ACM Conference, Pittsburgh, Richard Rimach Associates, Pittsburgh, pp. 127–133. May 1952.
   URBANO, and MUELLER: A Topological Method for the Determination of the Minimal Forms of a Boolean Function, IRE Transactions, vol. EC-5, pp. 126–132. 1952.
   KARNAUGH M.: The Map Method for Synthesis of Combinational Logic Circuits, Transactions AIEE, Communications and Electronics, vol. 72, pp. 593–599. Nov. 1953.
   BOOTH T. M.: The Vertex-Frame Method for Obtaining Minimal Proposition-Letter Formulas, IRE Transactions, vol. EC-11, No. 2, pp. 144–154. Apr. 1962.

Selected Bibliography

SERRELL R.: Elements of Boolean Algebra for the Study of Information-Handling Systems. Proceedings IRE, vol. 41, No. 10, pp. 1366–1380. Oct. 1953.
MEALY G. H.: A Method for Synthesizing Sequential Circuits, Bell Syst., Techn. Jour., vol. 34, pp. 1045–1079. Sept. 1955.
MOORE E. F.: Gedanken Experiments on Sequential Machines, Automata Studies. Princeton: Princeton University Press. 1956.
CALDWELL S. H.: Switching Circuits and Logical Design. New York: John Wiley and Sons. 1958.

HUMPHREY W. S., JR.: Switching Circuits. New York: McGraw-Hill. 1958.

PHISTER M.: Logical Design of Digital Computers. New York: John Wiley and Sons. 1958.

BERKELEY E. C.: Symbolic Logic and Intelligent Machines. New York: Reinhold Publishing Corp. 1959.

FLORES I.: Computer Logic. Englewood Cliffs: Prentice-Hall. 1960.

The following books are introductions to mathematical logic without emphasis on the application of the subject to electronic digital designs:

COOLEY J. C.: A Primer of Formal Logic. New York: The Macmillan Company. 1942.

ROSENBLOOM P.: The Elements of Mathematical Logic. Dover Publications. 1950.

QUINE W. V.: Mathematical Logic. Cambridge: Harvard University Press. 1951.

SUPPES P.: Introduction to Logic. Princeton: D. Van Nostrand. 1951.

LANGER S. K.: An Introduction to Symbolic Logic. New York: Dover Publications. 1953.

CHURCH A.: Introduction to Mathematical Logic. Princeton: Princeton University Press. 1956.

GARDNER M. A.: Logic Machines and Diagrams. New York: Mc Graw-Hill. 1958.

Following is a selection of publications concerned with the application of Boolean algebra to special circuits or requirements.

HUFFMAN D. A.: A Study of Memory Requirements of Sequential Switching Circuits, Technical Report 293, MIT Research Laboratory of Electronics, Cambridge. March 1955.

HUFFMAN D. A.: The Synthesis of Linear Sequential Coding Networks, Proceedings of the Third London Symposium on Information Theory. New York: Academic Press. 1956.

METZE G. A.: A Boolean Algebra with a Delay Operator etc., File No. 202, Digital Computer Laboratory, University of Illinois. 1956.

LOW, and MALEY: Flow Table Logic, Proceedings IRE, vol. 49, No. 1, pp. 221–228. Jan. 1961.

STRAM O. B.: Arbitrary Boolean Functions of N Variables Realizable in Terms of Threshold Devices, Proceedings IRE, vol. 49, No. 1, pp. 210–220. Jan. 1961.

AKERS, and ROBBINS: Logical Design with Three-Input Majority Gates, Computer Design, vol. 2, 4 parts appearing in the March, April, May and June issues, 1963.

AMAREL, COOKE, and WINTER: Majority Gate Networks, Transactions IEEE, vol. EC-13, No. 1, pp. 4–13. Feb. 1964.

WOOD P. E. Jr.: Hazards in Pulse Sequential Circuits, Transactions IEEE, vol. EC-13, No. 2, pp. 151–153. April 1964.

PERLIN A. I.: Designing Shift Counters, Computer Design, vol. 3, No. 2, pp. 12–16. Dec. 1964.

YOELI, and ROSENFELD: Logical Design of Ternary Switching Circuits, Transactions IEEE, vol. EC-14, No. 1, pp. 19–29. Feb. 1965.

GOLDBERG, and SHORT: Antiparallel Control Logic, Transactions IEEE, vol. EC-14, No. 3, pp. 383–393. June 1965.

UNGER S. H.: Flow Table Simplification — Some Useful Aids, Transactions IEEE, vol. EC-14, No. 3. pp. 472–475. June 1965.

The following reference is an introductory survey of the theory of finite automata, or Turing machines:

ELGOT C. C.: A Perspective View of Discrete Automata and Their Design, American Math. Monthly, pp. 125–134. Feb. 1965.

4. Logic Circuits

Computers, despite their complexity, are composed of surprisingly few types of basic circuits. Of course, circuit designs and circuit elements vary from machine to machine, but—disregarding auxiliary devices—the computer proper consists essentially of two types of functional elements: storage elements, and logic circuits. Storage elements retain information (numbers, instructions, signals, codes, etc.) for longer or shorter periods of time[1]. Logic circuits perform logic (or arithmetic) operations. They are sometimes also referred to as decision circuits, switching circuits or logical elements.

The action of logic circuits can be described well by symbolic logic. In turn, logic circuits may be considered as electronic implementations of Boolean algebra.

In order to associate the behavior of an electronic circuit with a logic function, it is necessary to associate individual states of the circuit with individual states of logic expressions. It is customary for pulse-type circuits to represent the presence of propositions by the presence of pulses on corresponding lines. The absence of a proposition is marked by the absence of pulses. When propositions are represented by levels of electric potentials, one level is assigned the logical meaning "1", another the meaning "0". For tube circuits, usually the higher level, say $+10$ volts, is the "1" level whereas the lower one, say -25 volts, is the "0" level. For transistor circuits, the assignment of "1" to the

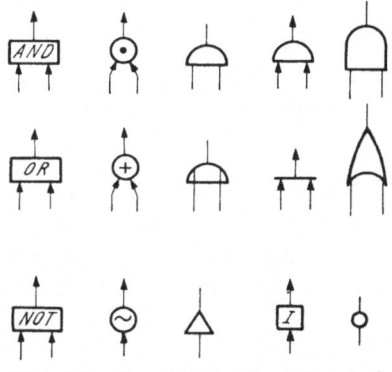

Fig. 4.1. Commonly Used Symbols for Logic Circuits

lower level, say -3 volts, and the assignment of "0" to the higher level, say zero volts, is as frequent as the opposite assignment. Really, the manner of assignment is of no major importance once it is defined in each instance.

---

[1] See chapter 5.

Logic diagrams avoid the possible confusion altogether since they show only the logic function of a circuit.

Fig. 4.1 shows several sets of commonly used symbols. We will only use the symbols given in the left column. Alternate symbols for the same circuits are indicated to the right.

The output of an AND-circuit will assume the "1" state (whatever the definition of the "1" state is) only if all inputs are simultaneously in the "1" state. The OR-circuit provides an output if one or more inputs are present and the NOT-circuit provides an output if there is no input and vice versa.

4.1. Diode Circuits

Practically all diodes in modern computer circuits are semi-conductor devices such as Germanium or Silicon diodes. Compared to vacuum tube diodes, which were once used extensively, they have the advantage of not needing filament supplies, they are smaller, faster, less expensive and they dissipate less heat. Although the theory of semi-conductor devices is fairly complicated, a very much simplified description of the diode behavior will suffice for the present purpose.

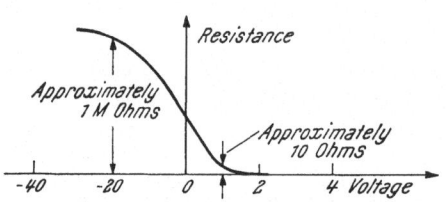

Fig. 4.2. The Resistance of a Semi-Conductor Diode as Function of the Applied Voltage

The resistance of a diode is a function of the applied voltage as indicated in Fig. 4.2.

With some idealization, we can say the diode is conducting (i.e. has a low resistance) if a voltage is applied in "forward" direction and it is non-conducting (i.e. has a high resistance) if a voltage is applied in reverse direction. The definition of forward and backward can be seen in Fig. 4.3.

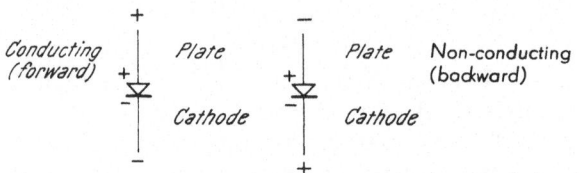

Fig. 4.3. Conventions Concerning Diode Symbols

4.1.1. Positive OR-Circuits

The symbol and the truth table for an OR-circuit are given in Fig. 4.4.

If we assume that the higher (the more positive) level represents a "1" and the lower (the more negative) level represents a "0", then the simple arrangement given in Fig. 4.5 will serve as a two-input OR-circuit.

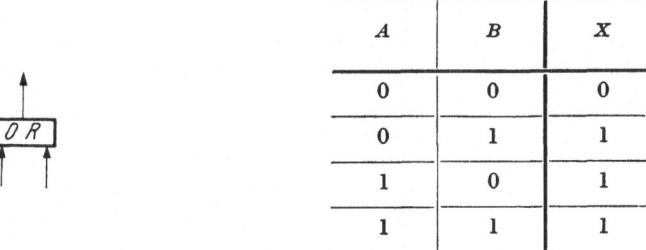

A	B	X
0	0	0
0	1	1
1	0	1
1	1	1

Fig. 4.4. Symbol and Truth Table for an OR-Circuit with Two Inputs

Let us suppose both inputs (*A* and *B*) are held to the higher of the two logical voltage levels, say +10 volts. Both diodes are conducting since their plates are positive with respect to their cathodes (which are connected via a resistor to some negative supply, say, —80 volts). The internal resistances of both diodes are negligible, and the output *X* assumes a level of practically +10 volts. If both inputs are held to the lower logic voltage

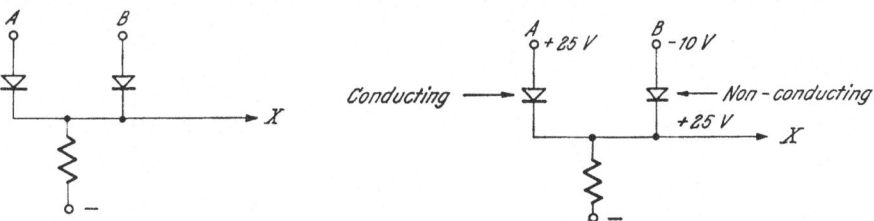

Fig. 4.5.   Diagram of a
Positive OR-Circuit with
      Two Inputs

Fig. 4.6. Diode OR-Circuit with One High and
                One Low Input

level, say —25 volts, the plates are still more positive than the cathode supply. The diodes are conducting and the output assumes a level of practically —25 volts. If one input is high, say *A*, and one input is low, say *B*, the diode connected to the high input is conducting, whereas the other one is not conducting because the plate is negative with respect to the cathode. The output assumes the high level as can be seen in Fig. 4.6.

The output voltage may be listed as function of the input voltage as shown in Table 4.1.

If we compare this listing with the truth table in Fig. 4.4, we recognize the behavior of an OR-circuit with the level of —25 volts representing a logical "0", and the level of +10 volts representing the logical "1"[1]. The

---

[1] We also have the behavior of an AND-circuit, if —25 volts represents the logic "1" and +10 volts represents the logic "0". Compare also paragraphs 4.1.5 and 4.4.

type of OR-circuit given in Fig. 4.5 is not restricted to the choice of logical voltage levels assumed here. Provided the more positive level represents the logic "1" and the more negative level represents the "0", almost any

Table 4.1.

Output Voltage as Function of the Input Voltage for the Circuit in Fig. 4.5

A	B	X
− 25	− 25	− 25
−25	+ 10	+ 10
+ 10	− 25	+ 10
+ 10	+ 10	+ 10

choice of voltages can be accommodated. The value of the resistor should be selected so that it is small compared to the internal resistance of a diode in reverse direction, and large compared to the internal resistance of a conducting diode. The supply voltage should be more negative than the lower logical level.

The circuit is also not restricted as far as the number of inputs is concerned. Fig. 4.7 shows an OR-circuit with four inputs and the corresponding block diagram symbol.

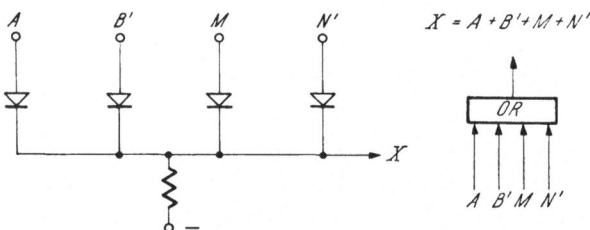

Fig. 4.7. OR-Circuit with Four Inputs

Problem 1: Suppose we have the condition $ABM'N'$. Which diodes in the circuit given in Fig. 4.7 are non-conducting? (Note that the presence of B implies the absence of B' etc.)

Problem 2: When diodes fail, they are usually "open" (i.e. they are not conducting in any direction) or "shorted" (i.e. conducting in both directions). Suppose you measure the voltages on inputs and outputs of an OR-circuit for different conditions, as listed below. Which diode is bad? Is ist shorted or open?

A	B	X
− 25	− 25	− 25
− 25	+ 10	− 25
+ 10	− 25	+ 10
+ 10	+ 10	+ 10

4.1.2. Positive AND-Circuits

The symbol and the truth table for an AND-circuit are given below:

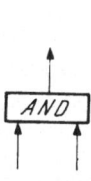

A	B	X
0	0	0
0	1	0
1	0	0
1	1	1

Fig. 4.8. Symbol and Truth Table for an *AND*-Circuit with Two Inputs

Assuming the same assignment of logical levels as in the last paragraph, we are looking for a circuit which gives a high output only if both inputs are high. The arrangement is not very complicated and is given in Fig. 4.9.

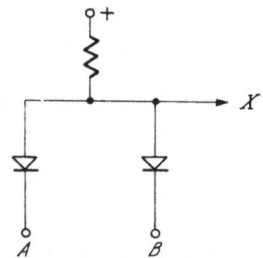

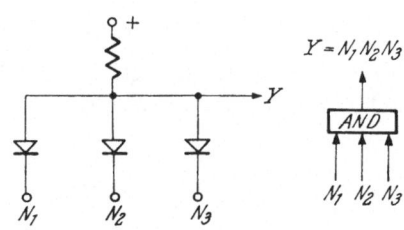

Fig. 4.9. Positive Diode *AND*-Circuit with Two Inputs

Fig. 4.10. Diode *AND*-Circuit with Three Inputs

If both inputs are high, the output assumes the high level. However, if one or more inputs are low, the corresponding diode or diodes pull the output down to the lower level. The output assumes, therefore, the high level, only if both inputs are high, or the circuit implements the truth table given in Fig. 4.8. Again, the circuit can be adapted for a variety of choices in logical levels and for several inputs. Fig. 4.10 shows an AND-circuit with three inputs.

Problem 3: Suppose the AND-circuit given in Fig. 4.10 is part of a computer in operation. The following oscillogram shows simultaneously the levels on all three inputs and on the output.

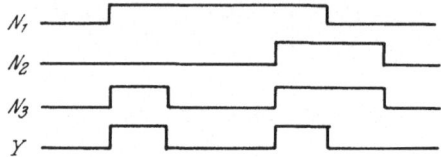

Assuming the possibility of open or shorted diodes only, find the one bad diode in this circuit.

Problem 4: Which diodes are conducting, if we have the condition $N_1 N_2 N_3'$, and the circuit given in Fig. 4.10.

4.1.3. Combinations of Positive OR- and AND-Circuits

Let us assume we have to design a circuit for the following function:

$$A = XY + UV \tag{4.1}$$

Realizing that the expression for A is a logic sum of two terms, we can write:

$$A = A_1 + A_2 \tag{4.2}$$

where:
$$A_1 = XY \tag{4.3}$$

$$A_2 = UV \tag{4.4}$$

Equation (4.2) requires an OR-circuit with two inputs, Equations (4.3) and (4.4) each require an AND-circuit.

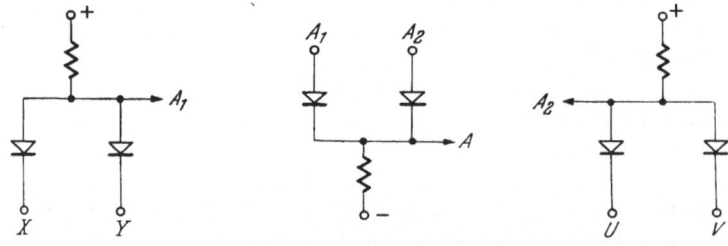

Fig. 4.11. Implementation of Equations (4.2) through (4.4)

The circuits for Equations (4.3) and (4.4) produce the outputs which are required as inputs for the circuit corresponding to Equation (4.2). Connecting outputs and inputs properly, we have a circuit which represents Equation (4.1).

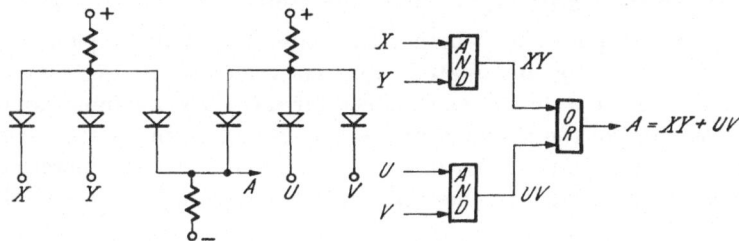

Fig. 4.12. Complete Circuit and Logic Diagram for Equation (4.1)

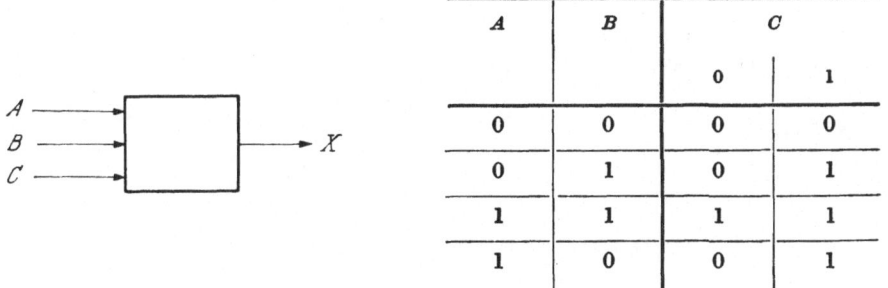

A	B	C	
		0	1
0	0	0	0
0	1	0	1
1	1	1	1
1	0	0	1

Fig. 4.13. Functional Diagram and Karnaugh Map for Sample Problem

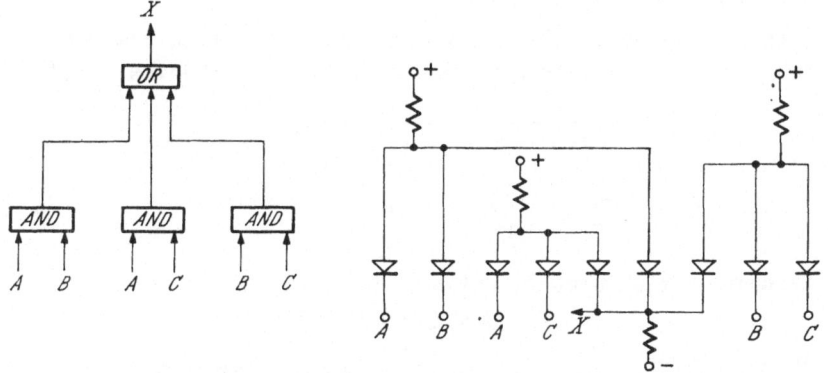

Fig. 4.14. Logic and Circuit Diagram for Sample Problem

Let us now attempt to solve a problem which is not yet stated in Boolean algebra. Suppose we want to construct a circuit which gives an output if there are two or more of its three inputs present.

4*

First, we construct a Karnaugh map. From the Karnaugh map we, subsequently, derive the following minimum sum:

$$X = AB + AC + BC \qquad (4.5)$$

The corresponding block and circuit diagrams are given in Fig. 4.14.

If we are not quite sure whether this circuit actually implements the function we originally anticipated, we can verify it by the method of perfect induction. We then have to check for *every* possible input combination, whether or not the circuit provides the correct output. Suppose one of the checks is made for the following condition: $A'B'C$. We can represent this unique condition in the logic diagram or in the circuit diagram as follows:

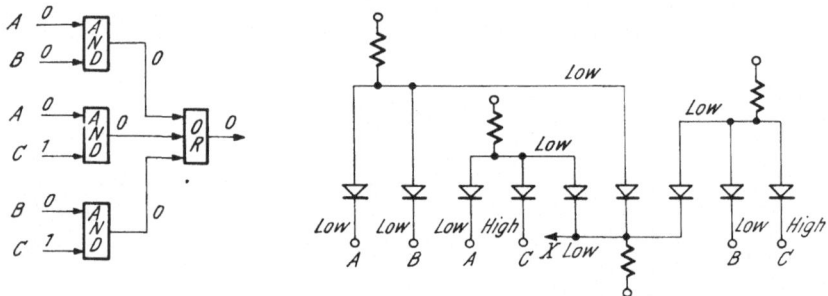

Fig. 4.15. Verification by Perfect Induction

The circuit provides the correct output for this combination. To be sure whether or not the circuit works for all combinations, a similar check on *all* conditions has to be made.

Problem 5: Show by the method of perfect induction, whether or not the following circuit is a representation of: $X = (A + B + C)(A + B')(B + C')$

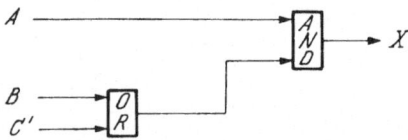

Problem 6: Implement the following function:
$$L = (XYZ + U + V)\ W$$

Problem 7: The circuit given in Fig. 4.14 contains nine diodes altogether Find an equivalent circuit using less diodes.

Problem 8: Implement the Karnaugh map given in Fig. 4.13 by a diode circuit representing the minimum product.

Problem 9: Find a simpler equivalent for the following circuit:

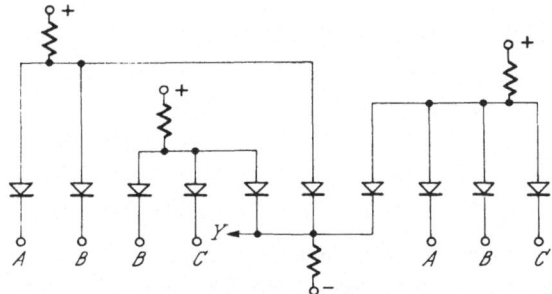

Perhaps, a word of caution is here appropriate. If a large number of OR- or AND-circuits are connected serially[1], the final output may deviate considerably from the correct voltage level defined for "1" or "0", due to the finite internal impedance of diodes. It is then harder to distinguish between 0's and 1's, and the circuit is more liable to cause malfunctions. Also, it becomes harder, if not impossible, to select appropriate values for the resistors in the circuit. Two levels of AND- or OR-circuits correspond to a standard or minimum sum or product, and are sufficient to represent any function. This is one of the reasons why these forms are attractive to computer designers. If for some reason, longer series of AND's or OR's are required, conservative designs restore voltage levels by active elements such as tubes or transistors.

For circuits which have to operate at high speeds or with high reliability, an active element is inserted after each diode AND- or OR-circuit.

4.1.4. Diode Matrices

"Decoding" circuits are given quite frequently in form of diode matrices. Let us suppose we want a circuit which can decode the decimal digits 0

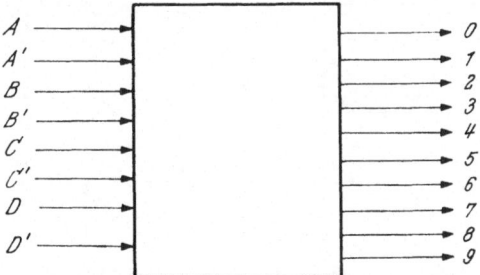

Fig. 4.16. Functional Diagram of a Decoding Circuit for the 8421 Code

through 9 from the 8421 code shown in Table 2.6. The circuit has ten outputs. Eight inputs are available.

---

[1] In Fig. 4.14 there are e.g. two circuits connected in series.

If proposition A has the weight 8, B the weight 4, C the weight 2 and D the weight 1, digits 0 through 9 are given by:

$$0 = A'B'C'D' \qquad\qquad 5 = A'BC'D$$
$$1 = A'B'C'D \qquad\qquad 6 = A'BCD'$$
$$2 = A'B'CD' \qquad\qquad 7 = A'BCD$$
$$3 = A'B'CD \qquad\qquad 8 = AB'C'D'$$
$$4 = A'BC'D' \qquad\qquad 9 = AB'C'D \qquad (4.6)$$

Each single digit can be decoded by a four-diode AND-circuit. Fig. 4.17 shows, for instance, the decoding circuit for the digit 3.

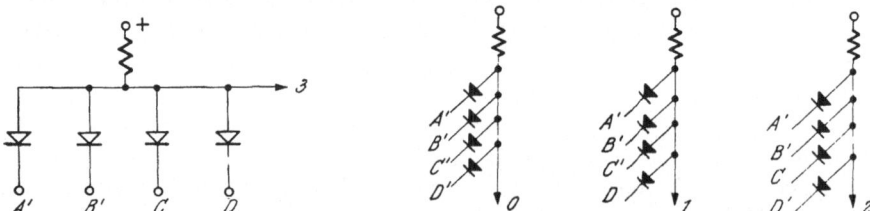

Fig. 4.17. Decoding of Digit 3   Fig. 4.18. Partial Matrix

For reasons which will be immediately apparent, let us draw the circuits for a few digits in a somewhat unconventional manner as shown in Fig. 4.18.

If we now draw all circuits and connect all points with the same label, we find the following diagram.

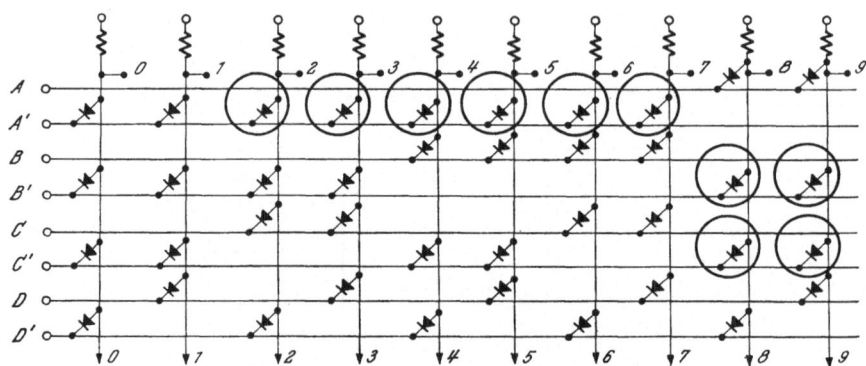

Fig. 4.19. Complete Decoding Matrix

Such an arrangement is called a diode matrix. Essentially, it is only a more compact way of drawing the circuit.

The ten diodes marked by circles in Fig. 4.19 could be omitted (see paragraph 3.5.3).

More complicated decoding circuits consist sometimes of several matrices. The following circuit can decode combinations of six variables (A, B, C, D, E, F). The circles in matrix I and II represent in simplified form, diodes as shown in Fig. 4.19. Circles in matrix III represent two diode AND-circuits as shown in the inset of Fig. 4.20.

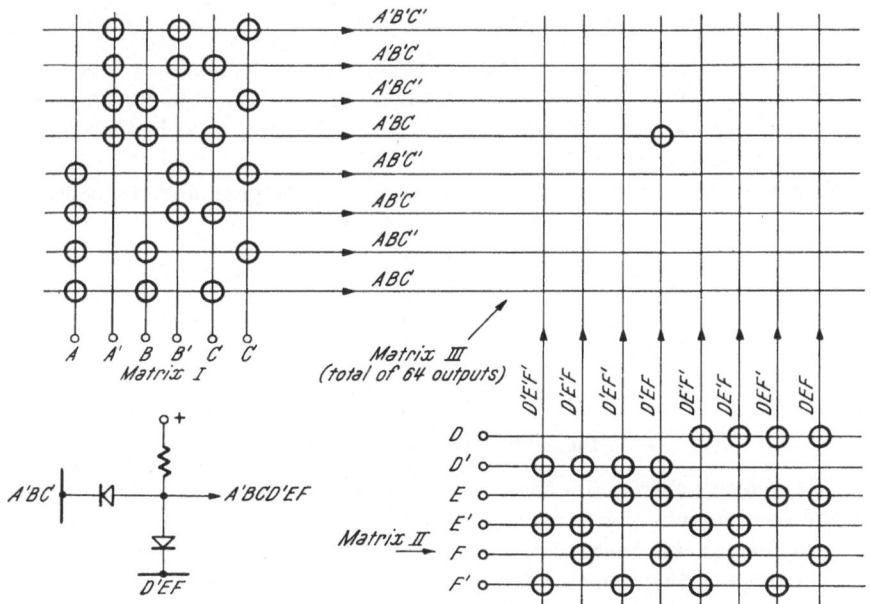

Fig. 4.20. Decoding Circuit for Six Variables

4.1.5. Negative AND- and OR-Circuits

The function of the diode circuits, shown up to now, has been defined if the more positive voltage level represents the logical "1", and if the more negative level represents the logical "0". Let us take the positive OR-circuit given in Fig. 4.5 and see how it will behave if the lower level is defined as the logical "1". From the list of outputs in Table 4.1, we see that it is now equivalent to an AND-circuit. The output is low (corresponding to a logic "1" according to our present definition) only if both inputs are low. We will call this a negative AND-circuit. Conversely, the positive AND-circuit in Fig. 4.9 is a negative OR-circuit. Its output is low if one or more inputs are low. A logic diagram may list either circuit as AND- or OR-circuit, depending upon the definition of logical levels and the function of the circuit.

Problem 10: Implement the functions given in problems 6 and 9 by diode circuits, assuming that the lower level of potential represents the logical "1".

Problem 11: Show a decoding matrix for the excess-3 code under the same assumption.

Problem 12 (Voluntary): Implement all functions of the collection started with problem 3—4 by positive and negative diode AND- and OR-circuits.

4.2. Transistor Circuits

For the understanding of transistor logic circuits, again a very simplified description of the basic element is adequate. The following remarks consider *pnp* transistors. Paragraph 4.2.7 will show changes in the concept if *npn*-types are considered.

A transistor has three electrodes: the emitter, the base, and the collector.

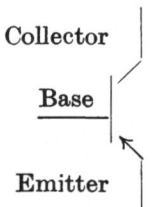

Collector

Base

Emitter

Fig. 4.21. Transistor Symbol

The collector current (i.e. the current from emitter to collector) is controlled by the base current (i.e. the current from emitter to the base). An increase in base current will increase the collector current. By decreasing the base current to zero, the collector current is decreased to practically zero. Since the amount of collector current is larger than the amount of base current required to control it, we speak of "current gain".

4.2.1. The Inverter or NOT-Circuit

With this relatively primitive knowledge of transistors, let us consider the circuit in Fig. 4.22.

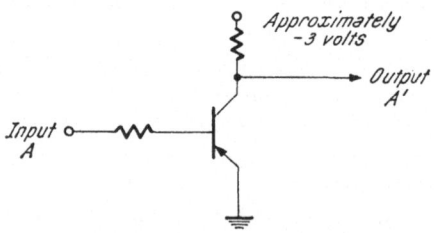

Fig. 4.22. Basic NOT-Circuit or Inverter

If the input is at ground potential, the emitter and the base have the same potential, and there will be no base current. The transistor is cut off and, consequently, the output assumes a potential of —3 volts. If the input is at a potential of —3 volts, there is a base current, the transistor is conducting and the output assumes (almost) ground potential.

We can now write the following relations:

Table 4.2. *Truth Table of the Circuit Given in Fig. 4.22*

Input	Output	A	A'
0 volts	— 3 volts	0	1
— 3 volts	0 volts	1	0

The output is high if the input is low and vice versa. The behavior is that of a NOT-circuit, regardless of the manner in which logical 0's or 1's are assigned to high or low potentials.

The value of the input resistor has to be small enough so that for an input of —3 volts there is enough base current to make the transistor fully conducting. On the other hand, the input resistor should be large so that an inverter represents a small load to the circuitry which drives the input. The load resistor (the resistor from the collector to the negative voltage supply) should be small so that the circuit can drive a large load. On the other hand, the value of this resistor should be high so that the internal impedance of the transistor, when fully conducting, is negligible in comparison. Actual design values constitute compromises between these conflicting requirements.

The circuits of actual transistor inverters are usually more complex than the basic circuit shown in Fig. 4.22. Additional circuit elements either increase the reliability of the inverter operation or they increase the operating speed. Fig. 4.23 shows an example of an actual high-speed transistor inverter.

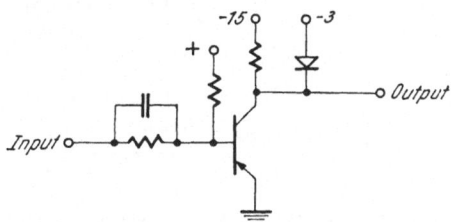

Fig. 4.23. Example of a High-Speed Transistor Inverter

The resistor from the base to the positive supply voltage pulls the base slightly positive for an input of approximately zero volts. Therefore, the transistor is cut off, even though there may be some noise on the input or the input may not quite reach zero volts due to the finite internal impedance of the input source. The capacitor across the input resistor couples momentarily the full swing of the input signal to the transistor base if the input changes states. The resistor alone—even though very important for steady state conditions—would couple only a fraction of the signal. (The input resistor together with the base impedance acts as voltage divider). The relatively large input signal accelerates the switching of the transistor (i.e. the transition from conducting to non-conducting or vice versa). The function of the clamping diode on the output can be best understood if we consider a capacitive load on the inverter output. (A capacitive load is always present, even if it is only the capacitance of the connected wiring.) Assume that the inverter in Fig. 4.23 is cut off by an input signal and that the load resistor is connected to —3 volts. The load resistor charges the capacitor on the output to —3 volts. The voltage will drop according to an exponential function, the time constant of which is determined by the product of R and C (trace a in Fig. 4.24).

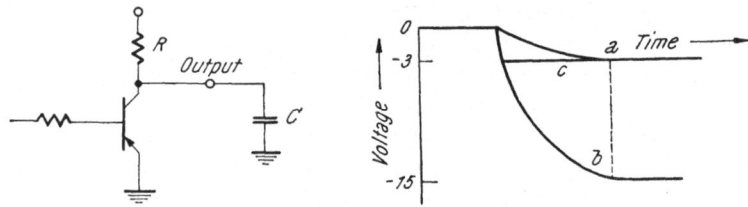

Fig. 4.24. Output Waveforms of a Transistor Inverter

Assume now that the load resistor is connected to —15 volts instead. The time constant remains the same, i.e. the output tends to assume a

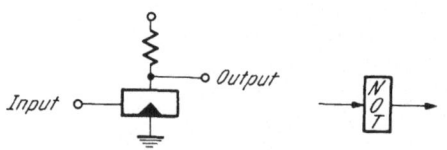

Fig. 4.25. Symbolic Representation of Transistor Inverters

voltage of —15 volts after the same time it would have otherwise taken to reach —3 volts (trace b). The clamping diode does not allow the output to assume a voltage lower than —3 volts, therefore the actual output follows trace c. We see that the transition in output voltage is speeded up considerably compared to trace a. The clamping diode also provides for a low effective output impedance since it allows the circuit to tolerate a wide variation in output loading without significantly changing the output voltage.

Incidentally, the single transistor shown in Figs. 4.22 and 4.23 is in some designs replaced by the *Darlington compound* of two transistors [1].

Since the actual circuitry of a transistor inverter is somewhat complicated, we shall from now on use the symbol given in Fig. 4.25a to represent it in circuit diagrams. For logic diagrams, we shall use the symbol given in Fig. 4.25b.

4.2.2. Combinations of NOT-Circuits with AND- and OR-Circuits

NOT-circuits combined with AND- and OR-circuits can perform any switching function[1]. Fig. 4.26 shows, for example, the function $X = A'B$ implemented by using the inputs A and B exclusively (not using A' or B').

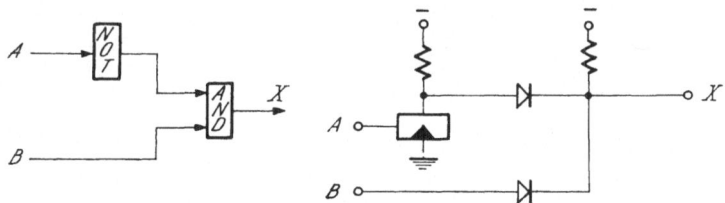

Fig. 4.26. Logic Diagram and Circuit Diagram for Sample Function

The circuit diagram is correct under the assumption that the more negative level (—3 volts) represents a "1". We shall from now on adhere to this convention for the remainder of this chapter unless otherwise noted.

Fig. 4.27 shows a logic diagram for the function $Z = (A'B + AB')(BC)'$ in which only the propositions A, B, C are used as inputs.

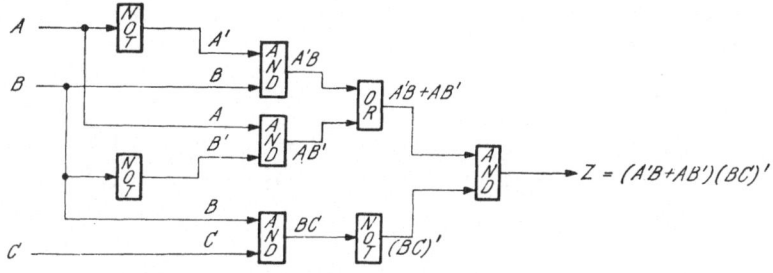

Fig. 4.27. Logic Diagram for $Z = (A'B + AB')(BC)'$

Problem 13: Show the simplest circuit you can find which is equivalent to the one given in Fig. 4.27. Draw the circuit diagram in addition to the logic diagram.

---

[1] The general theorem (see paragraph 3.4.2) states that any logic function can be represented by a sum of products, or a product of sums and, therefore, by logic AND, OR, NOT.

Problem 14: Design a circuit which gives an output if exactly two (any two) out of three inputs are present. Show only the logic diagram. Use only the inputs A, B, C (not A', B', C').

The following identities are frequently helpful in designing or simplifying circuits:

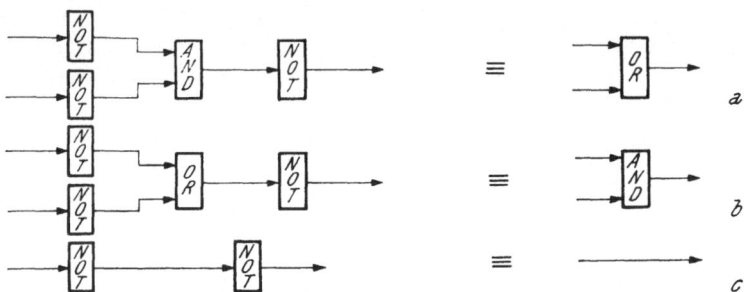

Fig. 4.28. Logic Equivalents

Problem 15: Prove the identities given in Fig. 4.28 by Boolean algebra.

From these identities we can deduce that AND-circuits plus NOT-circuits are sufficient to implement any logic function, since it is possible to make an OR-circuit from the other two. In the same manner we can deduce that OR-circuits plus NOT-circuits are self-sufficient.

Problem 16: Simplify the following circuits and show the resulting block and circuit diagrams.

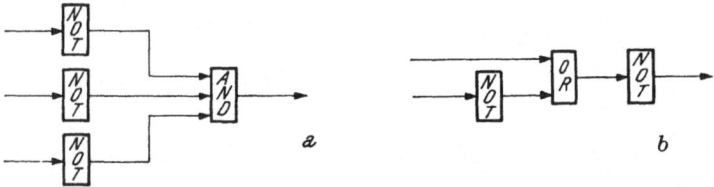

Problem 17 (Voluntary): Add the logic diagram and the circuit diagram to the collection started with problem 3—4. Use only the propositions A and B as inputs (not A' or B').

Diode circuits are frequently combined with transistor inverters in so-called "logic modules", i.e. plug-in cards or potted and sealed units. Fig. 4.29 shows the circuit diagram and a simplified schematic representation for such a circuit:

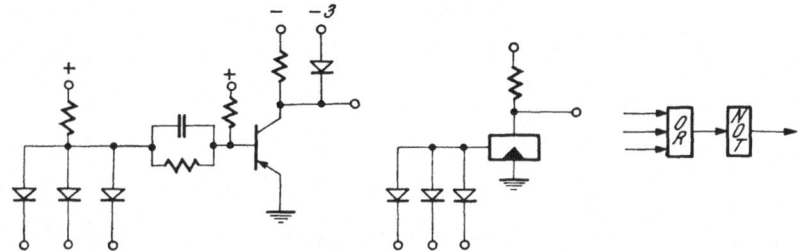

Fig. 4.29. Example of a NOR Logic Module

The circuit is equivalent to an OR-circuit in series with a NOT-circuit. It is, therefore, sometimes called a NOR- (=NOT OR) circuit. It is interesting from a theoretical and practical point of view to note that combinations of this circuit alone can perform any logic operation. For instance:

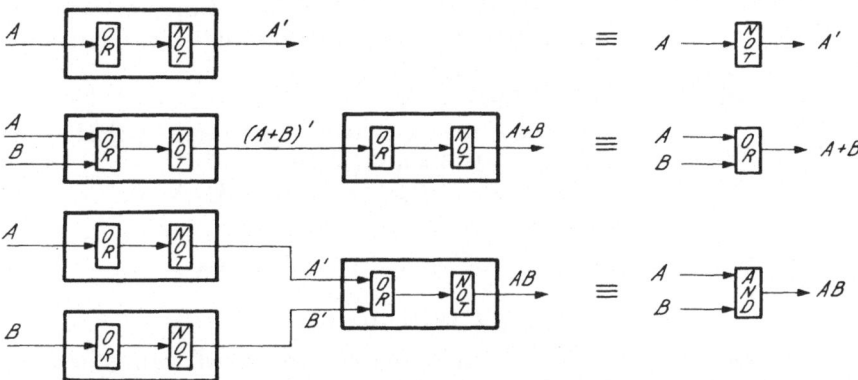

Fig. 4.30. The Implementation of NOT, OR, AND by NOR Modules

Since the three logical functions of AND, OR, NOT can be implemented, any function can be implemented[1]. Digital designs may consist of such modules exclusively. We speak then of NOR logic.

Problem 18: Implement the following block diagram by NOR modules exclusively. Use as few modules as possible.

Sometimes a negative AND-circuit is combined with a transistor inverter in a module as shown in Fig. 4.31.

The circuit is equivalent to an AND-circuit in series with a NOT-circuit. It is therefore sometimes called a NAND (=NOT AND) module.

---

[1] Two of these modules can also be combined to a flip-flop. See paragraph 5.1.1.

Problem 19: Show that the logic operations NOT, AND, OR can be implemented by NAND modules exclusively.

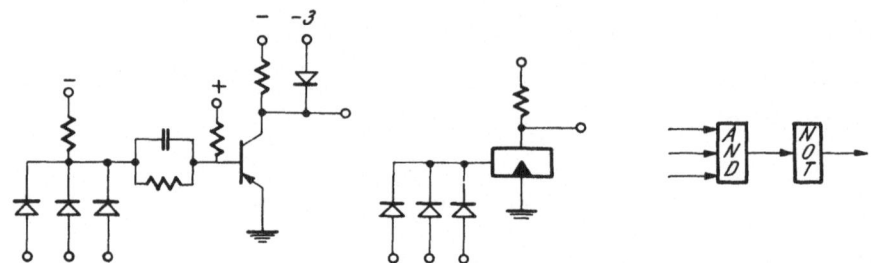

Fig. 4.31. Representations of a NAND-Module

Problem 20: Implement the following block diagram by NAND modules exclusively. Use as few modules as possible.

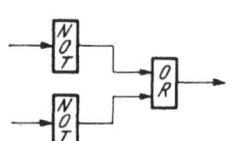

Problem 21: a) What logic function expressed in Boolean algebra is performed by the NOR and NAND modules?

b) What logic function is performed by these modules if the logic "0" is represented by the low and the logic "1" by a high potential?

State the output X as function of the three inputs M, N, P.

Problem 22 (Voluntary): Implement the logic functions of the collection started with problem 3—4 by NOR and NAND modules.

4.2.3. The Transistor Gate

The transistor gate can be considered as a modified transistor NOT-circuit. We notice that the transistor in Fig. 4.22 or 4.23 is able to pull the output to ground potential only if its emitter is at ground potential. If the emitter has a potential of —3 volts, the output will stay at a potential of —3 volts, regardless of what potential is applied to the base. If we connect the emitter to a second input, instead of grounding it, we obtain the circuit given in Fig. 4.32:

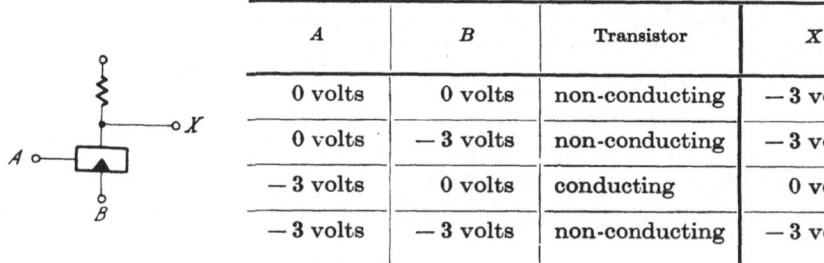

A	B	Transistor	X
0 volts	0 volts	non-conducting	— 3 volts
0 volts	— 3 volts	non-conducting	— 3 volts
— 3 volts	0 volts	conducting	0 volts
— 3 volts	— 3 volts	non-conducting	— 3 volts

Fig. 4.32. Transistor Gate

With our convention where —3 volts represent a logical "1" and 0 volts a logical "0", we obtain the following truth table.

Table 4.3. *Truth Table for Transistor Gate*

A	B	X
0	0	1
0	1	1
1	0	0
1	1	1

The behavior of the transistor gate may also be indicated by logic symbols or expressed in Boolean algebra:

$$X' = AB'$$
$$X = (AB')' = A' + B \tag{4.7}$$

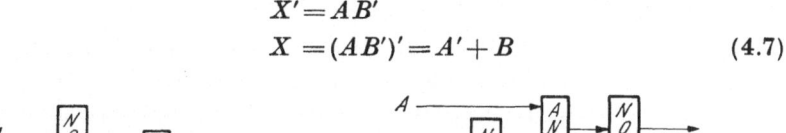

Fig. 4.33. Representations of the Transistor Gate

Problem 23: Show that the logic functions defined in Table 4.3 and Fig. 4.33 are equivalent.

4.2.4. The Implementation of Logic Functions by Transistors Exclusively

4.2.4.1. Parallel Connection of Transistor Gates: In the simplest case, two transistor gates are connected in parallel and both emitters are at ground potential.

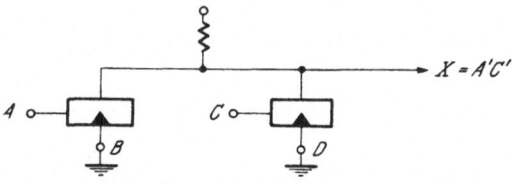

Fig. 4.34. Two Transistor Gates in Parallel

The output will assume ground potential if either one or both transistors become conducting, i.e., if either one or both bases become negative. This behavior may be shown in the following truth table:

Table 4.4. *Truth Table for the Circuit Given in Fig. 4.34*

A	C	X	A	C	X
0 volts	0 volts	− 3 volts	0	0	1
0 volts	− 3 volts	0 volts	0	1	0
− 3 volts	0 volts	0 volts	1	0	0
− 3 volts	− 3 volts	0 volts	1	1	0

The output, apparently, follows the equation:

$$X = A'C' = (A+C)' \tag{4.8}$$

Problem 24: What function is implemented by three transistors connected in parallel, all emitters being grounded?

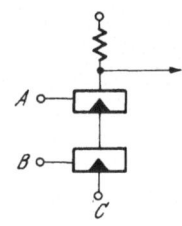

Fig. 4.35. Two Transistor Gates Connected in Series

Problem 25: The emitters (inputs B and D) of the circuit in Fig. 4.34 do not have to be grounded necessarily. Express the output X as function of the four inputs A, B, C, D.

4.2.4.2. Serial Connection of Transistor Gates: The serial connection of two transistor gates is shown in Fig. 4.35.

The chain of transistors can pull the output to ground potential only if both transistors are conducting, that is, if the input C is at ground potential *and* the inputs A and B are negative. We can express this statement as follows:

$$X' = ABC' \text{ or}$$
$$X = (ABC')' = A' + B' + C \tag{4.9}$$

Problem 26: What function is implemented by the following circuit:

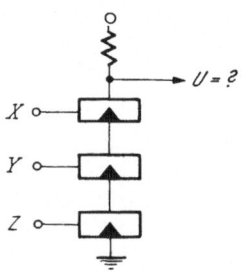

4.2.4.3. Arbitrary Connection of Transistor Gates: Transistorized switching circuitry usually consists of a combination of transistor gates and diode AND- and OR-circuits. However, it is possible to implement arbitrary logic functions by transistor circuits exclusively. Let us first show how to implement simple AND- and OR-circuits by transistor circuits.

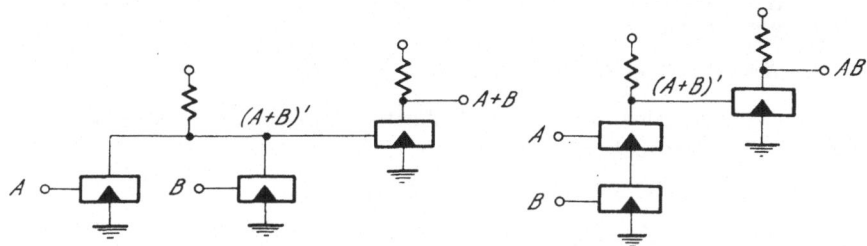

Fig. 4.36. Implementation of OR- and AND-Circuits by Transistors

An OR-circuit with only two transistors can be designed according to Fig. 4.37:

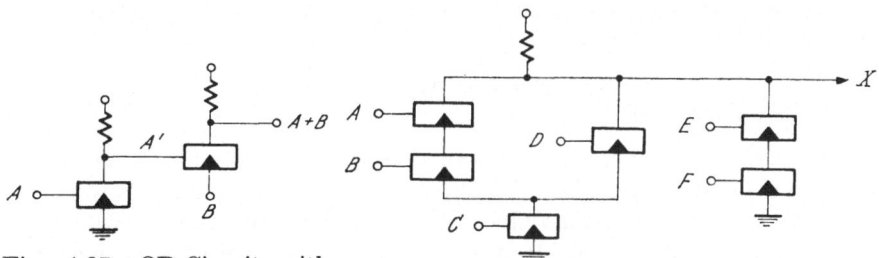

Fig. 4.37. OR-Circuit with Two Transistors

Fig. 4.38. Network of Transistor Gates

A disadvantage of the circuit is the relatively low impedance of input B[1].

Let us now take a fairly complicated network of transistor gates and try to find the logic function which it implements. (See Fig. 4.38.)

We will establish a current path if a string of transistors from the load resistor to ground becomes conducting, i.e. if we have one or more of the following conditions: ABC, or DC, or EF. For these conditions, the output will assume ground potential, i.e. will assume the logical "0" state. Therefore, we can write:

$$X' = ABC + DC + EF \qquad (4.10)$$

---

[1] An emitter input requires several times the driving current of a base input. The ratio of current required by a base input to that required by an emitter input is very nearly equal to the current gain of the transistor, beta. Representative values for beta fall in the range between 10 and 100.

or with De Morgan's Theorem:

$$X = (A' + B' + C') (D' + C') (E' + F') \qquad (4.11)$$

We could also have reasoned like this: We interrupt the current path (corresponding to a "1" on the output) if we have one or more of the following conditions: $A'D'E'$, $A'D'F'$, $B'D'E'$, $B'D'F'$, $C'E'$, $C'F'$. Correspondingly we can write:

$$X = A'D'E' + A'D'F' + B'D'E' + B'D'F' + C'E' + C'F' \qquad (4.12)$$

With a little practice, we can find shorter expressions directly from the diagram. For instance:

$$X' = C(AB + D) + EF, \text{ and thus}$$
$$X = C'(E' + F') + D'(A' + B') (E' + F') \qquad (4.13)$$

Problem 27: Implement the following functions by transistors exclusively:

a) $Y = (U + V' + W') D'$
b) $Z = (A + B) (C' + D')$

Problem 28: Implement the following logic functions by the most economical circuit you can find:

a) $X = A'B + AB'$ (use only the inputs A and B')
b) $Y = K + L + MN'$

The propositions A, B, K, L, M, N and their primes are available as inputs. Assume the cost for a diode to be \$2.—, and the cost for a transistor to be \$3.—. Other circuit elements like resistors or capacitors are free of charge. If you use diode AND- or OR-circuits, restore the level after each circuit.

Conservative designs connect not more than about four transistor gates in series or in parallel. Otherwise, the output will deviate too far from nominal potentials due to the finite internal impedance of transistors.

4.2.5. Pulse-Type Switching Circuits

Transistor gates are not only used for levels but also used extensively for pulse-type switching. The base is then "strobed" by negative pulses. The gate is "enabled" when the emitter is at ground potential. Input pulses produce then (positive) output pulses. If the gate is "inhibited" by a potential of —3 volts on the emitter, no output pulses occur.

Pulse gates may be represented in logic diagrams by AND, OR, NOT-circuits according to the logic function they perform. There are, however, several special symbols in use. We shall use the logic diagram symbol given in Fig. 4.39b. Pulse-type switching circuits frequently employ pulse transformers on their outputs. Fig. 4.40. shows, as an example, the circuitry of a transistor gate with pulse transformers[1].

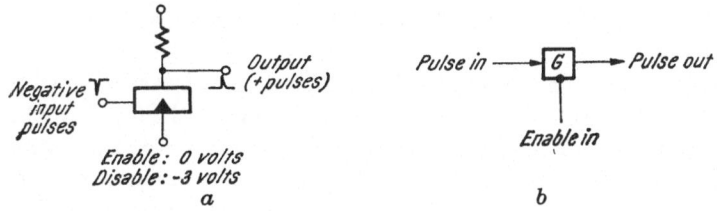

Fig. 4.39. Transistor Pulse Gate

Pulse outputs are completely decoupled as far as *DC* levels are concerned. The transformer ratios can be designed so that pulse amplitude requirements of different circuits are met. Furthermore, complete freedom is achieved as far as pulse polarities are concerned. For transmission of signals over long lines, the internal impedance of circuits can be matched to the line

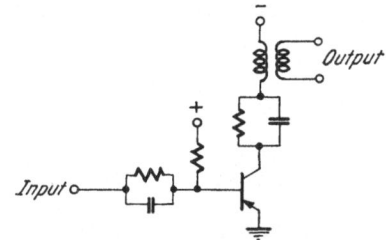

Fig. 4.40. Transistor Gate with Pulse Transformers

impedance. Using twisted pairs as transmission line, the noise level can be kept very low. Transistor pulse gates can be combined with transistor level gates in various ways. Fig. 4.41 shows several circuit diagrams to a transistor pulse gate enabled by the proposition $A + B$ and the equivalent logic diagram.

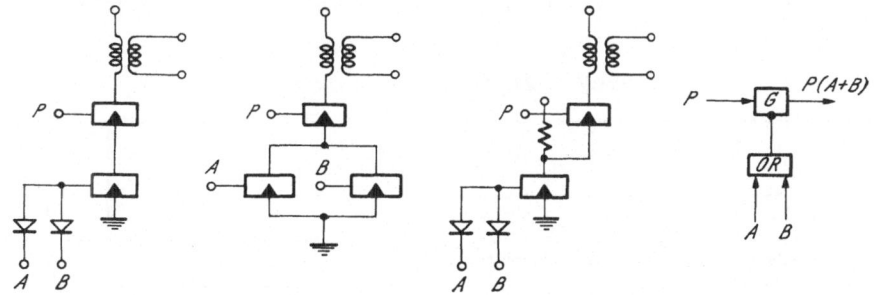

Fig. 4.41. Pulse Gate Enabled by the Proposition $A + B$

Problem 29: Design the circuitry for a pulse gate enabled by the proposition AB'. Show the logic diagram in addition to the circuit diagram.

---

[1] Pulse transformers are small iron ferrite transformers which are used to generate and shape pulses. They are useful for pulse widths in the approximate range between .05 and 50 μs.

4.2.6. The Emitter Follower

Emitter followers are not "logical" elements, but are sometimes used to obtain low-impedance outputs. The circuit of an emitter follower is given in Fig. 4.42.

We remember that a transistor is conducting if the base is more negative than the emitter. If we apply a ground potential to the input of the above circuit, the transistor becomes conducting, and the output potential drops from +15 volts to more negative values. If it has reached a potential of approximately 0 volts, the transistor begins to cut off and an equilibrium is reached. In a similar manner, the output will assume a potential of approximately —3 volts if the input is at that potential.

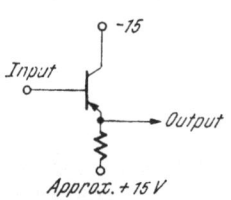

Fig. 4.42. Emitter
          Follower

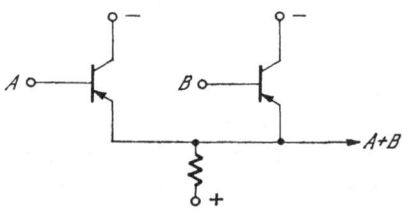

Fig. 4.43. Negative OR-Circuit with
          Emitter Followers

As we can see from Fig. 4.42, any voltage difference between the input and the output is applied directly to the base and the emitter of the transistor, that is, to the control terminals of a current amplifier. This causes the arrangement to oppose strongly any change in output voltage due to variations of the output load, and makes so for a very low impedance circuit.

Table 4.5. *Truth Table for the Circuit Given in Fig 4.43*

A	B	X	A	B	X
0 volts	0 volts	0 volts	0	0	0
0 volts	— 3 volts	— 3 volts	0	1	1
— 3 volts	0 volts	— 3 volts	1	0	1
— 3 volts	— 3 volts	— 3 volts	1	1	1

Two or more emitter followers may be combined to represent a negative OR-circuit as shown in Fig. 4.43.

The output assumes the potential of the more negative input.

The output, therefore, follows the equation

$$X = A + B.$$

4.2.7. Circuits with *npn*-Transistors

The *npn* transistor is a device complementary to the *pnp* transistor. Where the *pnp* transistor requires negative voltages for its operation, the *npn* transistor needs positive voltages. Consequently, all currents will have a direction opposite to that in a *pnp* transistor. Otherwise, the characteristics of both types of transistors can be considered the same. Before we see what consequences result for the interpretation of switching circuits in general, let us show the basic *npn*-type inverter.

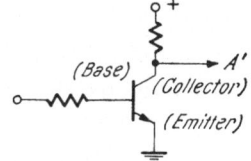

Fig. 4.44. Basic *npn*-Inverter

This circuit is identical to that in Fig. 4.22 except for the polarity of supply voltages. The transistor is conducting if the base is positive with respect to the emitter. If the base is at the same potential as or more negative than the emitter, the transistor is non-conducting.

A simplified *npn* transistor gate and its truth table are shown below:

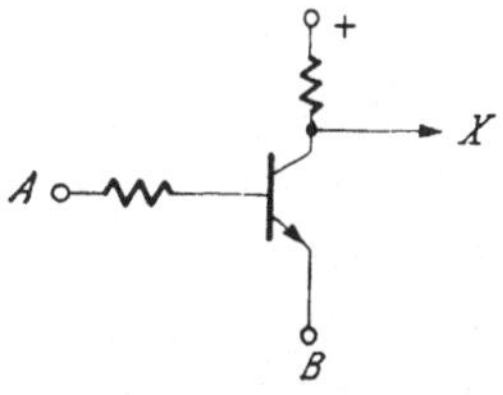

A	B	X	Transistor
0 volts	0 volts	+3 volts	Non-conducting
0 volts	+3 volts	+3 volts	Non-conducting
+3 volts	0 volts	0 volts	Conducting
+3 volts	+3 volts	+3 volts	Non-conducting

Fig. 4.45. *npn* Gate and Truth Table

If we compare Fig. 4.45 with Fig. 4.32, we see that the two circuits behave the same, except that there are now +3 volts where we used to have —3 volts. If we assign to the level of +3 volts the same logical meaning (i.e. "1") as previously to the level of —3 volts, the two circuits are logically identical. In the same manner, any circuit with *npn* transistors is equivalent to the same circuit with *pnp* transistors provided that the logical meaning of more positive and more negative voltage levels is interchanged.

Problem 30: Modify the circuit given in Fig. 4.34 for *npn* transistors and show that its logical truth table corresponds to the one given in Table 4.4.

4.3. Vacuum Tube Circuits

Practically no newly designed computer contains vacuum tube circuits as main elements. However, since vacuum tubes are still used to some extent in peripheral devices and have merits for some applications. a very short survey of logic vacuum tube circuits is included here. The logic function of tube circuits is described with the assumption that the more positive level corresponds to the logic "1" state.

4.3.1. Triode Circuits

A triode has three electrodes: cathode, grid and plate. The plate current is a function of the plate voltage (the voltage between plate and cathode) and the grid voltage (the voltage between grid and cathode). For the present purpose, it is sufficient to assume that the plate current is only a function of the grid voltage:

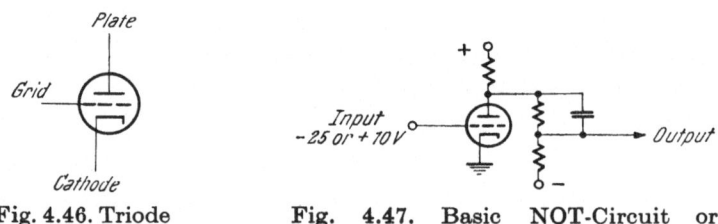

Fig. 4.46. Triode   Fig.   4.47.   Basic   NOT-Circuit   or
Symbol   Inverter

The plate current is increased by making the grid more positive. As the grid becomes more negative, the plate current decreases and, finally, ceases altogether.

4.3.1.1. NOT-Circuits with Triodes: By the proper choice of two grid voltages, we can make the triode conducting for one voltage, while the tube is non-conducting for the other.

In the above circuit (Fig. 4.47), either —25 or +10 volts are applied to the grid. In the first case, the tube is non-conducting. There is no plate current and, consequently, no voltage drop across the plate resistor. The plate assumes the voltage level of the positive supply. If +10 volts are applied to the grid, the tube is conducting. The plate current causes a voltage drop across the plate resistor and the output will assume a relatively low potential, the exact voltage depending upon the value of the plate resistor and the type of tube. Corresponding to the two input voltages we obtain two plate voltages. These two levels are usually made compatible to input levels by a proper voltage divider either on the input or on the output. We can now write the following list of inputs versus outputs:

Table 4.6. *Listing of Input and Output Voltages for the Circuit in Fig. 4.47*

Input	Output
— 25 volts	+ 10 volts
+ 10 volts	— 25 volts

The output is high, if the input is low and vice versa. We recognize the behavior of a NOT-circuit.

Vacuum tube NOT-circuits may be combined with diode AND- and OR-circuits in the same manner as transistor NOT-circuits.

Problem 31: Show the simplest circuit diagram equivalent to the logic diagram given below, using diode AND- and OR-circuits and triode NOT-circuits.

Problem 32 (Voluntary): Add the vacuum tube circuit diagram to the collection started with problem 3—4. Use only the propositions A and B (not A' and B').

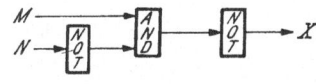

4.3.1.2. The Cathode Follower: The cathode follower, like the emitter follower, is in itself not a logic circuit. However, cathode followers are frequently used as low-impedance sources for logic propositions, and combinations of cathode followers can perform certain logic functions.

Let us assume we have a potential of —25 volts on the grid of the tube in the above circuit. The tube is conducting, since the grid potential is quite high compared to the cathode supply voltage. Consequently, the current through the

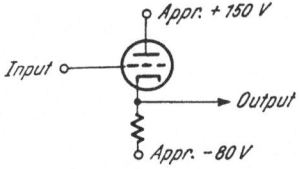

Fig. 4.48. Cathode Follower

tube causes a voltage drop across the cathode resistor. The cathode assumes a potential which is higher than —80 volts, and which is very nearly equal to the grid potential. If we apply a higher potential to the grid, we will have a higher current through the tube. The cathode will be at a potential which is more positive than previously, and once more nearly equal to the grid potential.

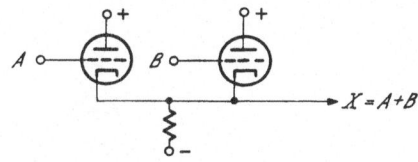

Fig. 4.49. OR-Circuit with Triodes

4.3.1.3. The OR-Circuit: Two cathode followers with a common load resistor constitute a vacuum tube OR-circuit:

If both inputs are at equal potentials, the output will assume the same potential. If one input is high, the other one low, the tube connected to the high input is conducting and pulls the output to the higher potential. The other tube is not conducting, since its grid is negative with respect to its cathode.

Using more than two triodes with a common load resistor, we can design OR-circuits with more than two inputs.

4.3.1.4. The Triode Gate: Two NOT-circuits with a common plate resistor constitute a triode gate.

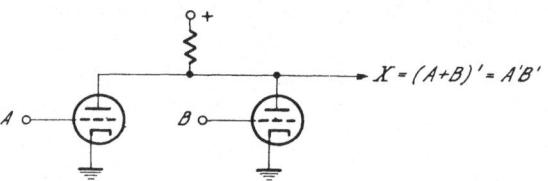

Fig. 4.50. Triode Gate

The output is high as long as the grids are low and neither tube is conducting. As soon as either one (or both) of the grids goes high, the corresponding tube conducts and the output goes low.

Problem 33: Show the truth table for this circuit and show that it is equivalent to the truth tables of the following circuits:

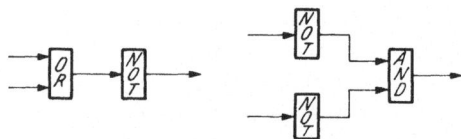

Problem 34: Show that it is possible to implement any logic function by triode gates exclusively.

Problem 35: What logic expression is implemented by the following circuit:

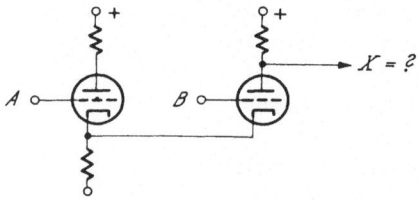

Assume that the input B uses slightly more negative logical levels than input A. (This can easily be accomplished by a proper voltage divider at the input of tube B.)

4.3.2. Pentode Circuits

A pentode has five electrodes: one cathode, three grids, and a plate:

For the understanding of switching circuits employing pentodes it is sufficient to know that both the control grid and the suppressor grid can cut off the tube, if the grid potentials are sufficiently low with respect to the cathode. The screen grid is usually connected to some fixed positive supply voltage, say $+100$ volts, and performs no logical function.

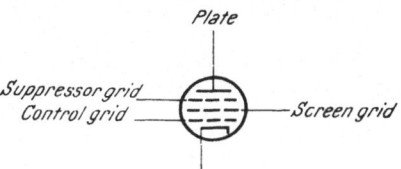

Fig. 4.51. Symbol for Pentode

4.3.2.1. The Pentode Gate: If we use two proper voltage levels, say $+10$ and -25 volts, a circuit can be designed so that the pentode is conducting only if both grids assume the more positive level and the tube is not conducting if either one, or both grids are at the lower level:

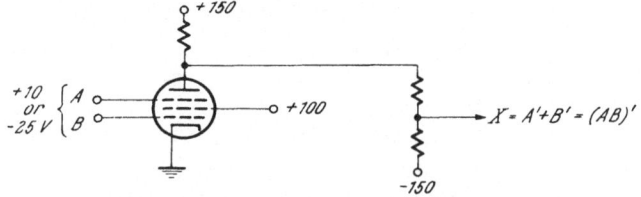

Fig. 4.52. Pentode Gate

The output of the circuit is made compatible to input levels by a voltage divider similar to the one used for triode NOT-circuits. The truth table of the circuit is given below:

Table 4.7. *Truth Table for Pentode Gate*

A	B	X	A	B	X
low	low	high	0	0	1
low	high	high	0	1	1
high	low	high	1	0	1
high	high ·	low	1	1	0

A pentode gate may be represented by either one of the following symbols:

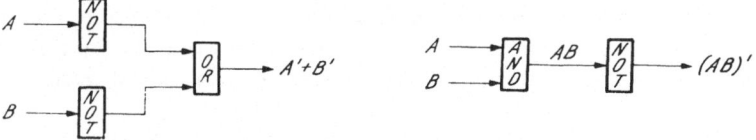

Fig. 4.53. Logic Equivalents of Pentode Gate

Problem 36: Show that the truth tables of the circuits given in Fig. 4.53 correspond to the truth table given in Table 4.7.

Problem 37: Show that the pentode gate is a self-sufficient circuit, i.e. that the three logic functions of AND, OR, NOT can be implemented by pentode gates exclusively.

Problem 38: What logic function is implemented by two pentodes which have a common plate resistor.

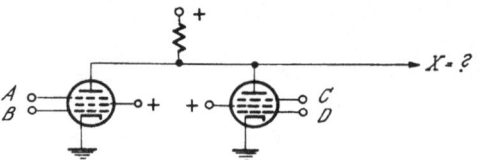

Show the (high or low) voltage on the output as function of (high or low) voltages on the inputs. Find corresponding logic expression.

Problem 39: Implement the expression $A+B+CD$ with the most economical circuit you can find. Assume costs for:

Germanium diode	$1.50
Pentode	$2.00
Double triode	$2.50

Resistors are free.

4.3.2.2. The Inhibit Gate: The inhibit gate is neither a truly pulse-type circuit nor a truly level-type circuit. It once has been used extensively in computers which are neither truly AC or DC machines[1].

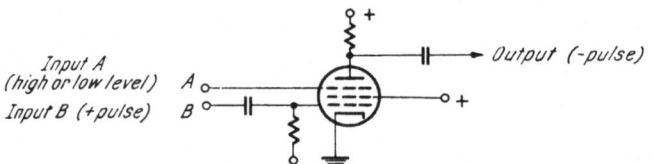

Fig. 4.54. Inhibit Gate

The circuit generates a (negative) output pulse only if there is a (positive) input pulse on the control grid and a high level on the suppressor grid. By applying a low potential to the suppressor grid, we can inhibit an output. Sometimes the following symbol is used for an inhibit gate:

---

[1] Machines with level-type circuits throughout are called DC machines. Machines with strictly pulse-type circuits are AC machines.

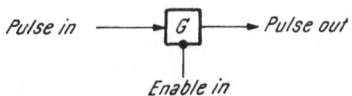

Fig. 4.55. Logic Symbol for Inhibit Gate

The "enable" (a high potential) opens the gate and lets pulses pass through. A low potential inhibits output pulses.

Inhibit gates have frequently pulse transformers on inputs and/or outputs. Fig. 4.56 shows such a circuit.

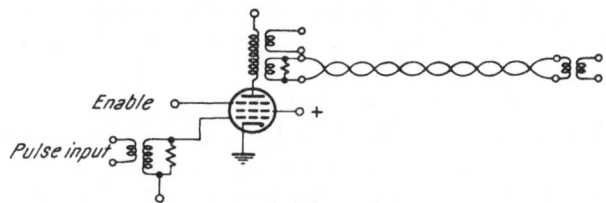

Fig. 4.56. Inhibit Gate with Pulse Transformers

Problem 40: Show the logic diagram and the circuit diagram for the function $X=(AB+C)p$. Assume that A, B, C are level inputs and that p is a pulse input.

4.4. Circuits with Inverted Levels

As we have seen in this chapter, the logic usage of high and low levels is not consistent. A high level may sometimes represent a logic "1" and sometimes a logic "0". Even within a particular design, levels may be sometimes reversed. It is, therefore, desirable to have general rules which allow us to deduce the logic function of a circuit for a certain definition of logic levels, if the function is known or defined for other assignments of levels.

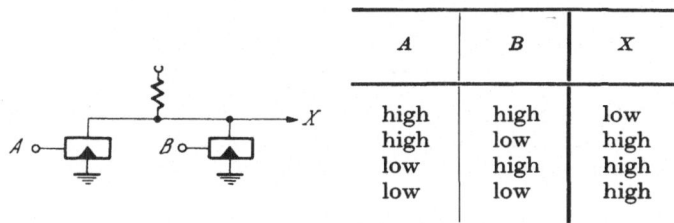

A	B	X
high	high	low
high	low	high
low	high	high
low	low	high

Fig. 4.57. Sample Circuit

Let us start with a specific, but relatively simple example. Fig. 4.57 shows two *pnp* transistor gates in parallel and a listing of output levels versus input levels.

Provided that the more negative level represents a logic "1", the circuit output is given by $X = A'B'$. If the high level represents the logic "1", the output function is $X = A' + B'$. Should we want to signify a logic 1 on the input by a low level, but on the output by a high level, then the circuit implements $X = A + B$. The representation of a logic 1 by a high level on the input but a low level on the output yields: $X = AB$. Depending upon the definition of logical levels, the identical circuit can perform: $X = AB$, $X = A + B$, $X = A'B'$ or $X = A' + B'$. In this particular example the various output functions can be derived relatively easily from the listing of high and low levels in Fig. 4.57 For an arbitrary case, we better resort to Boolean algebra.

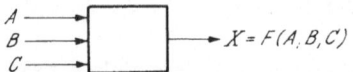

Fig. 4.58. Arbitrary Logic Circuit

Suppose we have an arbitrary logic circuit as given in Fig. 4.58.

The output is given by $F(A, B, C)$. If we interchange the definitions of logic levels only on the input, the output will then be given by $F(A', B', C')$ rather than by $F(A, B, C)$. The circuit performs identical logical functions upon the primes of previous inputs. In above example we obtain e.g. $X = AB$ rather than $X = A'B'$.

If we interchange the definition of the output only, we obtain $[F(A, B, C)]'$ instead of $[F(A, B, C)]$. In the above example we obtain $(A'B')' = A + B$ as output function.

If we interchange levels on both, inputs and output, we obtain $[F(A', B', C')]'$ instead of $F(A, B, C)$. In above example we obtain $(AB)' = A' + B'$ instead of $A'B'$.

Problem 41: What logic function will the following circuit implement if the definition of logical 0's and 1's is reversed for inputs and output? Show the appropriate new logic diagram.

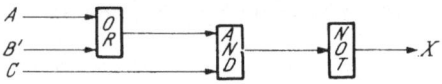

References

1. LINVILL J. G.: A New RC Filter Employing Active Elements, Proceedings Natl. Electronics Conference, vol. 9, pp. 342–352. 1953.
   SHEA R. F.: Transistor Circuit Engineering, pp. 130–133, New York: John Wiley. 1957.

Selected Bibliography

ANDERSON A. E.: Transistors in Switching Circuits, Proceedings IRE, vol. 40, No. 11, pp. 1541–1548. Nov. 1952.
Lo A. W.: Transistor Trigger Circuits, Proceedings IRE, vol. 40, p. 1535. Nov. 1952.

GLUCK, GRAY LEONDES, and RUBINOFF: The Design of Logical OR-AND-OR
Pyramids for Digital Computers, Proceedings IRE, vol. 41, No. 10, pp.
1388–1392. Oct. 1953.

LEBOW, and BAKER: Transient Response of Transistor Switching Circuits,
Proc. IRE, vol. 42, pp. 938–943. June 1954.

BRADLEY, BROWN, RUBINOFF, and BETTER: Surface Barrier Transistor
Computing Circuits, IRE Convention Record, part 4, pp. 139–145. 1955.

BRIGHT R. L.: Junction Transistors Used as Switches, Communications and
Electronics, No. 17, p. 111. March 1955.

YOKELSON, and ULRICH: Engineering Multistage Diode Logic Circuits,
Communication and Electronics, No. 20, pp. 466–475. Sept. 1955.

HUNTER L. P.: Handbook of Semiconductor Electronics. Section 15. New
York: McGraw-Hill. 1956.

MILLMAN, and TAUB: Pulse and Digital Circuits. New York: McGraw-Hill. 1956.

BOOTH, and BOTHWELL: Logic Circuits for a Transistor Digital Computer,
Transactions IRE, vol. EC-5, No. 3, pp. 132–138. Sept. 1956.

PROM, and CROSBY: Junction Transistor Switching Circuits for High-Speed
Digital Computer Applications, Transactions IRE, vol. EC-5, No. 4, pp. 192–
196. Dec. 1956.

CARROL J. M.: Transistor Circuits and Applications, chapter 7. New York:
McGraw-Hill. 1957.

DEWITT, and ROSOFF: Transistor Electronics, chapter 13. New York:
McGraw-Hill. 1957.

RICHARDS R. K.: Digital Computer Components and Circuits. Princeton:
D. Van Nostrand. 1957.

Transistor Circuit Engineering, chapter 10.5, edited by R. F. SHEA. New
York: John Wiley and Sons. 1957.

SMITH C. V. L.: Electronic Digital Computers. New York: McGraw-Hill. 1959.

NUSSBAUM, IRLAND, and YOUNG: Statistical Analysis of Logic Circuit Performance in Digital Systems, Proceedings IRE, vol. 49, No. 1, pp. 236-244.
Jan. 1961.

PRYWES, LUKOFF, and SCHWARZ: UNIVAC-LARC High-Speed Circuitry:
Case History in Circuit Optimization, Transactions IRE, vol. EC-10,
No. 3, pp. 426–438. Sept. 1961.

HOLST, P. A.: Bibliography on Switching Circuits and Boolean Algebra,
Transactions IRE, vol. E-10, No. 4, pp. 638-661. Dec. 1961.

CAN, HART, and RUYTER: The Junction Transistor as a Switching Device.
New York: Reinhold Publishing Corp. 1962.

HUSKEY, and KORN: Computer Handbook. New York: McGraw-Hill. 1962.

KEONJIAN E.: Microelectronics, Theory, Design, and Fabrication. New York:
McGraw-Hill, 1963.

CHUNG, and PALMIERY: Design of ACP Resistor-Coupled Switching Circuits,
IBM Journal of Research and Developement pp. 190–198. July 1963.

Special Issue on Integrated Electronics, Proceedings IEEE, vol. 52, No. 12.
Dec. 1964.

ATKINS J. B.: Worst-Case Circuit Design, IEEE Spectrum, pp. 152–161, March
1965.

BONGENAAR, and DE TROYE: Worst-Case Considerations in Designing Logical
Circuits, Transactions IEEE, vol. EC-14, No. 4. Aug. 1965.

5. Storage Elements

Logic elements alone—used in the straightforward manner shown in the previous chapter—are not sufficient to build a computer. It is necessary to have elements which perform the function of storage. This fundamental need can be illustrated by a basically simple example. Push-buttons with momentary contacts produce a certain output (opened or closed contacts) only as long as certain input conditions prevail (the button is pushed or not pushed). In this respect, they act like (and really are) logic elements. Using only such push-buttons, it will not be possible to design a circuit which turns a light on when a button is pressed, but leaves it on after the button is released. In order to accomplish this task, some storage element has to be incorporated into the circuit which stores (or "remembers") the fact that the button had been pressed. The storage element may be a relay as in a push-button motor control, or a simple mechanical device, as in a toggle switch which keeps the switch in the position into which it had been set last.

A similar need for storage exists in more complex mechanisms. A counter, for instance, has to "remember" a count; a computer has to "remember" at least the particular problem set-up and the particular operands. In general, any sequential machine requires storage of some sort[1].

The operating principle of storage elements incorporated in computers may be based upon almost any physical effect. There are mechanic, electric, magnetic, optic, acoustic, and cryogenic storage devices in use. Even the utilization of molecular and atomic effects is being investigated.

In this chapter, we shall consider only standard storage elements. Universal[2] or unusual elements will be discussed in chapter 11, and the organization of computer memories in chapter 8.

According to their mode of operation, we can distinguish two types of storage elements: static and dynamic. Static storage elements assume static conditions like on—off, high—low, left—right, etc., to represent the stored

---

[1] A mechanism which goes through a sequence of steps and, say, energizes three outputs in sequential order is a simple example of such a sequential machine requiring storage.

[2] A universal computer element is one which can perform the function of storage in addition to logic functions.

content. Dynamic storage elements represent their content by conditions like pulsating or not pulsating, oscillating with phase lag or phase lead with respect to a reference, etc. The choice of storage elements for a particular computer depends not only on technical considerations, but to some extent also upon the manufacturer's line of development, patents, etc. Ideally, storage devices should be inexpensive and reliable, they should consume little power and should be compatible with logic circuits as far as inputs and outputs are concerned. For some applications, one may look for certain additional characteristics. There are permanent storage devices which retain their content indefinitely, i.e. until their content is changed purposely, and temporary storage devices which can retain information only for a certain period of time[1]. Both permanent and temporary storage devices may make available their stored contents immediately or only after a certain elapsed time.

The stored information is sometimes destroyed when read out (destructive read). All stored information may be lost when a computer is turned off or the power fails (volatile storage), but for some applications it is essential to retain information even under those circumstances (nonvolatile storage).

Those storage elements which are used in computers can normally assume only one of two possible conditions. We may say they are in the "0" or in the "1" state, or their content is a "0" or a "1" at a certain time. In essence then, their "storage capacity" is one binary digit or one "bit" for short. Even devices with more than two positions can be treated as a combination of two-position or binary devices. In this manner the "bit" may be used as general measure for storage capacity and information content.

Problem 1 (Voluntary): Suppose a computer is built entirely of one type of circuit only (universal element). This element must be able to perform any logic function (e.g. AND, and NOT are sufficient) and the function of storage. Can you name one or two additional functions which such a circuit must be able to perform?

Problem 2 (Voluntary): a) How much information (how many bits) can be stored on a standard punched card?

b) How much information is transmitted by a 3-digit binary number, a 3-digit octal number, a 3-digit decimal number?

c) How much information can be "stored" in a 4-position rotary switch, a 5-position rotary switch?

---

[1] For instance, time delays rightfully belong in this latter category.

5.1. Static Storage Elements

5.1.1. The Static Flip-Flop

One may think of a flip-flop[1] as basically consisting of two NOT-circuits connected serially as shown in Fig. 5.1. If we assume binary variables on inverter inputs and outputs, the circuit must be in one of the two indicated states.

Fig. 5.1. Basic Flip-Flop Consisting of Two NOT-Circuits

These two conditions are stable (i.e. the flip-flop remains in a once assumed state) if the circuit is designed properly. The two conditions may be used to represent a stored "0" or "1" respectively. In order to define what we mean by the "0" or "1" state, let us label the output of one NOT-circuit with A, the output of the other with A' (see Fig. 5.2). We can now write the following truth-table:

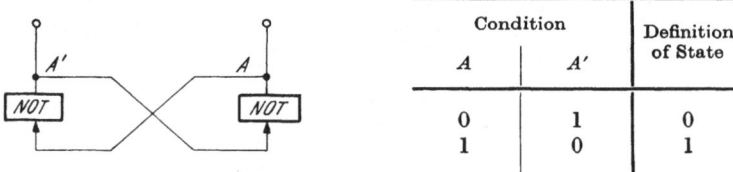

Condition		Definition of State
A	A'	
0	1	0
1	0	1

Fig. 5.2. Definition of Flip-Flop States

Frequently the terms off, cleared, reset or false are used for the "0" state, and the terms on, set or true for the "1" state.

Let us now look at the actual circuit diagram of a basic flip-flop.

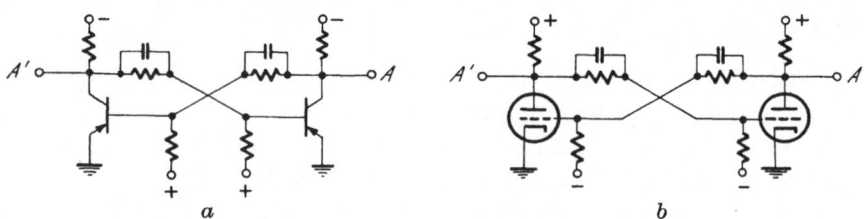

Fig. 5.3. Circuit Diagrams of Basic Transistor and Tube Flip-Flops

---

[1] Flip-flops are also known as Eccles-Jordan circuits or bistable multivibrators.

We recognize two transistor or tube inverters. In the given arrangement, always one tube or transistor is conducting while the other one is cut off. If, for instance, there is a negative potential on output A' in Fig. 5.3a, then the transistor on the right has a negative base potential and is conducting. Consequently, the output A assumes ground potential, and the base of the transistor on the left is slightly positive. The transistor is cut off, and allows the point A' to remain indefinitely at the negative po-

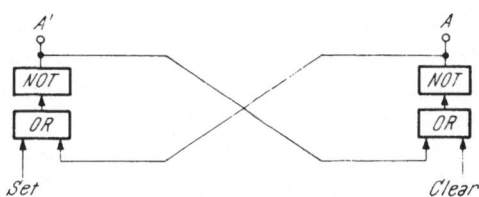

Fig. 5.4. Flip-Flop with Set and Clear Inputs

tential. Conversely, if A' is at ground potential, the transistor on the right is cut off. Output A is negative, and the transistor on the left is conducting.

So far, we have seen only the basic arrangement. Before we can make practical use of the circuit, we have to make several improvements. At least we must provide inputs so that we can "clear" or "set" the flip-flop by external signals. Suppose we add two OR-circuits to the basic arrangement as shown in Fig. 5.4.

As long as there are no inputs (i.e. the inputs stay in the "0" state), the circuit acts as shown previously, i.e. it remains in its state. If we apply, let us say, a clear input, the output A and, therefore, the flip-flop will assume the "0" state, no matter what state it had previously. Due to the previously explained flip-flop action, it will remain in this state even after we take off the input signal.

In effect, we have "cleared" the flip-flop. Conversely, if we apply a set input, the flip-flop will assume and remain in the "1" or set condition.

Problem 3: Flip-flops of the type shown in Fig. 5.4 are frequently constructed using two NOR modules (see paragraph 4.2.2). Show the circuit diagram for such a flip-flop. Label inputs and outputs. Indicate the polarity of set and clear inputs.

Problem 4: Design a "flip-flop" which consists of
a) transistor gates exclusively,
b) NAND modules exclusively.

Problem 5 (Voluntary): Express the output A as logic function of the inputs S and C. Refer to Fig. 5.4.

There is some danger in using this type of flip-flop: If we should happen to apply simultaneously set and clear inputs, both outputs of the flip-flop would assume the "0" state. The labelling A and A', which implies that one proposition is always the complement of the other, would then be incorrect. Furthermore, if we remove the clear and set input simultaneously, the flip-flop arbitrarily assumes one or the other state. This difficulty can be avoided in one of two ways.

By the proper logic design of the system in which the flip-flop is used, one can make sure that there are never simultaneous clear or set inputs. In this case, no further change of the flip-flop circuitry is necessary. We speak of flip-flops with DC inputs or DC flip-flops for short.

One can also design flip-flops in such a manner that simultaneous inputs make the flip-flop change or "complement" its previous state. Its behavior is then predictable and quite useful for many practical applications. Let us look at the circuit diagram of such a flip-flop:

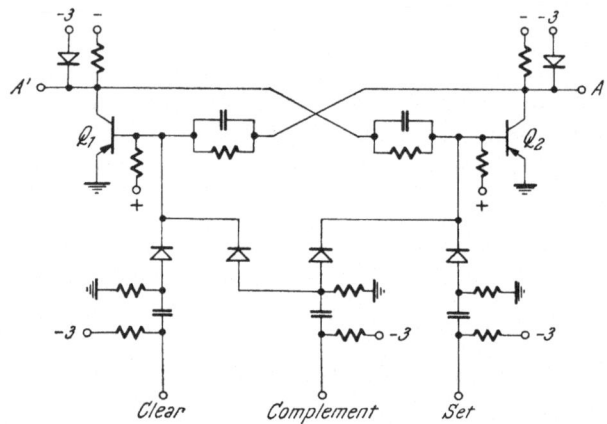

Fig. 5.5. Transistor Flip-Flop with Level or Pulse Inputs and Level Outputs

In the upper half, we recognize the basic flip-flop according to fig. 5.3. Set, clear and complement inputs are applied via the network in the lower half. The true OR-circuits of Fig. 5.4 are only functionally existent.

To understand the function of the input network, let us assume that at the moment the transistor Q_2 is conducting and that we apply a square pulse to the set input. The capacitor of the input network "differentiates" the waveform, and we obtain negative and positive spikes. Only the positive spike is transmitted to the base of Q_2 via the input diode (see Fig. 5.6). The positive spike cuts transistor Q_2 off. Its output assumes a low potential, and the flip-flop is set to the "1" state.

We see that the flip-flop is triggered by the positive transition of input signals. The existence of a signal itself (the low potential on an input terminal) does not affect the state of the flip-flop. For this reason, we may apply simultaneous signals to clear and set inputs (or to the complement input) without having both outputs of the flip-flop assume the zero state.

In order to understand why the flip-flop complements its state when signals are applied simultaneously to the clear and set inputs, it is necessary to comprehend the function of the coupling capacitors in more detail.

In one respect, the capacitors act like the speed-up capacitors of a transistor or tube inverter[1]. The total transition time of a flip-flop (i.e. the time it takes for the flip-flop to settle in its new state after an input pulse has been received which reverses its state) may be shortened by an order of magnitude if speed-up capacitors are added to the circuit. Since the speed of computers is largely determined by the speed of its flip-flops, this decrease in response time is very important. Let us investigate this speed-up in some detail. We note that, although the static or DC

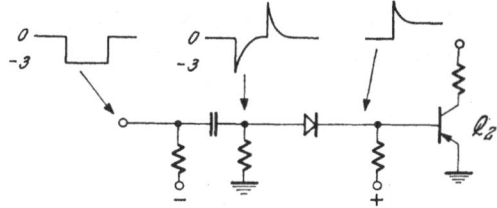

Fig. 5.6. Waveforms on the Input Network

behavior of the flip-flop is not changed by the coupling capacitors, the bases receive relatively large signals during any change of conditions. This produces a still larger change in collector voltages (due to the amplification of the transistors) which in turn is fed back to the bases. The effect is that, once the flip-flop starts to change its state, it continues to change until it is settled in its new state, even if the input signal is removed. An input signal which is so short or so weak that it would not trigger a resistor-coupled flip-flop, triggers a flip-flop with capacitor coupling reliably.

Let us now investigate what happens if we momentarily apply ˙short pulses to both the set and clear inputs. We note that the positive pulse on the base of the transistor which is already cut off produces no change, whereas the pulse to the conducting transistor starts to cut it off. The only immediate change is, therefore, in the direction of cutting off the previously conducting transistor. However, the collector of this transistor, which previously assumed ground potential, becomes more negative and feeds a negative going signal to the previously cut off transistor. This signal counteracts the input signal on its base. In other words, the flip-flop has begun to change state. If the input signal is now removed (by the differentiating action of the input capacitor), the flip-flop will continue to change state as explained previously.

A different way of looking at the complementation of a flip-flop is the following: Short pulses on both inputs of the flip-flop tend to cut off both

---

[1] See paragraph 4.2.1.

transistors momentarily. If we remove the inputs before the charges on the coupling capacitors have reached an equilibrium, the flip-flop will "come on" in the state determined by the amount of charges left in the coupling capacitors. It so happens that there was previously a larger potential drop across the capacitor from the collector of the non-conducting transistor to the base of the conducting transistor than across the other capacitor. Assuming a symmetrical circuit, the base of the previously conducting transistor will assume a more positive potential than the base of the previously non-conducting transistor. The previously conducting transistor will be cut off, and the previously non-conducting transistor will be turned on.

The complementation of a flip-flop may be induced by two separate but simultaneous signals to both, the clear and set input. However, many flip-flops for which a complementation is desired, have one or more separate complement inputs like the flip-flop in Fig. 5.5. The complement signal on this input is transmitted to both sides of the flip-flop. In this manner, complement signals reach by necessity both transistors or tubes simultaneously.

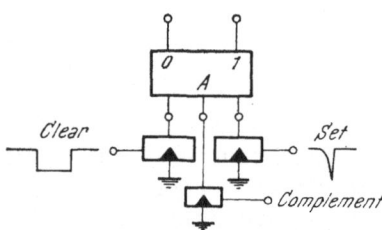

There are a few more details in connection with the circuit in Fig. 5.5 worth mentioning. It is not necessary to trigger the flip-flop by a square-shaped input signal i.e. by a change in input levels. Short pulses serve the

Fig. 5.7. Flip-Flop with Transistor Input Gates

same purpose. In one case we speak of level inputs, in the other of pulse inputs. Both pulse- and level-type inputs can be applied via transistor gates (see Fig. 5.7). The input transistors pull the flip-flop input to a ground potential while the signal is applied.

Fig. 5.7 shows a commonly used symbol for flip-flops. The outputs are either labelled by the "name" of the flip-flop (A or A' in above example), or the flip-flop is labelled by its name and the outputs are distinguished by "0" and "1".

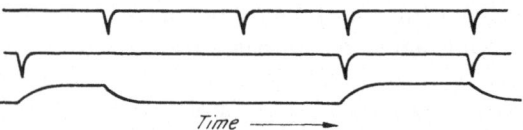

Fig. 5.8. Time Relation of Trigger Pulses and Output Levels of a Flip-Flop

Input pulses are normally rather short in duration compared with the response time of the flip-flop. Fig. 5.8 depicts this time relation.

In effect, the flip-flop has a built-in delay. (It takes a certain time before the new state of the flip-flop is available at the output, after an input pulse has triggered a change in state.) This delay time seems to be undesired at the first glance, but can be rather useful. It essentially allows to trigger a flip-flop and simultaneously "strobe" its output (=read out the previously contained information or state). As we shall see in chapter 6, this property will allow us to reduce the number of flip-flops required for certain applications to one half of the number otherwise required.

Let us make a final comment on the design of the OR-circuit indicated in Fig. 5.4. The flip-flop given in Fig. 5.5 does not show such a circuit. However, we note that the base of a transistor assumes a high potential if either the connected input goes positive or if the collector of the other transistor is positive. In effect, a transistor is cut off if the input goes high, or if the other transistor is conducting (or both). This is exactly the behavior symbolized in Fig. 5.4.

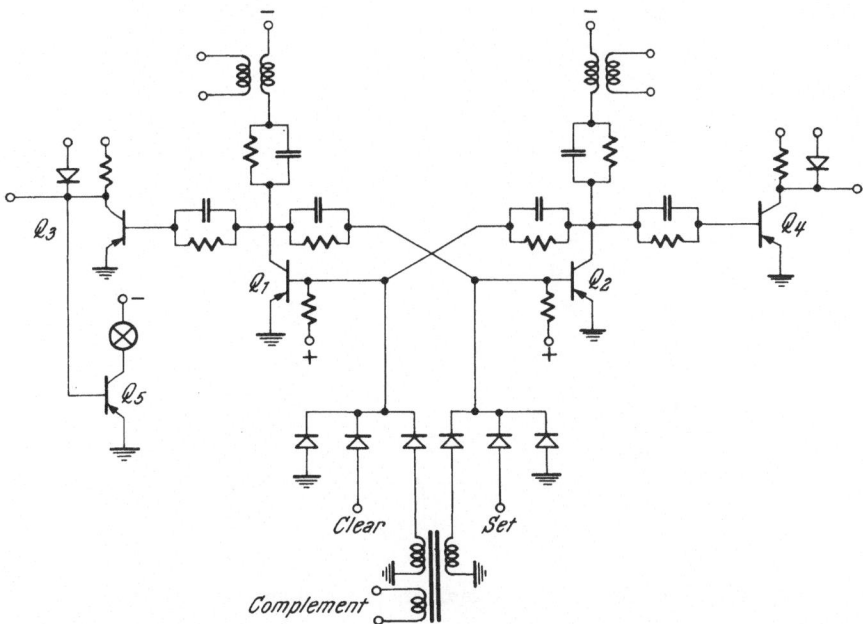

Fig. 5.9. Transistor Flip-Flop with Pulse and Level Outputs, Output Inverters and Indicator

Fig. 5.5 shows only one of many possible flip-flop circuits. Fig. 5.9 gives as further example another transistor circuit which has a pulse transformer on the complement input, pulse and level outputs, output inverters for levels, and an indicator to indicate the state of the flip-flop.

The output inverters separate the load from the flip-flop itself. This is desirable in two respects. First of all, an unsymmetrical load or a large load could make the operation of the flip-flop itself unreliable. Secondly, any load changes the response time of the flip-flop (it introduces additional

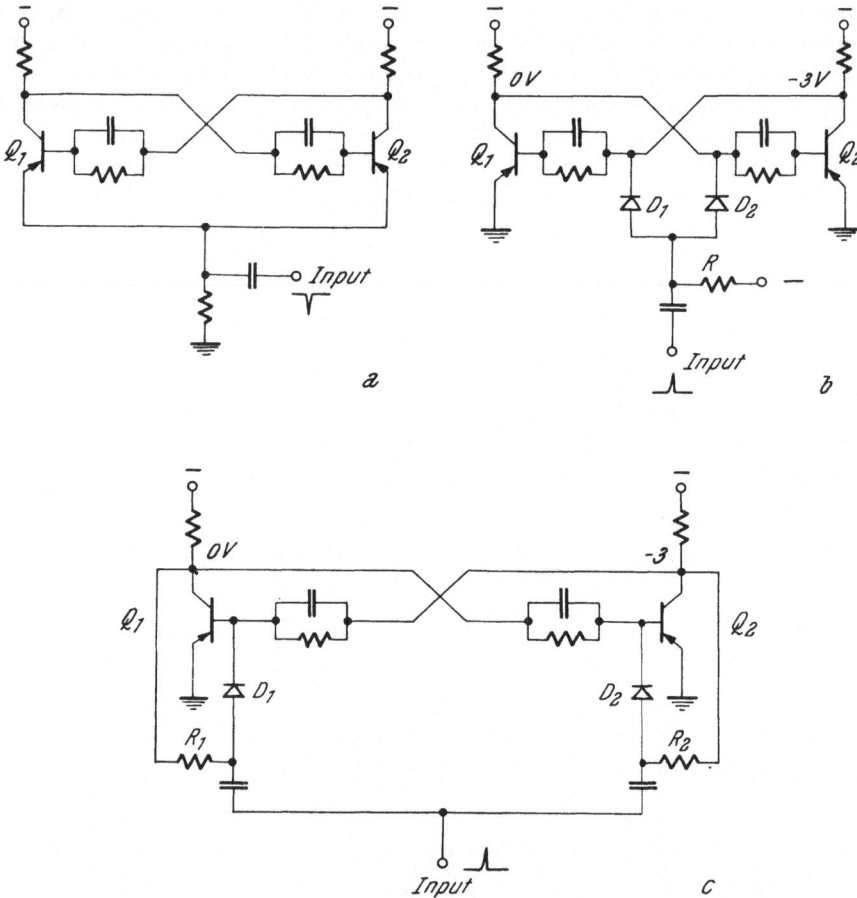

Fig. 5.10. Several Arrangements of Complement Inputs

capacitances, impedances, etc.). The computer design usually requires a rather tight tolerance on the response time, hence such variations are undesirable. Many flip-flop designs use emitter followers on the output, instead of inverters, to separate the load.

Problem 6: a) What logic state of the flip-flop is indicated by the light bulb connected to Q_5 in Fig. 5.9?

b) How are the outputs of Q_3 and Q_4 to be labelled?

c) Draw a rough time diagram of both level outputs and both pulse outputs for several complement inputs.

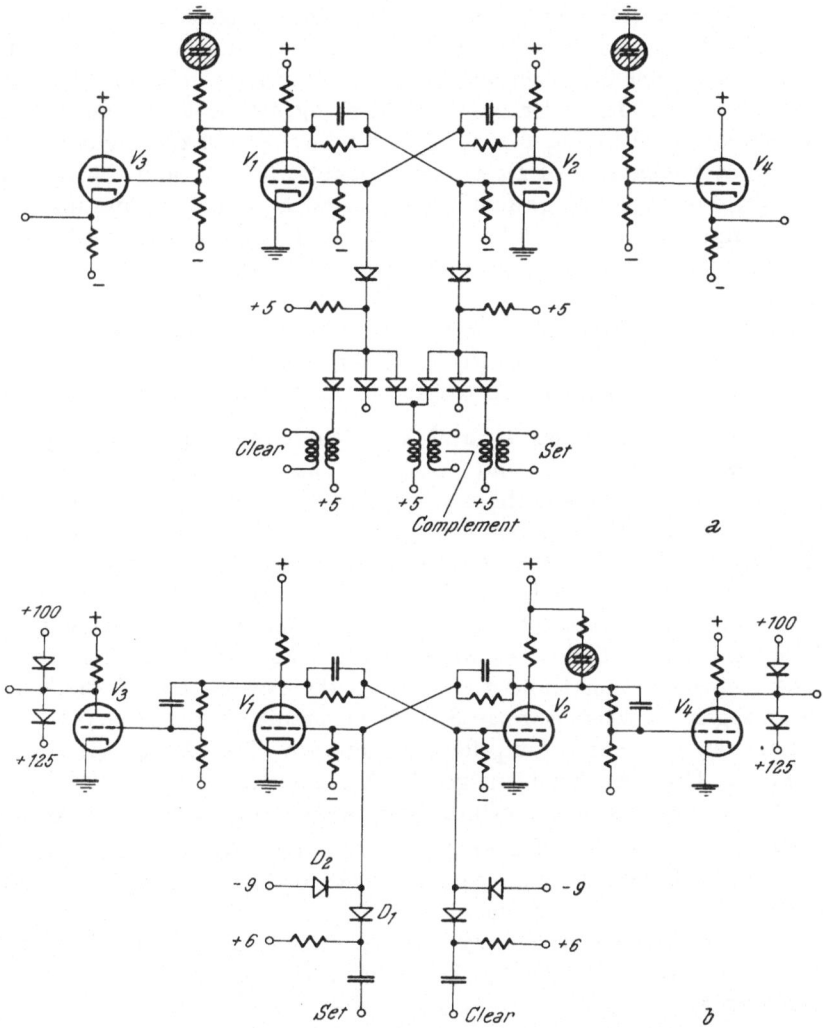

Fig. 5.11. Two Examples of Flip-Flop Circuits with Tubes

Fig. 5.10 shows three basic transistor flip-flop circuits with alternate arrangements of complement inputs.

A negative pulse applied to the input of the circuit in Fig. 5.10a tends to cut off both transistors by making the emitters more negative than before.

This starts a complementing action similar to the one previously explained. The complementing action continues even after the input signal is removed. The two circuits in Fig. 5.10b and 5.10c contain "steering diodes". The positive input signal is passed by one of the diodes to the presently conducting transistor (tending to cut it off), while the signal is blocked by the other diode from passing to the presently non-conducting transistor. This starts a complementing action which continues even after the input signal is removed. To understand the blocking action, let us consider the condition indicated in Fig. 5.10b. Q_1 is conducting and Q_2 is cut off. The potential at the cathode of D_2 is approximately 0 volts, and the potential at the plate of D_2 is approximately -3 volts. D_2 is not conducting and a positive input signal would have to exceed 3 volts in amplitude before diode D_2 starts to conduct. An input signal of smaller amplitude is, therefore, effectively blocked from reaching the base of Q_2. However, the input signal is not blocked by the diode D_1. Its plate has approximately the same potential as its cathode (-3 volts). A positive input signal finds the diode forward biased and can pass on to the base of Q_1 (and the collector of Q_2).

The cathodes of diodes D_1 and D_2 in the circuit shown in Fig. 5.10c always assume roughly ground potential (the cathode connected to the conducting transistors is perhaps a fraction of a volt negative). However, one of the plates assumes a potential of 0 volts (D_1 for the condition shown in Fig. 5.10c) while the other assumes -3 volts (D_2 in Fig. 5.10c). One diode is slightly forward biased (D_1) while the other is cut off (D_2). A positive input signal passes to the base of the conducting transistor (Q_1) via the forward biased diode (D_1) while it is blocked from reaching the base of the non-conducting transistor (Q_2) by the reverse biased diode (D_2). This arrangement, when compared with the circuit given in Fig. 5.10b, has the disadvantage that the flip-flop outputs are "loaded" by the resistors R_1 and R_2. However, smaller signals are sufficient to trigger the flip-flop reliably since input signals reach directly the base rather than the collector.

Fig. 5.11 shows two of many possible tube circuits. Both flip-flops can be set and cleared by (negative) pulses or levels. A (short) negative pulse is required to complement the flip-flop in Fig. 5.11a. The circuit in Fig. 5.11a has cathode followers to separate the load from the flip-flop proper. The circuit in Fig. 5.11b has clamped output inverters for the same purpose.

Problem 7: How is the output of V_4 and the indicator connected to V_2 to be labelled
    a) in Fig. 5.11a,
    b) in Fig. 5.11b.

There are many more variations of transistor and tube flip-flops possible.

For the practical design of computer circuits, it is important that we distinguish several types as far as their operation is concerned:

1. The *DC flip-flop* symbolized in Fig. 5.4. Its input is strictly a logic DC level. We cannot complement its state by simultaneous clear and set inputs, and it is not possible to set the flip-flop to a new state while simultaneously reading the old state.

2. *Flip-flops with AC coupled inputs.* The flip-flop can be set to a new state while the old state can be read simultaneously (built-in delay). Strictly speaking we can distinguish three types:

a) Only set and clear inputs are provided (*RS* flip-flop). The simultaneous application of set and clear inputs is not permissible and has to be avoided by the system designer.

b) Set, clear and complement inputs are provided (*RST* flip-flop). A simultaneous application of input signals to the set and clear inputs is not permissible but the application of an input signal to the complement input complements the flip-flop reliably.

c) Only set and clear inputs are provided but the application of simultaneous inputs to set and clear terminals complements the flip-flop reliably (*JK* flip-flop).

5.1.2. Magnetic Cores

Magnetic cores are the main elements of the internal memories of present-day computers[1]. Cores are very reliable components, they are inexpensive to manufacture and they can be made in very small sizes. The storage of information in a core is, in principle, a very simple process. It is shown schematically in Fig. 5.12.

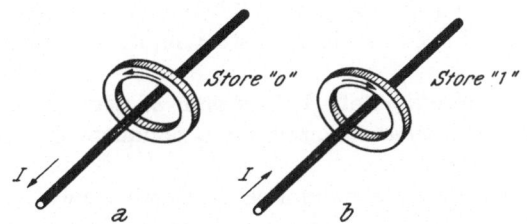

Fig. 5.12. Storage of a Zero and One in a Magnetic Core

The core is magnetized either in clockwise or counterclockwise direction by a fairly strong write current. This current is carried by one or more wires which are threaded through the aperture of the core. The material

---

[1] The organization of core memories is discussed in chapter 8, the use of cores for switching purposes in chapter 11.

of the core has a relatively high magnetic remanence, so that it remains practically saturated in one or the other direction even after the write current is turned off.

In principle, then, we have a binary storage element. One condition, say clockwise magnetization, is used to represent a "one", the other, say counterclockwise magnetization, to represent a "zero". Little use is made of a third possible state: no flux. This is due to the technical difficulties in producing this state.

In order to read, the core is set to the zero state by a current in the proper direction. If the core contains a "one" when this current is applied, the flux in the core is reversed and a signal is induced into the sense winding (see Fig. 5.13 b). If the core contained previously already a "zero", the flux of the core is essentially not changed and no signal is induced into the sense winding (see Fig. 5.13 a).

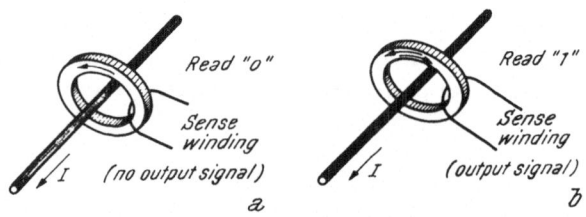

Fig. 5.13. Read-Out of a Magnetic Core

An output during read means then that there had been a "one" stored in the core; no output signifies the previous storage of a "zero". The read-out destroys the stored information. We speak, therefore, of a "destructive read". Computer memories usually have circuitry which automatically restores the destroyed information[1]. There are also several techniques by which a core can be read without destroying its content (non-destructive read)[2].

Cores require approximately 1/2 ampere of driving current and have a switching time (the time required for a complete flux reversal) in the order of a microsecond.

The design of large core memories is feasible only if cores exhibit a fairly square hysteresis loop. The approximate shape of the hysteresis loops is shown in Fig. 5.14 for ordinary magnetic materials and ferromagnetic materials as used for memory cores.

The hysteresis loop shows the relationship between the magnetic induction B (which is proportional to the magnetic flux Φ) and the magnetizing force H (which is proportional to the current I through the core).

---

[1] See paragraph 8.3.
[2] See paragraph 11.5.

The "1" and "0" states indicated in Fig. 5.14b correspond to the magnetization remaining in the core after the write currents have been turned off $(I=0)$. H_c is the coercive force.

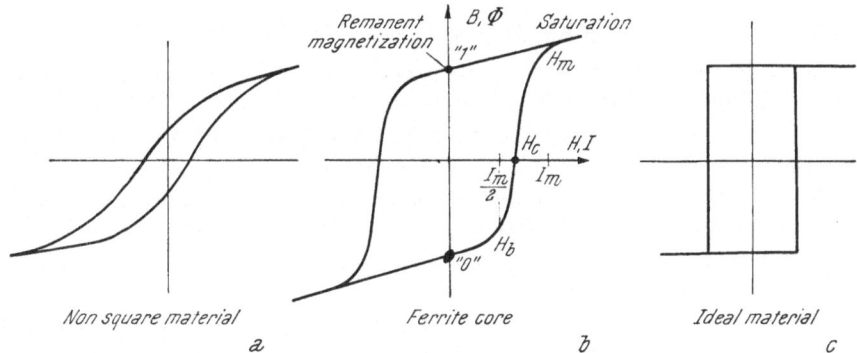

Fig. 5.14. Hysteresis Loops for Various Materials

The squareness of the loop is important in two respects. First of all, the "vertical" parts of the curve should be rather steep. It should be possible to select a current I_m on the curve which switches the core reliably, but one-half of this current should practically not change the flux in the core. The points H_c, H_b and H_m should, therefore, lie inside the range bounded by $\frac{I_m}{2}$ and I_m[1]. Secondly, the top and the bottom of the curve should be almost horizontal. To be more precise, a disturbance of the core by $\pm \frac{I_m}{2}$, or the driving of the core into actual saturation should not change the flux in the core significantly. The first requirement stems again from the selection scheme used in core memories, the second can be deduced from Fig. 5.13. The driving of a core in "zero" direction which contains already a zero shall produce practically no output signal on the sense winding (i.e. practically no flux change should take place).

The squareness of the loop is frequently expressed by a "squareness ratio", defined as the ratio of the remanent flux to the saturation flux. Squareness ratios of approximately .97 can be achieved.

Problem 8: Draw hysteresis loops roughly approximating that of Fig. 5.14b and indicate the traversal of the curve by heavy lines and arrows if the core is in the following initial states and a current of the following magnitudes is temporarily applied and then removed:

---

[1] This is important for the selection scheme used in core memories. See paragraph 8.3.2.

a) zero, $+I_m/2$ e) one, $+I_m/2$
b) zero, $+I_m$ f) one, $+I_m$
c) zero, $-I_m/2$ g) one, $-I_m/2$
d) zero, $-I_m$ h) one, $-I_m$

5.1.3. Magnetic Recording

Magnetic recording is used quite extensively for the storage of large amounts of information. The organization of some of these mass-storage devices[1] will be discussed in chapter 8. At the present, we will discuss only the principle upon which they are based.

The storage medium is usually a relatively thin layer of magnetic oxide which is attached to a magnetically inert support or carrier such as an aluminum drum or disk, or to a plastic tape. The support mainly facilitates the mechanical transportation of the storage medium. The oxide is (locally) magnetized by a write or record head. The magnetic flux patterns generated by the record head are retained by the material and can be detected by a read or reproduce head at some later time. Fig. 5.15 shows the basic arrangement.

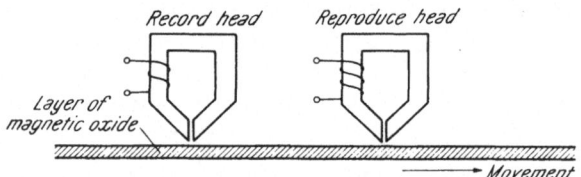

Fig. 5.15. Principle of Magnetic Recording

A current through the record head provides a relatively strong magnetic field in the vicinity of the recording gap. This field penetrates the magnetic oxide and saturates the material with a magnetic flux in a direction which is dependent upon the direction of the write current. If the direction of the write current is reversed while the storage medium passes the head, a part of the material will be magnetized in one direction, while another part is magnetized in opposite direction. The first two traces in Fig. 5.16 show this flux reversal schematically:

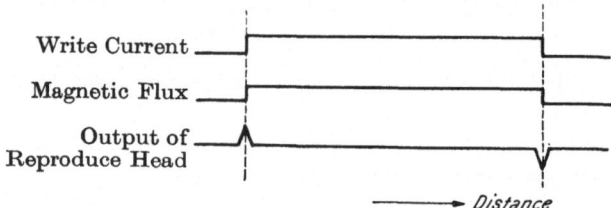

Fig. 5.16. Wave Shapes for Recording and Reproducing

---

[1] Magnetic drum, disk or tape storages.

The write current is reversed for a certain length of time. This causes the direction of the flux in the material to be changed for a corresponding distance. If such a piece of material passes the read head, it causes a small flux change in the iron core every time the flux in the material changes direction. The flux change in the core induces a voltage into the read winding as indicated in the bottom trace of Fig. 5.16. Essentially then, we can record and detect flux patterns of the magnetic material[1].

Five recording modes (i.e. the various representations of digital information by flux patterns) which are in frequent practical use are shown in Fig. 5.17. The recorded information is in all cases: 0100110.

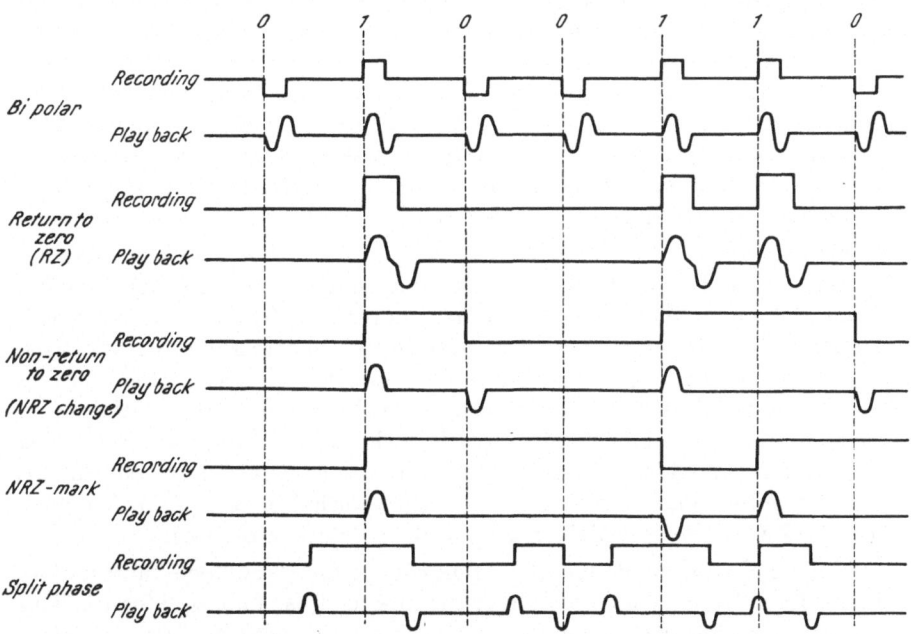

Fig. 5.17. Various Recording Modes for Digital Information

The *bi-polar* mode records pulses of opposite polarity for "0's" and "1's" while the flux remains neutral in between. This mode has the advantage that there can be no doubt where a digit begins or ends since there will be a pulse for every bit period. This property is important since the mechanical

---

[1] The principle is very similar to that used for home tape recorders. The only essential difference is that we usually try to saturate the magnetic material for the recording of digital information so that we get pulses of rather uniform amplitude during playback, whereas the recording of music or speech requires a faithful reproduction of all amplitudes.

speeds of a storage device may fluctuate so that timing alone is not sufficient
to determine how many digits there are in between consecutive but distant
pulses during playback. A disadvantage of the mode is the relatively low
amplitude (one-half of all other modes since we cannot more than saturate
the recording material). Furthermore, it is necessary to magnetically
neutralize the storage medium prior to recording.

The *return-to-zero* (*RZ*) mode records pulses of fixed polarity and duration for each "one" while the flux assumes the opposite polarity at all
other times. There is no necessity to erase since the recording medium can
be magnetically saturated by the record head in either one or the other
direction. However, the recording circuitry can be simplified for certain
applications if an erase head (powered by a simple DC supply) puts the
recording medium into the "zero" state before the record head records
only the pulses corresponding to "ones". The information recorded in the
return-to-zero mode is usually accompanied by a so-called sprocket. This
means that a pattern of all 1's is recorded on a separate track in synchronism
with the information. By looking at the output of this track it is simple to
determine the bit-timing of a recording.

The *non-return-to-zero* (*NRZ Change*) mode is the simplest mode as
far as recording is concerned. The flux remains in one or the other direction
during the entire bit period, depending upon what digit is to be recorded.
The advantage of this mode, compared to return-to-zero and bi-polar modes,
is a smaller number of flux changes for the same distance. This allows a
higher storage density on the recording medium and higher information
transfer rates at equal mechanical speeds of the device. Information recorded
in the *NRZ-C* mode is usually accompanied by a "clock" i.e. a continuous
string of pulses (one for every bit period) recorded on a separate channel.
The *NRZ-C* mode is widely used for drum storage devices.

The *NRZ-mark* mode records a flux-*change* for every "one" and no flux
change for a "zero". The direction of the flux itself is of no significance. This
has the advantage that a reversal of polarities does not affect the operating
condition of a storage device. Such a reversal may occur when the leads
to the write or read head are accidentally reversed, or when a magnetic
tape recorded at one installation is being read by another installation[1].
This feature, and the lower frequency of polarity changes compared to *RZ*
modes, make this mode of digital recording the most widely used. In practice,
information is usually recorded on several "tracks" simultaneously. A
"line" of information (i.e. all the information bits recorded simultaneously)
is frequently accompanied by a parity bit recorded on a "parity" track.

---

[1] The NRZ-M mode is also frequently used for the transmission of digital
information over wire lines or RF links. No matter whether inserted amplifiers
invert or do not invert polarities, the information is interpreted correctly.

The parity bit is such that there is a flux change in at least one track during each bit time. In this manner, the reconstruction of the clock signal poses no problem.

The *split-phase*[1] mode records two polarities for each bit. As shown in Fig. 5.17, a "zero" is represented by "down" followed by "up", while "up" and then "down" represents a "one". The average DC value of the signal in this arrangement is zero, which has the advantage that read amplifiers, or transmission elements in general, do not necessarily have to transmit DC.

The rate at which information can be recorded and reproduced depends upon the mode, the mechanical speed, and on achievable recording densities. Normal speeds at which digital tapes are presently recorded is up to 150 inches per second. The speed of magnetic drum devices can be somewhat higher. Recording densities are in the range of 100 to 1000 bits per inch. Commonly used densities for tape recordings are 200, 556 and 800 lines per inch. Non-saturation recording techniques promise substantially higher recording densities.

Problem 9: Assume that you have a tape transport which moves tape with a speed of 100 inches/sec. and that it is possible to record 200 flux changes per inch of tape on this device.

a) Show the possible recording densities in bits/inch for the five recording modes (see Fig. 5.17).

b) What are the transfer rates for these modes in bits/second for one track?

c) How much information can you store on a 3600 feet reel of magnetic tape if six information tracks are recorded simultaneously, and gaps of .75 inches are inserted after recording an average of 600 lines?

Magnetic recording devices show considerable latitude in engineering details. Fig. 5.18 shows schematically some head constructions.

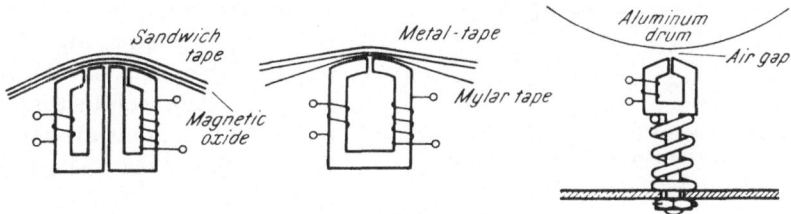

Fig. 5.18. Various Head Constructions

A head with two gaps which is frequently used for magnetic tape recording is shown in Fig. 5.18a. One side of the double head is used for recording the other for reproducing. This arrangement allows to read and to verify

---

[1] Also called bi-phase or Manchester mode.

information while it is recorded. Fig. 5.18 b shows a dual purpose head. It has record and reproduce windings. The same head can be used (alternately) as record and reproduce head. Several commonly used techniques to protect both, the head and the storage medium from mechanical abrasion are indicated in the same figure. Fig. 5.18 a shows a so-called "sandwich" tape. The layer of magnetic oxide is here enclosed inside mylar or acetate laminations which have relatively smooth surfaces. The separate mylar tape between the head and the metal tape in Fig. 5.18 b prevents the direct mechanical contact between the head and the storage medium. This mylar tape is subject to wear and is therefore slowly advanced by timer motors. Modern tape transports no longer require such protective devices, since friction is closely controlled by air pressure or vacuum arrangements. The head of the drum or disk storage devices is usually separated from the storage medium by an air gap. Fig. 5.18 c shows an adjustable head. There exist also "floating heads" which by aerodynamic principles keep a fixed distance from the storage medium, independent of possible expansions or contractions of the support.

5.2. Dynamic Storage Elements

5.2.1. One-Shots

One-shots (also called single-shots or monostable multivibrators) are temporary storage devices. Their internal circuitry is rather similar to that of flip-flops, the essential difference being that one of the cross-coupling networks between the two tubes or transistors is purely capacitive. Fig. 5.19 shows the basic arrangement with transistors.

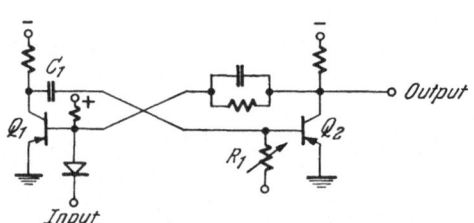

Fig. 5.19. Basic Circuitry of a Monostable Multivibrator with Transistors

In the quiescent or stable state, Q_2 is conducting (due to the negative bias on the base) and Q_1 is not conducting (due to the ground potential on the collector of Q_2 and the positive bias).

If the input goes momentarily negative, Q_1 becomes conducting. The rise of the collector potential is transmitted to the base of Q_2 via capacitor C_1, causing Q_2 to cut off. The fall in the collector potential of Q_2 is fed back to Q_1 so that its base stays negative, even after the input is removed. The "on" state of the circuit is sustained until capacitor C_1 is discharged, and Q_2 starts to conduct again. The resetting of the circuit is, from then on, accelerated by a flip-flop action. The delay time, i.e. the time it takes the circuit to reset to the "off" state, depends mainly upon the value of C_1 and R_1[1], the practical range being in the order of a fraction of a microsecond up to several seconds.

A one-shot as shown in Fig. 5.19 can accomplish two things: It can be used to produce a clean square pulse of desired duration from an input pulse which is short and distorted[2]. In this respect, the one-shot is an auxiliary device which may perform an essential function from an engineering point of view, but is then not really used as a storage device. However, the one-shot can store an event (an input signal) for a certain length of time. Its state may be interrogated and read out so that a one-shot is able to act as a (temporary) storage device.

One-shots frequently have pulse outputs in addition to level outputs. In the simplest case, a pulse transformer connected in series with the load resistor of Q_2 (see Fig. 5.19) serves as source of such an output pulse. The one-shot produces then a pulse at the end of its delay time. An input pulse is "delayed" by a certain length of time before it is reproduced on the output of the one-shot.

Unfortunately, there exists no standard symbol to represent one-shots in logic diagrams. The closest to a standard is probably the flip-flop like symbol in Fig. 5.20a for one-shots with level outputs and the symbol of Fig. 5.20b for delays with pulse outputs.

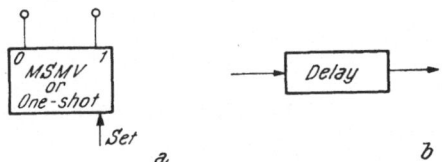

Perhaps the reason for considering the one-shot as a *dynamic* storage element should be explained. The "on" or "1" state is not a static condition (although it may be indicated

Fig. 5.20. Block Diagram Symbols for One-Shots

by a static condition, i.e. a level), but a time varying process i.e. the discharge of a capacitor. Also, the one-shot by itself returns to the zero state which, certainly, is a dynamic behavior.

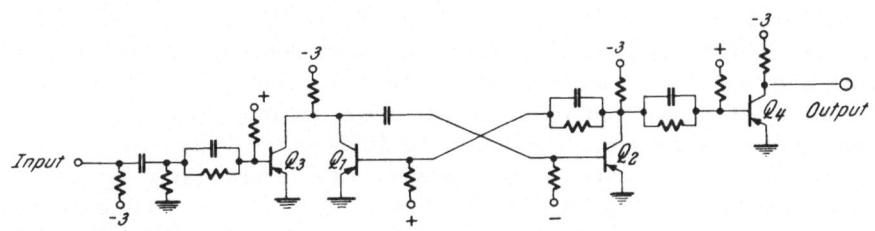

Fig. 5.21. One-Shot with Input Transistor and Output Inverter

---

[1] Q_2 is at this time reverse biased and draws no current. C_1 discharges solely via R_1.

[2] If equipped with an AC input it can also produce normalized pulses from long pulses or level changes.

One-shots, like flip-flops, show considerable latitude in engineering details. Fig. 5.21 shows, as an example, a one-shot with input transistor and output inverter.

The inputs and outputs of this circuit are electrically separated from the one-shot itself. This has the advantage that the impedances of the connected circuitry can in no way influence the delay time of the one-shot.

Problem 10: Draw a time diagram of the voltage on the base of Q_2 and the collector of Q_4 in Fig. 5.21 in relation to a positive square pulse on the input. How would you label the output in a block diagram? ("0" or "1").

The DC and AC couplings of a one-shot are not always as obvious as in the circuits of Fig. 5.19 and 5.21. Fig. 5.22 gives a frequently used tube circuit.

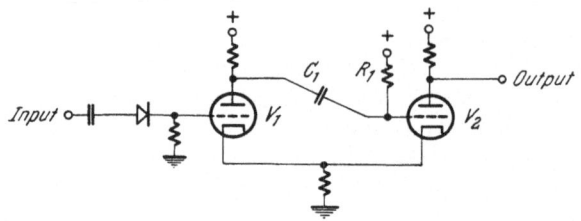

Fig. 5.22. Basic Circuitry of a One-Shot with Tubes

The DC feedback from V_2 to V_1 is here via the common cathode resistor. The bias of V_2 is adjusted so that V_2 is normally conducting. The cathodes of both tubes are, therefore, at a positive potential and V_1 is cut off. A sufficiently high positive input pulse turns tube V_1 on, which in turn produces a negative pulse on the grid of V_2 which turns off V_2. The one-shot stays on until C_1 is discharged through R_1 and returns then to the off state.

One-shots normally need some "recovery time". This means that they should be given a period of rest after they return to the zero state. If they are triggered too soon, the charges on the circuit elements will not have had sufficient time to reach an equilibrium, the circuit is not really in a quiescent state, and the delay time of the one-shot will be shortened. The recovery time has normally the same order of magnitude as the delay time, however, circuits exist which shorten the recovery time considerably by various measures.[1] "Integrating one-shots" will remain in the "on" state as long as input signals arrive which are spaced closer than the delay time. The integrating one-shot needs practically no recovery time.

---

[1] Essentially, the capacitor which determines the delay time is charged by a low impedance, but discharged by a relatively large impedance. Diodes, transistors or thyratrons may serve as such variable impedance devices.

5.2.2. Delay Lines

5.2.2.1. Electric Transmission Lines: Signals transmitted over electric transmission lines travel with a finite velocity. They experience therefore a certain delay. While a signal is travelling between the input and output of a transmission line, we may consider the signal to be "stored" in the line. The propagation velocity of normal transmission lines approaches the velocity of light. These lines would, therefore, have to be rather long in order to obtain appreciable delays. However, the speed at which a signal is propagated depends upon various parameters notably the inductance and capacitance per unit length of line[1]. If we make these parameters large, the signals travel slowly and short lines can produce relatively large delays.

In practice, there are two different methods employed to increase the capacitance and inductance of a line. The resulting lines are known as distributed parameter and lumped parameter delay lines.

A distributed parameter line has usually one of the conductors wound in the form of a spiral or double spiral. The mutual magnetic coupling between individual loops increases the inductance of the line. The second conductor presents an area as large as possible and as close as possible to the first conductor. In this manner, the capacitance of the line is increased. Fig. 5.23 shows two possible arrangements.

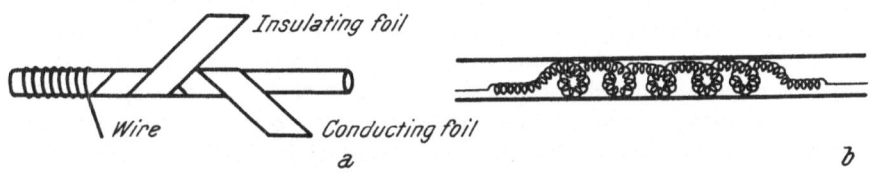

Fig. 5.23. Construction of Distributed Parameter Delay Lines

With delay lines of this type, it is feasible to produce delays up to a few hundred microseconds and to store up to 20 or 30 separate signals (bits) at any one time.

A lumped parameter delay line is shown schematically in Fig. 5.24.

The delay line consists of a lattice of individual inductances and capacitances simulating the inductances and capacitances of an actual trans-

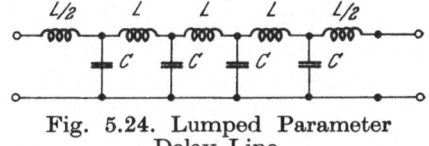

Fig. 5.24. Lumped Parameter Delay Line

mission line. Since they are concentrated or lumped, we speak of a lumped

---

[1] To be more exact, the group velocity is the speed with which a signal is propagated. It is approximately given by $1/\sqrt{LC}$, where L and C are the inductance resp. capacitance per unit length of a transmission line.

parameter delay line. Practically any reasonable delay can be achieved with this type of delay line[1].

Both, distributed and lumped parameter lines propagate the different frequency components of a signal with different speeds. This causes the signal to be distorted so that it usually has to be "reshaped" before it can be further used. Moreover, the characteristic impedance of the line has to be closely matched by the impedance of the input and output circuitry. Otherwise, reflections will occur at the ends of the line which distort the desired signals.

5.2.2.2. Magnetostrictive Delay Lines: The principle of magnetostrictive delay lines is similar to that of electric delay lines: A wave is propagated along a transmission line and needs a finite time to travel from the input to the output of the line. The wave is, however, not electric but mechanic or acoustic in nature. Fig. 5.25 shows such a magnetostrictive delay line.

Fig. 5.25. Magnetostrictive Delay Line

A nickel-alloy wire serves as the acoustic transmission line. The magnetic field, produced by a current pulse through the input coil, causes a momentary contraction of the nickel wire in the region of the transmitting coil. This contraction starts a longitudinal compression wave (i.e. an acoustic wave) traveling along the wire. Wherever the material in the wire is compressed (or expanded), the magnetic properties of the wire material are changed. If the wire is magnetized in the neighborhood of the output coil[2], the disturbance of the magnetic permeability produces a disturbance of the magnetic flux which is detected by the pickup coil. Since it is practically not possible to match the "mechanical impedance" of the line electrically by the input or output coil, mechanical dampers on the ends of the wire are required to suppress unwanted reflections.

Compared to electric delay lines, magnetostrictive delay lines have the advantage that the various frequency components of a signal are transmitted with more uniform speeds. Individual pulses are less broadened and it is technically feasible to store up to several thousand individual bits in a single line. Delay times in the order of milliseconds can be achieved.

---

[1] However, the feasible storage capacity expressed in bits is not larger than that of distributed parameter delay lines.

[2] Either by its intrinsic magnetism, or by a permanent magnet, or by a DC current through the output core.

Problem 11: Suppose that the sound waves in a magnetostrictive delay line are propagated with a velocity of 190000 inches per second and that the input pulse rate (the bit rate) is 200 kc.

a) What is the wave length of the resulting sound wave?

b) How long has the delay line to be in order to store 1000 bits at one time?

5.2.3. Dynamic Flip-Flops

A dynamic flip-flop consists basically of an amplifier, the output of which is fed back to the input via a delay line. Such an arrangement in itself can be in one of two possible states: It may be quiescent or in a state of oscillation. If precautions are taken that oscillations are not self-starting but that, on the other hand, once excited oscillations are sustained, the device can be used as a storage element. The quiescent state may, for instance, be used to represent a logical "0", and the oscillating or recirculating state to represent a "1".

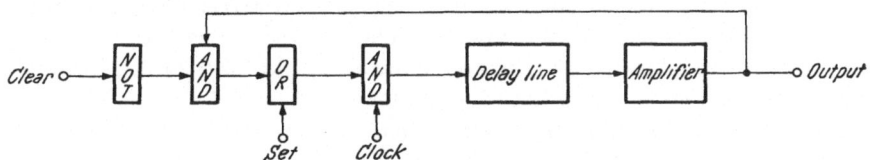

Fig. 5.26. Functional Diagram of a Dynamic Flip-Flop

Fig. 5.26 shows the functional details of such a dynamic flip-flop. The feedback loop connecting the delay and the amplifier contains several logic elements so that an oscillation or recirculation of pulses can occur only if certain conditions are met.

The AND-circuit connected to the input of the delay line synchronizes any oscillations of the circuit with a master clock. Obviously, it lets only input pulses pass which occur simultaneously with a clock pulse. The delay time itself is adjusted closely to the time of one clock cycle. Any output pulses appear, therefore, almost simultaneously with a clock pulse. In this manner all flip-flops in a system are forced to oscillate with a common frequency and phase (if they oscillate at all), and the output pulses of one flip-flop can be used as input pulses to other flip-flops.

Oscillations of the circuit are excited by an input pulse on the set input. This pulse passes through the OR- and AND-circuit. It is then delayed and amplified. From then on, it recirculates via the AND-, OR-, AND-circuits, and is also available on the output once during every clock cycle. A single pulse on the set input is, therefore, sufficient to place the circuit into the oscillating or "1" state.

The recirculation can be suppressed by a "1" signal to the clear input. This causes the output at the NOT-circuit to assume the "0" state, which in turn prohibits any output of the following AND-circuit, or in effect, opens the recirculation loop. Again a single pulse (in synchronism with the clock) is sufficient to "clear" the flip-flop.

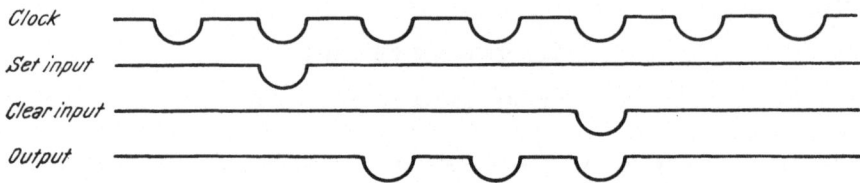

Fig. 5.27. Input and Output Wave Forms

Fig. 5.27 shows a time diagram of inputs and outputs for a dynamic flip-flop. A transistorized circuit is assumed in which the more positive voltage level (0 volts) represents a logical "0" and the more negative level (say —3volts) represents a "1". To be more exact: the logical "1" is represented by a negative level at the time when the clock is negative. We see from the time diagram that circuit outputs are delayed with respect to inputs by one clock cycle. This allows us to "read" the old state of the flip-flop simultaneously with setting it to a new state.

Dynamic flip-flops, like static flip-flops, may have several clear and set inputs or complement inputs. This requires a more complicated logic circuitry while the basic storage element remains the same. Fig. 5.28 shows the circuitry necessary to complement a dynamic flip-flop.

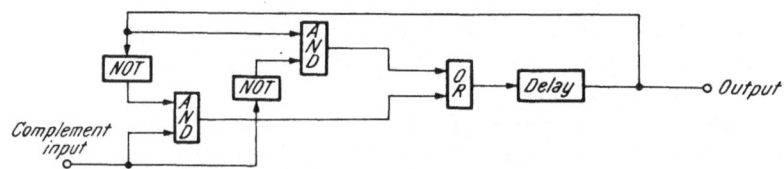

Fig. 5.28. Dynamic Flip-Flop with Complement Input

The storage element itself is symbolized by a delay. It should be understood that it contains an amplifier and the synchronization gate in addition to the delay line.

Problem 12: Show the wave forms on the output of the flip-flop and on the two inputs of the OR-circuit in relation to clock pulses and complement pulses in a time diagram. Assume that there is a complement pulse with every third clock pulse and that the dynamic flip-flop is originally in the "0" state. Refer to Fig. 5.28.

Problem 13: Draw the block diagram of a dynamic flip-flop with two set, two clear, and one complement input.

Selected Bibliography

BEGUN S. J.: Magnetic Recording. New York: Murray Hill Books. 1949.

WANLASS C. L.: Static-Dynamic Design of Flip-Flop Circuits, IRE Transactions, vol. EC-1, pp. 6–18. Dec. 1952.

ELBOURN, and WITT: Dynamic Circuit Techniques Used in SEAC and DYSEAC, Proceedings IRE, vol. 41, No. 10, pp. 1380–1387. Oct. 1953.

COHLER E. V.: Transistor Flip-Flops for High-Speed Digital Computer Applications, Proceedings of WESCON, pp. 38–43. Aug. 1954.

BROWER D. F.: A "One Turn" Magnetic Reading and Recording Head for Computer Use, IRE Convention Record, pp. 95–100, part 4. 1955.

McMAHON R. E.: Designing Transistor Flip-Flops, Electronic Design, vol. 3, pp. 24–27. Oct. 1955.

MILLMAN, and TAUB: Pulse and Digital Circuits. New York: McGraw-Hill. 1956.

SURAN, and REIBERT: Two-Terminal Analysis and Synthesis of Transistor Multivibrators, Transactions IRE, vol. CT-3, p. 26. 1956.

THOMPSON, and LYON: Analysis and Application of Magnetostriction Delay Lines, Transcations IRE, vol. UE-4, pp. 8–22. Aug. 1956.

BAY, and GRISAMORE: High-Speed Flip-Flops for the Millimicrosecond Region, Transactions IRE, vol. EC-5, No. 3, pp. 121–125. Sept. 1956.

RICHARDS R. K.: Digital Computer Components and Circuits. Princeton: D. Van Nostrand. 1957.

Transistor Circuit Engineering, edited by R. F. SHEA, chapter 10.6. New York: John Wiley and Sons. 1957.

MORLEIGH S.: A Survey of Delay Lines for Digital Pattern Storage, Electronic Engineering, vol. 30, pp. 380–387. June 1958.

CARROLL J. M.: Modern Transistor Circuits, chapter 5. New York: McGraw-Hill. 1959.

SMITH C. V. L.: Electronic Digital Computers. New York: McGraw-Hill. 1959.

Digital Applications of Magnetic Devices, edited by ALBERT J. MEYERHOFF. New York: John Wiley and Sons. 1960.

HOAGLAND, and BACON: High-Density Digital Magnetic Recording Techniques, Proceedings IRE, vol. 49, No. 1, pp. 258–267. Jan. 1961.

ROTHBART A.: Bibliography on Magnetostrictive Delay Lines, Trans. IRE, vol. EC-10, No. 2, p. 285. June 1961.

HUSKEY, and KORN: Computer Handbook. New York: McGraw-Hill. 1962.

GOLDSTICK, and KLEIN: Design of Memory Sense Amplifiers, Transactions IRE, vol. EC-11, No. 2, pp. 236–253. Apr. 1962.

KUMP H. J.: The Magnetic Configuration of Stylus Recording, Transactions IRE, vol. EC-11, No. 2, pp. 263–273. Apr. 1962.

DUNDON T. M.: Specifying Magnetostrictive Delay Lines for Digital Applications, Computer Design, vol. 2, No. 1, pp. 14–25. Jan. 1963.

BARKOUKI, and STEIN: Theoretical and Experimental Evaluation of RZ and NRZ Recording Characteristics, Transactions IEEE, vol. EC-12, No. 2, pp. 92–100. Apr. 1963.

STEIN I.: Generalized Pulse Recording, Transactions IEEE, vol. EC-12, No. 2, pp. 77–92. Apr. 1963.

SHEW L. F.: Discrete Tracks for Saturation Magnetic Recording, Transactions IRE, vol. EC-12, No. 4, pp. 383–387. Aug. 1963.

GILLIS J. E.: A Method for Achieving High Bit Packing Density on Magnetic Tape, Transactions IEEE, vol. EC-13, No. 2, pp. 112–117. Apr. 1964.

6. Computer Circuits

This chapter presents computer circuits and sub-units in which storage and logic elements are inter-connected, so that they perform particular logic or arithmetic operations. All subunits are treated individually, i.e. they are considered separately from other units of the computer. Chapter 8 will then show how the different subunits of a computer work together within the over-all concept. For the following discussions, we shall free ourselves from the burden of detailed schematic diagrams and use logic or block diagram symbols as far as possible.

6.1. Registers

Registers are storage devices for those pieces of information which the computer uses in its current operation[1]. A register consists of one or more storage devices to retain information and an arrangement of logic circuits which permits the input, output, and, possibly, the modification of information.

6.1.1. Flip-Flop Registers

The simplest register consists of an array of flip-flops as shown in Fig. 6.1

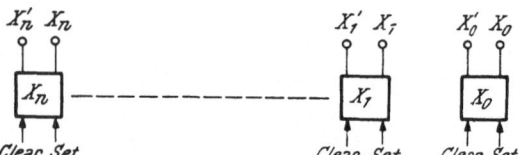

Fig. 6.1. Flip-Flop Register

Each flip-flop stores one bit of information. The storage capacity of the register is, therefore, given by the number of its flip-flops[2]. Any binary

---

[1] Larger amounts of information which are not immediately being used can be stored in internal or external "memories" (see chapter 8).

[2] Information as used in computers usually consists of "words", i.e. a for a machine fixed number of binary digits. Many registers have a storage capacity or "length" of one computer word, although longer or shorter registers are not uncommon.

configuration can be placed into the register by energizing appropriate inputs. If we, for instance, energize the "set" inputs for X_n, ...X_4, X_1, X_0, and the "clear" inputs for X_3, X_2, the X-register will assume the state: $X_n.....X_4X_3'X_2'X_1X_0$ or, in other words, store the binary number $1.....10011$. The information stored within the register may be read out by detecting the logic state of its outputs. Suppose, for example, the information contained in an A-register is to be transferred to B-register. Assuming static flip-flops with pulse inputs, the transfer pulse "A to B" will "read" the information contained in A, and place it into B as shown in Fig. 6.2. Such a transfer of "zeros" and "ones" is called a "forced" transfer.

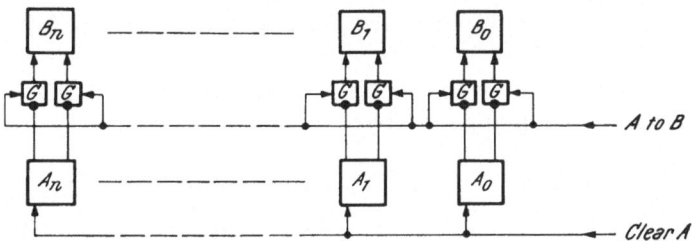

Fig. 6.2. Transfer of Information from one Register to Another

It is to be noted that the transfer pulse causes a copying or duplication of the contents without destroying the original information. However, if desired, a "clear A" signal can be given simultaneously with the transfer command.

In order to reduce the amount of hardware required for a transfer, very frequently only "1's" are transferred while the register to be copied into is cleared (i.e., set to zero) in advance. We then speak of a "one's" transfer as compared to a forced transfer.

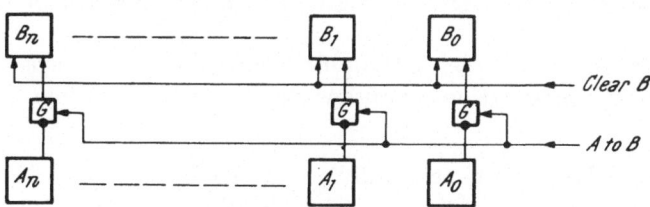

Fig. 6.3. Transfer of 1's only

This type of transfer takes longer (two pulse times instead of one) but requires only half the number of transfer gates.

If we consider flip-flops with AC coupling but with level inputs and outputs, the pulse gates in Fig. 6.2 or 6.3 are replaced by AND-circuits.

As can be seen from Fig. 5.6, the transfer of information takes place at the trailing edge of the transfer signal. Again, B may be cleared before the transfer, and only 1's transferred.

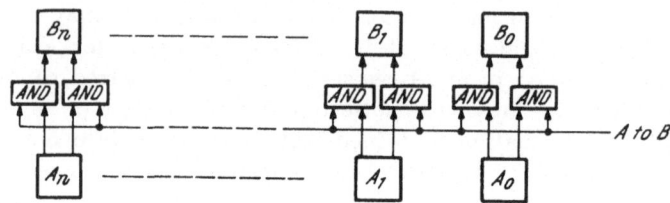

Fig. 6.4. Transfer of Information for Registers with Level Inputs

An arrangement for transferring information between registers consisting of dynamic flip-flops is given in Fig. 6.5.

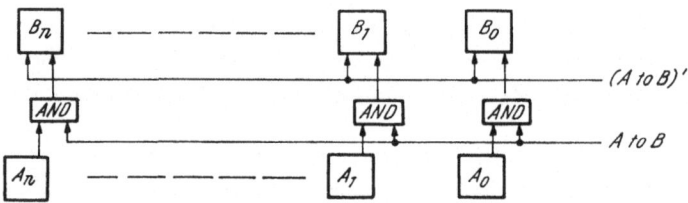

Fig. 6.5. Transfer of Information for Registers with Dynamic Flip-Flops

The transfer signal A to B is applied for one pulse time. During this time, the recirculation of pulses within the B flip-flops is inhibited by the complement of the transfer signal. Pulses enter only those B flip-flops whose corresponding A flip-flops contain one's. The AND-circuits shown on the input to the B-flip-flops can be considered an integral part of the B-flip-flops.

In all cases, except when DC flip-flops are used, the A-register may be set to a new configuration simultaneously with the readout of the old information.

Problem 1: Registers have usually provisions for the input of information from several sources as well as provisions for the output to several destinations. Show the logic diagram of an X-register including the input and output circuitry required for all of the following provisions:

a) Upon a pulse A to X, the content of A is placed into X.

b) Upon a pulse B' to X, the complement of the information contained in B is placed into X.

c) Upon a pulse X to A, the content of X is placed into A.

Assume static flip-flops with pulse inputs. Assume further that all registers are cleared before a transfer of information takes place.

Problem 2 (Voluntary): Show the circuits for problem 1 under the assumption of
a) in flip-flops with AC coupling and level inputs,
b) dynamic flip-flops. Show all details of the flip-flop input circuitry.

6.1.2. Recirculating Registers

The flip-flop registers shown so far are found quite extensively in "parallel" machines, where the transfer of information (as well as any arithmetic or logic operation) is in parallel, i.e. concurrent in all digits of a word. Serial machines, which transfer and operate upon one bit of information at one time, frequently use delay lines as an adequate and economic storage medium for registers. The basic arrangement of such a register, is shown in Fig. 6.6.

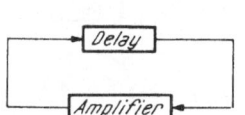

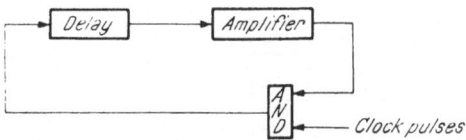

Fig. 6.6. Basic Arrangement of a Recirculating Register

Fig. 6.7. Synchronization of Information in a Recirculating Register

Pulses which enter the delay are available at the output after a certain time. If the delay time is long compared to the pulse time, several pulses may be "stored" within the delay at any one time. The output of the delay is fed back to the input via an amplifier or pulse shaper. As a result, a pulse once introduced continues to recirculate. The information is represented as a pulse "train". "One's" or "zeros" correspond to the presence or absence of pulses within the train. The storage capacity of the register is determined by the total time which is required for any one pulse to complete a cycle and by the frequency with which consecutive pulses may follow each other. Suppose the total cycle time (i.e., the delay time of all circuitry required to complete the loop) is 30 μs in a particular example and pulses may follow each other with 1 μs distance and still be individually distinguishable, then the total storage capacity of the register is 30 bits.

Recirculating Registers usually have provisions for the synchronization of pulses with the computer "clock", a pulse source with which also all other operations of the computer are synchronized. A typical synchronization circuit is shown in Fig. 6.7.

Pulses at the output of the delay (which are attenuated and distorted) are first amplified and then "AND-gated" with clock pulses. The resulting output pulses are of normalized amplitude and duration, and they are in synchronism with clock pulses.

Like flip-flop registers, recirculating registers may have various circuits for input and output attached. Fig. 6.8 shows an arrangement for the transfer of information from an A-register to a B-register.

Amplifiers and synchronization circuits are considered part of the registers. When the "A to B" signal is not present, both registers recirculate as previously described. When the "A to B" signal is applied, the recirculation of the B register is interrupted and the contents of the A register are routed into B.

The transfer signal is usually applied for exactly one cycle time. Consequently, the old information is still available for other parts of the computer, while new information is entered into the register. If required, register A may be cleared simultaneously with the transfer by preventing its recirculation at the input.

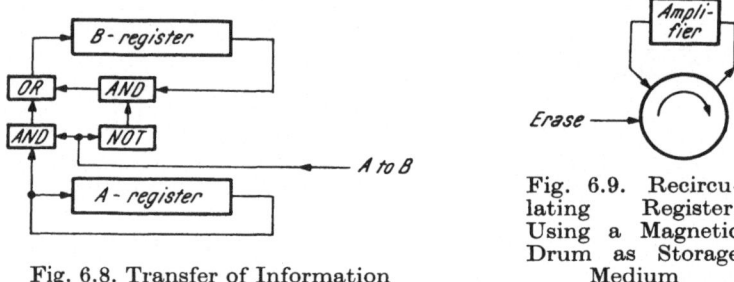

Fig. 6.8. Transfer of Information

Fig. 6.9. Recirculating Register, Using a Magnetic Drum as Storage Medium

Problem 3: Draw the logic diagram of a circuit which allows the transfer of information from the recirculating register A to the recirculating register B while simultaneously the content of B is transferred to A.

Problem 4: Try to find a reason why in recirculating registers information bits are almost always recirculated in the order of their significance, i.e., least significant bit first, then the next significant, and finally the most significant.

Recirculating registers may be built with any type of delay lines discussed in paragraph 5.2.2. It may be interesting to note that also magnetic drums can be used to construct recirculating registers. Fig. 6.9 shows the basic arrangement.

Information recorded by the write head travels with the drum surface to the read head, before it is read. By the appropriate physical spacing of read and write heads, the desired delay time, i.e. storage capacity, can be achieved.

Registers of this type are frequently provided with an "erase" head which makes certain that the drum surface is cleared of all previous information before a new pattern is recorded. Since, in this mode of operation, the drum surface is erased i.e. set to the "zero" state, the write head has to record only "one's". The clock pulses used for the synchronization of this type of registers are usually derived from a clock channel which is permanently recorded on one of the drum tracks. In this manner, the information rate stays in synchronism with the clock rate, even if the speed of the drum varies. Since a register usually needs only a small part of the drum circumference for the required delay, several registers may be accomodated by a single track. Such an arrangement can be seen in Fig. 6.10.

Recirculating registers using magnetic drums as storage devices usually record information in the NRZ-C mode[1]. Pulses produced by the read head

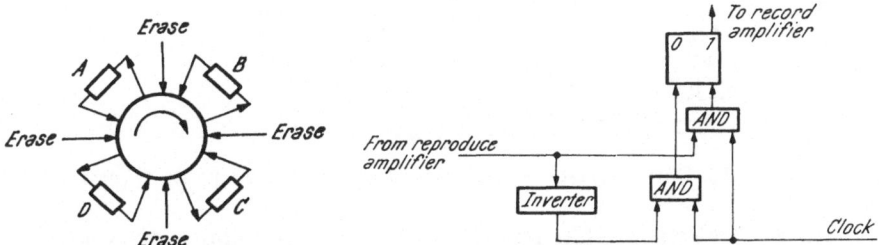

Fig. 6.10. Multiple Use of   Fig. 6.11. Conversion of Output Pulses to
a Single Track   Output Levels

are converted to levels before they are re-recorded. A flip-flop set by pulses of one polarity (positive, according to Fig. 5.17) and cleared by pulses of opposite polarity (negative according to Fig. 5.17) accomplishes this rather conveniently. Fig. 6.11 shows one possible arrangement.

The level output of the flip-flop can not only be used for the re-recording of information, but it is also the appropriate representation of the information for a computer using level-type circuitry.

6.1.3. Shift Registers

The registers shown so far, have no provisions for arithmetic or logic operations other than the storage and transfer of information. However, most actual registers are able to perform additional operations on the contained information. One of the simplest operations of this type is shifting. Suppose a register contains the following binary number:

$$\boxed{1011 \ \ldots \ 0101}$$

---

[1] See Fig. 5.17.

The same number shifted one place to the left, appears like:

$$011 \ldots\ldots 01010$$

A shift register performs such a shift, independent of the particular information contained in the register.

6.1.3.1. Shifting with Flip-Flop Registers: Flip-flop registers which are able to shift, have more elaborate switching circuitry than the simple registers shown in paragraph 6.1.1. The particular type of circuitry depends upon the type of flip-flops used.

Suppose we have a register as shown in Fig. 6.1. Two pulse gates per flip-flop make it a shift register as indicated in Fig. 6.12.

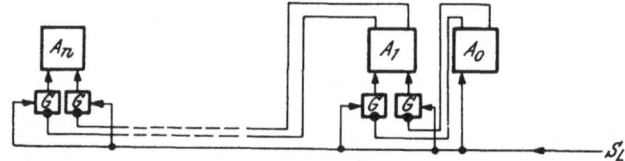

Fig. 6.12. Shift Register with Static Flip-Flops and Pulse Inputs

The application of a shift pulse (S_L) causes the flip-flop A_1 to be set if A_0 contains a "1", and it causes A_1 to be cleared if A_0 contains a "0". In other words, A_1 will assume the previous state of A_0, or the bit of information previously contained in A_0 is shifted into A_1. In the same manner, the content of A_1 is shifted to A_2, and so on. Finally, the content of A_{n-1} is shifted into A_n. In this manner, all bits of the information originally contained in A are shifted one place to the left. By the repeated application of shift pulses, the information may be shifted an arbitrary number of places.

Usually, the least significant bit is set to zero with the shift pulse (in this way, zeros are shifted in) whereas the most significant bit is "shifted out", i.e., lost. In some instances, registers have provisions for a "circular" or "end around" shift, where the bits of information which are shifted out of one end of the register, enter on the other end. Fig. 6.13 shows the first and the last stage of such a register.

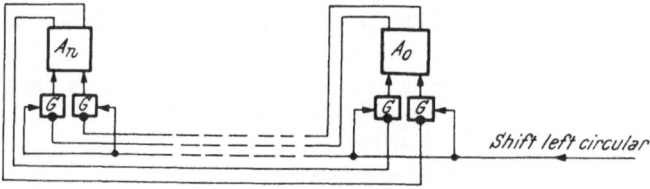

Fig. 6.13. Circular Shift

Problem 5: Show the details of the logic circuitry for a transistorized flip-flop shift register. The register shall perform a left circular shift or a right circular shift, upon the receipt of "shift left" or "shift right" pulses. Assume flip-flops with pulse or level inputs. Shift pulses are negative.

If a register consists of flip-flops with level inputs, the pulse gates in Fig. 6.12 and 6.13 are replaced by AND-circuits.

Note how important the possibility of simultaneous setting and reading the flip-flops is for shift registers. If this is not feasible[1], single bits of information could be shifted by a series of appropriately spaced shift pulses as schematically indicated in Fig. 6.14.

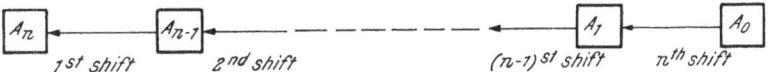

Fig. 6.14. Shifting by Staggered Shift Pulses

Such a shift requires n pulse times instead of the one pulse time required for the scheme given in Fig. 6.12. A faster operation can be achieved by the use of an auxiliary storage register as indicated in Fig. 6.15.

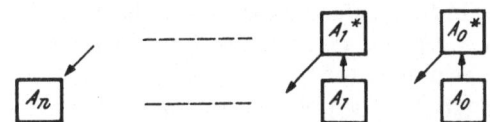

Fig. 6.15. Use of Auxiliary Storage for Shifting

Information is first transferred from A into the auxiliary storage A^*, and then transferred back into A, shifted by one position. The scheme needs two pulse times for a shift by one place[2].

Fig. 6.16 shows the diagram of a shift register with dynamic flip-flops:

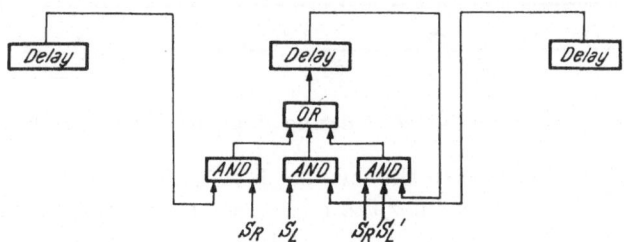

Fig. 6.16. Shift Register with Dynamic Flip-Flops and Provisions for Left and Right Shifts

---

[1] When, for instance, DC flip-flops are used.

[2] This type of shift is used exclusively in some strictly DC computers.

S_L pulses (shift left) cause a flip-flop to be set if the next lower flip-flop contains a "1". The complement of the shift pulse $(S_L)'$ prevents the recirculation in a flip-flop during the transfer, so that it will be cleared if the next lower flip-flop stage contains a "0". The "shift right" is performed in an analogous manner.

The information contained in a shift register has to be shifted frequently by a certain fixed number of binary positions. For instance, a computer which uses a binary coded decimal system may have certain registers which shift information by four binary places, i.e., by one decimal position. Such a shift can be facilitated by groups of four shift pulses or by special circuitry. Fig. 6.17 shows a shift register which shifts four places for each shift pulse.

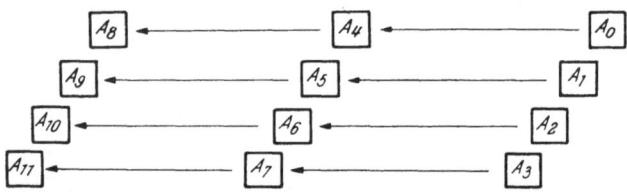

Fig. 6.17. Shift by a Fixed Number of Binary Places

6.1.3.2 Shifting in Recirculating Registers: The significance of a bit which is recirculating in a register is given by its position relative to a marker which signifies the beginning (or the end) of a recirculation cycle[1]. Fig. 6.18a shows the time diagram of a pulse train representing the binary number 01...011001011 in relation to the mark pulse. Fig. 6.18b shows the same

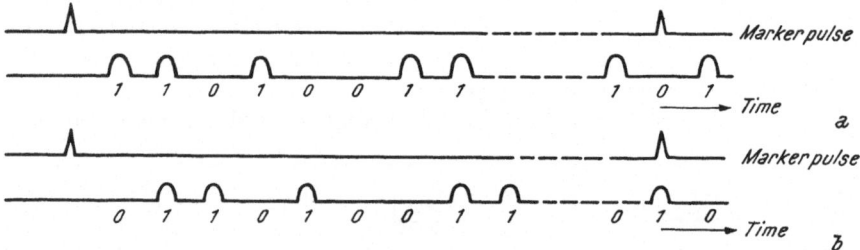

Fig. 6.18. Information in a Recirculating Register Relative to a Mark Pulse

number shifted one place to the left, i.e. 1...011010110. It is to be noted that the time diagram assumes the recirculation of a pattern in which the least significant bit comes first, then all other bits in the order of their

---

[1] If the register is synchronized with a computer clock, the mark can be derived from this clock. For instance, a register which recirculates 42 bits requires a mark pulse every 42 clock cycles. Using a modulo 42 counter (see paragraph 6.2.2) we can select every 42nd clock pulse and pass it as mark pulse.

significance, the most significant bit being the last. In this manner, the bits of a number appear in Fig. 6.18 in reversed order, i.e., the least significant bit to the left and the most significant bit to the right. For the same reason, a left shift of information corresponds to a right shift in Fig. 6.18.

We see that a shift is equivalent to a change in the time relation between the information and the mark pulse. Since we cannot shift the mark pulse (all registers in the computer will be synchronized with it and a shift in the mark pulse would correspond to a shift of information in all registers), we have, in some way, to shift the information with respect to the mark pulse. As we see from Fig. 6.18, a left shift by one bit corresponds to a delay of information by one pulse time. There-

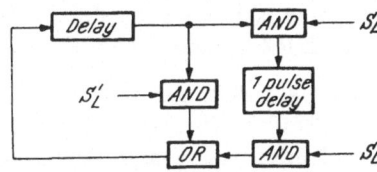

fore, the introduction of an additional delay of one pulse time into the recirculation loop of the register will shift the information one bit to the left.

Suppose we apply the shift signal in synchronism with a mark pulse. The regular recirculation loop is interrupted. The first bit which enters the main

Fig. 6.19. Recirculating Register with Provisions for a Left Shift

delay after the mark pulse is a "zero" since the one-pulse delay has been previously at rest. The following bits correspond to the information previously stored in the main delay line. As we can see, they enter one pulse time late with respect to their original position. If the shift signal is removed in synchronism with the next mark, the previously most significant bit will be *trapped* in the one pulse delay and lost, whereas all other bits recirculate in the normal manner. In effect, we have performed a straight left shift. Shifts by more than one position can be accomplished by repeated shifts of one position.

If a circular shift is desired, we can modify our circuit as shown in Fig. 6.20.

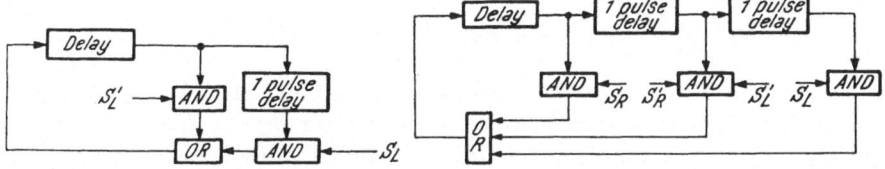

Fig. 6.20. Recirculating Register with Provisions for a Circular Shift

Fig. 6.21. Recirculating Register with Provisions for Left and Right Shift

The first bit which enters the delay line after the application of the shift signal is the previously most significant bit which, at the time of the mark pulse, is stored in the one pulse delay. It will reappear as the least significant bit.

A right shift in a recirculating register can be performed in two different ways. One possibility is a repeated left circular shift until the information appears in the desired place. The principle of the second possibility is indicated in Fig. 6.21.

The delay for a normal recirculation consists of a long delay and one additional one-pulse delay. For a right shift, the additional delay is by-passed, whereas two additional delays are used for a left shift.

Problem 6: Frequently, flip-flops are used as convenient delays for one pulse time. Design the logic diagram of a recirculating register which uses a magnetic drum as main delay. Additional delays required for left or right (circular) shifts are flip-flops with level inputs. The recording on the drum is in NRZ-C mode. The computer clock is available as input.

6.1.3.3. Shift Registers as Serial-to-Parallel and Parallel-to-Serial Converters: Shift Registers are frequently used to convert parallel information to serial information and vice versa.

By parallel information we mean information which is available or transferred in parallel. For instance, the register given in Fig. 6.1 presents parallel information. All bits are available at the same time, but at different terminals. Serial information is available or transferred one bit at a time. (The recirculating register of Fig. 6.19 presents e.g. serial information.) All bits are available at the same terminal, but during consecutive time intervals[1]. It is customary to speak not only of parallel or serial information, but also of parallel or serial registers.

The basic arrangement for the use of a shift register as parallel-to-serial converter is shown in Fig. 6.22.

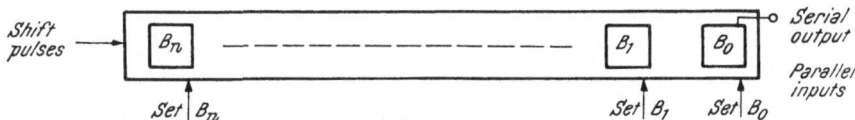

Fig. 6.22. The Shift Register as Parallel-to-Serial Converter

B_0 to B_n are the flip-flops of a shift register. Information is transferred to the shift register in parallel via the set and clear inputs. If we now apply shift pulses to the register, the information is shifted to the right and always one bit at a time is available on the output of the shift register, i.e. at the output of flip-flop B_0. The output is serial.

A similar register may be used to convert from serial to parallel information.

---

[1] Many computers use either entirely serial or entirely parallel representation of information. We speak then of serial or parallel machines.

The input to the shift register is in serial form. If we apply shift pulses to the register synchroneously with the information, then the information is shifted into the register and is later available in parallel on the outputs of the B flip-flops.

Serial input

Shift pulses

Fig. 6.23. The Shift Register as Serial-to-Parallel Converter

Information is not always represented in purely serial or purely parallel form. There are instances when part of the information is stored or transferred partly in serial and partly in parallel. One example of such serio-parallel information is information recorded on magnetic tape. Here, some bits of a computer word may be recorded in parallel but, altogether, a computer word may comprise several lines (e.g. the 36 bits of a computer word are recorded in six lines with 6 bits each). The conversion of serio-parallel information to truly serial or truly parallel information can be performed by shift registers similar to that indicated in Fig. 6.17.

Problem 7 (Voluntary): Design the logic diagram for a 12-bit shift register which can accept serial information, parallel information and serio-parallel information, 4 bits at a time. It shall have the further capability of delivering information in these three formats. Show the necessary input and output circuitry. List the number and sequence of control signals to be applied for the input and output of each of these data formats.

6.2. Counters

6.2.1. Binary Counters

Electronic counters are circuits which count the number of inputs (of electronic pulses) they receive. Counters can be made to count in any number system, the binary system being the most convenient one.

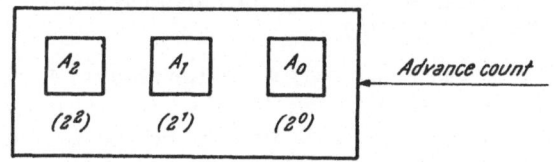

Advance count

Fig. 6.24. Basic Arrangement of a Binary Counter

Counters ordinarily consist of a number of storage elements which store the current count, and of associated logic circuitry which causes the count to be increased every time a count pulse is received. Fig. 6.24 shows the basic arrangement of such a counter. It is assumed that it counts in the binary system and that the storage elements are flip-flops.

The consecutive states of the flip-flops A_0 through A_2 follow a straight binary count and are indicated in the table below.

Table 6.1. *Table of Consecutive Flip-Flop States for a Binary Counter*

A_2	A_1	A_0	State
0	0	0	At beginning
0	0	1	After 1st Pulse
0	1	0	After 2nd Pulse
0	1	1	After 3rd Pulse
1	0	0	After 4th Pulse
	etc.		

6.2.1.1. Counters with Complement Flip-Flops: In order to find the required switching circuitry systematically, let us first observe the state of the flip-flop A_0. We see that it must change state for every advance pulse. Connecting the advance pulse to the complement input, we have found the required circuitry for the first stage of the counter.

Flip-flop A_1 has to change its state when (to be exact: if and only if) there is an advance pulse *and* the low order flip-flop A_0 contains a "1".

Fig. 6.25. First Stage of a Binary
Counter

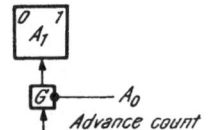

Fig. 6.26. Second Stage of a
Binary Counter

Similarly, A_2 is complemented by the advance pulse if both, A_0 *and* A_1 contain "1's". (See Fig. 6.27).

We could have used a single pulse gate on the input to A_2 which is enabled by A_0 *and* A_1. However, if we notice that the gate which is enabled by A_0 and strobed by the advance pulse is already incorporated into the second stage, we can simplify the overall diagram.

We see that all stages contain identical circuitry. The output of the third stage can be used as input to further stages or can be used to indicate that the capacity of the counter has been exceeded.

Problem 8: Show the details of the logic circuitry for a typical counter stage. Assume transistor flip-flops with pulse inputs. If desired, pulse gates with pulse transformers and two secondary windings are available to obtain simultaneously positive as well as negative pulse polarities.

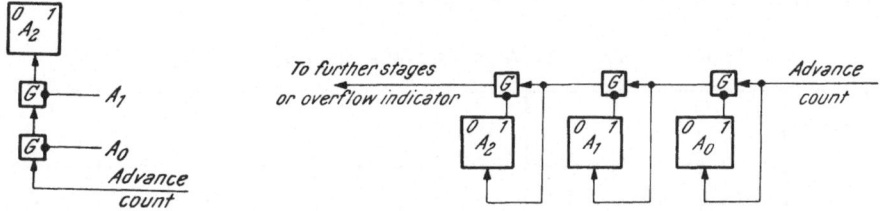

Fig. 6.27. Third Stage of a Binary Counter

Fig. 6.28. Complete Three-Stage Counter

Problem 9 (Voluntary): Repeat problem 8, assuming tube circuitry.

Sometimes counters are capable of increasing or decreasing their count dependent upon whether "count up" or "count down" pulses are received. Such counters are usually designated as up-down counters.

The circuitry for increasing the count stays essentially the same as shown in Fig. 6.28. For the count down, Table 6.1 shows the consecutive states of flip-flops A_0 through A_2 from bottom to top. We see that a particular digit has to be complemented if all lower order stages contain zeros and a count down pulse is received[1]. Fig. 6.29 shows the logic diagram of such an up-down counter.

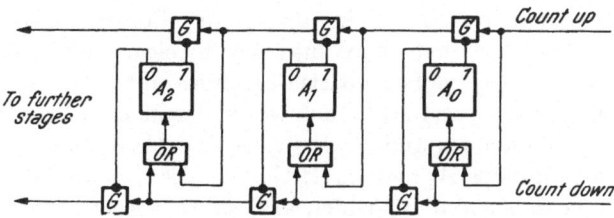

Fig. 6.29. Binary Up-Down Counter

Problem 10: Show in a time diagram the state of all flip-flops in a three-stage up-down counter. Assume a repeated sequence of three "count up" and two "count down" signals.

6.2.1.2. Counters with Set/Reset Flip-Flops: So far, we have assumed flip-flops with complement inputs. If flip-flops have no such provisions, they have to be cleared and set separately. Consequently, the design of counter

---

[1] For a count up, the state of a flip-flop is complemented if all lower order stages contain ones, as we have seen.

circuits become more cumbersome. Let us derive the circuitry required for such a counter in detail.

Table 6.2 lists the consecutive counter states and the required inputs (set and clear signals) for a three stage binary counter. Set signals have the prefix 1 and clear signals the prefix 0.

Table 6.2. *Truth Table for a Three Stage Binary Counter*

Count	A_2	A_1	A_0	$_1a_2$	$_0a_2$	$_1a_1$	$_0a_1$	$_1a_0$	$_0a_0$
0	0	0	0	0	×	0	×	1	0
1	0	0	1	0	×	1	0	0	1
1	0	1	0	0	×	×	0	1	0
3	0	1	1	1	0	0	1	0	1
4	1	0	0	×	0	0	×	1	0
5	1	0	1	×	0	1	0	0	1
6	1	1	0	×	0	×	0	1	0
7	1	1	1	0	1	0	1	0	1
0	0	0	0						

The right-hand side of the table shows the inputs which, in conjunction with the advance count signal, are required to switch the counter from a particular state (indicated in the left half on the same line) to the state representing the next higher count (indicated in the left half on the next line). For instance, in order to go from the top line ($A_2'A_1'A_0'$) to the second line ($A_2'A_1'A_0$), the flip-flop A_0 has to be set. The setting of this flip-flop is indicated by the "1" in column $_1a_0$. Columns $_0a_0$, $_1a_1$ and $_1a_2$ show "0's" to indicate that there may be no clear pulse for A_0 (which would be simultaneously applied with the set pulse and is therefore forbidden), and no set pulses for A_1 and A_2 (which would result in an erroneous count). The crosses in columns $_0a_1$ and $_0a_2$ indicate that there may or may not be clear pulses for flip-flops A_1 and A_2. If these pulses are applied, they would clear the flip-flops A_1 and A_2 which are set to zero anyhow.

Let us now derive the logic expressions which govern the count from the truth table. Listing standard sums, we find:

$$_1a_0 = (A_2'A_1'A_0' + A_2'A_1A_0' + A_2A_1'A_0' + A_2A_1A_0') C \tag{6.1}$$

$$_0a_0 = (A_2'A_1'A_0 + A_2'A_1A_0 + A_2A_1'A_0 + A_2A_1A_0) C \tag{6.2}$$

$$_1a_1 = (A_2'A_1'A_0 + A_2A_1'A_0) C \tag{6.3}$$

$$_0a_1 = (A_2'A_1A_0 + A_2A_1A_0) C \tag{6.4}$$

$$_1a_2 = (A_2'A_1A_0) C \tag{6.5}$$

$$_0a_2 = (A_2A_1A_0) C \tag{6.6}$$

The term C represents the advance count signal. If we simplify these expressions to minimum sums we can write:

$$_1a_0 = A_0'C \tag{6.7}$$

$$_0a_0 = A_0C \tag{6.8}$$

$$_1a_1 = A_1'A_0C \tag{6.9}$$

$$_0a_1 = A_1A_0C \tag{6.10}$$

$$_1a_2 = A_2'A_1A_0C \tag{6.11}$$

$$_0a_2 = A_2A_1A_0C \tag{6.12}$$

Equation (6.7) states that flip-flop A_0 has to be turned on, if it is turned off when an advance pulse is received. Equation (6.8) states that A_0 has to be turned off, if it is turned on when an advance pulse is received. This separation of clear and set pulses is common to all stages and constitutes the essential difference between this counter and the one given in Fig. 6.28.

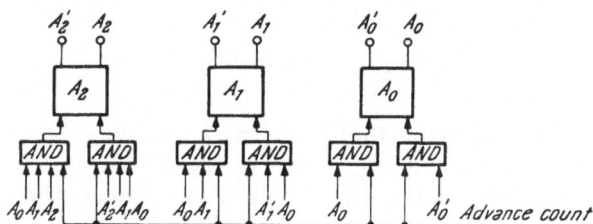

Fig. 6.30. Binary Counter with Separate Clear and Set Pulses

For simplicity, connections from flip-flop outputs to AND-circuit inputs are not shown. Some savings in circuitry could be achieved by cascading the various AND-circuits. Whether this is possible and feasible depends upon the specific design. Incidentally the crosses in Table 6.2 have been replaced by "0's" [Equation (6.1) through (6.12)]. If we would use "1's", the resulting equations would be unnecessarily complicated[1].

Problem 11: Show the logic equations for the fourth and fifth stage of a counter similar to the one given in Fig. 6.30.

6.2.1.3. Counters with DC Flip-Flops: All counter circuits discussed so far, require flip-flops which can be simultaneously read out and changed. However, if a counter is to be built from DC flip-flops,[2] two

---

[1] Shown on a Karnaugh map, the crosses are neighboring to zeros, so that extra terms have to be introduced if crosses are replaced by ones.

[2] Category 1, page 89.

flip-flops per bit must be provided. One rank of flip-flops holds the old count, and the other rank is set to the new count. Fig. 6.31 shows one possible solution.

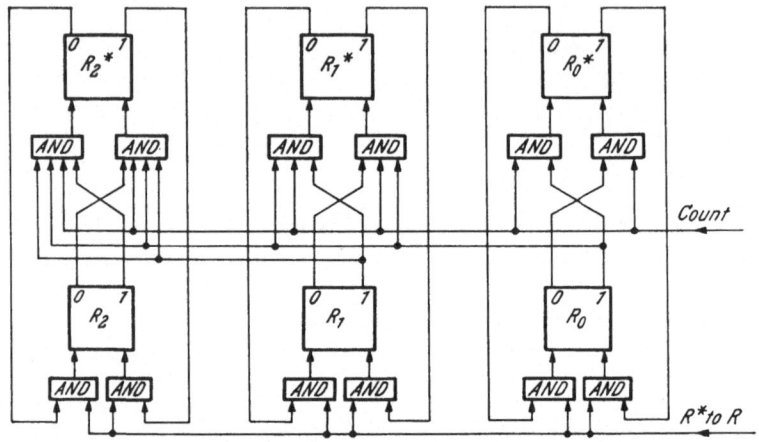

Fig. 6.31. Binary Counter with DC Flip-Flops

To start with, let us assume that both, the R and the R^* storages contain the current count. The count signal now changes the state of the flip-flops R^* as necessary to obtain the new count. For instance, the state of R_0^* is changed regardless of any further considerations. The state of R_1^* is changed by the count signal only if the lower order flip-flop R_0 contains a "one" and the state of R_2^* is changed if both, R_0 and R_1 contain "ones". The complementation of a flip-flop is accomplished by separate clear and set inputs, and only one or the other but never both are applied simultaneously.

As we can see, the count signal sets the R^* flip-flops to the new count, while the old count remains stored in the R flip-flops. In order to prepare the counter for a new count input, the current count contained in R^* is duplicated in R by a signal "R^* to R". Each consecutive count in this scheme requires a sequence of two signals: a count signal, and then (somewhat delayed) a transfer signal "R^* to R".

One might think of several variations of this scheme, but, in general, counting is much more cumbersome with DC flip-flops than with other types. Of course, there is some justification for using DC flip-flops. For one, the fact that all signals in a system are represented by levels and nowhere by pulses or dynamic conditions may give some aesthetic satisfaction. But it also simplifies maintenance and, in the case that only one type of logic module is used throughout a system, the spare parts requirements. Moreover, the flip-flops are less expensive than other types, and the two separate storages might possibly be made use of for sundry other purposes.

Problem 12: Design the logic diagram for a four-stage binary "countdown" counter with DC flip-flops.

6.2.1.4. Counters with Dynamic Flip-Flops: So far we have assumed static flip-flops for all counter circuits. Fig. 6.32 shows the first two stages of a binary counter with dynamic flip-flops and Fig. 6.33 the corresponding time diagram of advance pulses and flip-flop outputs.

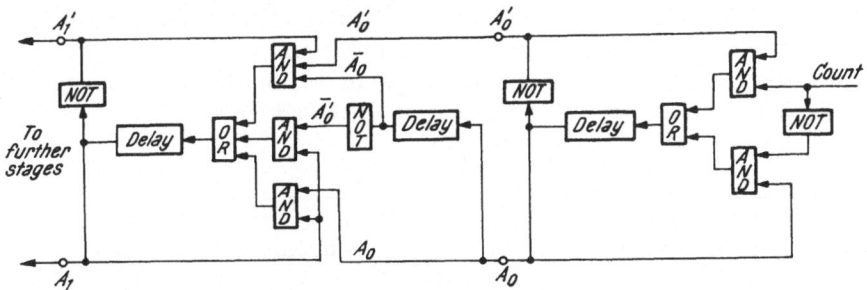

Fig. 6.32. Binary Counter with Dynamic Flip-Flops

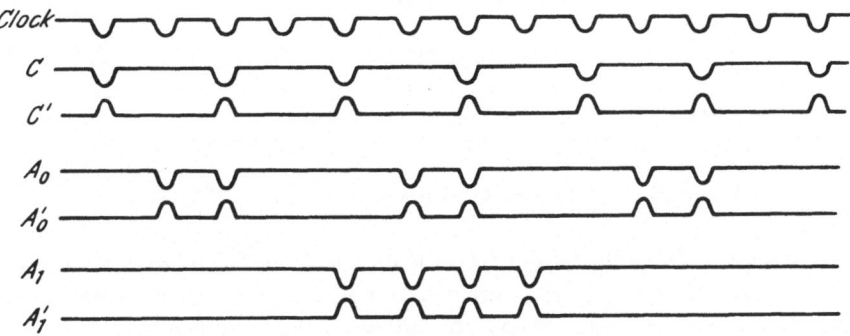

Fig. 6.33. Time Diagram for Above Counter

The first stage is complemented by every advance pulse in a manner which is typical for dynamic flip-flops[1]. Its outputs A_0 and A_0' are available to the second stage. By means of an additional delay, the propositions $\overline{A_0}$ and $\overline{A_0'}$ are derived. These propositions represent the state of the first stage

---

[1] See paragraph 5.2.3.

delayed by one pulse time. In effect then, both, the present state of the first stage (A_0 and A_0') and the previous state ($\overline{A_0}$ and $\overline{A_0'}$) are available to the second stage flip-flop A_1 during a particular pulse time.

The second stage is complemented if the first stage changes from a "1" to a "0". To be more detailed: The second stage is *set* if it contains a "zero" (A_1') *and* if the first stage contains a "zero" (A_0') *and* if the previous content of the first stage was a "one" ($\overline{A_0}$). The second stage is *cleared*, i.e. its recirculation is prohibited, if the first stage contains a "zero" (proposition A_0 is in the "0" state) *and* if the previous content of the first stage was a "one" (proposition $\overline{A_0'}$ is in the "0" state).

In an arrangement as given in Fig. 6.32, the higher stages of a counter are identical. The fact that it takes several pulse times before the correct count is available, is a disadvantage for certain applications. Fig. 6.34 shows a counter with dynamic flip-flops in which the correct count is available one pulse time after application of the advance pulse.

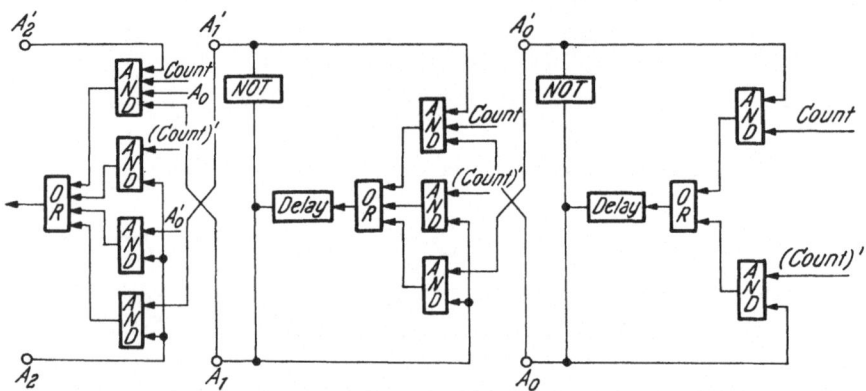

Fig. 6.34. Fast Binary Counter with Dynamic Flip-Flops

The second stage flip-flop A_1 is set if it contains a "zero" *and* A_0 contains a "one" when the advance pulse arrives. It is cleared, when it contains a "one" *and* A_0 contains a "one" when the advance pulse arrives[1].

For the second stage we can write the following input equation:

$$a_1 = A_1'A_0C + A_1A_0' + A_1C' \qquad (6.13)$$

The first term of the right hand member in Equation (6.13) corresponds to the "set" input in the diagram, the other two terms take care of the recirculation. One input equation is sufficient to describe the behavior

---

[1] A_1 will recirculate for $A_0' + C'$. Therefore, it will *not* recirculate or be cleared for: $(A_0' + C')' = A_0C$.

of the flip-flop since, being a dynamic flip-flop, it will be cleared if there are no inputs at any one pulse time.

Similarly, we can write for the third stage:

$$a_2 = A_2'A_1A_0C + A_2A_1' + A_2A_0' + A_2C' \qquad (6.14)$$

Effectively, this equation states that the third stage is to be complemented if all lower order stages contain a "1" when the advance pulse arrives.

Problem 13: Draw a time diagram similar to the one in Fig. 6.33 for the counter given in Fig. 6.34.

6.2.1.5. Counters with Recirculating Registers: Up to now, we have seen counters which contain flip-flops as storage elements. In a sense, we can regard these counters as consisting of flip-flop registers and associated circuitry permitting the counting. However, not only flip-flop registers but also recirculating registers can be used to construct counters. Fig. 6.35 gives the basic arrangement.

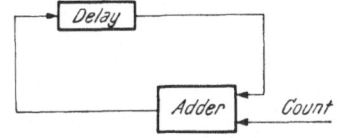

Fig. 6.35. Recirculating Register as Counter

The recirculation loop contains an "adder". Since adder circuits will be discussed to quite some extent in paragraph 6.3, let it suffice to say here only that the register contains the current count and that for every count input, a "one" is added arithmetically to its contents.

6.2.2. Counters for Other Number Systems

6.2.2.1. Decade Counters: Electronic counters which count in a number system other than the binary system, normally use binary codes to represent numbers. As an example of such a counter, let us derive the circuitry required for a decade of a decimal counter which uses the 8421 code[1].

The basic arrangement is given in Fig. 6.36.

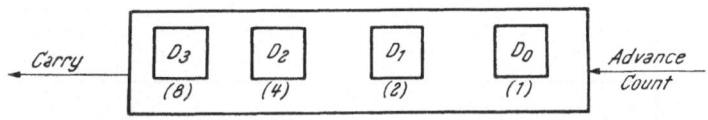

Fig. 6.36. Basic Arrangement of a Decade Counter Using the 8421 Code

The count given by the states of flip-flops D_0 through D_3 follows the straight binary count up to 1001 which is the equivalent of the decimal digit 9. However, with the next advance pulse, the counter assumes the state

---

[1] See Table 2.6.

0000 (equivalent to the decimal digit 0), and produces a carry or overflow which may be used to advance a second decade, counting in units of ten.

Table 6.3. *Truth Table for Decade Counter*

State	D_3	D_2	D_1	D_0	d_3	d_2	d_1	d_0
0	0	0	0	0	0	0	0	1
1	0	0	0	1	0	0	1	1
2	0	0	1	0	0	0	0	1
3	0	0	1	1	0	1	1	1
4	0	1	0	0	0	0	0	1
5	0	1	0	1	0	0	1	1
6	0	1	1	0	0	0	0	1
7	0	1	1	1	1	1	1	1
8	1	0	0	0	0	0	0	1
9	1	0	0	1	1	0	0	1

Table 6.4. *Karnaugh Maps for the Complementation of Flip-Flops D_0 through D_3*

		D_1	0	1	1	0
D_3	D_2	D_0	0	0	1	1
0	0		1	1	1	1
0	1		1	1	1	1
1	1		×	×	×	×
1	0		1	×	×	1

d_0

		D_1	0	1	1	0
D_3	D_2	D_0	0	0	1	1
0	0		0	0	1	1
0	1		0	0	1	1
1	1		×	×	×	×
1	0		0	×	×	0

d_1

		D_1	0	1	1	0
D_3	D_2	D_0	0	0	1	1
0	0		0	0	1	0
0	1		0	0	1	0
1	1		×	×	×	×
1	0		0	×	×	0

d_2

		D_1	0	1	1	0
D_3	D_2	D_0	0	0	1	1
0	0		0	0	0	0
0	1		0	0	1	0
1	1		×	×	×	×
1	0		0	×	×	1

d_3

Table 6.3 constitutes the truth table for this counter.The right hand side indicates the actions which are required to advance the counter from the present state to the next sequential state. Its last line shows the actions required for the return to the zero state. Rather than listing separate columns for the clear and set inputs of individual flip-flops, the right hand columns in Table 6.3 show when a certain flip-flop is be complemented. In effect, it is assumed that we will use flip-flops with complement inputs to build the counter.

The "$1's$" and "$0's$" in Table 6.4 correspond to those in Table 6.3. Crosses have been entered into places for which no specifications are given by the truth table. Since these entries correspond to conditions which the counter never assumes, we may enter "$0's$" or "$1's$" as we see fit to simplify the counter circuitry.

Looking for patches of "$1's$" (or patches of "$1's$" and crosses) that are as large as possible, we find the minimum functions encircled in Table 6.4. The condition 0111 for the proposition d_3 is given as a single entry. This implementation is somewhat simpler than the combination of 0111 with 1111 to a double patch since it can be easily derived from the existing proposition d_2. Similarly, d_2 is given as a patch of two, rather than as a patch of four entries, since it can in this form easily be derived from d_2.

Labelling the "advance count" pulse with the letter C, we have the following input equations:

$$d_0 = C \tag{6.15}$$

$$d_1 = D_3'D_0C \tag{6.16}$$

$$d_2 = D_3'D_1D_0C \tag{6.17}$$

$$d_3 = D_3'D_2D_1D_0C + D_3D_0C \tag{6.18}$$

We notice that:

$$d_2 = D_1d_1, \text{ and} \tag{6.19}$$

$$d_3 = D_2d_2 + D_3D_0C \tag{6.20}$$

Finally, the carry to the next decade should be produced if the counter is in the state 1001 and an advance pulse is received. This condition can be expressed as:

$$\text{Carry} = D_3D_2'D_1'D_0C \tag{6.21}$$

or simpler as[1]:

$$\text{Carry} = D_3D_0C \tag{6.22}$$

Fig. 6.37 gives the implementation of the above equations and the complete logic diagram of the counter.

Following the same line of thinking, counters for any other number systems can be designed.

---

[1] Compare paragraph 3.5.3.

Problem 14: Design the logic diagram for a decade counter using:
a. the 2421 code
b. the excess 3 code

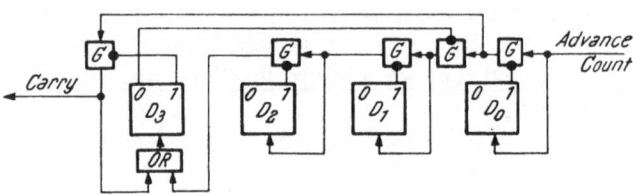

Fig. 6.37. Logic Diagram of Decade Counter

Problem 15: Design the logic diagram for a modulo 6 counter (counting from 000 to 101, then resetting and producing a carry).

If we cannot use flip-flops with complement inputs, the set and clear pulses have to be applied separately to a flip-flop (similar to the binary counters in paragraph 6.2.1).

Problem 16: Design the logic diagram for a decade counter using the 8421 code. Assume that you have to use separate level inputs for clear and set, and flip-flops of the category 2a, page 89.

Problem 17 (Voluntary): Design a 8421 decade counter
a. with flip-flops of the category 1, page 89
b. dynamic flip-flops.

Problem 18 (Voluntary): Design a 8421 decade up-down counter which has provisions to deliver count-up and count-down carries to the next decade.

6.2.2.2. Gray Code Counters: A special case, in some respects, are counters for Gray codes. Let us derive here, as an example, the circuitry of a counter employing a specific Gray code, the inverted binary code[1]. In order to find its circuitry, we first list the consecutive counter states (columns G_3, G_2, G_1, G_0, in Table 6.5) and the instances when a particular stage has to be complemented (columns g_3, g_2, g_1, g_0,).

If we want, we may indicate the transition from one state to the next in the Karnaugh map shown in Table 6.5b. From either Table 6.5a or b, we can derive the Karnaugh map given in Table 6.6, which indicates under which conditions the low order stage G_0 has to be complemented.

The implementation of this Karnaugh map may be straightforward, but it is not simple. We see that all "one's" have to be implemented as single

---

[1] See paragraph 2.4.2.

Table 6.5. *Consecutive States of the Gray Code Counter*

State	G_3	G_2	G_1	G_0	g_3	g_2	g_1	g_0
0	0	0	0	0	0	0	0	1
1	0	0	0	1	0	0	1	0
2	0	0	1	1	0	0	0	1
3	0	0	1	0	0	1	0	0
4	0	1	1	0	0	0	0	1
5	0	1	1	1	0	0	1	0
6	0	1	0	1	0	0	0	1
7	0	1	0	0	1	0	0	0
8	1	1	0	0	0	0	0	1
9	1	1	0	1	0	0	1	0

etc.

a

		G_1	0	0	1	1
G_3	G_2	G_0	0	1	1	0
0	0		● → ● → ● → ●
0	1		● ← ● ← ● ← ●
1	1		● → ● → ● → ●
1	0		● ← ● ← ● ← ●

b

Table 6.6. *Required Inputs to the Low-Order Stage of the Gray Code Counter*

		G_1	0	0	1	1
G_3	G_2	G_0	0	1	1	0
0	0		1	0	1	0
0	1		0	1	0	1
1	1		1	0	1	0
1	0		0	1	0	1

entries. For the first stage of a 4-stage counter we would, therefore, have to have eight AND-circuits with four inputs each. (We would find similar results if we would investigate the required inputs for higher order stages. Let us, therefore, see whether we can find a simpler arrangement by a different approach.

From Table 6.5a, we can see that G_0 has to be complemented for every second input. Fig. 6.38 shows the circuitry to accomplish this.

The circuit contains an additional auxiliary flip-flop A, but the total

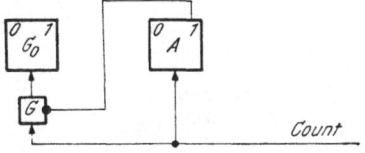

Fig. 6.38. Input Circuitry for the Low Order Stage of the Gray Code Counter

circuitry for the first stage of the Gray counter is less complicated (and probably less expensive) than the implementation of the Karnaugh map given in Table 6.6.

Since we now have an additional flip-flop, let us list its state together with the states of G_3, G_2, G_1, G_0 in a new truth table, before we derive the circuitry required for the higher stages.

Table 6.7. *Truth Table for the Gray Code Counter with Auxiliary Flip-Flop*

State	G_3	G_2	G_1	G_0	A
0	0	0	0	0	1
1	0	0	0	1	0
2	0	0	1	1	1
3	0	0	1	0	0
4	0	1	1	0	1
5	0	1	1	1	0
6	0	1	0	1	1
7	0	1	0	0	0
8	1	1	0	0	1
9	1	1	0	1	0

We see that G_1 has to be complemented if, and only if G_0 is in the "1"-state *and* A is in the "0" state. Similarly, G_2 is complemented if and only if G_1 is in the "1"-state *and* both, G_0 and A are in the "0"-state. In effect, a particular flip-flop is complemented if, and only if the flip-flop immediately to its right contains a "1", but all other flip-flops further to the right contain zeros. This behavior is indicated by arrows in Table 6.7. Let us now construct the applicable circuitry.

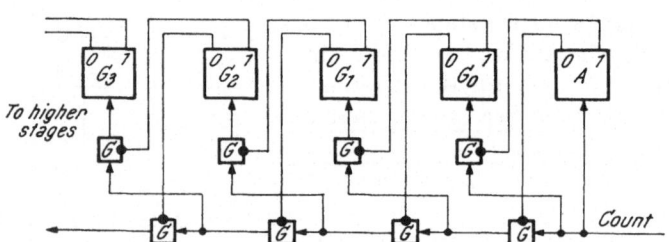

Fig. 6.39. Gray Code Counter

The gates at the bottom of Fig. 6.39 pass a count pulse as long as all low order flip-flops are in the zero-state. If there is a "1" in any particular stage, the count pulse is diverted to the next higher stage as complement input.

Here, we have deduced the required circuitry entirely by logic reasoning. However, starting with Table 6.7 we could have used Karnaugh maps or algebraic methods and derived the same results. The desirability of the auxiliary flip-flop A in the Gray code counter could not have been found by the methods of Boolean algebra. These methods allow us to find the required switching circuitry for a particular case, provided that the number of storage elements is known and the meaning of their states is defined.

Incidentally, it is interesting to note that it is possible to build counters for the inverted binary code with only one flip-flop per bit, even if DC flip-flops are used. The reason for this is that only one flip-flop changes its state at any one time (disregarding the auxiliary flip-flop) and that its input signal is independent of its output.

Problem 19 (Voluntary): Show the Karnaugh maps for the inputs to G_0, G_1, G_2, G_3, in above counter. This is a five-variable problem (A, G_0, G_1, G_2, G_3, are input variables). Show two sets of four-variable Karnaugh maps (one for $A = 0$, and one for $A = 1$).

Problem 20: Design a three-stage Gray code counter without auxiliary flip-flop.

Problem 21 (Voluntary): Design a Gray code counter with separate clear and set inputs to each flip-flop.

Problem 22 (Voluntary): Design a Gray code up-down counter.

6.2.3. Counters with Special Features

Sometimes it is desired to have flexible counters, i.e. counters which progress from one state to the next not in a fixed sequence determined by their internal logic circuitry, but which alter their sequence dependent upon external signals or conditions. Such counters frequently constitute the main part of computer control units[1]. Here, different actions have to be taken dependent upon conditions external to the control unit. For instance, a multiply operation requires a sequence of control signals different from that for a divide operation. A flexible counter, going through one particular sequence of states for the multiply operation and going through another particular sequence of states for a divide operation, can be the source of such control signals.

A relatively simple example of a flexible counter is shown in Fig. 6.40.

The counter itself is a straight binary counter. Advance count signals increase the count. A "reset counter" signal clears all flip-flops of the counter,

---

[1] See paragraph 8.2.

The highest count of the counter is selectable by switches. With the switches set as indicated in Fig. 6.40, the highest count is 011. If the counter is in the state corresponding to the highest selected count, the output of the AND-circuit assumes the logic "1" state, advance count pulses are inhibited, and the next count input resets the counter to zero. The reset pulse serves simultaneously as a carry. In the above example, the counter assumes consecutively the states 000, 001, 010, 011. It then resets to 000, and starts a new count cycle.

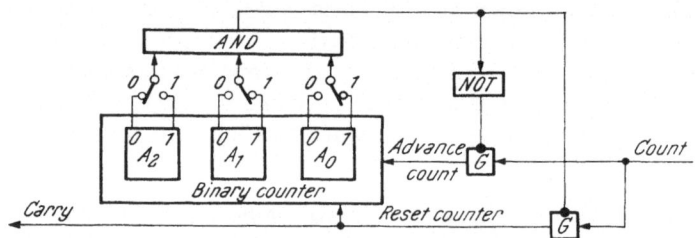

Fig. 6.40. Counter with Externally Selectable Modulo

Since the counter goes through a repeated sequence of four different states we can speak of a modulo 4 counter[1]. By proper setting of the switches any other modulus between 1 and 8 can be selected. Of course, we may use counters with more stages and construct in this fashion flexible counters with higher moduli.

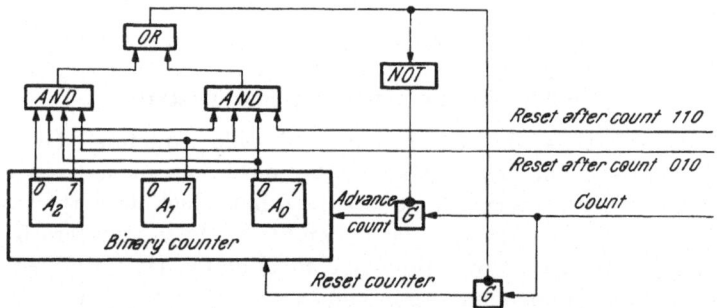

Fig. 6.41. Counter with Selectable Modulo

Slightly different circuitry is required when the modulus of a counter is not selected by switches but by input signals. Fig. 6.41 gives an example of such a counter.

As long as there are only count inputs, the circuit behaves like an ordinary 3-stage binary counter, i.e. it counts modulo 8. However, if there is a "reset after count 110" condition, the AND-circuit on the right produces

―――――――――

[1] For a more detailed explanation of the term modulo, see paragraph 8.1.1.1.

a "1" output for the state 110, so that the next count pulse resets the counter to zero. In effect, the counter counts modulo 7. In a similar manner, a "reset after 010" input causes the counter to count modulo 3.

Problem 23: Design the complete logic diagram for a flexible counter which counts either modulo 10 (a decade, in the 8421 code) or modulo 8 (equivalent to one octal position). Assume flip-flops of the category 2, page 89 and separate set and clear inputs.

So far, we have seen counters with a rather limited degree of flexibility. They simply reset after they have reached a certain count. Now imagine a more flexible counter which ordinarily counts in the binary system, but, which upon external signals, can skip a few intermediate states, go back to a previous state, or perhaps remain in a state until the external signal is removed. Let us indicate the specific behavior of such a counter in the flow-chart-like diagram given in Fig. 6.42.

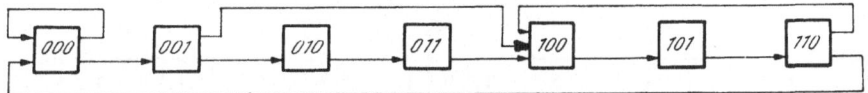

Fig. 6.42. Flow Diagram for Sample Counter

We assume a 3-stage binary counter, normally counting from 000 to 110. These normal transitions from one state to another are indicated by the arrows in the lower part of Fig. 6.42. However, in the presence of an external "skip" command, the counter shall follow the transitions indicated by arrows in the upper part of the same figure. In this particular example, we

Table 6.8. *Input Requirements for Sample Counter*

State			Count						Skip					
A_2	A_1	A_0	$_1a_2$	$_0a_2$	$_1a_1$	$_0a_1$	$_1a_0$	$_0a_0$	$_1a_2$	$_0a_2$	$_1a_1$	$_0a_1$	$_1a_0$	$_0a_0$
0	0	0	0	×	0	×	1	0	0	×	0	×	0	×
0	0	1	0	×	1	0	0	1	1	0	0	×	0	1
0	1	0	0	×	×	0	1	0	×	×	×	×	×	×
0	1	1	1	0	0	1	0	1	×	×	×	×	×	×
1	0	0	×	0	0	×	1	0	×	×	×	×	×	×
1	0	1	×	0	1	0	0	1	×	×	×	×	×	×
1	1	0	0	1	0	1	0	×	×	0	0	1	0	×
1	1	1	×	×	×	×	×	×	×	×	×	×	×	×

prescribe that the counter progresses from 001 to 100 if a skip command is present (rather than going to 010). In the same manner, the counter shall go back from 110 to 100, or remain in the state 000 if there is a skip command.

In order to find the required circuitry, let us list all possible counter states and the inputs to each flip-flop required for counting and skipping. Let us further assume that the flip-flops have no complement inputs so that we have to apply separate set and clear inputs. (See Table 6.8.)

Table 6.9. *Karnaugh Maps for Set and Clear Inputs of Sample Counter*

A_2 A_1 $\quad$ A_0 / S	0 / 0	0 / 1	1 / 1	1 / 0
0 0	0	×	1	1
0 1	0	×	×	1
1 1	×	×	×	×
1 0	0	×	×	1

$_0a_0$

A_2 A_1 $\quad$ A_0 / S	0 / 0	0 / 1	1 / 1	1 / 0
0 0	1	0	0	0
0 1	1	×	×	0
1 1	0	0	×	×
1 0	1	×	×	0

$_1a_0$

A_2 A_1 $\quad$ A_0 / S	0 / 0	0 / 1	1 / 1	1 / 0
0 0	×	×	×	0
0 1	0	×	×	1
1 1	1	1	×	×
1 0	×	×	×	0

$_0a_1$

A_2 A_1 $\quad$ A_0 / S	0 / 0	0 / 1	1 / 1	1 / 0
0 0	0	0	0	1
0 1	×	×	×	0
1 1	0	0	×	×
1 0	0	×	×	1

$_1a_1$

A_2 A_1 $\quad$ A_0 / S	0 / 0	0 / 1	1 / 1	1 / 0
0 0	×	×	0	×
0 1	×	×	×	0
1 1	1	0	×	×
1 0	0	×	×	0

$_0a_2$

A_2 A_1 $\quad$ A_0 / S	0 / 0	0 / 1	1 / 1	1 / 0
0 0	0	0	1	0
0 1	0	×	×	1
1 1	0	×	×	×
1 0	×	×	×	×

$_1a_2$

Crosses are inserted in positions where we "don't care" what the input will be. For example, crosses are indicated for the state 111. Since the counter, according to the specifications, will never be in this state, we do not care how the counter gets out of it. A similar justification applies to the entries for skipping in the states 010 through 101: The flow chart in Fig. 6.42 does not specify particular actions.

Let us now indicate the inputs $_1a_2$ through $_0a_0$ in Karnaugh maps so that possible simplifications become more apparent. The entries for which $S=1$, correspond to skipping. $S=0$ indicates counting. (See Table 6.9.)

From the Karnaugh maps we can derive the following input equations:

$$_0a_0 = A_0 C \tag{6.23}$$

$$_1a_0 = (A_2' + A_1') A_0' S' C \tag{6.24}$$

$$_0a_1 = (A_2 + A_0) A_1 C \tag{6.25}$$

$$_1a_1 = A_1'A_0S'C \qquad (6.26)$$

$$_0a_2 = A_1A_0'S'C \qquad (6.27)$$

$$_1a_2 = (A_1+S)\,A_0C \qquad (6.28)$$

The term C in above equations indicates the advance count input. Fig. 6.43 gives now the complete counter diagram.

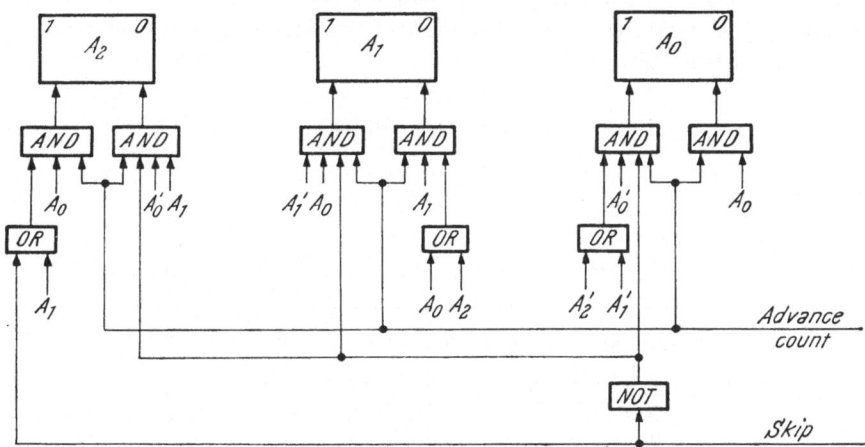

Fig. 6.43. Complete Sample Counter

The diagram does not take into consideration further simplifications possible by deriving certain input propositions from already existing ones.

Problem 24: Draw a flow-chart similar to the one in Fig 6.42. Show for all possible states of the actual counter given in Fig. 6.43 the transitions for count and skip.

Problem 25: Design a flexible counter for the flow chart given in Fig. 6.42 using only complement inputs to all flip-flops.

At the end of this paragraph on counters, it is appropriate to mention that flip-flops may assume random states when the equipment power is turned on. It is, therefore, possible that the flip-flops of a decade counter using the 8421 code might assume the illegal state 1111. Whenever the possibility exists that a counter assumes an illegal state, it becomes necessary to provide some means to force the counter into a "legal" state. This is frequently accomplished by a "master clear" signal which sets all flip-flops to the zero state. For some applications, however, it may be simpler to investigate each individual illegal state and to assure that solely the input of normal count pulses gets the counter into a legal state. This may require the addition of inputs according to logic terms which may seem to be rather "illogical", when one considers only the usual operation.

6.3. Adders

Adders are circuits which perform the arithmetic operation of addition. If adders operate on one digit at a time (as we do in a paper and pencil calculation) they are called serial adders. Parallel adders work on all digits of a number simultaneously.

6.3.1. Binary Adders

Table 6.10 shows the truth table for the addition of two individual binary digits[1].

Table 6.10. *Truth Table for the Addition of Two Binary Digits*

Addend (A)	Augend (B)	Sum (S)	Carry (K)
0	0	0	0
0	1	1	0
1	0	1	0
1	1	0	1

From this table, we can derive the following equations:

$$S = A'B + AB' = A \oplus B \tag{6.29}$$

$$K = AB \tag{6.30}$$

A "half adder" is an implementation of these equations and may be shown in block diagrams by the following symbol:

Fig. 6.44. Symbol for Half Adder

Fig. 6.45 shows two equivalent logic diagrams for a half adder.

Problem 26: Show the logic diagram of a half adder consisting of NOR modules exclusively. Try to find as simple circuits as possible.

---

[1] See also Table 2.4.

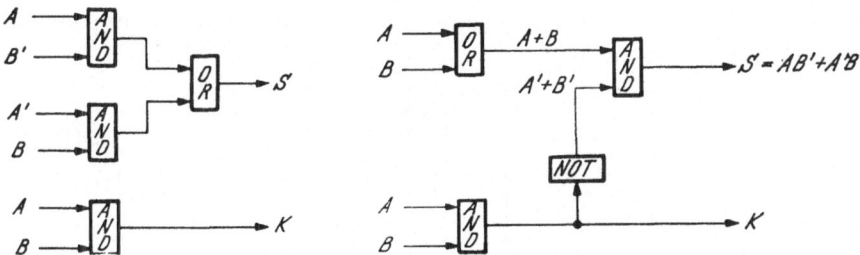

Fig. 6.45. Logic Diagrams for Half Adder

A half adder can add two binary digits, but it cannot take care of carries which might be originated in lower order positions when numbers with more than one digital position are added. The truth table of a "full adder" which takes such carries into consideration is given in Table 6.11.

Table 6.11. *Truth Table for a Full Adder*

A	B	K*	S	K
0	0	0	0	0
0	0	1	1	0
0	1	0	1	0
0	1	1	0	1
1	0	0	1	0
1	0	1	0	1
1	1	0	0	1
1	1	1	1	1

The full adder has three inputs: the addend A, the augend B, and the previous carry K^*. It has two outputs: the sum S, and the carry K.

From the truth table, we derive the following equations:

$$S = ABK^* + AB'K^{*'} + A'B'K^* + A'BK^{*'} \qquad (6.31)$$

$$K = AB + BK^* + AK^* \qquad (6.32)$$

These equations may also be written in the following form:

$$K = AB + BK^* + AK^* \qquad (6.33)$$

$$S = ABK^* + (A + B + K^*) K' \qquad (6.34)$$

Equation (6.34) states that the sum is equal to "1" if there are three 1's on the input of the full adder (ABK^*), or if there is at least one "1" ($A + B + K^*$) but less than two 1's (K').

Fig. 6.46 shows the logic diagram of a full adder which implements Equations (6.33) and (6.34):

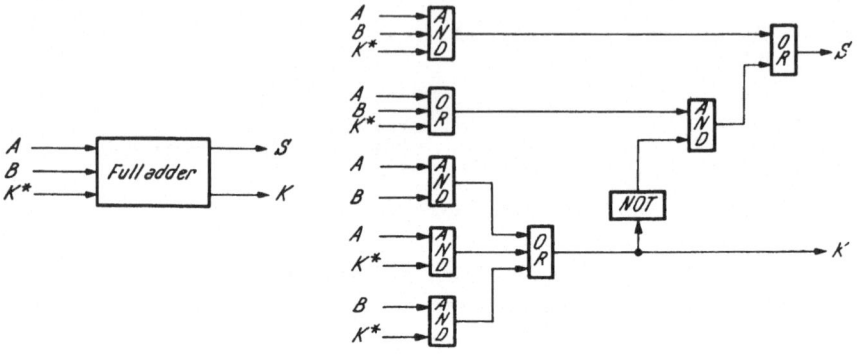

Fig. 6.46. Full Adder

Problem 27: Convince yourself that the logic diagram given in Fig. 6.46 is that of a full adder. Try to find a simpler diagram.

A full adder can be designed using two half adders:

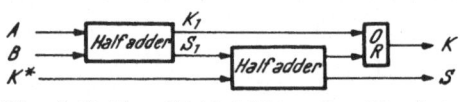

Fig. 6.47. Two Half Adders Combined to a Full Adder

In this arrangement, the addend A and augend B are added first to generate the partial sum S_1. The second half adder derives then the sum of S_1 and the previous carry K^*. Carries from either the first or second half adder appear as final carry.

6.3.1.1. Serial Addition: For the serial addition of two binary numbers, only one full adder is used and the digits of the two numbers to be added enter the adder in the sequence of their significance, i.e. the least significant bit first, and the most significant bit last.

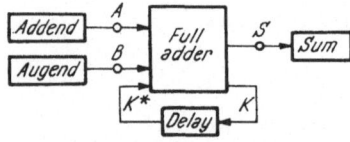

Fig. 6.48. Serial Addition

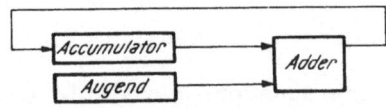

Fig. 6.49. Accumulative Addition

Let us suppose that the addend and the augend are contained in recirculating registers. For the addition of the two numbers, the outputs of both registers are connected to the adder inputs. The adder generates the sum, S, which is entered in serial form into the sum register. The carries which are generated during the addition of a certain digit are "remembered" in a pulse delay and re-enter the adder simultaneously with the next significant bits of augend and addend.

Problem 28: Draw a time diagram of the conditions prevailing at the points labeled A, B, S, K, K^* (see Fig. 6.48) during the addition of the two binary numbers ...01001, and ...00101.

Under certain conditions, we may obtain a sum which requires one digital position more than any of the operands (e.g. $101 + 100 = 1001$). Sum registers may or may not have provisions for the storage of such an "overflow". If the capacity of the sum register is exceeded, usually an overflow indication is given. (The condition of the one pulse delay at the end of an addition is a very simple criterion for an overflow alarm. If the delay contains a carry pulse, there is an overflow. Conversely, if it does not contain a carry, there is no overflow.)

Frequently, a single register is used to hold both the addend and the sum. Such a register is then referred to as an "accumulator". (See Fig. 6.49.)

The recirculation loop of the accumulator contains the adder. If the augend is zero, (or if it is disconnected), the contents of the accumulator are recirculated without change. However, if the augend register contains a finite number, the contents of the accumulator will be augmented by this number. The augend register is connected to the adder only when its contents are to be added to the number contained in the accumulator, and only for the time required to perform this addition.

The arrangement for serial addition given in Fig. 6.48 tacitly assumes a pulse-type representation of information. However, in many cases, the addend,

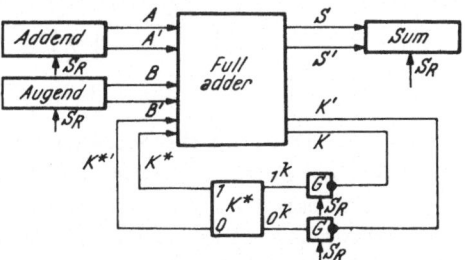

Fig. 6.50. Serial Adder with Level-Type Circuitry

augend, and sum-registers, or accumulator and augend registers are flip-flop shift registers with level outputs. While this level-type representation of information requires no change in the design of the adder itself, the temporary storage for the carry has to be modified.

In order to perform an addition, shift pulses are applied simultaneously to all registers. The adder derives the sum for each digit from the level inputs A, A', B, B'. The level outputs S and S' are available as inputs to the shift register containing the sum. Simultaneously with the shifting of information out of A and B, the sum is shifted into S.

In this arrangement, a flip-flop K^* is used to "remember" previous carries. It will be set if the output K of the adder is in the "1" state and it will be cleared if K is in the "0" state. Since it will retain its state until the next shift pulse is applied, it serves as a memory for the previous carry from one shift pulse to the next.

Problem 29: The output K of the adder follows equation (6.32). The inputs to the flip-flop K^* in Fig. 6.50 can therefore be written as:

$$_1k^* = (AB + BK^* + AK^*)\, S_R$$
$$_0k^* = (AB + BK^* + AK^*)\, 'S_R$$

Show that the following simpler equations (corresponding to a simpler adder circuitry) are sufficient as input equations for the K^* flip-flop:

$$_1k^* = ABS_R \qquad\qquad _0k^* = A'B'S_R$$

6.3.1.2. Parallel Addition: Parallel addition requires as many adders as there are digits in the two numbers which are to be added. Fig. 6.51 shows the basic arrangement:

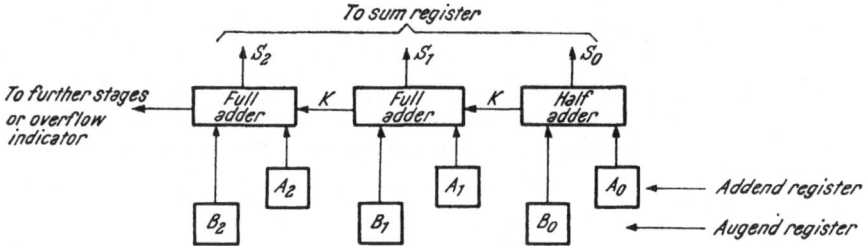

Fig. 6.51. Parallel Addition

The adders set up the correct sum for each digital position, depending upon the digits of the operands and propagated carries. The sum can be read out in parallel and written into the sum register. There is no storage for carries required, because all digits are added simultaneously. The final carry may be used to indicate an overflow.

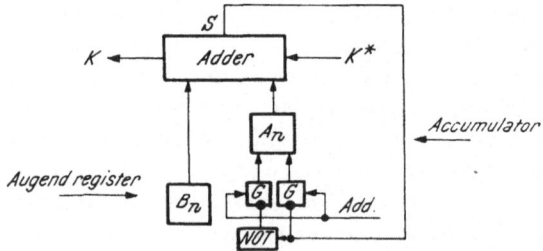

Fig. 6.52. Typical Stage for an Accumulative Addition

As for serial adders, a single register (the accumulator) may be used to store both, the addend and the sum. Fig. 6.52 gives the diagram for a typical stage.

The sum output of the adder is used as input to the accumulator. Upon an add pulse, the addend in the accumulator is replaced by the sum.

The number of circuit elements required for a parallel addition scheme as indicated in Fig. 6.51 or Fig. 6.52 is considerable. One can reduce the number of elements if the addition is performed in two steps: First, the two corresponding digits of addend and augend are added and then, in a second step, carries are generated and propagated. Let us suppose we apply the scheme to accumulative addition. The accumulator A shall contain the addend and the B-register shall contain the augend. The states of the accumulator after the first step are indicated in the truth table below[1]:

Table 6.12. *Truth Table for a Modulo 2 Addition*

		A (original)	
		0	1
B	0	0	1
	1	1	0

The first entry can be interpreted as: the accumulator shall contain a "0" after the first step of the addition, if the accumulator contained previously a "0" and the augend is a "0". Any other entry can be interpreted similarly. For instance: if the accumulator contains a "1" prior to this step, *and* the augend is a "1", then the accumulator shall contain a "0" as result of the partial addition. Simplifying these statements, we can say: <u>The state of the accumulator is to be changed if the augend is a "1"</u>. Using level type flip-flops with pulse inputs we find a rather simple circuitry:

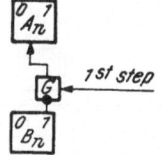

Fig. 6.53. Partial Addition Shown for a Typical Stage

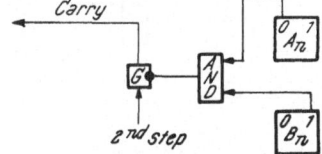

Fig. 6.54. Second Step, Generation of Carries

The accumulator flip-flop A_n is complemented if the augend flip-flop B_n contains a "1" and a pulse arrives which signifies the first step of the addition.

---

[1] This table is equivalent to the sum column in Table 6.10.

As a second step, we have to take care of the generation and propagation of carries. A carry has to be *generated* when in any digital position both operands contain a "1"[1]. One condition for the generation of a carry is, therefore, that the augend is equal to "1" or, simpler, $B=1$. Since the second operand, originally contained in A, may have been changed during the first step of the addition, the second condition $A=1$ is no longer correct. We know, however, that the state of the accumulator has been complemented, if the augend contained (and still contains) a one. Consequently, the correct condition for the generation of a carry is that B contains a "1" <u>and A contains a "0" which can be expressed as: $A'B$.</u>

A carry which is generated in a certain digital position should be *propagated* and increase the contents of the accumulator by a "1" to the left of the stage which generated the carry. Effectively then, the accumulator should have the properties of a counter.

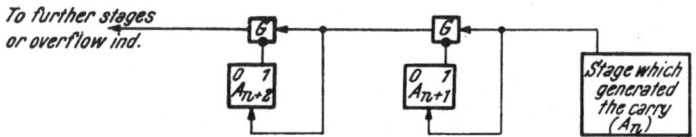

Fig. 6.55. Propagation of Carries

If we now combine Fig. 6.53 through 6.55, we obtain the complete diagram for this type of addition:

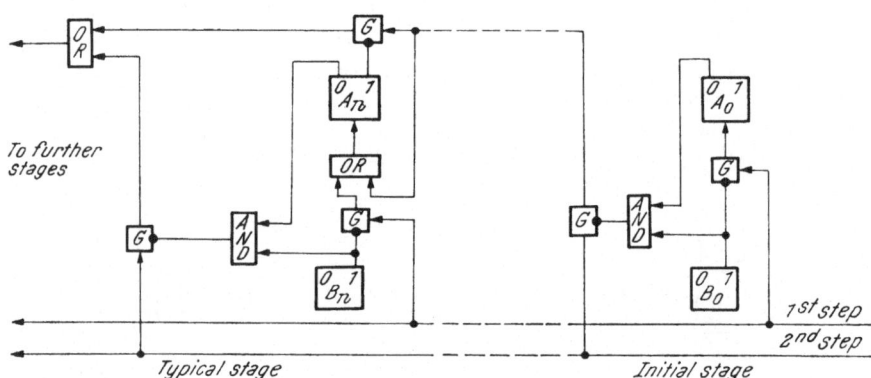

Fig. 6.56. Complete Diagram

Problem 30: Show the contents of the accumulator after the first and after the second step of the addition if the following two numbers are added:

$$(A)=\ldots 0110110$$
$$(B)=\ldots 0011100$$

[1] See also Table 6.10.

The diagrams for accumulative addition given in Figs. 6.52 and 6.56 can be modified for flip-flops with level inputs (but AC coupling) and for dynamic flip-flops. However, they cannot be applied to strictly DC flip-flops since it is not possible to read and set such flip-flops simultaneously.

Accumulators with DC flip-flops usually contain a dual rank A-register (A and A^*). The sum of A and B is first recorded in A^*. The sum is then brought back to A before a new addition of A and B takes place. This procedure is schematically indicated in Fig. 6.57.

This scheme requires a transfer of operands into the A and B registers before an addition can take place. Fig. 6.58 shows a

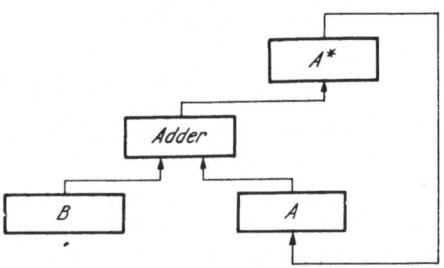

Fig 6.57. Basic Arrangement of a Dual Rank Accumulator

rather interesting variation of the basic scheme. Here, the transfer is not straightforward but implemented in such a manner that part of the logic operation required for an addition is performed by the transfer itself. The operands A and B are not brought into separate A- and B-registers, but into X- and Y-registers in such a manner that X contains the logic sum $A + B$,

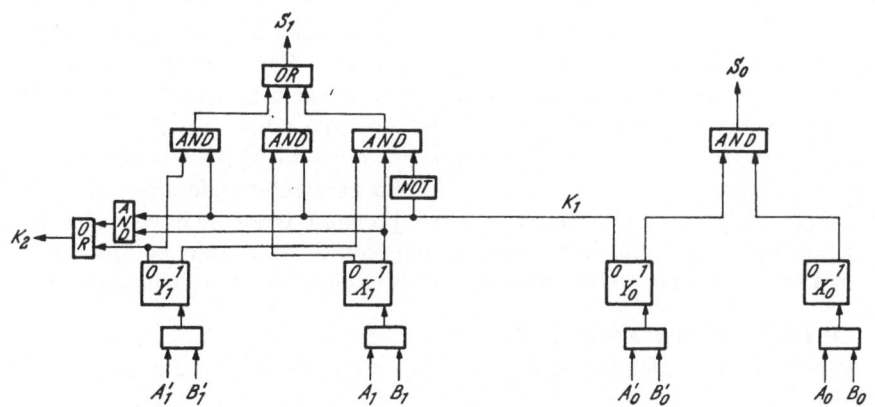

Fig. 6.58. Addition Scheme in which a Part of the Required Logic Operation is Performed by the Transfer

and Y contains the quantity $A' + B'$ prior to the addition. The fact that these two quantities, rather than the quantities A and B, are available as inputs to the adder make the adder circuitry itself less complex than the circuitry of a straightforward parallel adder. This can be easily verified by comparing the circuitry deriving S_0 and K_1 with the half adder shown in Fig. 6.45, and by comparing the circuitry deriving S_1 and K_2 with the full adder shown in Fig. 6.46.

Problem 31:   a) Express S_1 and K_2 as a function of X_1, Y_1, and K_1.
                  b) Express S_1 and K_2 as a function of A_1, B_1, and K_1.
                  c) Show that the resulting functions are equivalent to those of a full adder.

The propagation of carries through a parallel adder may take a fairly long time. In the worst case, a carry generated in the least significant position has to travel through all other stages before the correct sum is established. Each of these stages may place several logic circuits in the carry propagation paths. Frequently we speak of a carry "rippling" through the adder and also of ripple-adders. In any case, the carry propagation severely restricts the speed of parallel addition and many schemes for high-speed addition, which speed up the carry propagation, have been proposed[1].

6.3.2. Decimal Adders

Computers which work in the decimal system represent (decimal) numbers by certain binary codes[2]. An adder for such a computer not only has to derive the proper arithmetic result, but also has to present the result in an acceptable code. For the understanding of binary adders it was sufficient to consider one bit at a time. For decimal adders it is necessary to treat one decimal digit, i.e. several bits, simultaneously. Of course, there will be a wide variety of possible adder circuits dependent upon the various codes used to represent decimal digits. For the illustration of the problems involved, let us take a particular code, the 2421 code, which is neither very simple nor very complicated. Once we have understood the basic problems and how they are overcome, we should be able to design adders for other codes, or improve the circuits which we will derive here.

6.3.2.1. Parallel Addition: Before we set out, let us first indicate in a rough functional diagram what we want to accomplish with our decimal adder.

The adder shall produce the sum S of addend A and augend B considering the carry K^* which may have been produced by the next lower decade. The adder shall further generate a carry K to the next higher decade. A, B, and S are represented in the 2421 code. The propositions A_3, B_3, S_3 have the weight 2; A_2, B_2, S_2, have the weight 4; A_1, B_1, S_1, the

---

[1] See paragraph 8.1.1.3.
[2] See Table 2.6.

weight 2; and A_0, B_0, S_0 the weight 1. Basically we have here a 9-variable logic problem. The five output variables $S_3 \ldots S_0$, K are functions of the nine input variables $A_3 \ldots A_0$, $B_3 \ldots B_0$, K^*. This is quite a formidable problem. If we should try to list all possible different input conditions, we would end up with a truth table containing $2^9 = 512$ entries. The minimization of the output functions would be an almost impossible task.

Let us see then whether we can simplify the problem. One possible point of attack is the following: Not all conceivable combinations of input states are of interest. Some of them will never occur according to the definition of the 2421 code. As a matter of fact, the augend and addend can each assume only 10 different states (whereas four binary variables in

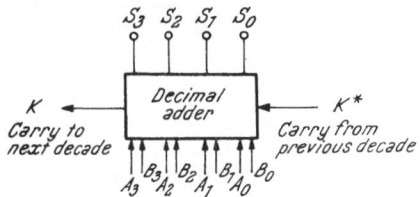

Fig. 6.59. Functional Diagram of a Parallel Decimal Adder

general may assume 16 different states). The number of different input states for augend and addend together is, therefore. 10 times 10, or 100. But with the possibility of a carry or no carry from the lower order decade we still have to consider 200 different input states. Even the size of this reduced problem is too large for our liking. Let us see then whether we can split the problem into several smaller ones. The least significant bit has the weight 1 and essentially signifies whether a number is odd or even. Now, it should be possible to determine whether the sum is odd or even by looking at the least significant bit of augend, addend and the carry. In fact, a binary full adder is exactly the circuit to accomplish this. Having perceived this fact, we can restate our problem.

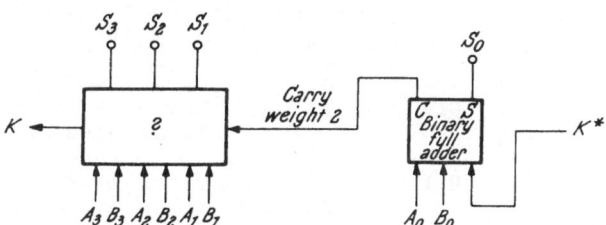

Fig. 6.60. Partial Diagram of an Adder for the 2421 Code

What we are left with is a 7-variable problem. In general, we would have $2^7 = 128$ different input conditions. However, considering the peculiarity of the 2421 code, the variables A_3, A_2, A_1 can assume only 5 different states. The same holds for B_3, B_2, B_1. The carry from the low order bit can assume two states. So altogether, we have now to consider $5 \times 5 \times 2 = 50$

Table 6.13. *Straight Binary Sums and Required Corrections*

a) No Carry from Low Order Bit

Augend

Addend	Decimal Digit	High Order Bits	0 000	2 001	4 010	6 011	8 111
	0	000	0) 000 ok	0) 001 ok	0) 010 ok	0) 011 ok	0) 111 ok
	2	001	0) 001 ok	0) 010 ok	0) 011 ok	0) 100 0) 111	1) 000 ok
	4	010	0) 010 ok	0) 011 ok	0) 100 0) 111	0) 101 1) 000	1) 001 ok
	6	011	0) 011 ok	0) 100 0) 111	0) 101 1) 000	0) 110 1) 001	1) 010 ok
	8	111	0) 111 ok	1) 000 ok	1) 001 ok	1) 010 ok	1) 110 1) 011

(b Carry from Low Order Bit

Augend

Addend	Decimal Digit	High Order Bits	0 000	2 001	4 010	6 011	8 111
	0	000	0) 001 ok	0) 010 ok	0) 011 ok	0) 100 0) 111	1) 000 ok
	2	001	0) 010 ok	0) 011 ok	0) 100 0) 111	0) 101 1) 000	1) 001 ok
	4	010	0) 011 ok	0) 100 0) 111	0) 101 1) 000	0) 110 1) 001	1) 010 ok
	6	011	0) 100 0) 111	0) 101 1) 000	0) 110 1) 001	0) 111 1) 010	1) 011 ok
	8	111	1) 000 ok	1) 001 ok	1) 010 ok	1) 011 ok	1) 111 ok

different input conditions. Unfortunately, we find no additional output variables which are functions of relatively few input variables. So, in one way or another we have to live with the problem to provide the correct outputs for any one of 50 input conditions.

We could now list the five desired outputs for each of the 50 different input conditions, derive the logic expressions, and try to simplify them as much as possible. However, let us consider the problem from a new point of view. We have to add two numbers in the range from one to ten in a representation which corresponds to a large extent to a true binary representation (8 codes are true binary equivalents and only two, the codes for 8 and 9 are not). If we would perform a true binary addition upon these codes, we, probably, would find the correct result for a large number of the 50 possible input conditions. Perhaps it will be possible to correct the wrong results in some simple manner. Looking a little more closely, the problem does not seem hopeless at all. The codes for 8 and 9 are binary representations of the decimal digits which are too large by 6. With some optimism we might expect that also the binary sums of augend and addend are off by the binary equivalent of 6.

Before we get too deep into speculation, let us see what really happens. Since we have already taken care of the least significant bit, we have to consider only the three most significant bits and the carry from the low order bit. Let us list the true binary sum of the two operands and the correct sum according to the 2421 code in Table 6.13.

Let us see what we have here. For the 50 combinations of input variables which are of interest (5×5 entries in each of the two tables), we obtain 33 times the correct result (indicated by "*ok*" below the binary sum) and 17 times an incorrect result (indicated by the correct sum below the binary sum). This is not bad, provided the necessary correction for the 17 incorrect cases is not too difficult and we are able to find relative simple rules telling when or when not to correct a result.

Well, let us first see what corrections are necessary. In all cases, the binary result is off by the binary equivalent of 6, as we had hoped. That is to say, in 16 cases, we have to add 0110 (011 if we consider only the three most significant bits) in binary fashion to the true binary sum to derive the correct result. For one case (the entry in the lower right hand corner of Table 6.13a) we have to subtract 0110.

Before we try to find the rules which govern whether or not a certain correction is to be applied, let us again restate our problem as we see it now.

In order to derive the sum of two digits in the 2421 code, we first produce the straight binary sum. The least significant bit of the binary sum, S_0, is also the correct bit of the sum expressed in the 2421 code. The remaining bits of the binary sum (P_3, P_2, P_1) and the carry (C_p) are correct only for certain combinations of digits. For others, the binary sum has to be modified. Hopefully, we have reduced the original 9-variable problem to the 4-variable problem indicated in Fig. 6.61 which we should be able to handle by normal methods.

Let us now derive when and how we have to modify the straight binary sum. For this purpose, we draw a Karnaugh map for the four variables C_p, P_3, P_2, P_1, and indicate when a particular state corresponds to the desired sum, when the state is not the correct sum and how this state is

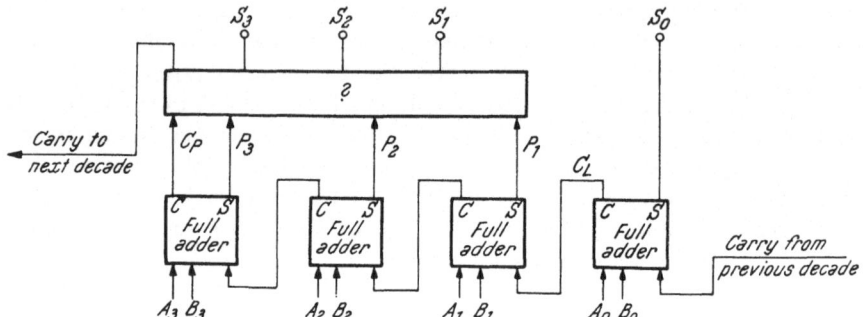

Fig. 6.61. Partial Diagram of 2421 Adder

then to be modified. Part of the work is already accomplished since the entries in Table 6.13 show the straight binary sum and the correct sum. The Karnaugh map which we want to draw is, therefore, essentially an extraction of Table 6.13.

Table 6.14. *Required Corrections as Function of the Straight Binary Sum*

C_p	P_3	P_2 P_1	0 0	0 1	1 1	1 0
0	0		ok	ok	ok	ok
0	1		add 6	add 6	ok add 6	add 6
1	1		×	×	ok	subtract 6
1	0		ok	ok	ok	ok

The crosses in Table 6.14 represent conditions which will not occur. (No matter what decimal digits constitute the inputs to the straight binary adder, it will never produce the binary sum 1)100 or 1)101[1]. Unfortunately, there are two entries for the condition 0111. If we investigate where they came from, we find that the entry "*ok*" comes from Table 6.13a and the entry "add 6" from Table 6.13b. This means essentially that we were overly

---

[1] See Table 6.13.

optimistic. The problem is not a four-variable problem as we indicated in Fig. 6.61 but really a five-variable problem. Our action does not only depend upon the four variables C_p, P_3, P_2, P_1, but also whether we have, or have not a carry C_L from the low order bit.

Well, a five-variable problem is not too difficult, especially since we can deduce quite a bit about it from Table 6.13. First of all, modifications are necessary only if P_3 is in the "1"-state. If we would show our five-variable problem in two four-variable Karnaugh maps, one for $P_3=0$ and one for $P_3=1$, we would find that all entries in the map for $P_3=0$ are "ok". Apparently then the map for $P_3=1$ is the one we should look at. Let us show this in a new table.

Table 6.15. *Required Corrections as Functions of* C_L, C_p, P_2, P_1

C_L	C_p	P_2 P_1	0 0	0 1	1 1	1 0
0	0		add 6	add 6	ok	add 6
0	1		×	×	×	subtract 6
1	1		×	×	ok	×
1	0		add 6	add 6	add 6	add 6

From this figure we deduce that we can subtract 6 for the condition $C_p P_1'$, and that we have to add 6 for the condition C_p', unless we have simultaneously the condition $C_L' P_2 P_1$. If we further take into consideration that we have to modify only if $P_3=1$, we can express the following rules:

Table 6.16. *List of Corrective Actions*

Condition	Action
$C_p' P_3 (C_L' P_2 P_1)'$	Add 6 to binary sum
$C_p P_3 P_1'$	Subtract 6 from binary sum
All others	Do not modify binary sum

It should not be too difficult to implement these rules. The addition of 6 to the straight binary sum can be accomplished by an additional binary adder. If we disregard for the moment the subtraction of 6 required for one special case, we can draw a new functional diagram.

The adder as it stands now, takes care of all cases, except the one case when both operands are equal to 8 and there is no carry from the lower decade. The subtraction of 6 from the straight binary sum for the latter case is somewhat difficult. However, we have to do it only for a single one of the 50 possible input conditions. Let us see what this special case is. From table 6.13a we see that we have to modify 1)110 into 1)011. We could design a circuit which, in this special case, applies the correct code on the output rather than the output of the auxiliary adder. However, since we have already an auxiliary adder, let us try to make use of it. Suppose we would

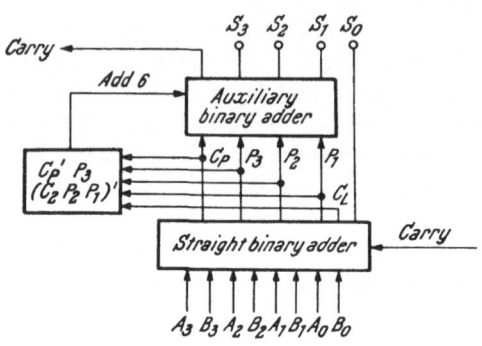

Fig. 6.62. Decimal Adder Incorporating an Auxiliary Binary Adder

add 101 to the straight binary sum 1)110 in the auxiliary adder. The result is the desired configuration 011 and a new carry[1]. If we combine the carry from the auxiliary adder with the carry from the basic adder in an OR-circuit, and send this combination as carry to the next decade, only one carry will be sent, even if both adders overflow. Let us now show the complete diagram in Fig. 6.63.

What we have found here is one possible circuit for a 2421 adder. But let us consider it critically. Essentially, it constitutes an engineer's solution to a rather complex logic problem. We have no way of knowing whether the circuit represents the simplest or the least expensive solution. A theoretical approach which would allow us to minimize, requires the treatment of the 7-variable problem indicated in Fig. 6.60. The treatment is not simple but should probably be attempted if cost and complexity are important design criteria. On the other hand, it seems certain that a decade adder for the 2421 code is necessarily more complex than a 4-bit straight binary adder. The true "minimum" solution to the problem (if we are able to find one at all) may be simpler than our circuit, but probably not much more so.

One possible approach to further minimization of our solution might be the theoretical treatment of the five-variable problem stated in Fig. 6.61. (As we have found, the state of C_L has to be considered in addition to the states of C_p, P_3, P_2, P_1). In this case, we would accept the straight binary adder as necessary but derive the additional circuitry in a minimum form.

---

[1] This is a subtraction by the addition of the complement. (See paragraph 8.1.1.)

We discussed here only an adder for the 2421 code. Other codes will require different adder circuits. Dependent upon the code, decimal adder circuits may be simpler than that for the 2421 code, but the circuits required

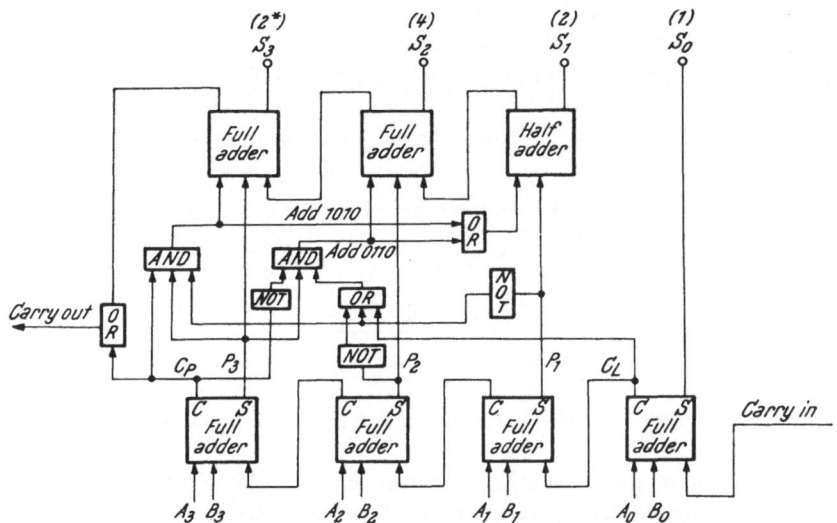

Fig. 6.63. Complete Diagram of 2421 Decimal Adder

for operations other than additions are then probably more complicated. In general, the design of decimal computers is more complicated than the design of binary machines, and the computer designer will select that decimal code which he thinks is best in his particular case.

Problem 32: Design the logic circuitry of a parallel decimal adder for the 8421 code. Note: the circuitry will be more straightforward than the circuitry for a 2421 adder, since the codes resemble the straight binary representation more closely.

6.3.2.2. Serial Addition: To start with, let us again indicate the requirements of a serial decimal adder in a functional diagram.

We assume that addend, augend and sum are represented in coded form and, in particular, in the 2421 code for the purposes of this example. The addend and augend enter the adder in serial form. The output S shall

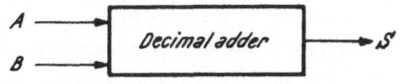

Fig. 6.64. Functional Diagram of a Serial Decimal Adder

produce the sum of both also in serial form. The decimal adder, in contrast to a serial binary adder, cannot produce the sum of addend and augend, by using the states of the inputs A and B during only one binary position.

As we have seen during the discussion of the parallel 2421 adder, the state of certain sum bits depends not only upon the state of the input positions with the same weight and the carry, but also upon the state of bits which are up to two positions removed from the position under consideration. The serial decimal adder must, therefore, by necessity contain a delay of at least two bit times. This gives sufficient time to receive two bits (and the possibility to look simultaneously at three bit positions) before an output is derived.

One could think of a solution where the inputs are stored as indicated in Fig. 6.65.

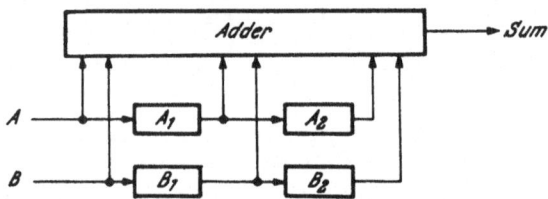

Fig. 6.65. Functional Diagram of a Serial Decimal Adder

The inputs A and B are shifted into storage elements so that always three bits of augend and addend are available simultaneously. The required adder circuit would then be essentially a parallel decimal adder as described in the previous paragraph. However, four storage elements (six storage elements if the outputs have to be delivered in true serial form) plus a parallel adder seem to be too high a price for serial addition.

The alternate approach indicated in Fig. 6.66 seems to be more reasonable.

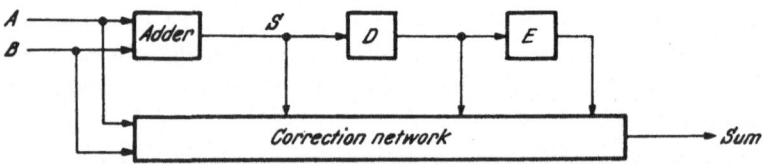

Fig. 6.66. Alternate Functional Diagram of a Serial Decimal Adder

Rather than storing the operands, we store the binary sum. Inputs from three bit positions are available for the correction logic.

Let us now investigate the requirements for this arrangement at the different bit times in more detail. There are four bit times per decade. Let us agree to call them T_1, T_2, T_3 and T_4. Let us further agree to call the weights respectively 2*, 4, 2, 1. At the time T_1 the least significant bit (weight 1) is present on the input and we have the situation indicated in Fig. 6.67a.

Flip-flop C contains the carry from the previous decade with the weight 1. Flip-flops D and E contain two bits of the decimal sum (produced by the correction network) for the previous decade with the weights 2^* and 4. The output S of the binary adder has the weight 1, and the output K the weight 2.

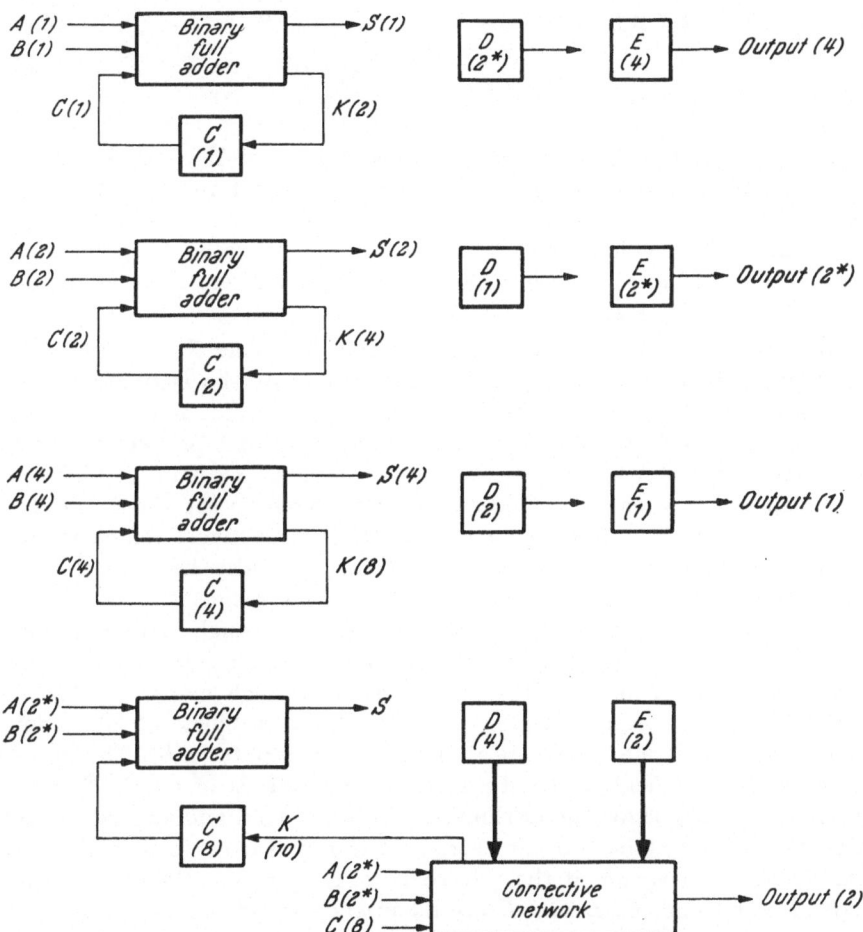

Fig. 6.67. Functional Diagram for the Four Bit-Times

During time T_2, the inputs have the weight 2, C has the weight 2, D has the weight 1 and E has the weight 2. Similar situations exist at times T_3 and T_4.

If we inspect the transitions from T_1 to T_2, T_2 to T_3, and T_3 to T_4 we find the requirements for a binary adder with carry flip-flop (C) and for a two bit shift register (D and E). However, during the transition from T_4 to T_1 a number of changes have to take place.

The logic circuitry required during the times T_1, T_2 and T_3 is described, without difficulty, by the following equations:

$$S = ABC + A'B'C + A'BC' + AB'C' \tag{6.35}$$

$$_1c = AB \qquad\qquad _0c = A'B' \tag{6.36}$$

$$_1d = S \qquad\qquad _0d = S' \tag{6.37}$$

$$_1e = D \qquad\qquad _0e = D' \tag{6.38}$$

The binary sum, S, is derived in the usual manner. The on/off terms to the carry flip-flop ($_1c$ and $_0c$) are in their shortest form (see also problem 29). The on/off terms for the D and E flip-flops indicate the behavior of a shift register (D follows S, E follows D). Of course, each of the on/off equations should also contain a clock or shift pulse term, but let us disregard this for the moment.

Let us now study the requirement during the time T_4 in detail. The flip-flops C, D and E have eight possible states at the beginning of T_4 (depending upon the least significant bits of augend and addend). During the time T_4 itself, we may have no input, one input or two inputs to the adder. In essence then, we have to consider 3×8 or 24 different possibilities. For each of the possibilities, we have to set the C, D, and E flip-flops to their correct state, representing respectively the carry into the new decade, the 2* bit, and the 4 bit. We also have to produce the correct output F, representing the bit with the weight 2.

This situation is indicated in Table 6.17. The Karnaugh maps on the left contain eight fields corresponding to the eight possible states of C, D and E at the beginning of T_4. There are three Karnaugh maps, one for the possibility of no input ($A'B'$), one for the possibility of one input ($A'B + AB'$), and one for the possibility of two inputs (AB). The entries in the individual fields show the required new states of C, D, E, F. In other words, they show the arithmetic sum of addend and augend in each case. For instance, the top left entry in the top Karnaugh map is 0)000. This is the correct sum if there have been no previous "ones" in either addend or augend (C, D and E are all in the zero state) *and* there is no input at the time T_4 (both inputs A and B are in the zero state). In effect, the two operands are decimal 0's and the result should be a decimal 0 with no carry to the next decade[1]. The entry in the bottom right of the bottom map reads 1) 011. This is the representation of a carry and a decimal 6, which is the correct sum for $8+4+2+2$ (C, weight 8; D,

---

[1] Remember that the least significant bit of the sum (weight 1) has already been taken care of. It may have been a "one".

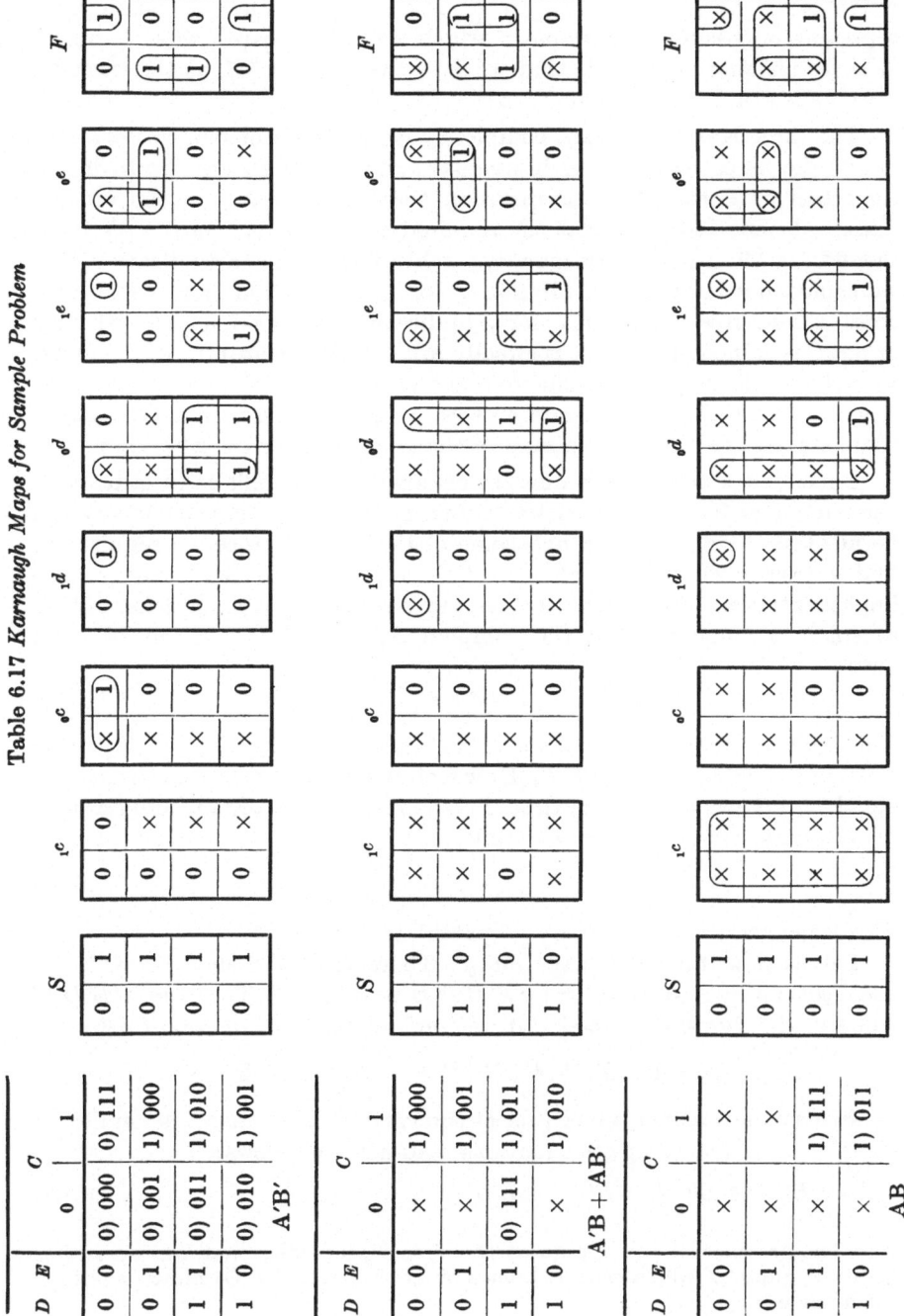

Table 6.17 Karnaugh Maps for Sample Problem

weight 4; A, weight 2; B, weight 2). The crosses in the Karnaugh maps represent impossible conditions in which we "don't care" what happens. It is, for instance, impossible that we have two inputs (AB) at the time T_4, if we have also the condition $C'DE$[1].

The three Karnaugh maps completely define the problem. We can now derive the maps representing individual actions as they are shown in the right half of Table 6.17. For instance, the top left entry in the original Karnaugh map indicates that the required state of C is "zero". We enter therefore a "0" in the corresponding field of the map $_1c$ (indicating that we should not turn on the flip-flop C for this condition), and we enter a cross in the corresponding field for $_0c$ (indicating that we don't care whether or not C is turned off, since it is already in the "zero" state). In this manner, we indicate the required actions for each of the 24 possibilities. Where the original Karnaugh maps contain crosses, we retain crosses. For the sake of completeness, we also draw a map of the state of S for the 24 possibilities.

There exists now a large number of different possibilities to implement the individual functions. Various "patches" of ones or zeros or crosses can be identified. Let us go through some of the mental processes required to find a reasonable implementation. Let us start with the term $_1c$. The implementation $_1c = 0$ seems to be the simplest implementation. If we accept it, we have to write the complete equation for $_1c$ in the following form:

$$_1c = ABT_4' \tag{6.39}$$

This indicates that $_1c$ is equal to AB at times other than T_4 (see Equation 6.36) but is equal to zero at T_4. However, if we notice that we can make $_1c$ equal to AB also during the time T_4 (the map for $_1c$ contains only crosses for the condition AB), we can simplify to

$$_1c = AB \tag{6.40}$$

This latter term is indicated by a "patch" in Table 6.17.

Let us now take the maps for $_0c$. There is only a single "one". We could cover it by $A'B'CD'E'$ or by $A'B'D'E'$ or by $SD'E'$. The term $A'B'D'E'$ will give the shortest expression if we combine it with Equation (6.36):

$$_0c = A'B'T_4' + A'B'D'E' = A'B'(T_4' + D'E')$$

In a similar manner, we can select the various other "patches" indicated in Table 6.17. The complete equations read now as follows:

$$S = ABC + A'B'C + A'BC' + AB'C' \tag{6.41}$$

---

[1] The presence of two inputs at the time T_4 indicates that we are adding eights or nines, in which case at least a binary sum of 12 (CD) must have been accumulated.

$$_1c = AB \qquad\qquad\qquad _0c = A'B'(T_4' + D'E') \qquad (6.42)$$

$$_1d = S\,(T_4' + D'E') \qquad\qquad _0d = S' + T_4D\,(E' + AB') \qquad (6.43)$$

$$_1e = D\,(T_4' + S' + A + B) + D'T_4SE' \qquad _0e = D'(T_4' + S' + E) \qquad (6.44)$$

$$F = E\,(T_4' + S' + A + B) + T_4SE' \qquad\qquad\qquad (6.45)$$

Although we have made an effort to produce short terms, the equations not necessarily produce the simplest implementation. We note, for instance, that the term $(T_4' + S' + E)$ is the complement of the term T_4SE'. If we derive one of them, we can obtain the other one fairly simply by complementation. Also the term $D'E'$ could, in this particular instance, be replaced by the longer term $D'T_4SE'$ which we have to derive anyway. Whether or not such rearrangements reduce the number of required circuits and the cost of them, depends very much on the type of circuits to be used and the number and cost of individual inputs/or outputs. It depends also upon some requirements which have to be specified for an actual design. For instance, whether or not the logic complements of inputs are available and whether or not the complement of the output is to be derived, can make quite some difference in our optimization procedure.

Let us conclude the treatment of the serial decimal adder with some general observations. If we compare our result to the parallel decimal adder derived in paragraph 6.3.2.1, we see a number of similarities. We first add the four bits of the decimal code in binary fashion. We then correct the binary sum. In both cases, this correction is a five-variable logic problem. In case of the parallel adder, we accepted two additional binary full adders and one additional binary half adder as necessary to produce the correct sum. In case of the serial adder, we took the trouble to find the exact requirements.

Problem 33 (Voluntary): Design a serial decimal adder for the 8421 code. Use the length of logic expressions as criterion for minimization.

6.3.3. Subtracters

The operation of subtraction, in its degree of complexity, is almost identical to the operation of addition. We shall find that practically all

Table 6.18 *Truth Table for a Binary Half-Subtracter*

Minuend A	Subtrahend B	Difference D	Borrow K
0	0	0	0
0	1	1	1
1	0	1	0
1	1	0	0

of the discussed addition schemes can be easily modified for subtraction.

Corresponding to a binary half-adder we may define a binary half-subtracter. Its truth table is given in Table 6.18.

The binary half-subtracter generates the difference, D, and the borrow, K, from the state of the minuend, A, and the subtrahend, B. From the truth table we can derive the equations:

$$D = A'B + AB' \tag{6.46}$$

$$K = A'B \tag{6.47}$$

Similarly, we can define a full subtracter which produces the difference, D, and the borrow, K, from minuend, A, subtrahend, B, and previous borrow, K^*.

Table 6.19 *Truth Table for Binary Full Subtracter*

Minuend A	Subtrahend B	Previous Borrow K^*	Difference D	Borrow K
0	0	0	0	0
0	0	1	1	1
0	1	0	1	1
0	1	1	0	1
1	0	0	1	0
1	0	1	0	0
1	1	0	0	0
1	1	1	1	1

The corresponding equations are:

$$D = ABK^* + AB'K^{*'} + A'B'K^* + A'BK^{*'} \tag{6.48}$$

$$K = A'B + A'K^* + BK^* \tag{6.49}$$

One of the possible implementations of a binary full subtracter is by two half subtracters as shown in Fig. 6.68:

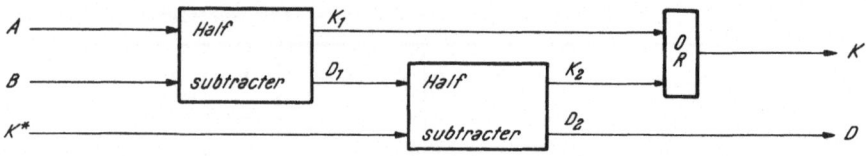

Fig. 6.68. Implementation of a Full Subtracter by two Half-Subtracters

One can also use two half adders and the arrangement shown in Fig. 6.69.

Problem 34: Prove the validity of the implementations given in Figs. 6.68 and 6.69.

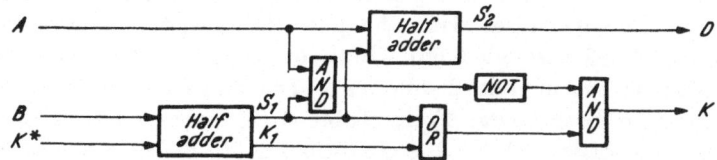

Fig. 6.69. Implementation of a Full Subtracter by Two Half Adders

Another possible implementation is according to the following equations:

$$D = ABK^* + (A + B + K^*)(AB + AK^* + BK^*)' \qquad (6.50)$$
$$K = (B + K^*)(BK^* + D) \qquad (6.51)$$

The resulting logic diagram is similar to the one shown in Fig. 6.46 in so far as only one inverter is needed to produce the difference and the borrow, even if the complements of the inputs are not available.

If we compare Tables 6.11 and 6.19, we notice that the conditions for which the difference assumes the "one" state are identical to those for which the sum assumes the "one" state. In other words, the only essential difference between an adder and subtracter lies in the generation of carries (respectively borrows). All schemes for binary addition can, therefore, be used for subtraction, provided that the circuits for carry generation are appropriately modified.

Let us take the serial addition scheme shown in Fig. 6.50 as an example. We immediately can design the corresponding subtraction scheme:

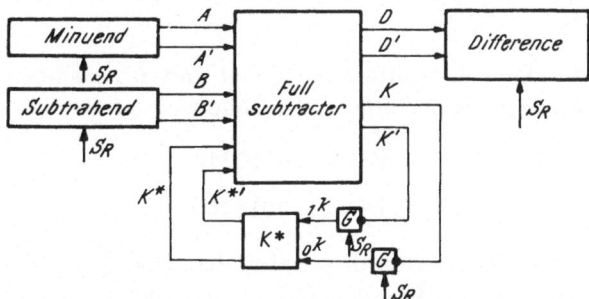

Fig. 6.70. Serial Subtracter Corresponding to the Serial Adder in Fig. 6.50

Similarly as for the adder (see problem 29), the inputs to the carry flip-flop can be simplified. Here we obtain:

$$_1k = A'B\,S_R; \quad _0k = AB'\,S_R$$

Problem 35: Show that above equations are sufficient as input equations for the serial subtracter given in Fig. 6.70. (See also problem 29).

The modification of the binary parallel addition schemes for subtraction should pose no difficulties with the possible exception of the two-step addition (Fig. 6.53 through 6.56). However, we know that the modulo 2 sum is equal to the modulo 2 difference. The first step should, therefore, be the same in both cases. From Table 6.19, we see that a borrow is originated in a stage only for the conditions $A'B$. Since A has been complemented during the first step if B contains a *"one"*, we have the condition AB for the generation of a borrow. A borrow is propagated where A contains a "zero" (down-counter). A typical stage for two-step binary subtraction (corresponding to the typical stage for addition in Fig. 6.56) will therefore look like the one given in Fig. 6.71.

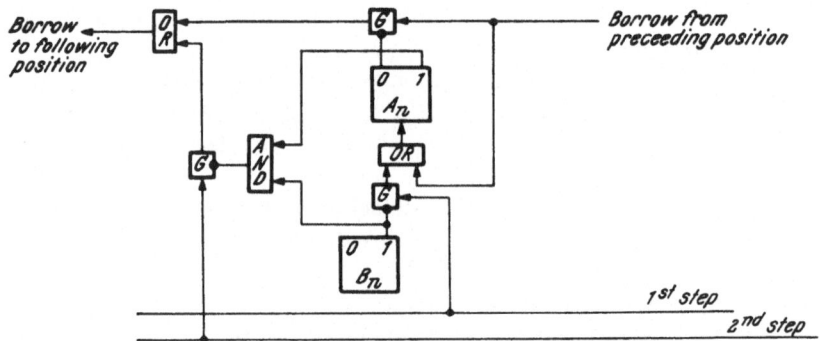

Fig. 6.71. Typical Stage for a Two-Step Subtraction

Problem 36: Show the contents of the A and B registers after the first and second step of the subtraction if the following numbers are subtracted:

$$(A) = \dots 0110110$$

$$(B) = \dots 0011100$$

In the case of decimal parallel or serial adders, the required modifications are by no means obvious. Here it may be better to start the design problem from scratch, following the line of thinking already discussed in detail for the adder.

Problem 37 (Voluntary): Design a serial decimal subtracter for the 2421 code. Follow the line of thought expressed in paragraph 6.3.2.2 as close as possible.

Selected Bibliography

Special Issue on Computers, Proceedings IRE, vol. 41, No. 11. Oct. 1953.

RICHARDS R. K.: Digital Computer Components and Circuits. New York: D. van Nostrand. 1957.

PHISTER M.: Logical Design of Digital Computers. New York: John Wiley and Sons. 1958.

HUSKEY, and KORN: Electronic Digital Computers. New York: McGraw-Hill. 1959.

Special Issue on Computers, Proceedings IRE, vol. 49, No. 1. Jan. 1961.

PERLIN A. I.: Designing Shift Counters, Computer Design, vol. 3, No. 12, pp. 12–16. Dec. 1964.

7. The Basic Organization of Digital Computers

So far, we have concerned ourselves with the details of logic design, that is, we have developed the skills and the techniques to implement individual computer subunits. But no matter how well we understand the details of these circuits, the overall picture of a computer will be rather vague as long as we do not perceive the systematic organization of its components. If we want not only to recognize the structure of a specific machine, but also attempt to judge the significance of variations in the layout of different machines, it is further necessary to comprehend the philosophy of their design.

During our discussion, let us keep in mind that the organization of digital computers is still in a state of evolution. We do not yet have the ideal computer (if there is an ideal computer at all). Designers continuously conceive ideas which change the basic structure of computers in one respect or another. Where future developments may lead, is by no means settled in the mind of experts. In order to give a fair but comprehensible account of this situation, we will first try to put into the proper perspective what might possibly be accomplished by computers. Then we will determine what structural elements are essential to this end. In later chapters, when the basic layout common to present machines has become a familiar concept, we shall discuss individual peculiarities and attempt to grasp the implications of more unorthodox concepts.

7.1. Design Philosophy

The most fundamental question which may be asked about any system is: what can it do? Even though the question is rather simple, it is not easily answered in the case of a digital computer. The answer depends to some extent upon our point of view, and also somewhat on the meaning which we associate with the terms used to describe the capabilities. Definitions of and limitations to these capabilities have been proposed but, at least some of them, are the subjects of heated arguments.

One undisputed capability of digital computers, although not necessarily a limitation, is the automated numerical computation. Before we attempt to define this capability more clearly, it may be appropriate to investigate why such a capability is desirable. There must be some incentive for building digital computers. The computer must be an improvement in some respect. It is either more effective or more reliable, more powerful or less expensive, or perhaps more convenient to use than other computational devices.

Let us begin by examining the operation of a familiar computational device, the desk calculator. A desk calculator performs, essentially, the four basic arithmetic operations following relatively simple rules implemented in its hardware. The operator enters numbers and prescribes the arithmetic operations to be performed. Compared to a paper and pencil calculation, the desk calculator has taken over some of the more mechanical tasks, namely the performance of the four basic arithmetic operations. A calculation with the aid of a desk calculator is more convenient, faster and produces less errors. In general, the combination of man and machine is more effective than the man alone.

Suppose we now try to improve the basic concept. Observing the actions of the human operator closely, we are able to identify additional "mechanical" tasks. Therefore, it should be possible, at least in principle, to shift even more tasks from the man to the machine. Assume that a certain calculation requires a repeated and fixed sequence of addition, subtraction, transfer of numbers from one register to another, etc. In such cases, the operation of a desk calculator becomes mechanical and may be considered as the pushing of buttons according to a *pre-determinable* schedule. Certainly, it should then be possible to construct a mechanism which executes just such a predetermined sequence. If we are able to provide different sequences for the different types of calculations which we want performed, and enable the operator to select them, we have the concept of a primitive and crude computer.

We find now some significant differences in the operation of our hypothetical computer versus the operation of a desk calculator. Clearly, the machine is more powerful than a desk calculator. We might also say that the operator controls the machine on a "higher level". He specifies now not single arithmetic operations, but the touching of a single button may cause the machine to evaluate an entire algebraic formula. An example might be the computation of a power series to find the value of sin x for a given x. With some imagination we might even see the possibility of having machines which are capable of solving any mathematical problem, provided that the problem can be stated in the form of an algorithm[1] and that the machine is elaborate enough to handle rather long sequences of individual operations.

We could now go and try to construct such a machine. But let us save

---

[1] Algorithm, as used here, means a specific prescription or procedure for a computation.

the technical problems for later and find out what we could expect if we were to succeed.

The accuracy of results would probably be the same as for a paper and pencil calculation. Humans can carry as many significant digits in a calculation as required and so can, we should expect, properly designed computers.

The reliability, i.e. the probability that correct results will be delivered, should be increased. True, failures of mechanisms, like errors of humans, will occur, but the rate of failures should be relatively small if the device is properly designed. Moreover, computers should be insensitive to adverse environmental conditions like noise, and unaffected by nervous strain which influences the performance of humans.

The speed of any calculation should be increased tremendously. The desk calculator which is a mechanical device adds or subtracts much faster than a human. The speed of an electronic device could be many orders of magnitude larger, but we have to be careful in one respect: Although the computation time itself may be short, the overall time for solving a problem on the computer could be long, if it takes painstaking efforts to set up the computer for a specific problem. Let us investigate this latter point more closely. Particularly, let us find out what work humans have to do before the computer is able to solve a specific problem.

The computer designer and manufacturer have to do a large amount of creative work before the computer is in existence. The problem analyst has to state the problem in mathematical language and a mathematician has to find an algorithm for its solution. The computer programmer breaks the algorithm down into individual operations like add, subtract, etc. He further specifies the exact sequence of operations necessary for the solution and the exact numerical values for necessary parameters, constants, etc. The computer operator handles operational details such as "loading" the program, starting the machine, replenishing the paper for the printing of results etc. Finally, the maintenance technician diagnoses malfunctions and repairs the computer. Altogether, a considerable effort is required and, in some instances, we may find that it is faster to solve a problem by hand, than to set up and run the computer.

Well then, do we save any work at all by using the computer? Yes, we do. True, in order to solve one problem, it is necessary to expend a large amount of work (both mental and physical), but much of it is expended only once, and we are able to capitalize on it repeatedly. (This is the essence of human progress in general). If there are large amounts of computations to be performed, the efforts of the designer and manufacturer are well spent. If there is a large amount of reasonable similar or identical calculations, the efforts of the problem analyst and programmer are worth their while. The performance of the computer operator can then be very effective. By pushing a few buttons, he is able to do large amounts of computation in a very short time.

All this may seem not sufficient incentive for building the very complex computers of our time. But really there is not much more, except perhaps the encouragement to perform calculations of such complexity that they would not be attempted without computers. A well advertised problem in this category is the (hopefully) accurate weather prediction from an enormous number of individual weather observations. The calculations are not too difficult, but the masses of data to be evaluated for this purpose are so large that if humans evaluated them, results could not be obtained in time for a forecast. In such cases, the computer delivers more accurate results (in a given time) than humans can.

There is another more or less philosophical point which would be considered here: the intelligence of computers. We probably would never say that a desk calculator has intelligence of its own. But when referring to a computer we might not be so sure. Let us first argue in the following manner:

No matter how sophisticated a computer (i.e. a mechanism) is, the designer of the machine who establishes and implements the rules of its mechanism should be able to predict or duplicate its performance. Certainly then the machine is not more intelligent than its designer. Well then, does it have any intelligence of its own? Let us take an example. A mathematically incompetent operator is able to solve complicated differential equations simply by inserting parameters and initial conditions into the computer. Are his accomplishments due to the intelligence of the computer? Well no, a mathematically competent programmer has previously set up the computer to solve the equations. The computer itself simply follows instructions. In this example, we are justified in taking the position that the computer is extremely "stupid". It has to be told exactly and in all detail what to do. (The only point which we have to concede is the fact that it is sufficient to tell the computer only once what to do. It will not forget).

But suppose now that it will be possible to design a "learning machine", that is a computer which modifies its own program according to the statistical distribution of data, or according to the results of previous trials. Suppose further, this machine has "learned" to translate an ancient text in a hitherto unknown language, or has "learned" to play chess so well that it beats its designer every time. (Both are feats which are well within the range of foreseeable capabilities of digital computers.) Will it now be necessary to attribute "artificial intelligence" to computers? This is a debatable point on which people sometimes vehemently disagree. One side argues that intelligence is obvious. The other side, in order to be consistent, is willing to claim that even humans accomplishing comparable feats are not necessarily intelligent. For our purposes here, let us take a more practical point of view: The argument is useless, as long as there is no commonly accepted definition of the term intelligence.

Problem 1: Review problem 3 in chapter 1.

a) Will it always be possible to write a straightforward scheme for any numeric calculation or can you think of cases where you cannot unequivocally prescribe a predetermined sequence of operations?

b) Try to state a minimum set of instructions which the operator has to understand in order to perform *any* numeric calculation.

7.2. The Basic Internal Functions of an Automatic Digital Computer

Knowing approximately what we want to accomplish, namely an automated numerical computation, let us try to identify the various necessary internal functions which a computer must perform. As far as possible, we shall indicate them in functional diagrams, but keep in mind that such diagrams not necessarily bear a resemblance to the block diagram of an actual computer.

The most obviously required function of a digital computer is the performance of arithmetic operations like add, subtract, etc. Let us indicate this capability by a functional unit, the "arithmetic unit". We may tentatively assume its capabilities as approximately those of an electronic replica of a desk calculator.

Something must guide or control the arithmetic unit, i.e. make it perform the specific sequence of arithmetic operations necessary to solve a specific problem. In other words, we need a sequencing mechanism. If we want our computer to be able to solve various different problems, then a sequencer with a fixed sequence is inadequate. The sequencing mechanism must necessarily contain provisions for the changing of the "program". If we study this requirement a little more closely, we see the necessity of some storage in which we can store the specific sequence of operations before a computation is started. It will be helpful later if we separate the sequencer into two functional parts, the "program storage", and the "control unit". The functions of the program storage are already sufficiently defined for our present purpose. We may imagine the control unit to be "sensing" certain conditions external to it, and issuing "commands" to other parts of the machine. For example, the control unit sends commands to the arithmetic unit which initiates arithmetic operations, or sends commands to the program storage which cause the "lookup" of the next instruction. The control unit senses such conditions as the completion of an arithmetic operation in the arithmetic unit, or the start and stop signals initiated by the computer operator.

Before we go any further, let us draw the functional diagram of the computer as we see it now. Additional needs may then be easier to identify.

The machine, as it stands now, can perform any sequence of arithmetic operations, but we recognize immediately some oversights. We have not yet made provisions for the input and output of numerical operands (numerical values to be used in the computation) to and from the arithmetic unit, for the storage of these operands, and for the storage of partial results (so to speak a scratch pad for the machine). We can solve this problem by adding an "operand storage" and a "temporary storage" to the diagram given in Fig. 7.1. The transfer

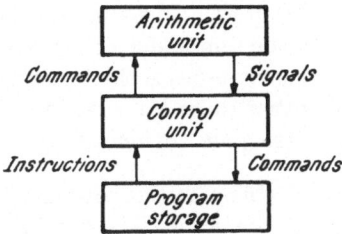

Fig. 7.1. Partial Functional Diagram of an Automatic Digital Computer

of operands or, in general, of information is indicated by heavy lines in Fig. 7.2. The transfer can be initiated by the control unit in very much the same manner as the initiation of arithmetic operations or the fetching of the next instruction.

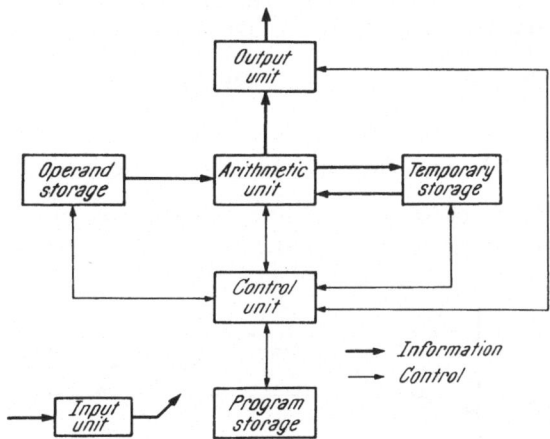

Fig. 7.2. Functional Diagram of an Automatic Digital Computer

There are essentially two more capabilities which must be incorporated: provisions for the input, and the output of information. An output is obviously necessary, even if it is only for the output of the results of a computation. Let us, therefore, tentatively connect an output unit to the arithmetic unit. The understanding should be that the control unit can issue commands which, at appropriate times during the computation, transfer the results contained in the arithmetic unit to external devices like a display, printer or plotter. The need for an input unit may not be so obvious. However, if we regard our computer as a system in itself which has to communicate with the outside world, the necessity for input provisions becomes apparent, even if it is only to change the contents of the program and operand storage.

At present, we may not be so sure how to incorporate the input feature into the remainder of the functional diagram. Just so that we do not forget it, let us indicate the function by a separate box, and leave the connections to other units for later. Fig. 7.2 shows now all the major functions which we are able to identify at the moment.

Problem 2: Suppose a computer is set up to compute $A \div (B - C)$ and print the result. Make a listing of the conditions to be sensed, and of the sequence of commands to be issued by the control unit for this problem. Assume the functional diagram given in Fig. 7.2. The first three entries should approximately read:

Condition	Command
1. Start signal	Fetch first instruction
2. Instruction requests the transfer of B into the arithmetic unit	Transfer B from the operand storage to the arithmetic unit
3. Transfer completed	Fetch next instruction

7.3. The Layout of Early Computers

We could now proceed and see how modern computers are organized around the basic functions indicated in Fig. 7.2. However, we will gain

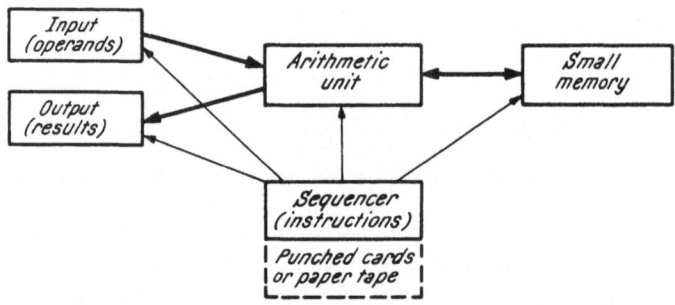

Fig. 7.3. Layout of an Externally Programmed Computer

a much better understanding of modern machines if we first briefly review earlier computer layouts, giving special attention to their shortcomings.

We will also be in a better position to recognize the modern computer as a link in a continuing process of evolution.

The earliest automatic computers were *externally programmed*. Fig. 7.3 gives a representative layout of such a machine.

The control unit and the program storage of Fig. 7.2 are here shown as a single unit, the sequencer. It consisted of one or more mechanisms to read punched cards or punched paper tape. The program itself (i.e. the sequence of individual operations to be performed) was coded and punched into a paper tape or in a deck of punched cards[1]. The sequencer advanced the tape or the cards. Sensing and interpreting the combinations of holes in the paper, it sent signals to the remaining units, making them perform appropriate operations.

The numerical values of the operands are read as needed from cards or tape, and transferred under control of the sequencer to the arithmetic unit via the input unit. Arithmetic results are transferred in a similar manner from the arithmetic unit to a printer or punch via the output unit. A small memory keeps intermediate results for later reference.

It is not hard to see that the externally programmed computer, as it stands, can perform calculations for which the sequence of individual operations can be predetermined. It is an automatic computer in the sense that it, by itself, proceeds with a computation, once it is set up. The change from one problem to the next is rather easily accomplished. Only the program (the program tape or card deck) and the operands (another tape or card deck) have to be replaced, and the computer is ready for an entirely different task. Further flexibility and convenience can be derived from the fact that the operands and the program are on separate storage media. In order to repeat a computation with different operands, it is only necessary to change the operand tape or deck. Once prepared programs can, therefore, be used over and over again[2]. Existing programs, say for the numerical integration of a third order differential equation, written for a particular application can be used as a part of a program for an entirely new problem. Conceivably, "program libraries" will be able to furnish a wide variety of existing routines so that only a fraction of the otherwise necessary original effort has to be spent in programming a new problem. The mechanical duplication of punched paper tapes or punched card decks required for this purpose presents no problems.

---

[1] If we want, we can consider the paper tape or the deck of punched cards as "program storage" and the reading mechanism and its associated circuitry as "control unit".

[2] Conceivably, instructions and operands might be (and have been) entered by the same tape or card deck. Each re-run of a computation with different operands requires then the preparation of an entirely new program tape or card deck.

Altogether we have a workable concept but, of course, it has its limitations. Programs may become voluminous by the repetition of identical instructions. Suppose for instance, that the computation for a particular problem requires time and again the calculation of the square root of operands. The program for such a problem may be rather long and conceivably consist mainly of a monotonous repetition of the sequence of instructions which make the arithmetic unit extract the square root by more elementary arithmetic operations. Inconveniences of this type can be repaired by incorporating an auxiliary sequencer into the system. The auxiliary sequencer, in our example, would contain a paper tape loop with the particular instructions necessary for the extraction of the square root. This program is regarded as a "subroutine". Every time a square root is to be extracted, the main program would contain a transfer or "jump" instruction which causes the main sequencer to transfer the control to the auxiliary sequencer. The auxiliary sequencer extracts then the square root and transfers control back to the main program. The paper tape loop for the subroutine in the auxiliary sequencer would be exactly as long as needed. It would, therefore, be back in the starting position when the subroutine has been executed. In this manner, the subroutine can be executed as many times as required. Of course, the concept is not limited to relatively simple subroutines, but any specific sequence of operations which has to be executed repeatedly during a computation can be accomodated. Nor is the concept limited to a single subroutine. By providing several auxiliary sequencers, a number of different subroutines can be referenced.

A more serious problem is the handling of situations where different actions have to be taken depending upon results computed by the machine itself. The simplest illustration of such a situation is perhaps the determination of the absolute value, say $|x|$. If x is the result of a preceding computation, the sign of x may not be known beforehand and, therefore. it cannot be predetermined whether or not the sign of x has to be reversed in a particular instance. In other situations, the necessary operations in a computation may depend upon whether or not a variable is larger than a certain limit. In all such cases, the computer has to make a "decision", and perform either one or the other sequence of operations. Such a program is said to be "branching". At the first glance it seems as if our computer would not be able to handle such situations. However, if the machine contains auxiliary sequencers, a solution can be found. For instance, in order to determine the absolute value of x, the main program might contain an instruction which transfers control to the auxiliary sequencer *only if* the content of the arithmetic unit (i.e. x) is negative. The auxiliary sequencer would then change the sign of x, and return control to the main program. If the content of the arithmetic unit were positive, the control would not be transferred to the auxiliary sequencer, and the sign would not be changed.

In any event, the main program can assume for any following computations that the arithmetic unit contains a positive number, i.e. the absolute value of x. Transfers which depend upon certain conditions, such as the one described, are called "conditional" jumps or transfers, in contrast to the previously discussed "unconditional" jump.

The above example is a special case. One branch of the program contains only operations in addition to the operations required for the other branch. But a solution can be found also for the more general case that entirely different operations have to be performed in the two branches of the program. In such cases, the control may be conditionally transferred to one or the other auxiliary sequencer. When the instructions for one or the other branch have been executed, the control is returned to the main program.

With the provisions for branching, the computer can (theoretically) solve any problem as long as an algorithm for its solution can be found. With its capabilities, the externally programmed computer is vastly superior to any previous aid for computation. Surprisingly, the construction of these machines required neither a technological breakthrough nor new scientific discoveries. The idea to mechanize computations is not new and the first automatic computers consisted of previously known components like relays and card or paper tape punches and readers. We have to regard their construction mainly as an engineering feat stimulated by the desire or the requirement to attack complex computations.

Problem 3: Try to state some of the practical limitations of externally programmed computers.

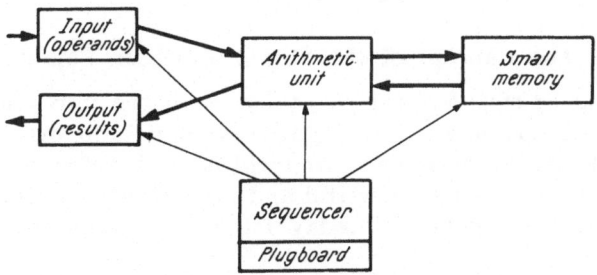

Fig. 7.4. Layout of a Plugboard Programmed Computer

An improvement over the externally programmed machine was the *plugboard programmed* computer. Let us consider it briefly. Fig. 7.4 shows the basic layout.

The layout is very similar to that of Fig. 7.3. The only essential difference is the replacement of the previously used mechanical program storage (a card deck or paper tape) by an electrical program storage (i.e. a plugboard). The sequence of instructions for a particular program is no longer punched into cards or tapes, but specified by the wiring of the plugboard. The completely wired plugboards can be easily interchanged for the various computations to be performed.

The advantages of the plugboard programmed computer compared to the externally programmed computer are the increased speed and flexibility. The computer is faster because it is no longer necessary to manipulate the program storage mechanically. Consecutive instructions can, therefore, be "looked up" very rapidly[1]. More important, however, is the increased flexibility in handling subroutines and program branches. The program board contained a certain number of positions in which particular operations could be specified. In principle, it made no difference how these positions were divided between the main program and various branches. The programmer had large freedom in specifying conditional or unconditional jumps throughout the available program storage.

It may seem that the limited number of operations which could be specified (limited by the size of the plugboard) is a step back from the unlimited length of program which can be allowed for externally programmed computers. However, the problems which were thought worthwhile to set up and run on the computer at this time, were mainly of the sort for which identical sequences of operations were to be performed over and over on a large amount of operands or input data. Furthermore, even plugboard programmed computers, in many instances, contained additional card readers for long programs. In this manner, the partial routines wired on the plugboard could be very effectively scheduled by a relatively small input of instructions via the card reader.

7.4. The Concept of the Stored Program Computer

A tremendous step forward was taken with the introduction of the stored program computer. Its designers attempted to overcome simultaneously all disadvantages and inconveniences of earlier machines. The resulting concept proved so powerful and convenient that even the most advanced digital computers of today follow essentially the same basic layout.

Before we draw this layout, let us review the reasoning which probably led to the design of the stored program machine. The fundamental considerations may have been the following: The designers wanted to incorporate a much larger storage than prior machines had. This allows problems requiring the storage of large amounts of intermediate results to

---

[1] In addition, plugboard programmed computers contained to a limited extent electronic components which allowed a faster internal operation.

be attacked in a fully automatic manner. They wanted an all-electronic computer for reasons of increased speed and reliability, and they wanted as much freedom as possible especially as far as the number of subroutines, and the number of branches in a program were concerned. All further considerations may be considered as consequences of these wishes and the desire to provide the best possible answer to them.

The all-electronic machine had to include by necessity an electronic storage for operands and intermediate results, since the speed of a fast computer is useless if it cannot quickly access the numerical values required for its calculation. Rather than providing separate storages for operands, temporary results, and the program, the designers felt that it would be advantageous to store all three in one large "memory". This, in itself, is not unreasonable if provisions for fast input means (like inputs from magnetic tapes) are made, so that the program and the operands can be quickly changed. Moreover, if the assignment of operand storage, temporary storage and program storage is flexible, the programmer can allocate the available storage space as it fits a particular problem best.

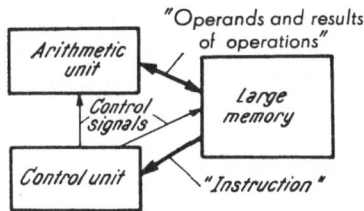

Fig. 7.5. Partial Layout of a Stored Program Computer

Let us now draw a partial layout of the stored program computer which reflects this arrangement.

The functions defined in Fig. 7.5, except for input and output which we disregard for the moment, can still be related to earlier concepts if we remember that the memory contains the operand, temporary and program storages. The control unit looks up the sequence of instructions in the memory, it initiates arithmetic operations, and causes operands or intermediate results to be transferred between the arithmetic unit and the memory, as required by program instructions.

The layout has the advantage over previous machines that all operations can be performed at electronic speeds. Further implications of the scheme are not so obvious, but let us try to derive them one at a time.

In the new concept, operands can be used in "random" order, whereas older computers required the operands to be "stored" in the exact order in which they were to be used. We may speak of "random access" to operands in contrast to the "sequential access" of earlier machines. The computer program obtains operands by "addressing". A program instruction may, for instance, read: transfer the contents of storage location 526 to the arithmetic unit, or: multiply the contents of the accumulator in the arithmetic unit by the contents of storage location 317. Addressing allows to retain a valuable feature of previous machines: programs can be written in such a manner that they can be re-run with different operands but no changes, other than the replacement of the operands themselves.

Random accessing of operands may seem a minor point, but it is of great importance. If we would ask the programmer of a modern computer to order all operands in the sequence in which they are used, he would probably reply that this is impossible. Even though it may not be truly impossible, it, certainly, would be a great inconvenience. Just think of a case where certain operands have to be used only for one branch of the program while the alternate branch uses different operands or, perhaps, the same operands in a different sequence. The addressing scheme not only allows the ordering of operands to be disregarded, but also allows the same operands to be used repeatedly without storing them repeatedly. This, of course, reduces the capacity requirements of the operand storage.

The addressing scheme has a further implication. Instructions, like operands, can be addressed. The program may, for instance, ask the control unit to take the instruction in a certain memory location as the next instruction. In effect, the computer has performed a jump. Such jumps may be conditional or unconditional. With the given scheme, there are no restrictions as far as the number of jumps is concerned, other than the memory capacity itself. Conceivably, every instruction in the memory could be a jump instruction. Some computers actually use each instruction as a jump instruction, i.e. each instruction specifies the address of the next instruction to be performed. With such an arrangement, not only operands but also instructions can be stored in random order in the memory. Most computers, however, store instructions in the sequence in which they are normally executed and programs specify jumps only if the computer has to deviate from this sequence. In any event, the programmer is free to incorporate as many program branches into his program as he likes.

The features of the stored program computer which we have discussed so far, make the machine much more flexible and versatile than its predecessors but we have not yet touched the real reason for its power.

Operands, i.e. numbers, are stored in the same memory as instructions. Both can be obtained by addressing. Theoretically it is then possible to use (or misuse) instructions as numbers and numbers as instructions. At the first glance, this may seem undesirable and even dangerous. But let us not be hasty. Even though right now, we may not grasp all possible implications, let us here at least derive, in general terms, the uses we can make of it.

To treat an instruction as a number means to perform an operation upon it which we, normally, would perform only upon a number. Such an operation may be a transfer or an arithmetic operation[1]. Transfers of

---

[1] Later we shall see that computers can also perform logic operations so that we might also perform a logic operation upon an instruction.

instructions are desired for the input and output of computer programs, i.e. for the loading or "dumping" of programs[1]. In these cases, instructions are transferred like any other "information". Let us show here also at least one specific example where the transfer of a single instruction within the computer memory is useful. Suppose we have to use a subroutine repeatedly during the execution of a main program. It is no problem to transfer the control from several places in the main program to the subroutine. The main program simply contains at the appropriate places a jump instruction to the location of the memory which contains the first instruction of the subroutine. The problem is, however, to make the subroutine transfer the control back to different places in the main program. Obviously, the last instruction of the subroutine must be a jump instruction back to the main program. Since, the subroutine, by itself does not "know" the appropriate place in the main routine to go back to, the main program has to "insert" the appropriate jump into the subroutine. This inserting may be done by transferring the appropriate jump instruction into the subroutine, before the subroutine is entered by the main program[2].

Let us now look at an example in which an arithmetic operation is performed on an instruction. Suppose an add instruction, located somewhere in the computer memory, is coded as: 350001. The 35 shall specify an add operation, and 0001 shall specify the address of storage location 0001. If the instruction were executed as it stands, the content of storage location 0001 would be added to the contents of the accumulator. However, if we should happen to add the number 000001 to our instruction before it is executed, the instruction is modified to read: 350002, which specifies that the contents of storage location 0002 (rather than the contents of storage location 0001) shall be added to the contents of the accumulator[3].

The examples we have discussed so far were concerned with the modification of addresses. However, by appropriate operations we can also change one kind of instruction into another. For instance, an add instruction (code 35) could be changed to a subtract instruction (say code 36), or any other instruction by the addition of an appropriate numerical value. We recognize here the potential capability of the computer not only to modify, but also to generate entire programs (let us say from specifications given perhaps in a semi-english symbolic language), and the potential

---

[1] The dumping of a program means the printing of a program (e.g. for trouble-shooting), or the temporary storage of a program on some external means, like a magnetic tape.

[2] This process is called the "setting of the return link". There are various ways to accomplish this. In some computers, the return link is set automatically by the computer hardware.

[3] Modifications of this nature can be used very well to perform identical operations upon different sets of operands, to "rewrite" a program for differently located operands, or to re-locate a program itself.

capability to "learn", that is to generate and modify its program according to successes or failures with previous versions of the "same" program.

We are now in a position where we have some feeling for the capabilities of the concept of a stored program computer. A full understanding of all details and implications requires more than a superficial knowledge of programming. The problem of finding a mechanism which implements a desired operational feature cannot very well be separated from the problem of programming such a mechanism. A programming course, however, is well beyond the scope of this book. We will, therefore, restrict ourselves to the more detailed study of the computer hardware, only occasionally mentioning the programming aspects of a certain scheme or layout. If at all possible, it is recommended that the reader gain some practical experience in programming an actual computer before, or simultaneously with, the study of the following paragraphs and chapters.

7.5. A Model Layout of a Digital Computer

In the preceeding paragraphs of this chapter, we attempted to derive and justify the concept of the stored program computer. Let us now see in more detail what internal structure is necessary to implement the concept. In short, let us look at the stored program computer as a system. Suppose we start with the model layout shown in Fig. 7.6.

This layout does not intentionally represent any existing computer; rather, we should consider it as one possible implementation which helps us to define more clearly the functions of the individual units and to recognize the interface requirements.

The general functions of the four units have been previously defined roughly as follows:

The memory stores instructions, operands, results and any other information which may be needed in the course of a computation.

The arithmetic unit performs arithmetic operations (such as add, subtract, etc.) or logic operations (such as shift, mask, etc.).

The input/output unit provides communication paths between the computer and external or peripheral equipment (such as magnetic tape units or card readers).

The control unit governs the internal operation of all units and the transfer of information throughout the system.

Let us now state more detailed specifications for our model layout.

The *memory* is divided into a number of storage locations or cells, each having a unique address and a storage capacity of one word. The circuitry of the memory performs essentially three functions: selecting a cell according to the address contained in the memory address register (MAR); reading a stored word and putting it into the memory information register (MIR);

storing the word contained in MIR. These functions are in short: select; read; write.

The interface of the memory consists of: lines for the input and output of information (originating or terminating at MIR); lines for the input of commands initiating read or write operations (terminating at some control circuits which are not specifically shown in Fig. 7.6); lines for the output of status indications such as ready or busy (originating in the internal control circuitry of the memory).

The control unit governs the operation of the overall system. In particular, it translates instructions obtained from the memory into specific sequences of internal commands. These commands are distributed to all units and initiate there appropriate operations or information transfers (e.g. a read operation in the memory, an add operation in the arithmetic

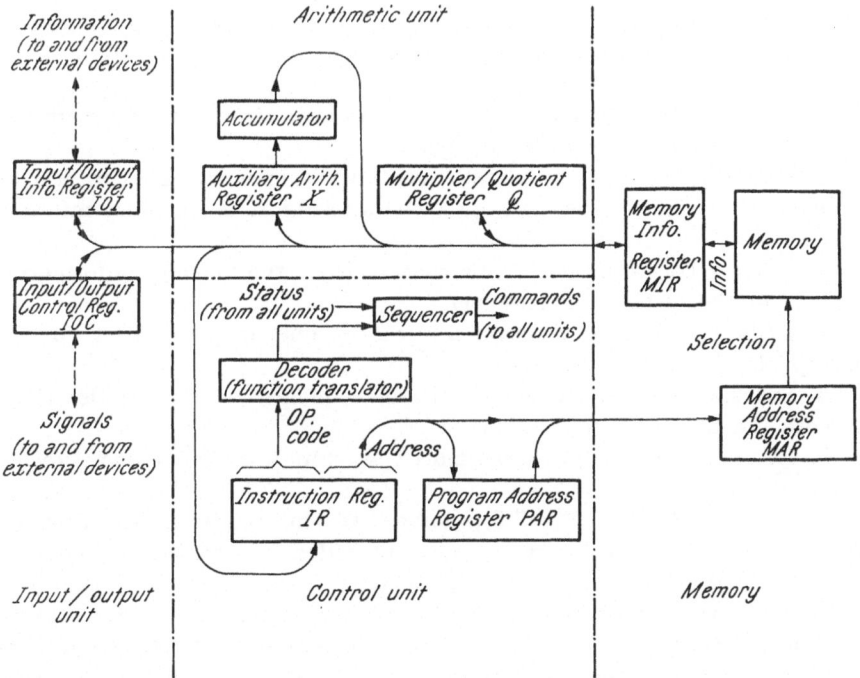

Fig. 7.6. A Model of a Stored Program Computer

unit, or an information transfer between the memory and the input/output unit). The control unit also assures that the timing of commands is commensurate with the operational speed and the operational status of all units. For this purpose, it receives indications of specific conditions (such as "addition complete" or "memory cycle in progress") from all units of the overall system.

All these functions are performed by the circuits represented in the left half of the control unit, as shown in Fig. 7.6. The circuits in the right half perform a function which has not yet been specifically mentioned, i.e. the fetching of instructions from the memory and keeping track of the program address. To accomplish this, the control unit contains a "program address register". For the "look-up" of a program instruction, its contents are sent to the memory, and a read cycle is initiated. At the conclusion of the read cycle, the contents of the accessed storage location are transferred to the instruction register. The program address is normally increased by one before the next instruction is looked up. The internal functions of the control unit are performed in the same manner as any other operation in the system, i.e. each operation is initiated by a command from the sequencer. In fact, the commands pertaining to the fetching of instructions are usually an integral part of the overall sequence of commands executing an instruction. A typical sequence of commands issued for the execution of an instruction might be:

1. PAR → MAR (transfer program address from program address register to the memory address register).

2. Initiate read (initiate a read cycle in the memory. The instruction obtained as a result of the read operation appears in the memory information register).

3. MIR → IR (transfer the instruction from the memory information register to the instruction register).

4. PAR+1 (increase program address by one, in preparation for the pickup of the next instruction).

5. Specific sequence (issue a specific sequence of commands which execute the obtained instruction).

6. Reset sequencer (reset sequencer to step 1, to fetch and execute the next instruction).

In the control unit of Fig. 7.6, we notice two information paths which we have not yet discussed. There is a path provided to send the address part of an instruction to the memory address register, and another to send it to the program address register. The first of the two is used to send the address contained in an instruction to the memory (for any memory operation other than an instruction lookup). The second path is used to modify the contents of PAR if a jump instruction is executed. In this case, the address contained in the jump instruction is sent to PAR, so that the next instruction is not looked up in sequence but from the specified address[1].

---

[1] From then on, the instruction look-up is sequential until another jump instruction is encountered.

The functions of the control unit can be expressed in short as: keeping track of the program address; addressing the memory; generating a sequence of commands consistent with the instruction to be executed, the speed of all units, and the status of all units.

The interface may be defined as follows: lines for the input of instructions (terminating at the instruction register), lines for the output of addresses (originating at IR and PAR), lines for the output of commands (originating at the sequencer), lines for the input of status signals from all units (terminating at the sequencer).

The *arithmetic unit* consists of a number of registers and associated circuitry connected in such a manner that arithmetic and logic operations can be performed. In particular, the contents of the auxiliary arithmetic register (X) can be added to the contents of the accumulator (A) or can be subtracted from it. A multiplication is performed as a repeated addition, wherein the multiplicand is contained in X, the multiplier is contained in Q, and the product is derived in A. A division is performed as a repeated subtraction, wherein the dividend is contained in A, the divisor in X, and the quotient is derived in Q[1]. All three registers can be cleared; the A and Q registers can be shifted left and right.

The interface of the arithmetic unit consists of: lines for the input of operands (terminating at X and Q); lines for the input of commands initiating arithmetic and logic operations (terminated at control circuits which are not shown); lines for the output of status indications such as "content of the accumulator negative", or "addition completed" (originating in the logic circuitry associated with the arithmetic unit).

The *input/output unit* consists essentially of two registers, the input/output information register (IOI), and the input/output control register (IOC). The first transmits information between the external equipment and the computer system; the latter transmits command codes which select the proper equipment and initiate operations such as read, write, print, etc. in the external equipment. It also accepts status information from the external equipment such as "cycle complete", "not ready", etc. An important function of the I/O unit is the buffering of signals. The transfer of any information between the external equipment and the I/O registers is usually timed by the external equipment, whereas the transfer between the I/O registers and the rest of the computer is timed by the computer control unit.

The interface between the I/O unit and the remainder of the computer system consists of: lines for the input and output of information (originating or terminating at IOI); lines for the input of command codes (terminating at IOC); lines for the output of status indications concerned with the status

---

[1] Paragraphs 8.1.2 and 8.1.3 show these processes in detail.

of external equipment (originating at IOC); lines for the input of command signals (terminating at same control circuits which are not shown in Fig. 7.6); lines for the output of status signals concerned with the status of the I/O unit itself (originating in the control circuitry associated with the I/O unit).

Problem 4: Show a model sequence of commands to be issued by the control unit in order to execute:

a) a transmit instruction, say, from a memory cell to the Q-register.

b) an add instruction, adding the contents of a memory cell to the contents of the accumulator.

c) an unconditional jump.

Problem 5 (Voluntary): Try to define a minimum set of instruction which the model computer must be able to execute in order to solve any mathematical problem. Assume that subtractions are programmed as sign changes and additions, and that multiplications and divisions are programmed as a sequence of add subtract and test instructions.

Problem 6 (Voluntary): Try to define a minimum set of control signals which the sequencer must be able to issue in order to execute the minimum set of instructions defined in problem 5.

The definitions of internal functions and interface requirements stated above apply, strictly speaking, only to our model computer. However, the characteristics of the model computer are typical enough to give us at least a basic understanding of the overall operation before we now discuss individual units in more detail.

Selected Bibliography

AIKEN H.: Proposed Automatic Calculating Machine, 1937. Reprinted in IEEE Spectrum, vol. 1, No. 8, pp. 62–69. Aug. 1964.

Staff of the Computation Laboratory Harward, A Manual of Operation for the Automatic Sequence Controlled Calculator, Cambridge. 1946.

BURKS, GOLDSTINE, and VON NEUMAN: Preliminary Discussion of the Logical Design of an Electronic Computing Instrument, 1946. Reprinted in Datamation, September and October issues, 1962.

KELLY J. L., JR.: Sophistication in Computers: A Disagreement, Proceedings IRE, vol. 50, No. 6, pp. 1459–1461. June 1962.

SERRELL, ASTRAHAN, PATTERSON, and PYNE: The Evolution of Computing Machines and Systems, Proceedings IRE, vol. 50, No. 5, pp. 1040–1058. May 1962.

ARMER P.: Attitudes Toward Intelligent Machines, Datamation, March and April issues, 1963.

Special Issue on Computers, American Mathematical Monthly, vol. 72, No. 2, part II. Feb. 1965.

8. The Functional Units of a Digital Computer

8.1. The Arithmetic Unit

In chapter 7, the function of the arithmetic unit has been defined loosely as the performance of arithmetic operations. As such, the capabilities of the arithmetic unit have been compared to those of a desk calculator. Although this analogy is valid in a general sense, the capabilities of arithmetic units exceed those of the desk calculator: in addition to arithmetic operations, certain logic data manipulations can be performed. Moreover, the particular manner in which operations are performed is influenced by the electronic design. In the following paragraphs we shall discuss three types of operations: fixed-point arithmetic operations, logic operations, and floating-point arithmetic operations. Incidental to this discussion, we shall see structures required for the implementation of the individual operations. In conclusion, several sample layouts of arithmetic units are indicated in which the individual requirements are combined.

8.1.1. Addition and Subtraction

Earlier, in chapter 6, we have already seen the basic mechanisms performing additions and subtractions. At that time we also discussed the integration of adders and subtracters with arithmetic registers. Remaining for our present discussion are then essentially three topics: a method to perform subtractions by additions, or, conversely additions by subtractions; rules for algebraic additions and subtractions; and certain high-speed techniques.

8.1.1.1. Subtraction by the Addition of the Complement: Let us illustrate the principle of this method by a few specific examples in the decimal system. Suppose we have a counter (similar to the mileage counter in an automobile) which displays two decimal digits and resets from 99 to 00 (or from 00 to 99 in reverse direction). We can make the counter display any desired two-digit number by driving it either forward or backward from an arbitrary initial state. If the counter presently displays, for instance, the number 95, we can make it display the number 58 either by driving it backward 37 units, or driving it forward 63 units. Corresponding regular computations would show:

$$95 - 37 = 58 \tag{8.1}$$

or:
$$95 + 63 = 158 \tag{8.2}$$

The counter disregards any carries into the third digital position. We say: it counts (or adds and subtracts) "modulo 100". We obtain identical results by driving it B positions backward (subtracting B), or driving it $(100-B)$ positions forward (adding $100-B$). We can express this behavior by the following equation:

$$A-B = A+(100-B) \pmod{100} \tag{8.3}$$

Example: $95-37 = 95+(100-37) = 95+63 = 58 \pmod{100}$ (8.4)

The numbers B and $(100-B)$ are said to be complementary as far as our counter is concerned. One number complements the other to 100 or is the complement "modulo 100" to the other number. If there is no doubt about the modulus, we can simply speak of the "complement". In the above numerical example we may, for instance, say that 63 is the complement of 37.

Problem 1: What is the complement modulo 100 to: 7, 96, 0, 100 ?

Problem 2: What is the complement modulo 10^3 to above numbers ?

Problem 3: What is the complement modulo 2^5 to: 101_2, 1011_2, 0 ?

Problem 4 (Voluntary): Write an equation modulo 2^N, equivalent to Equation (8.3).

Equation (8.3), potentially, gives us a method to perform subtractions by the addition of the complement. Of course, this method is practically useful only if we can find a simple way to derive the complement of a number, i.e. a way to obtain the complement of a number without actual subtraction.

The complement of a number, in our example, is given by: $(100-B)$. We may write this in the following manner:

$$(100-B) = (99-B)+1 \tag{8.5}$$

The left hand side of Equation (8.5) represents the true complement. The term in parentheses on the right hand side of Equation (8.5) is known as the 9's complement. The 9's complement is generally easier to derive than the true complement. Let us demonstrate this with our numerical example:

True complement:	9's complement:
100	99
—37	—37
63	62

Nothing short of a complete subtraction will give us the true complement. However, in deriving the 9's complement, it is sufficient to subtract within one digital position. Since the minuend consists of all 9's, there will never be a borrow from another digital position. In fact, the 9's complement can be written down immediately. We simply inspect a number and write down those digits which complement the digits of the original number to 9.

Example:

Number:	37	00	93
9's complement:	62	99	06

The derivation of the 9's complement by a computer can be extremely simple if an appropriate number code is used[1]. In this case, it is only necessary to invert the binary 1's and 0's.

Let us now see where our method stands. Combining Equations (8.3) and (8.5), we obtain:

$$A - B \quad = \quad A \quad + \quad (99 - B) \quad + 1 \pmod{100} \quad (8.6)$$
$$\uparrow \qquad\qquad \uparrow \qquad\qquad \uparrow \qquad\qquad \uparrow$$

subtraction   addition   complementation   addition

We see that subtraction is replaced by addition and complementation. Hence by the use of complementation we can make a computer subtract without incorporating subtracters. The fact that results are "modulo" results is no drawback of our particular method, rather, it is a property of any machine calculation. Even additions are, by necessity, modulo operations. The reason is very simple: There cannot be an unlimited number of digital positions in the machine. The "capacity" of the machine will be exceeded in some instance, that is some register will eventually "overflow" when we keep adding numbers.

Problem 5 (Voluntary): Write an equation modulo 2^N equivalent to Equation (8.6).

Let us now apply our method to a few numerical examples:

Example 1:   66 — 13 = ?

The 9's complement of 13 is 86, so we obtain:

$$66 - 13 = 66 + 86 + 1 = 1)52 + 1 = 53$$

Example 2: $3726 - 1367 = 3726 + 8632 + 1$
$$= 1)2358 + 1 = 2359$$

So far, we have considered only cases where the minuend is larger than the subtrahend. Let us see what happens if this is not the case:

---

[1] See paragraph 2.4.1.

Example 3: $13-66 = 13 + 33 + 1$

$$= 0)46 + 1 = 47$$

We see that there is no overflow or "end-carry", and also that the result is not correct. In order to interpret it, let us return to the counter analogy. Suppose the counter indicates "13". When we drive it back by 66 counts, it will display "47" which is exactly the result we obtained in example 3. "47" is then really the result of the subtraction 13—66 "modulo 100". (The counter disregards borrows from the third digital position in the same manner as it disregards carries to the third digital position.) Suppose we would step the counter back slowly, one count at a time. We would then obtain:

Correct Result:	Counter State:
13	13
12	12
.	.
.	.
.	.
02	02
01	01
00	00
—01	99
—02	98
.	.
.	.
.	.
—53	47

By comparing the two columns, we see that the counter displays true complements instead of negative numbers. The interpretation of the result in example 3 is then not very difficult. Instead of "47", we take its true complement, "53", as result and attach the minus sign. If desired, we can derive the true complement of 47 again by taking the 9's complement and increasing it by 1.

Let us repeat the previous example in detail:

Example 3: $13-66 = 13 + 33 + 1 = 0)46 + 1 = 47$
(repeated)

We note that there is no end-carry, so we take the complement of 47:

$$(99-47) + 1 = 52 + 1 = 53$$

We further attach the minus sign. The correct result is then: —53.

Let us now state the rules for the subtraction by the addition of the complement in the form of a flow chart:

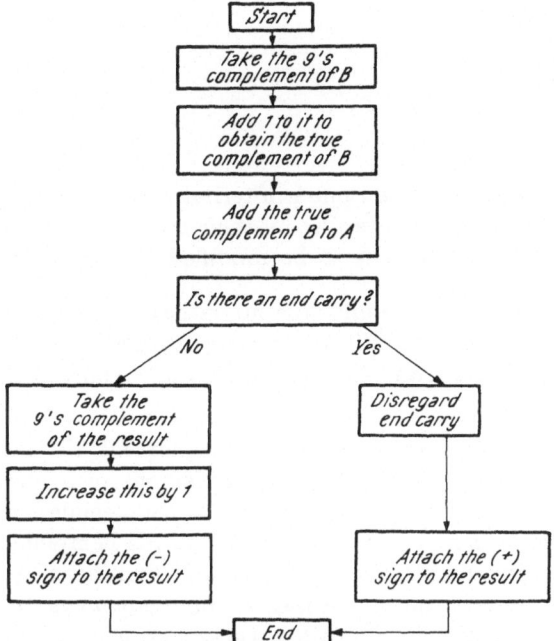

Fig. 8.1. Flow Chart for the Calculation of $A-B$ by the Addition of the Complement

Example:
$$\begin{array}{r} 37\,631 \\ -29\,867 \\ \hline \end{array} = \begin{array}{r} 37\,631 \\ 70\,132 \\ +\quad 1 \\ \hline 1)07\,764 \end{array}$$

There is an end-carry. The result is, therefore, 07764 with a $(+)$ sign attached, or: $+07764$.

Example:
$$\begin{array}{r} 29\,867 \\ -37\,631 \\ \hline \end{array} = \begin{array}{r} 29\,867 \\ 62\,368 \\ +\quad 1 \\ \hline 0)92\,236 \end{array}$$

There is no end-carry. The result is, therefore, the 9's complement of 92236 (i.e. 07763) increased by 1 and with a $(-)$ sign attached, or: -07764.

Problem 6: Calculate the following differences by the addition of the complement:

  a) 7329—4987
  b) 4987—7329
  c)   37—5634
  d) 6784—   17
  e)   43—   43

Problem 7: Suppose a computer internally represents numbers to be subtracted and also negative results by their true complement. How would the flow chart of Fig. 8.1 have to be modified to take this into account?

Problem 8 (Voluntary): Some computers have only subtracters and no adders. They can perform addition by the subtraction of the complement. Find the applicable rules and show them in a flow chart similar to Fig. 8.1.

Up to now, we have shown the applicable rules only for the decimal system. However, there is no difficulty in adapting the method to other number systems. In deriving the 9's complement of a number, we complement digit by digit to 9, the largest digit in the decimal system. In order to find the complement corresponding to the 9's complement in another number system, we complement each digit of a number to the largest digit in that number system. Since the largest digit is always one smaller than the base of the system, we may call all these complements the "base minus one" complements or $B-1$ complements for short. The $B-1$ complement is in all cases the true complement decreased by 1. The flow chart of Fig. 8.1 is generally applicable to all number systems if we replace the term "9's complement" by the term "$B-1$ complement". If we refer specifically to the binary (base 2) number system, we may speak of the 1's complement instead of the $B-1$ complement. Also, perhaps not always quite correctly, we sometimes speak of the 2's complement instead of the true complement.

Problem 9: Show the true complement and the $B-1$ complement for the following numbers: 376_8, 2011_3, 0100110_2.

Problem 10: Perform the following subtractions by the addition of the complement:

  a) $110010_2 - 101101_2$
  b) $01001_2 - 11001_2$

Problem 11 (Voluntary): How is the flow chart of Fig. 8.1 to be modified if negative numbers are represented by their $B-1$ complement? Simplify the resulting flow chart as much as possible (note that for no end-carry, there is actually no need to add 1 twice. Identical results would be obtained if, for this case, nothing would be added to the $B-1$ complements. See also example 3, repeated). Use the simplified flow chart to solve problems 6 c, d, e.

8.1.1.2. Algebraic Addition and Subtraction: The method described in paragraph 8.1.1.1 allows us to perform addition and subtraction in a computer which has only adder circuits[1]. The method itself may seem not very attractive since the necessary operations are much more involved than the straightforward subtraction with subtracter circuits. However, before we discard the method, we have to take several things into consideration. First of all, subtracters in addition to adders will require a considerable amount of additional hardware. The hardware necessary to control the sequence of operations shown in Fig. 8.1 will probably be less expensive than additional subtracters. Secondly, we shall see that it is possible to simplify the flow chart of Fig. 8.1 considerably when computers represent negative numbers by their complement[2]. Most important of all, however, is the requirement that a computer must be able to perform algebraic additions and subtractions, i.e. add and subtract numbers of arbitrary sign. In order to do so, even computers which have both, adders and subtracters, must operate according to relatively complicated flow charts, so that the additional complication required for the subtraction by the addition of the complement becomes of minor concern.

While discussing the mechanization of algebraic additions and subtractions in computers, let us assume a binary machine. However, we shall always speak of true and $B-1$ complements (instead of 2's and 1's complements) so that it will be simple to adapt the given rules also for a decimal machine. Furthermore, let us clearly specify the number representation we refer to, since the procedures are quite different for the different representations.

Before we start, it may be well to list the three most commonly used representations for binary numbers in a table so that we can refer to it later. We will assume here a word-length of four bits. The state of the leftmost bit serves usually as a sign indication. Positive numbers are identified by a "zero" in this position, negative numbers by a "one". Positive numbers are normally represented by their magnitude but negative numbers can be represented either by (the sign and) the magnitude, or by their true complements, or by their $B-1$ complements.

The procedures for algebraic addition and subtraction using each of these number representations will be discussed below. The implications and advantages or disadvantages of the various specific number representations will then become apparent. As a very general observation we may state

---

[1] A modification of the original method enables us to add and subtract in a machine which has only subtracter circuits.

[2] Inspecting Fig. 8.1, we see that negative quantities are always complemented or re-complemented. This complementation or re-complementation is not necessary in a machine which represents all negative numbers by their complement.

here that the representation of negative numbers by sign and magnitude is closest to the familiar usage for paper and pencil calculations. The representation of negative numbers by their complements is more convenient for the internal operations of a computer which subtracts by the addition of the complement (or adds by the subtraction of the complement).

Table 8.1. *Table of Commonly Used Number Representations in Binary Computers*

Decimal Equivalent	Negative Numbers represented by		
	Magnitude	True Complement	B—1 Complement
+7	0111	0111	0111
+6	0110	0110	0110
. . . .	. . .	. . .	. . .
+2	0010	0010	0010
+1	0001	0001	0001
+0	0000	0000	0000
−0	1000		1111
−1	1001	1111	1110
−2	1010	1110	1101
. . .	. . .	. . .	. . .
−6	1110	1010	1001
−7	1111	1001	1000
−8	non-existent	1000	non-existent

It may be worthwhile to note that Table 8.1 shows by no means all possible binary number representations. For instance, one might consider the possibility to represent positive numbers by their complement and negative numbers by their magnitude.

Problem 12: Convince yourself that negative numbers and zeros are correctly represented in Table 8.1 according to the column headings. In particular, determine the decimal equivalent of the bit configuration 1100 for each of the three representations. Make your assignment not by letting the given configuration "fall into the pattern" of the table, but justify your choice in terms of the weights of bit positions and in terms of the particular true and B—1 complements.

Problem 13: Assume that the leftmost bit of the binary number 1 011 001 indicates the (—) sign. Show the representation of the corresponding positive quantity if 1 011 001 is a number represented:
  a) by the sign and the magnitude;
  b) by the sign and the 2's complement;
  c) by the sign and the 1's complement.

Table 8.2. *Rules for the Algebraic Addition of two Numbers, when Negative Numbers are Represented by Their Sign and Magnitude, and when Subtractions are Replaced by the Addition of the Complement*

Problem	$(+A)+(+B)$	$(+A)+(-B)$	$(-A)+(+B)$	$(-A)+(-B)$
Actual Operation	$A+B$	$A+\overline{B}$	$\overline{A}+B$	$\overline{(A+B)}$

(Note: $\overline{A}$ and $\overline{B}$ indicate here the true complements of A and B respectively)

If there is no End-Carry:

Sign of Result	+	—	—	—
The Sum represents	Magnitude of Result	True Compl. of Result	True Compl. of Result	Magnitude of Result
Corrective Action	None	Decomplement	Decomplement	None

If there is an End-Carry:

Sign of Result	+	+	+	—
The Sum represents	Magnitude of Result (overflowed)	Magnitude of Result	Magnitude of Result	Magnitude of Result (overflowed)
Corrective Action	Set overflow Alarm	None	None	Set overflow Alarm

Algebraic Addition of Two Numbers, when Negative Numbers are Represented by Sign and Magnitude: When we add two numbers algebraically, we may distinguish four different cases, dependent upon the sign of the operands. These four cases are listed in the four columns of Table 8.2. The second line of the table gives the most straightforward operations for the solution when numbers are represented by their sign and magnitude[1] and when all subtractions are performed by the addition of the true complement. (Table 8.2.)

The rules given for the first and last cases are obvious. The rules for the second case have been derived in paragraph 8.1.1.1. The rules for the third case are the same as the rules for the second case (in each case one operand is positive, and the other operand is negative).

Let us now translate these rules into a flow chart.

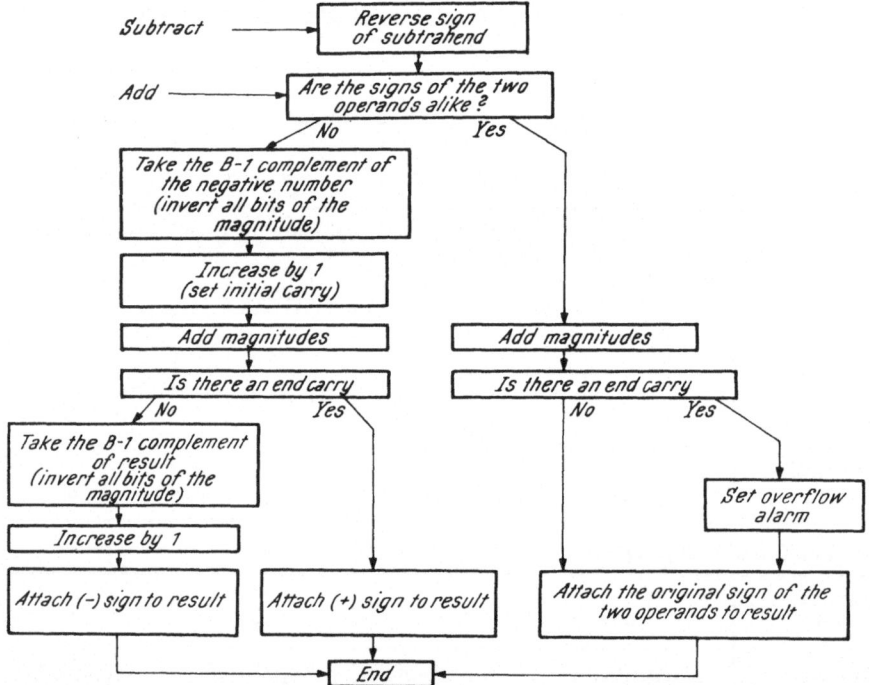

Fig. 8.2. Flow Chart for the Algebraic Addition and Subtraction of Two Numbers, when Negative Numbers are Represented by Their Sign and Magnitude, and when Subtractions are Replaced by Additions of the Complement

The given flow chart constitutes one of several possible interpretations of the rules defined in Table 8.2. Alternate flow charts may differ in the manner in which decisions are made, and in the arrangement of different paths for the different cases but, in one way or another, all operations

---

[1] See second column in Table 8.1.

shown in Fig. 8.2 have to be performed. It is to be noted that a computer may perform simultaneously several of the operations shown in separate boxes. For instance: It is possible to invert all bits in the (negative) operand and, simultaneously, to set the initial carry.

The flow chart in Fig. 8.2 includes algebraic subtraction. For this operation, the sign of the subtrahend is changed, and then the applicable paths for algebraic addition are followed.

Example: Add 1 011 to 0 010

$$
\begin{array}{c}
0\ 010\ (+2) \\
+1\ 011\ (-3) \\
\hline
\end{array}
\rightarrow
\begin{array}{c}
010 \\
100 \\
+\quad 1 \\
\hline
0)\ 111
\end{array}
\left\lceil
\begin{array}{c}
\rightarrow 000 \\
1 \\
\hline
-001
\end{array}
\right.
\rightarrow 1 \quad 001 \quad (-1)
$$

Example: Subtract 1 100 from 0 010

$$
\begin{array}{c}
0\ 010\ (+2) \\
-1\ 100\ (-4) \\
\hline
\end{array}
\rightarrow
\begin{array}{c}
0\ 010 \\
+0\ 100 \\
\hline
\end{array}
\rightarrow
\begin{array}{c}
010 \\
+100 \\
\hline
0)\ 110
\end{array}
\rightarrow 0\ 110\ (+6)
$$

Problem 14: Perform the following calculations by using the flow chart given in Fig. 8.2. Show all intermediate results.

a) Add 0 100 to 0 010
b) Add 0 100 to 1 010
c) Subtract 1 010 from 1 011
d) Subtract 1 011 from 0 000
e) Add 0 101 to 0 110
f) Add 1 100 to 1 101
g) Subtract 0 100 from 0 010
h) Subtract 1 101 from 1 011
i) Subtract 0 011 from 0 011

Assume that the leftmost bit in above binary numbers represents the sign and the remaining three bits the magnitude (Second column in Fig. 8.2).

Problem 15 (Voluntary): Use the flow chart in Fig. 8.2 to calculate the sums and differences of the following decimal numbers.

a)   17+32
b)   17—32
c)   32+17
d)   32—17
e)   86+45
f) —36—29
g) —76—29

Algebraic Addition of Two Numbers, when Negative Numbers are Represented by Sign and True Complement; The applicable rules are shown in Table 8.3.

Table 8.3. *Rules for the Algebraic Addition of Two Numbers, when Negative Numbers are Represented by Their Sign and True Complement, and when Subtractions are Replaced by Additions of the Complement*

Problem	$(+A)+(+B)$	$(+A)+(-B)$	$(-A)+(+B)$	$(-A)+(-B)$
Actual Operation	$A+B$	$A+\bar{B}$	$\bar{A}+B$	$\bar{A}+\bar{B}$

(Note: $\bar{A}$ and $\bar{B}$ indicate here the true complements of A and B respectively)

If there is no End-Carry:

Sign of Result	$+$	$-$	$-$	$-$
The Sum represents	Magnitude of Result	True Compl. of Result	True Compl. of Result	True Compl. of Result (overflowed)
Corrective Action	None	None	None	Set Overflow Alarm

If there is an End-Carry:

Sign of Result	$+$	$+$	$+$	$-$
The Sum represents	Magnitude of Result (overflowed)	Magnitude of Result	Magnitude of Result	True Compl. of Result
Corrective Action	Set Overflow Alarm	None	None	None

The corrective actions listed for the second and third case are different from those given in Table 8.2 (A decomplementation of negative results is neither necessary, nor desired). The rules for the fourth case have very little similarity to those of Table 8.2. The reason is that we compute here $\overline{A+B}$ instead of $A+B$. This is preferable since the two operands are represented by their complements. The correctness of the rules listed for this case is most easily verified by calculating a few numerical examples[1].

---

[1] One can also show that $(-A)+(-B)=(2-A)+(2-B)=\bar{A}+\bar{B}$. (mod. 2)

Before we attempt to draw the applicable flow chart, it is worthwhile to note some peculiarities of the listed operations. The addition of operands by itself gives the complement of the result when the result is negative. A comple-mentation or decomplemen-tation of numbers is, there-fore, not required. One can use different approaches to de-termine the sign of the result. The simplest approach is probably a binary "addition" of the sign bits of the two operands and of the end-carry from the arithmetic addition. We can see from Table 8.3 that such an "arithmetic" addi-tion of the signs (0's for positive operands and 1's for negative operands) generates the proper sign of the result for all cases, except when an

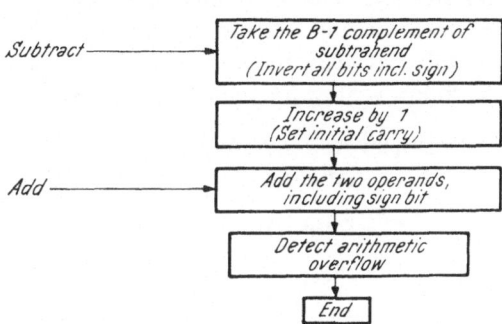

Fig. 8.3. Flow Chart for the Algebraic Addi-tion and Subtraction of Two Numbers, when Negative Numbers are Represented by Their Sign and True Complement, and when Sub-tractions are Replaced by Additions of the Complement

arithmetic overflow occurs. If we disregard for the moment the detailed operations necessary to detect such an overflow and are satisfied with the wrong sign of overflowed numbers (which, in any case, are not very meaningful as they are), we can draw the flow chart shown in Fig. 8.3.

Let us now apply the flow chart to two numerical examples.

Examples: Add 1 101 to 0 010   Add 1 010 to 0 111

$$\begin{array}{ll} 0\ 010 & (+2) \\ +1\ 101 & (-3) \\ \hline 1\ 111 & (-1) \end{array} \qquad \begin{array}{ll} 0\ 111 & (+7) \\ +1\ 010 & (-6) \\ \hline 1)0\ 001 & (+1) \end{array}$$

Arithmetic overflows can be detected in several ways[1]. For instance, we can see from Table 8.3 that an overflow occurs if both operands are positive and there is an end-carry, or if both operands are negative and there is no end-carry[2]. Alternately, an overflow is to be indicated if both operands are positive and the sum is negative, or if both operands are negative and the sum is positive. (An equivalent statement is the following: the accumu-lator changes from a positive to a negative quantity if a positive quantity

---

[1] Not all computers do, however, detect arithmetic overflows. It is then entirely the programmers responsibility to assure that operands are scaled in such a fashion that overflows cannot occur.

[2] The end-carry, as defined here, is the carry *into* the sign position. The carry *from* the sign position is disregarded.

is added, or the accumulator changes from a negative to a positive quantity if a negative quantity is added). As a third possibility, an overflow has occured when the signs of the two operands are alike, but the sum has the opposite sign. (An equivalent statement is the following: The signs of the accumulator and augend register have been alike before the addition but are different after the addition). The actual design of the overflow detector will follow that approach which is most convenient under the circumstances.

Example: Addend (Accumulator)   1 010   (—6)
Augend (Augend Register) 1 001   (—7)

Sum (Accumulator)   0 011   (+3)   Overflow!

There has been an overflow, since either:

a) The operands are both negative but no carry has occured from the most significant numerical bit position into the sign position; or

b) The accumulator has changed from a negative sign to a positive sign while a negative quantity has been added; or

c) The accumulator and augend register contained numbers of like sign to start with, but contain different signs after the addition.

Problem 16: Perform the arithmetic operations specified in problem 14 a) through i). Use the flow chart given in Fig. 8.3 and assume that negative numbers are represented by their true complement (third column in Table 8.1).

Problem 17 (Voluntary): Repeat problem 15, using the flow chart in fig. 8.3.

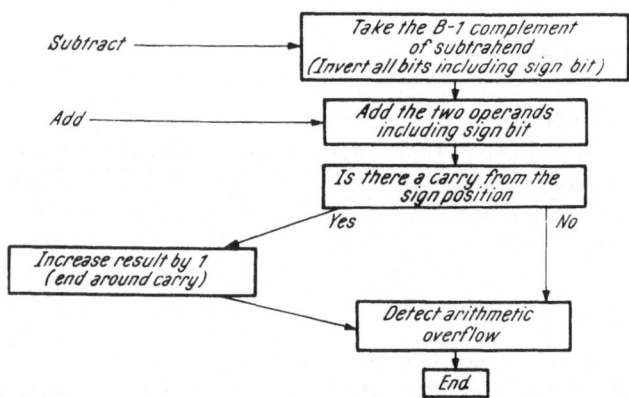

Fig. 8.4. Flow Chart for the Algebraic Addition and Subtraction of Two Numbers, when Negative Numbers are Represented by their Sign and $B—1$ Complement, and when Subtractions are replaced by Additions of the Complement

Algebraic Addition of Two Numbers, when Negative Numbers are Represented by Sign and B—1 Complement: The applicable rules are shown in Table 8.4.

Table 8.4. *Rules for the Algebraic Addition of Two Numbers, when Negative Numbers are Represented by their Sign and B—1 Complement, and when Subtractions are Replaced by Additions of the Complement*

Problem	$(+A)+(+B)$	$(+A)+(-B)$	$(-A)+(+B)$	$(-A)+(-B)$
Actual Operation	$A+B$	$A+\overline{B}$	$\overline{A}+B$	$\overline{A}+\overline{B}$

(Note: $\overline{A}$ and $\overline{B}$ indicate here the B—1 complements of A and B respectively)

If there is no End-carry:

Sign of Result	$+$	$-$	$-$	$-$
The Sum represents	Magnitude of Result	B—1 Complement of Result	B—1 Complement of Result	B—1 Complement (overflowed) (decreased by 1)
Corrective Action	None	None	None	Increase by 1, Set overflow alarm

If there is an End-Carry:

Sign of Result	$+$	$+$	$+$	$-$
The Sum represents	Magnitude of Result (overflowed)	Magnitude of Result (decreased by 1)	Magnitude of Result (decreased by 1)	B—1 Complement (decreased by 1)
Corrective Action	Set Overflow Alarm	Increase by 1	Increase by 1	Increase by 1

The results of the additions are rather similar to those shown in Table 8.3. That is, the signs of the result and overflows can be detected in the same manner. Also, negative results are represented by a complement of sorts. Here, however, arithmetic operations are required, in some cases, to generate the correct magnitude (for positive results) or the correct B—1 complement (for negative results). The logic required for this correction is rather straightforward. We can deduce from Table 8.4 that the result of the addition is to be increased by 1 in all cases where the binary addition produces a carry in the sign position (two negative numbers or one negative number and an end carry into the sign position). We then simply take this carry and add it as "end-around carry" into the least significant bit position. The resulting flow chart is shown in Fig. 8.4.

It is to be noted that the addition of the end-around carry into the least significant bit position can be performed simultaneously with the addition of the two operands, at least, when parallel adders are used.

Let us show the propagation of the end-around carry in a numerical example:

Example:

$$
\begin{array}{ll}
1\ 110 & (-1) \\
+1\ 101 & (-2) \\
\hline
1)\ 1\ 011 & \\
\vdash\!\!\longrightarrow 1 & \\
\hline
1\ 100 & (-3)
\end{array}
$$

Problem 18: Perform the arithmetic operations specified in problem 14a) through i). Use the flow chart given in Fig. 8.4 and assume that negative numbers are represented by their B—1 complement (fourth column in Table 8.1).

Problem 19 (Voluntary): Repeat problem 15, using the flow chart in fig. 8.4.

Comparing the procedures for algebraic addition and subtraction, we can find advantages and disadvantages for each of the three number representations. However, the fact that all three representations are in practical use indicates already that none of them has truly outstanding characteristics.

The representation of negative numbers by sign and magnitude requires slightly more hardware due to the somewhat more complex flow chart. On the other hand, the fact that this representation is most similar to the notation we are used to, can be considered an advantage. However—at least for programming—this advantage becomes less and less important with the modern trend to use symbolic notations. Perhaps one could say with some justification that at least decimal computers (which go so far as to use the electronically more inconvenient decimal notation for their internal operations in order to be more compatible with humans) should use also the representation by sign and magnitude for negative numbers.

The representation of negative numbers by their true, or B—1 complements requires approximately the same amount of hardware. A slight advantage of the B—1 notation is the easier complementation of quantities. The true complement notation, instead, is more natural for counting from negative to positive quantities or vice versa. The distinction between negative and positive zeros may or may not be a desirable property of the B—1 notation in a particular instance.

Problem 20 (Voluntary): Design the logic diagrams of parallel adders for algebraic addition and subtraction for each of the three commonly used number representations. Assume a word-length of four bits, including the sign bit. Use the two-step addition scheme discussed in paragraph 6.3.1. Show the sequence of control signals to be applied. Provide for overflow detection.

8.1.1.3. High-Speed Addition Techniques: The speed of parallel adders is limited by the carry propagation time. A carry which is generated in the least significant digital position may have to propagate through all remaining positions before the final result of the addition is available. Normally, we have to allow sufficient time for the worst case before we assume that a result is valid. We speak of ripple carry adders or simply of ripple adders.

Efforts to reduce the time from the beginning of an addition until it is certain that the addition is complete. follow two lines of attack. The first approach speeds up the carry propagation by minimizing the number of circuit elements in the propagation path. We speak then of carry by-pass, or carry look-ahead adders. The second approach indicates in each individual case when an addition is complete. Here, then we have to wait only as long as actual carries are propagating and not as long as the worst case may take. We speak of completion recognition adders.

The principle of carry by-pass adders is shown in Fig. 8.5.

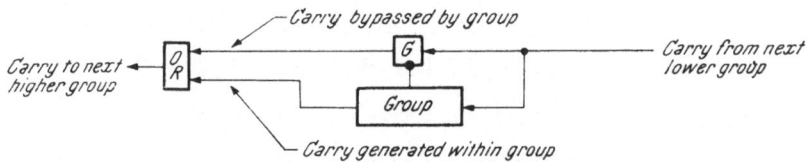

Fig. 8.5. Principle of a Carry By-Pass Adder

Several digital positions are combined to a "group". The group has a single carry by-pass gate so that a carry which is generated by a lower order group and which has to be propagated to a higher order group has to pass only one level of logic circuits, instead of being propagated through the individual stages of the group.

The conditions for by-passing the carry by a group are not difficult to derive. An individual binary position *stops* a propagating carry when both operands are "zeros". A binary position in which both operands are "ones" *generates* its own carry. It is therefore not necessary to propagate an arriving carry beyond this position. A carry is *propagated* through a binary position only if one operand is a "one" and the other a "zero". A group has, therefore, to propagate a carry only if in each position within the group one operand is a "one" and the other a "zero". If we designate the individual stages within one group by the indices 0 through n, we obtain the following condition for the carry by-pass enable of the group:

"Carry by-pass enable" $=$
$$= (A_0 B_0' + A_0' B_0)(A_1 B_1' + A_1' B_1) \cdots (A_n B_n' + A_n' B_n) \tag{8.7}$$

The basic approach can be modified in several respects. Obviously, one can use small or large groups. The use of small groups reduces the time between the generation of a carry within a group and its appearance at the output of that group. With larger groups, one shortens the overall propagation time by reducing the total number of groups. The optimum group size depends upon the length of operands. One may also provide carry by-passes for groups of groups. This approach is a straightforward extension of the principle shown in Fig. 8.5. A group of groups enables a common by-pass gate if all group by-pass enables are present.

A different approach is the provision of a carry look-ahead for carries generated within a group. Suppose we have the stages 0, 1 and 2 within a group. A carry from this group has to be produced if stage 2 generates a carry, *or* if stage 1 generates a carry and stage 2 propagates it, *or* if stage 0 generates a carry and both, stage 1 and 2 propagate the carry. The following equation shows these conditions:

"Carry produced by the group" $=$
$$
\begin{aligned}
&= A_2 B_2 + A_1 B_1 (A_2 B_2' + A_2' B_2) + A_0 B_0 (A_2 B_2' + A_2' B_2)(A_1 B_1' \\
&+ A_1' B_1) = A_2 B_2 + A_1 B_1 A_2 B_2' + A_1 B_1 A_2' B_2 + A_0 B_0 A_2 B_2' A_1' B_1 \\
&+ A_0 B_0 A_2 B_2' A_1 B_1' + A_0 B_0 A_2' B_2 A_1 B_1' + A_0 B_0 A_2' B_2 A_1' B_1
\end{aligned} \tag{8.8}
$$

Clearly, the scheme can be extended to larger groups if desired. If we are willing to use enough circuitry, the carry from a group can be generated by two levels of logic circuits (AND's and OR's) no matter how large the group is.

The various discussed approaches can be combined, but the comparison of individual combinations is rather difficult. An evaluation should consider not only the speed and the cost of particular addition schemes but the increase in speed and cost for the overall computer system[1].

Problem 21: Design a carry by-pass for the two-step addition scheme discussed in paragraph 6.3.1. Assume a typical group of five binary positions.

---

[1] For a more detailed discussion of computer evaluations see chapter 10.

Problem 22: Consider the three low-order stages of a binary parallel adder. What is the minimum number of logic levels (not the minimum number of circuits) required to produce the three sum bits and the carry? What is the minimum number of logic levels required for n low-order stages? How many levels has a n-bit ripple adder?

Problem 23 (Voluntary): State the condition for a carry by-pass by a decade in an 8421 adder.

Let us now have a look at completion recognition adders. If each individual stage of an adder would provide a completion signal when the addition in that stage is complete, we could detect the actual end of an addition as is schematically indicated in Fig. 8.6.

Since the average "length" of a carry, i.e. the number of positions through which a carry is propagated, is much shorter than the total length of the adder, the average time for additions is greatly decreased.

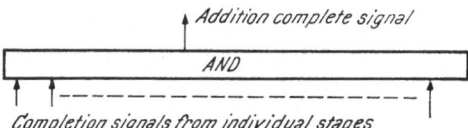

Fig. 8.6. Principle of a Completion Recognition Adder

A slight difficulty lies in the generation of the completion signal for the individual stage. The stage has to "know" whether or not a carry will arrive before it can "say" that it is ready. It must therefore wait until it receives a "carry" or a "no-carry" signal from the previous stage. Consequently, the adder must propagate a "no-carry" signal, in addition to the normal carry signal.

Let us restate what we already know about the carry signal. A carry is generated in positions, where both operands are "ones". It is propagated through positions where one operand is a "zero" and the other a "one". The propagation is stopped in positions where the operands are both "zeros" or both "ones". Let us now investigate the no-carry signal. Positions where both operands are "zeros" will neither generate nor propagate a carry. We can therefore generate in these positions a no-carry signal. We can propagate this signal through those positions to the left which contain one "zero" and one "one" as operands. We want to stop the propagation of the no-carry signal in positions where both operands are "ones" or "zeros", since these positions generate their own carry or no-carry signals.

Table 8.5. *Generation and Propagation of Carry and No-Carry Signals*

Carry	←←← ↑	←← ↑
Augend	1 0 1 1 0 1 0 0 0 1 1 0	
Addend	0 1 0 1 1 0 0 0 1 0 1 1	
No Carry		←← ↓ ↓

Table 8.5 shows the generation and propagation of "carry" and "no-carry" signals for a numerical example. Positions which generate a "carry" or "no-carry" signal are marked by vertical arrows, and positions which propagate these signals are marked by horizontal arrows. We see that all positions carry an arrow of some kind no matter what the operands are. The only exception to this rule is the least significant position. Following our previous line of thinking, we have to assume that an artificial external "no-carry" signal enters this position from the right so that it propagates a "no-carry" signal unless it generates a "carry" or "no-carry" signal of its own accord[1]. We may now consider an addition as a sequential process. Positions which carry vertical arrows generate immediately a "carry" or "no-carry" signal. The signals then ripple to the left through places where horizontal arrows are shown. If all positions are marked with either a carry or no-carry signal, the addition is complete.

The logic expression for the completion signal of an individual stage can be stated as follows:

$$\text{"Addition complete"} \quad = C_n + N_n, \tag{8.9}$$

$$\text{where: } C_n = A_n B_n + (A_n B_n' + A_n' B_n)\, C_{n-1}$$

$$\text{and: } N_n = A_n' B_n' + (A_n B_n' + A_n' B_n)\, N_{n-1}$$

C_n is the carry signal. N_n is the no-carry signal. We obtain a carry signal if both operands are "ones" $(A_n B_n)$ *or* if one operand is a "zero", the other a "one" $(A_n B_n' + A_n' B_n)$ *and* there is a carry signal from the next lower stage (C_{n-1}). Conversely we obtain a no-carry signal if both operands are "zeros" $(A_n' B_n')$ *or* if one operand is a "zero", the other a "one" $(A_n B_n' + A_n' B_n)$ *and* there is a no-carry signal from the next lower stage (N_{n-1}).

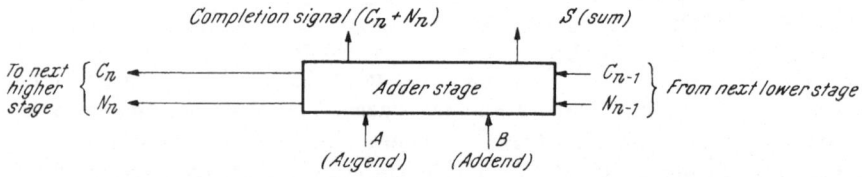

Fig. 8.7. Inputs and Outputs for One Stage of a Completion Recognition Adder

Equation (8.9) together with the expression for the sum (e.g. Equation 6.31) completely determines the circuitry of a stage of the binary completion recognition adder. Fig. 8.7 shows all inputs and outputs of one stage.

---

[1] Sometimes it is desired to enter an artificial carry (e.g. end-around carry, see paragraph 8.1.1.2.) from the right.

Problem 24: Design a logic diagram for a typical stage of a binary completion recognition adder.

Problem 25 (Voluntary): Try to estimate the speed increase which can be achieved by a completion recognition adder versus a ripple adder. Take the carry propagation time through one stage, T, as basis of your comparison.

Problem 26: Design a carry completion detector for the two-step addition discussed in paragraph 6.3.1.

In our discussion of the completion recognition adder we have, so far, been not quite exact. We have said that an addition is complete when each stage produces a carry or a no-carry signal. However, an addition is complete not when a carry signal is produced, but only after the carry has been added to the next higher stage (This stage by itself may generate an early carry or no-carry signal). Therefore, there is some danger that the carry complete signal may be produced slightly too soon. This can be corrected by delaying the "addition complete" signal by a small amount of time. Also, while new operands are being transferred to the arithmetic registers, there is the danger of the addition complete signal being present (indicating the completion of the previous addition). Another small time delay suppressing the completion signal at the beginning of the addition corrects this situation.

The principle of the completion recognition adder is well suited for asynchronous machines. Here, the next operation can be immediately started after the completion of the current operation. In synchronous machines, the next operation will be started on the first clock pulse[1] following the "addition complete" signal. Thus, part of the time saved in this scheme is lost waiting for the clock pulse so that the full potential is not realized. If the word-length is short, it probably will not pay to incorporate a completion recognition adder into a synchronous machine.

The principle of carry completion recognition can be combined with carry by-passes. It can also be applied to decimal adders, subtracters and counters. In addition to carry by-pass, carry look-ahead, and addition complete techniques, there exists a high-speed "carry save" scheme. This addition technique is specifically suited for multiplication and is described in connection with high-speed multiplication techniques[2].

---

[1] In some designs, two clock pulses are required to re-synchronize an asynchronous signal.

[2] See paragraph 8.1.2.3.

8.1.2. Multiplication

8.1.2.1. Binary Multiplication: Computers normally perform a multiplication by a series of additions. However, a number of variations are in use. Details depend upon the number system and the number representation used, whether only single-length or also double-length products are to be derived, whether integers or fractions are to be multiplied, and

Table 8.6. *Example of a Binary Paper and Pencil Multiplication*

```
            1 0 1 1 0              multiplicand
   ×        1 0 0 1 1              multiplier
            _____
            1 0 1 1 0              1st partial product
          1 0 1 1 0                2nd partial product
        0 0 0 0 0                  3rd partial product
        0 0 0 0 0                  4th partial product
    + 1 0 1 1 0                    5th partial product
    _____
    1 1 0 1 0 0 0 1 0              product
```

Table 8.7. *Accumulation of Partial Products*

```
            1 0 1 1 0              multiplicand
   ×        1 0 0 1 1              multiplier
   _____
   0 0 0 0 0 0 0 0 0 0            (accumulator at the beginning)
        + 1 0 1 1 0               1st partial product
   _____
   0 0 0 0 0 1 0 1 1 0            (accumulator after addition of the first partial
                                  product)
        + 1 0 1 1 0               2nd partial product
   _____
   0 0 0 1 0 0 0 0 1 0            (accumulator after addition of the second
                                  partial product)
      + 0 0 0 0 0                 3rd partial product
   _____
   0 0 0 1 0 0 0 0 1 0            (accumulator after addition of the third
                                  partial product)
    + 0 0 0 0 0                   4th partial product
   _____
   0 0 0 1 0 0 0 0 1 0            (accumulator after addition of the fourth
                                  partial product)
 + 1 0 1 1 0                      5th partial product
   _____
   0 1 1 0 1 0 0 0 1 0            product (accumulator after addition of the
                                  fifth partial product)
```

whether the computer is a parallel or a serial machine. Moreover some of the schemes are especially designed for high-speed multiplication. Let us begin here with the straightforward paper and pencil multiplication of two binary numbers. Table 8.6 shows the conventional computation for a specific numerical example.

The product is the sum of the partial products. In this example, we add all five partial products in one step. However, since it is easier for machines to add only two numbers at a time, let us re-write the same computation in a slightly different form as shown in Table 8.7.

In this approach, we "accumulate" the partial products individually toward the final result. We know that the partial products are either equal to zero or equal to the multiplicand. We can, therefore, state the following rule for this type of multiplication: "Inspect individual bits of the multiplier. Add the multiplicand into the accumulator for a "1" in the multiplier, but do not add for a "0". Shift the multiplicand for each step by one bit as in a conventional multiplication."

Let us now try to find an implementation of this rule. To start with, assume that we have a multiplicand register, a multiplier register and an accumulator. Suppose the multiplicand register and the multiplier register are shift registers as indicated in Fig. 8.8.

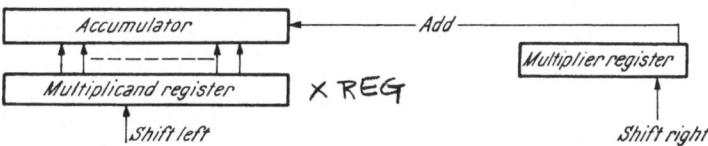

Fig. 8.8. Functional Layout of a Simple Multiplication Scheme

The multiplicand register and the accumulator are double-length registers. The multiplicand is placed into the right half of the multiplicand register and the accumulator is cleared at the beginning of the multiplication. If the least significant position in the multiplier register contains a "one", the contents of the multiplicand register are added to the contents of the accumulator. The accumulator contains now the first partial product. We now shift the contents of the multiplicand register one bit to the left (shifting a zero into the vacated position) and the contents of the multiplier register one bit to the right. Again, we add the contents of the multiplicand register to the accumulator if the least significant position in the multiplier register contains a "one". By repeating this procedure, we multiply two n bit numbers by performing n additions and obtain a $2n$-bit result in the accumulator. We speak of a double-length product.

The indicated layout is a direct implementation of the computation indicated in Table 8.7, and results not necessarily in the simplest implementation. We notice, for instance, that the multiplicand register is a double-length register, but that it holds only a single-length operand. This suggests the arrangement shown in Fig. 8.9 which requires only a single-length register.

Here, the contents of the accumulator, rather than the contents of the multiplicand register are shifted. The first partial product is added into the left half of the accumulator, but will finally end up in the right half due to the repeated shifts during the multiplication. The arrangement can be varied in a number of ways.

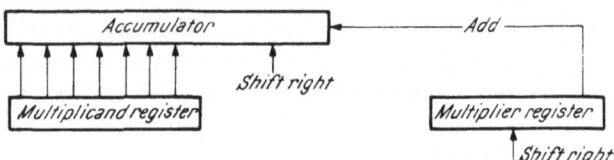

Fig. 8.9. Alternate Multiplication Scheme

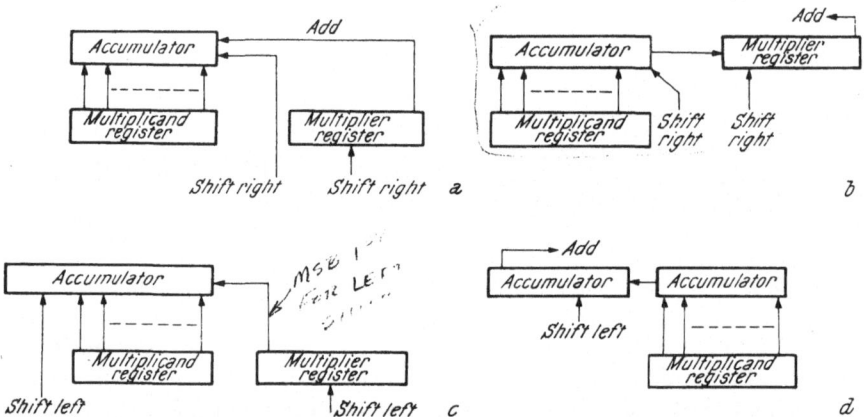

Fig. 8.10. Variations of the Multiplication Scheme Shown in Fig. 8.9

Fig. 8.10a is identical to Fig. 8.9 except that only a single-length accumulator is used. Therefore, only a single-length product (the most significant half) can be obtained. The arrangement given in Fig. 8.10b allows a double-length product to be derived in spite of the fact that only a single-length accumulator is used. The least significant part of the product is shifted into the multiplier register while the multiplier is shifted out during the process of multiplication. Fig. 8.10c shows an arrangement where the most significant partial product is added first and then shifted to the left. This requires the expense of a double-length accumulator (to propagate carries through high order positions). However, once incorporated, the double-length accumulator may provide the capability of double-length additions or subtractions at practically no cost. The arrangement given in Fig. 8.10d is equivalent to that in Fig. 8.10b except that the left half of the accumulator is used to store the original multiplier.

Problem 27: Show the contents of all registers during the consecutive steps of a multiplication for each of the arrangements shown in Figs. 8.8, 8.9 and 8.10. Assume the multiplicand to be 0 101 and the multiplier to be 1 001.

Problem 28: What is the rule for rounding a double-length product to a single-length number?

Problem 29: Additions can produce overflows. What can you propose to do about overflows during a multiplication?

Problem 30: How long does it take to multiply two 36-bit numbers? Assume a clock rate of 1 Mc and the two-step addition scheme discussed in paragraph 6.3.1.

Probably each one of the shown arrangements has been implemented in one or another computer. None of the arrangements possesses any fundamental advantage or disadvantage. The selection of a particular scheme is based on its compatibility with other aspects of the overall system. In any event, we notice that multiplication requires only slightly more hardware than addition. If an arithmetic unit is already layed out for addition, essentially only one more register, — the multiplier register — and provisions for shifting are required.

The schemes for multiplication given in Figs. 8.9 and 8.10 can be applied to both, parallel and serial designs. Parallel multipliers are fairly straightforward implementations of the schemes as they are shown. The design of serial multipliers requires a careful consideration of timing signals so that all bits of the operands are operated upon in the proper sequence.

Fig. 8.11 may serve as a representative functional diagram of a serial multiplier:

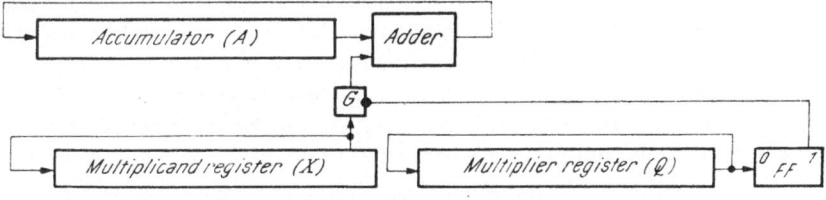

Fig. 8.11. Functional Diagram of a Serial Multiplier

All three arithmetic registers are recirculating registers. The contents of the multiplicand register can be added serially to the contents of the accumulator. A flip-flop has been provided for the storage of one bit of the multiplier. (We notice that a multiplier bit has to be available for a full word time, but that this multiplier bit stays for only one bit time in a specific position of the recirculating multiplier register).

Let us now study the requirements for shifting (not recirculating) the contents of the arithmetic registers. If we want to implement a scheme equivalent to the one given in Fig. 8.10a, both, the accumulator and the multiplier register must have provisions for a right shift. From what we have found in paragraph 6.1.3, we know that we have to "shorten" a register by one bit position in order to perform such a shift. Let us indicate this requirement in Fig. 8.12.

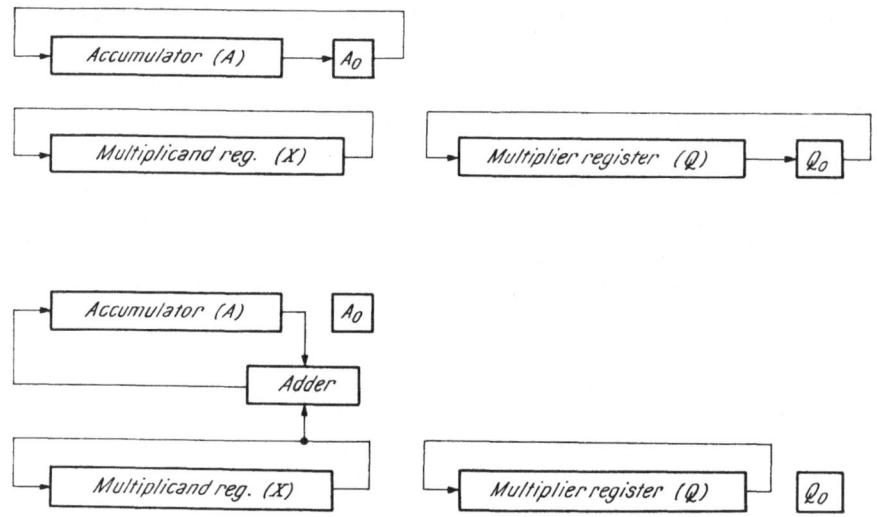

Fig. 8.12. Recirculation and Shifting

The top half of the diagram shows all three registers in their normal recirculating configuration. The right-most bit positions of the accumulator and the multiplier register are shown separately. The bottom half of Fig. 8.12 shows the registers during multiplication. The recirculation loops of the accumulator and the multiplier register have been "shortened" by one bit, so that their contents are shifted right one bit during each word time.

Figs. 8.11 and 8.12 tell us quite a bit about the logic circuitry we have to provide, but it is not yet clear just when we should switch from one configuration to the other. In order to find the exact timing conditions, let us construct a flow chart or a time sequence diagram for individual bits as is shown in Fig. 8.13.

We assume 4-bit words. A word time (W) consists, therefore, of four bit times, (labelled P_0, P_1, P_2, P_3, on the left hand margin). The four individual bits of the multiplicand (MC_0, MC_1, MC_2, MC_3) recirculate through the individual storage locations of the multiplicand register (X_0, X_1, X_2, X_3) as it is indicated in the column labelled "multiplicand". The bits of the multiplicand appear in sequential form on the output of the register (X_0)

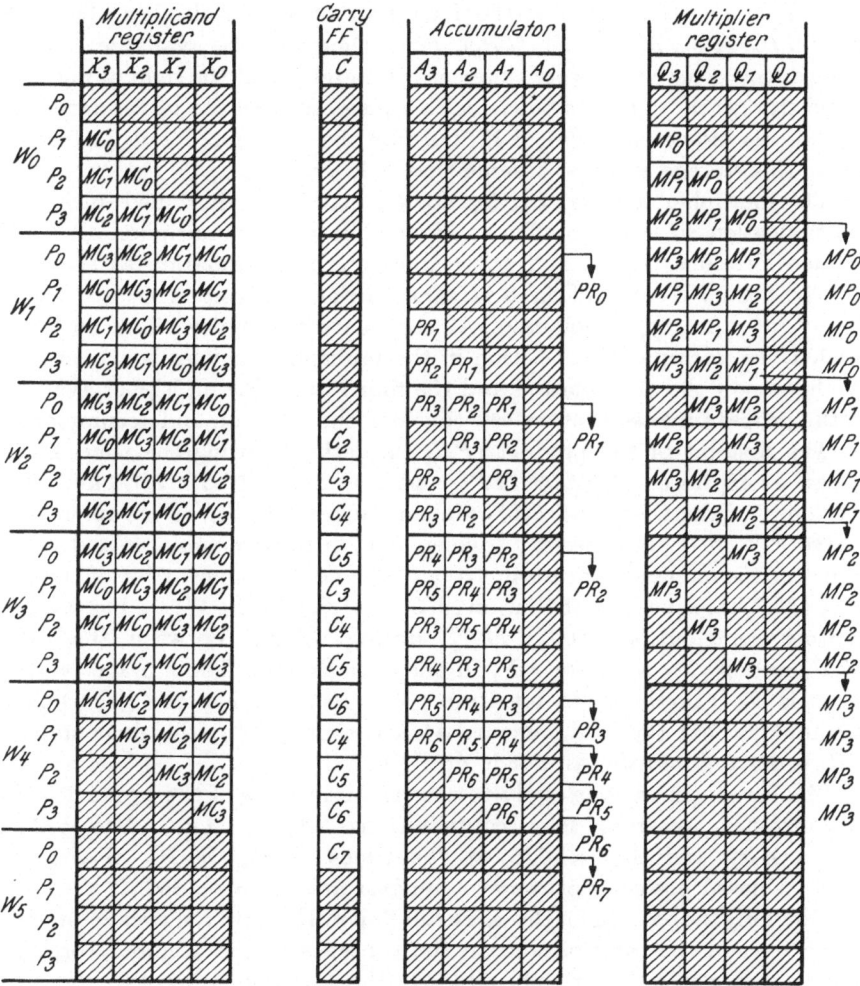

Fig. 8.13 Minimum Storage Requirements During the Multiplication

so that the least significant bit (MC_0) appears at the time P_0, the next significant bit (MC_1) appears at the time P_1, etc. Fig. 8.13 shows only four recirculation cycles (corresponding to the four times an addition of the multiplicand has to be performed during the multiplication of four-bit numbers). Only the minimum length of time during which the multiplicand has to stay in the register has been shown, although the listing could be easily extended. While the multiplicand appears at the output of the multiplicand register four times during the word times W_1, W_2, W_3, and W_4, it may be added or not be added to the contents of the accumulator depending upon the state of the four bits of the multiplier. The four bits (MP_0, MP_1, MP_2, and MP_3), must therefore, be available continuously during consecutive word times, as it is indicated at the right of Fig. 8.13. Some storage must be provided for this purpose, perhaps in the form of a flip-flop as indicated in Fig. 8.11, but, for the moment, let us not worry about this storage and just assume that it is available.

The multiplier register (Q_3, Q_2, Q_1, Q_0) has to be shortened during the multiplication so that we effect a right shift of the multiplier. Only the storage locations Q_3, Q_2 and Q_1 recirculate the information as shown in the right hand column of fig. 8.11. The entries for position Q_0 are shaded to indicate that Q_0 is not used. As soon as a multiplier bit is no longer needed, its corresponding position in the multiplier register is also shaded. We have now the multiplicand and the four bits of the multiplier available at the proper times so that we can consider the addition and accumulation.

The first bit of the product (PR_0) is derived during the time W_1P_0. It can be expressed as $MC_0 \times MP_0$, and is a "one", if both MC_0 and MP_0 are "ones". Otherwise it is "zero". Since no further additions are required to derive PR_0, it is not necessary to keep PR_0 in the accumulator and we assume for the moment that we put it out. The next significant bit of the product (MP_1) receives a contribution ($MC_1 \times MP_0$) during W_1P_1, but receives another contribution ($MC_0 \times MP_1$) during W_2P_0. Between the times W_1P_1 and W_2P_0, we store PR_1 in the accumulator. The storage requirement is for exactly three bit times. We observe that the "shortened" accumulator fits the requirement. At the time W_2P_0, the final state of PR_1 is derived and we can put it out.

Similar considerations apply to the remaining bits of the product. PR_3, for instance, receives four contributions: $MC_3 \times MP_0$, $MC_2 \times MP_1$, $MC_1 \times MP_2$, and $MC_0 \times MP_3$ at the times W_1P_3, W_2P_2, W_3P_1, and W_4P_0. Unused positions of the accumulator are again shaded.

Now only the column for the carry flip-flop in Fig. 8.13 requires comments. When we add the multiplicand to the contents of the accumulator, there may be carries which have to be added to the next higher bit of the product. The entries in the column for the carry flip-flop indicate the position of the product into which the carry has to be added. For instance, during the time W_2P_0, the product bit PR_1 is derived. A possible carry (C_2) into the product position PR_2 is set into the carry flip-flop, so that during the time W_2P_1 all three PR_2, MC_1, and C_2 are available in order to derive the current state of PR_2.

Fig. 8.13, as is, shows the minimum storage requirements for the three arithmetic registers during a multiplication of four-bit numbers. We may now make use of the shaded areas as we see fit in order to simplify the circuitry, conserve additional storage space, or improve the operation.

Let us begin with the multiplicand register. There is nothing we can do during the multiplication itself; all positions are used. At the beginning, during the time W_0, we may load the multiplicand into the register. However, we can do this also during some earlier time, but without knowing more about the design of the remainder of the machine, it is not possible to make a sensible decision. At the end, during the times W_4 and W_5, it may be desirable to fill the register with "zeros", i.e., to clear the register in preparation for some new operation.

There are a number of things we can do with the multiplier register. We notice, for instance, that the position Q_0 is not at all used during the multiplication, but that, on the other hand, we need an additional storage location for one bit of the multiplier, as indicated to the right. The obvious solution is, of course, the use of Q_0 for the storage of the currently used bit of the multiplier. We notice also that more and more positions within the train of information recirculating through Q_1, Q_2, Q_3 become available during the multiplication. On the other hand, we need positions to store the bits of the product as they are derived. An ideal solution would be the shift of the four least significant bits of the product into the multiplier register, and to keep the four most significant bits of the product in the accumulator. In effect, we would then implement the scheme indicated in Fig. 8.10b. Unfortunately, there is no position in the multiplier register available at the time PR_n is derived, so that we have to provide a temporary storage somewhere else.

We notice that the position A_0 in the accumulator is not used, so we can use it for this purpose. In effect, we can use A_0 for the storage of PR_0 during the times W_1P_1, W_1P_2, W_1P_3, and then transfer PR_0 into the position Q_3 which is empty at the time W_2P_0. In a similar manner, we can use A_0 to store PR_1, PR_2 and PR_3 until a second, third, and fourth position in Q becomes available.

As far as the remainder of the accumulator is concerned, we should retain the bits PR_4, PR_5, PR_6, and PR_7. Thus the most significant half of the product is contained in the accumulator after a multiplication. Also, in order to simplify the required circuitry, we ought to assume that the accumulator is cleared at the beginning of a multiplication.

Let us reflect all these ideas in Fig. 8.14.

The timing corresponds to that of Fig. 8.13. At the beginning of the multiplication, that is at the time W_1P_0, multiplicand and multiplier registers contain the operands in their natural order. The accumulator is cleared. The multiplicand recirculates during three word times and is replaced by "zeros" during W_4. Individual bits of the multiplier recirculate up to three word times in the shortened multiplier register. During W_4, only the three least significant bits of the product recirculate in the shortened multiplier register. During W_5 (and possibly during later word times),

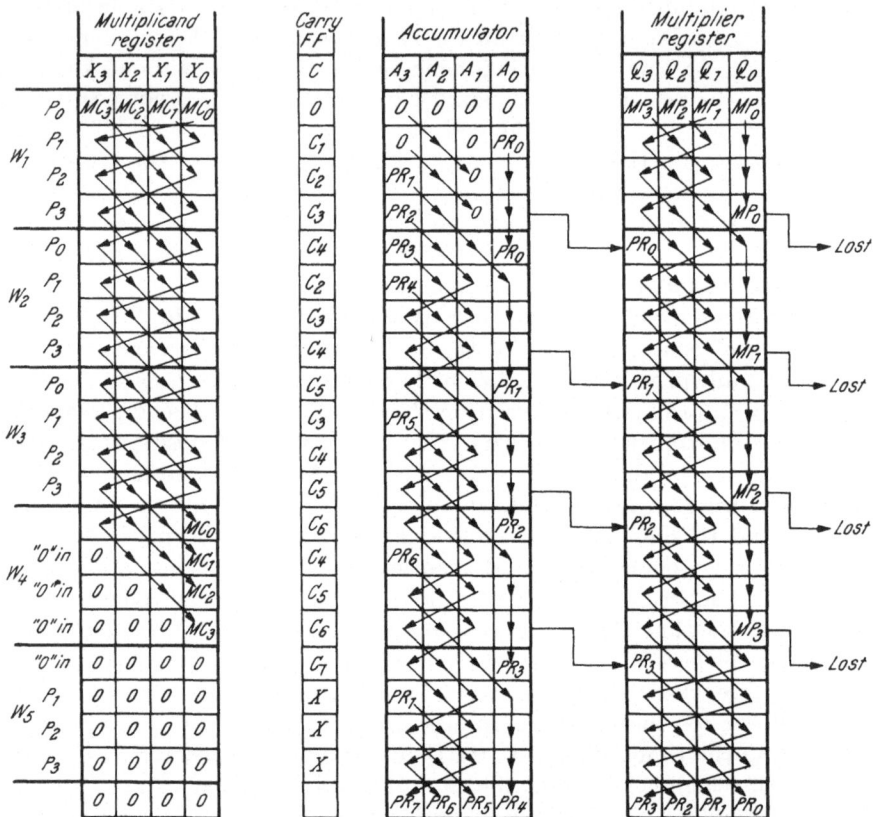

Fig. 8.14. Actual Use of the Available Storage During a Multiplication

the four least significant bits of the product recirculate in the full-length multiplier register. The accumulator is shortened during five word times corresponding to five right shifts. Four of these shifts (during W_1, W_2, W_3, W_4) are required to derive the individual bits of the product. The fifth one is required to let the four most significant bits of the product appear in their natural order at the beginning of a word time (W_6P_0).

Fig. 8.14 completely defines the use of the storage elements during the multiplication. As an exercise, let us derive from it the logic equations for the required circuitry.

Suppose we begin with X_3. The state of X_3 follows the state of X_0 during W_1, W_2, W_3, and W_5 but is to be set to zero during W_4. We can express this as follows:

$$_1x_3 = X_0(W_1 + W_2 + W_3 + W_5) \tag{8.10}$$

$$_0x_3 = X_0'(W_1 + W_2 + W_3 + W_5) + W_4$$

The remaining three positions of the multiplicand register behave in the manner of a shift register:

$$_1x_2 = X_3 \; ; \quad _1x_1 = X_2 \; ; \quad _1x_0 = X_1 \; ; \tag{8.11}$$

$$_0x_2 = X_3' ; \quad _0x_1 = X_2' ; \quad _0x_0 = X_1' ;$$

Similarly, we can derive for the multiplier register:

$$_1q_3 = Q_1(W_1 + W_2 + W_3 + W_4)P_3' + A_0(W_1 + W_2 + W_3 + W_4)P_3 + Q_0W_5 \tag{8.12}$$

$$_0q_3 = Q_1'(W_1 + W_2 + W_3 + W_4)P_3' + A_0'(W_1 + W_2 + W_3 + W_4)P_3 + Q_0'W_5$$

$$_1q_2 = Q_3 \quad _1q_1 = Q_2 \tag{8.13}$$

$$_0q_2 = Q_3' \quad _0q_1 = Q_2'$$

$$_1q_0 = Q_1(W_1 + W_2 + W_3 + W_4)P_3 + Q_1W_5 \tag{8.14}$$

$$_0q_0 = Q_1'(W_1 + W_2 + W_3 + W_4)P_3 + Q_1'W_5$$

Let us now derive the applicable expression for the adder. One of the inputs is A_1, another is X_0Q_0 (we add X_0 only if the multiplier bit in Q_0 is a "one"). The third input, C (the carry), is to be added, except during the times P_0. At this time, the least significant bit of the multiplicand is to be added into the accumulator (there is no carry into this position), but the carry flip-flop contains a carry into a higher order bit. The sum output of the adder becomes consequently:

$$S = [A_1(X_0Q_0)' + A_1'(X_0Q_0)]P_0 + \tag{8.15}$$

$$+ [A_1(X_0Q_0)'C' + A_1'(X_0Q_0)C' + A_1'(X_0Q_0)'C + A_1(X_0Q_0)C]P_0'$$

The input equations to the carry flip-flop become:

$$_1c = A_1(X_0Q_0) \tag{8.16}$$

$$_0c = A_1'(X_0Q_0)' + [A_1(X_0Q_0)]'P_0$$

The second term in the equation for $_0c$ clears the carry flip-flop during P_0, unless both operands are "ones". (Remember that a new cycle of addition starts at P_0 and previous carries have to be cleared out, unless a new carry is generated.) The input to the accumulator follows the sum output of the adder, S, except for the time P_0 when the sum is recorded in A_0 and only the previous carry is recorded in A_3:

$$_1a_3 = S(P_1 + P_2 + P_3) + CP_0 \qquad (8.17)$$
$$_0a_3 = S'(P_1 + P_2 + P_3) + C'P_0$$

A_2 and A_1 behave in the manner of a shift register:

$$_1a_2 = A_3 ; \qquad _1a_1 = A_2 ; \qquad (8.18)$$
$$_0a_2 = A_3' ; \qquad _0a_1 = A_2' ;$$

The equations for A_0 become:

$$_1a_0 = SP_0 \qquad (8.19)$$
$$_0a_0 = S'P_0$$

Equations (8.15) through (8.19) apply during the five word times: W_1, W_2, W_3, W_4, W_5. We note that a true addition is required only for four word times, but the split-timing would unnecessarily complicate all expressions. We may apply the equations during five word times since the accumulator is cleared when we start, and since the multiplicand register is cleared during the time W_5.

Let us now represent above the equations by a simplified logic diagram:

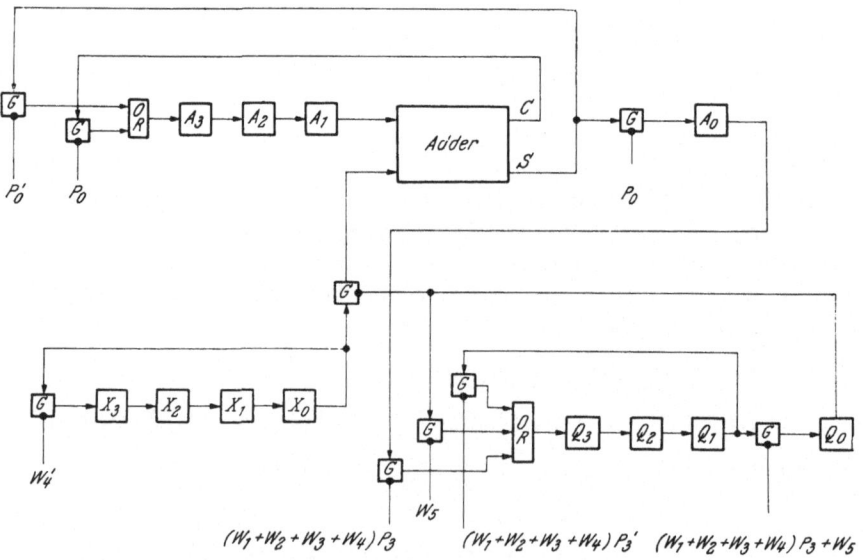

Fig. 8.15. Simplified Logic Diagram for the Serial Multiplier

The diagram is simplified in two respects: Only one line for the transfer of both, "zeros" and "ones" is shown, and clock signals effecting transfers are omitted.

Problem 31 (Voluntary): Assume the multiplicand to be 0111, and the multiplier to be 0101. Show the actual contents of all registers and of the carry flip-flop . Follow the diagram given in Fig. 8.14.

Problem 32: How many clock cycles does it take to multiply two 36-bit numbers in a serial multiplier. Calculate the number of cycles from the time when all operands are available until the time when the result in its final form becomes available.

What we have derived is one possible implementation of a serial multiplier. However, we can think of many variations. First of all, the multiplier as it stands is an implementation of the scheme given in Fig. 8.10b, but we might just as well have implemented one of the other schemes given in Fig. 8.8, 8.9, or 8.10. Even if we restrict ourselves to the scheme of Fig. 8.10b, we may obtain different solutions depending upon the exact number representation. The bit position to the left of the most significant position may be an empty position, an overflow position, or a sign position. Depending upon the actual use, further simplifications (or complications) are possible. Another possible requirement which we have not considered is that for algebraic multiplication. Depending upon the number representation we may have to remove the sign (or mask it out) or complement numbers if they have a negative sign. Furthermore, if our computer contains only subtracters and no adders, we may have to reduce multiplications to a repeated subtraction of the complement rather than to repeated additions.

In conclusion of this paragraph on binary multiplication, a few remarks on the operational differences between integral and fractional machines are in order. The multiplication of two single-word operands produces, in general, a double-length product. If such a result is desired, the most significant and the least significant halves of the product appear frequently in two separate words in the following form:

$$\boxed{\pm\, XX \cdots X} \qquad \boxed{\pm\, XX \cdots X}$$

The sign of the least significant half duplicates the sign of the most significant half. Since the bits of the two halves are arithmetically contiguous when they are derived, a position in the second word must be vacated before the sign can be inserted. This can be accomplished by the additional shift of one of the halves, or by connecting "around" the sign position during the accumulation and the repeated shifting of the product while it is derived.

The applicable procedures are identical for integral and fractional machines. If, however, the sign is not repeated in the least significant half, the placement of the bits in the product is different. In an integral machine, the least significant bit of the product (having the weight 1) appears in the least significant position of the least significant word:

$$\boxed{\pm OX \cdots X} \qquad \boxed{XXX \cdots X.}$$

In a fractional machine, the most significant bit of the product (having the weight $1/2$) appears to the right of the sign position:

$$\boxed{\pm . XX \cdots X} \qquad \boxed{XX \cdots XO}$$

We see, the final placement of the products is different by one bit position. Integral and fractional machines require, therefore, slightly different control sequences.

If only a single-length result is desired, fractional machines use the most significant bits of the double-length product arranged as follows:

$$\boxed{\pm . XX \cdots X}$$

Integral machines use the least significant bits:

$$\boxed{\pm XX \cdots X.}$$

The fractional machine may round the most significant half, while the integral machine may detect an arithmetic overflow of the least significant half.

Incidentally, if the arrangement of arithmetic registers is such that only a single-length product can be computed, altogether different procedures may be applicable. The layout shown in Fig. 8.10a, for instance, allows only the computation of the most significant half. (The least significant bits are shifted out of the accumulator and are lost.) The layout is, therefore, only applicable to a fractional machine. It would have to be modified for an integral machine by the provision of left shifts instead of right shifts, and by using the most significant position of the multiplier register to determine whether or not an addition of the multiplicand takes place.

A final remark about accumulative multiplication may be appropriate. The schemes as we have discussed them, assume a cleared accumulator at the beginning of a multiplication. The final content of the accumulator is, therefore, simply the product of multiplicand and multiplier. If, however, the accumulator contained a value B at the beginning, the final result would be the sum of B and the product. In effect we would have an accumulative multiplication. Such an accumulative multiplication could be quite valuable if we had to compute an expression like: $a \cdot b + c \cdot d + e \cdot f + \ldots$ Instead of storing the individual products and adding them later, we could multiply and add the product in one operation. An accumulative multiplication is easily implemented if there is a double-length accumulator (e.g. Fig. 8.8, 8.9, or 8.10c). In all other cases, it would require an additional register which holds B (or part of B), and we would have to shift B into the (single-length) accumulator while the multiplication is performed.

In practice, only a fraction of all computers has provisions for accumulative multiplication. Most of those have a double-length accumulator and have a two-address instruction code, so that one instruction is sufficient to specify both, multiplicand and multiplier.

8.1.2.2. Decimal Multiplication: Decimal multiplication is more complicated to perform than binary multiplication. We note that a decimal multiplier digit may have any value between "0" and "9", whereas a binary multiplier digit has only the two possible states "0" and "1". The basic difference between binary and decimal multiplication lies, therefore, in the derivation of individual digits of the partial product. The addition of partial products can follow the procedures outlined for binary multiplication except that, of course, a decimal rather than a binary adder has to be used.

Let us concentrate here on the multiplication by a single digit, rather than on deriving the complete implementation of a decimal multiplier. The simplest solution is, probably, a repeated addition of the multiplicand. Dependent upon the multiplier digit, we would add the multiplicand from zero to nine times as indicated in Fig. 8.16.

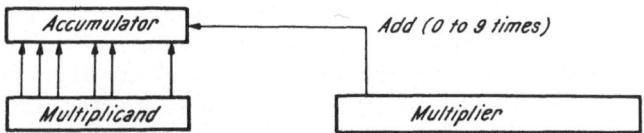

Fig. 8.16. Decimal Multiplication by Repeated Addition

In order to conserve time, we could vary the scheme so that we would add only for multiplier digits equal to, or smaller than 5. For multiplier digits from 6 to 9, we could add ten times the multiplicand (shifted one decade) and subtract then the multiplicand by an appropriate number of times. Of course, we realize that it would be easier to subtract first, and perform the addition after the shift (which is required in any event, before we can operate with the next significant multiplier digit).

A different approach is illustrated in Fig. 8.17.

A logic network derives the appropriate multiple of the multiplicand before the partial product is added (once) into the accumulator. Such a logic network may be rather complex, but certainly not impossible to build.

One possible implementation is by a nine-input adder. Dependent upon the multiplier digit, we would energize from zero to nine inputs of the adder with the bits of the multiplicand.

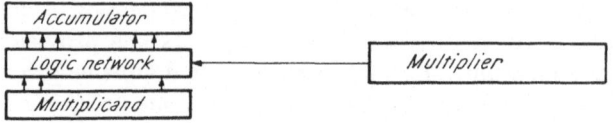

Fig. 8.17. Immediate Decimal Multiplication

Unless high speed is of utmost importance, it probably will not pay to implement such a scheme. If we are satisfied with a few additions instead of one, we can separate the multiplier digit into its, let us say four, binary positions and operate with one binary position of the multiplier at a time. Suppose the 2421 code is used to represent a multiplier digit. If we perform four additions, the logic network has to derive only the following multiples of the multiplicand: 0, 1, 2, 4. Depending upon the individual bits of the multiplier digit, we would add: zero, the multiplicand, twice the multiplicand, or four times the multiplicand. The logic network should be much simpler than that indicated in Fig. 8.17. Of course, we need the time for four, rather than one individual addition.

While the 2421 code requires only a doubling and quadrupling of the multiplicand, other number codes may demand other multiples. Each number representation will require its own particular network. The derivation of multiples can be combined with the shifting of operands. Furthermore, the technique of deriving multiples can be combined with subtraction techniques. For instance, a multiplication by "three" can be performed as an addition of the quadrupled multiplicand and a subtraction of the simple multiplicand. With the possibilities of serial or parallel designs, pulse or level type logic, there is an almost infinite variety of possible solutions. However, specialized schemes seldom incorporate any basically new ideas and are challenging mainly as minimization problems.

Problem 33: Design the logic diagram for the doubling of a decade in the 8421 code. Assume the five inputs A_0, A_1, A_2, A_3 with the respective weight of 1, 2, 4, 8 and the input C (from the doubler of the next lower decade). Provide the outputs D_0, D_1, D_2, D_3 with the respective weights of 1, 2, 4, 8, and the carry (weight 10) to the next decade. Minimize the solution.

8.1.2.3. High-Speed Multiplication Techniques: The time required for multiplication can be decreased by employing high-speed components, or by employing high-speed addition and shifting techniques. In addition, there are a number of high-speed multiplication methods which, even though they are expensive, are being used more and more frequently. These techniques aim at the reduction of the number of individual additions required for a multiplication.

Shifting Across Zeros: An obvious reduction in the number of "additions" can be achieved if no addition cycles are initiated for "zeros" in the multiplicand, but shift cycles are initiated immediately. This technique is employed in practically all parallel computers and causes variations in the execution times for multiply instructions. (The execution time depends upon the number of "ones" in the particular multiplier.) A further reduction can be achieved if provisions for a shift with variable shift-length are made. With such an arrangement, a single shift could shift "across" a whole string of zeros.

Problem 34: Assume that one clock cycle is required for a "shift" and two clock cycles are required for an "add". Try to estimate the time for the multiplication of two average 36-bit binary numbers, if:

a) a shift and an add is initiated for each bit in the multiplier;

b) adds are initiated only for "ones" in the multiplier, but a shift is initiated for each bit in the multiplier;

c) adds are initiated only for "ones" in the multiplier and shifts over an arbitrary number of binary positions are possible within one clock cycle.

Multiplication by Addition and Subtraction: If a multiplier contains a series of neighboring "ones", one may replace the individual additions for the series by a single addition and a single subtraction. Let us take an example. Suppose the multiplier is the binary number: 0011110. The computation: multiplicand × 0011110 requires normally four additions. We note, however, that we obtain an identical result by computing: (multiplicand × 0100000)—(multiplicand × 0000010). In other words, we say 0011110 is equal to 0100000—0000010. This scheme requires only one addition and one subtraction. The rules for such a multiplication could be expressed as follows:

Inspect the bits of the multiplier, starting with the least significant bit.

1. If you encounter the first "one" in a series of "ones", subtract the multiplicand from the contents of the accumulator;

2. if you encounter the first "zero" after a series of "ones", add the multiplicand to the contents of the accumulator;

3. if you encounter a single "one", add the multiplicand to the contents of the accumulator.

Of course, the multiplicand has to be shifted an appropriate number of places, as in a normal multiplication procedure. Again, much time can be saved if provisions are made for shifts with a variable number of places.

Problem 35: Show the contents of the multiplicand register and the accumulator during the multiplication of the two binary numbers 01001×01110. Assume a double-length accumulator, provisions for variable shift length, and a representation of negative numbers by sign and magnitude.

Problem 36: Estimate the average time required to multiply two 36-bit binary numbers with the given multiplication scheme. Make the same assumptions as in problem 34.

Problem 37: Find a set of rules for a multiplication scheme equivalent to the one above but with the inspection of the multiplier starting with the most significant bit.

The multiplication by addition and subtraction can be developed slightly further. There are instances when the scheme results in neighboring additions and subtractions. In such cases, the two operations can be replaced by a single operation. Let us demonstrate this in a specific example.

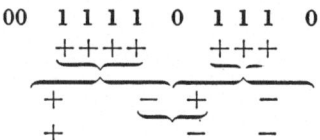

The first line indicates the multiplier. The second line indicates the operations required for a straightforward multiplication, i.e. an addition for each single "one" in the multiplier. The third line indicates the operations required for the previously discussed high-speed multiplication scheme, i.e., an addition for a single "one" in multiplier, and a subtraction and an addition for each string of "ones" in the multiplier. The fourth line shows the combination of neighboring additions and subtractions to a single subtraction.

All previously discussed high-speed multiplication schemes require a variable shift length feature if they are to be used to their full advantage. However, such shift matrices or similarly flexible arrangements are fairly expensive, so that sometimes arrangements with limited flexibility are employed. A possible implementation of such a scheme might be an arrangement in which a shift by a number of positions between, say, 1 and 6 is possible. Shifts by more than 6 places have to be performed by repeated shifts of six or less places.

Problem 38 : Try to estimate the speed of a high-speed multiplication scheme in which the shift length is variable only in the range from one to six bits. Take the results of problem 36 as basis of the comparison.

Multiplication by Uniform Multiple Shifts: A different approach to avoid the complexity of a completely flexible variable shift is the use of uniform multiple shifts. Let us explain the principle with the specific example of uniform shifts of two.

The multiplier is inspected in groups of two bits. Depending upon the results of this inspection, different actions are taken. Table 8.8 represents a set of possible actions.

Table 8.8. *A Sample Set of Rules for Multiplication by Uniform Shifts of Two Binary Positions*

Multiplier	Action
00	do not add
01	add multiplicand
10	add twice the multiplicand
11	add three times the multiplicand

The first two of the indicated four possible actions are easily implemented. The third action could be easily accomplished if the multiplicand register had an additional output which presents the multiplicand shifted by one place to the left (or multiplied by two). The fourth action could be performed with two additions. However, we could also subtract the multiplicand and remember during the handling of the next higher group (when the multiplicand is shifted two places or multiplied by four) to add one extra time. In effect, we would add four times the multiplicand and subtract it once. An arrangement in which it is possible to add or subtract one or two times the multiplicand would then be sufficient to perform a multiplication by groups of two bits. A possible set of applicable rules is given below:

Table 8.9. *Alternate Set of Rules*

Marker	Multiplier	Action	Marker	Multiplier	Action
0	00	—	1	00	add 1 ×m'cand & clear marker
0	01	Add 1 ×m'cand	1	01	Add 2 ×m'cand & clear marker
0	10	Add 2 ×m'cand	1	10	Subtract 1 ×m'cand & set marker
0	11	Subtract 1 ×m'cand & set marker	1	11	set marker

The inspection of the multiplier starts with the least significant pair of bits. If the multiplier is zero, one, or two, the proper multiple of the multiplicand is added. If the multiplier is three, the multiplicand is subtracted and the marker is set. The state of the marker is examined together with the higher-order bit-pairs of the multiplier. If the marker is not set, the action proceeds as for the least significant pair of bits. If the marker is set, an action is initiated which normally would be appropriate for a multiplier which is (by one) larger than the actual multiplier. As we can see from the table, there is no arithmetic operation performed if a long string of "zeros" or a long string of "ones" is found in the multiplier. In effect, the method becomes equivalent to the previously discussed addition and subtraction method, except that shifts are always in groups of two bits.

Problem 39: Compare the execution time for this multiplication scheme to those of the previously discussed methods.

The scheme, as it stands, has a slight disadvantage. It requires a separate storage element for the marker. Table 8.10 gives a set of rules which avoids this inconvenience.

Table 8.10. *Third Set of Rules*

Multiplier	Action	Multiplier	Action
0 — 00	—	1 — 00	Subtr. 4 × m'cand
0 — 01	Add 2 × m'cand	1 — 01	Subtr. 2 × m'cand
0 — 10	Add 2 × m'cand	1 — 10	Subtr. 2 × m'cand
0 — 11	Add 4 × m'cand	1 — 11	—

The multiplier is shifted in groups of two bits, but is inspected three bits at a time. In effect, the least significant bit of each pair is inspected twice. From the left half of the table, we see that all actions produce either the proper result, or a result which is too large by one times the multiplicand. A result which is too large is derived only when the least significant bit of pair is a "one". We correct for this during the inspection of the next lower ordered pair. If we find that the higher pair is odd we produce an action which corrects for the error which we have made or will make. The following example may serve to illustrate the application of these rules.

$$\begin{array}{ccccc|ccccc}
00 & 11 & 11 & 01 & 10 & 00 & 11 & 11 & 01 & 10 \\
 & +4 & & -2 & -2 & + & & - & + & - \\
+ & & - & - & & + & & & - & - \\
\end{array}$$

The first line on the left shows the multiplicand separated into pairs of bits. The second line shows the appropriate action according to Table 8.10. The third line shows the equivalent actions in our previously used notation. The right-hand side indicates the actions for the previously discussed high-speed multiplication scheme. We see that the two methods produce identical results. In summary, the scheme allows multiplication to be performed in groups of two bits. It is only necessary to add and subtract either two or four times the multiplicand. Providing two sets of outputs for the multiplicand register (one giving two times the multiplicand and the other giving four times the multiplicand), the scheme is easily implemented. Of course, more operations are required than in the previously discussed high-speed schemes with variable shifts, but the shifting of operands is greatly simplified. It is also interesting to note that the scheme works for both, an inspection starting with the least significant pair of bits, and an inspection starting with the most significant pair, whereas the previous scheme works only for an inspection starting with the least significant pair.

Problem 40: Compare the average number of addition and shift cycles of the above to previously discussed schemes.

Problem 41 (Voluntary): For the given set of rules, beginning and/or end corrections may be necessary. Try to derive the rules for these corrections. Assume that the multiplier is inspected in pairs,
  a) starting at the least significant end;
  b) starting at the most significant end.

We have shown here the method of uniform multiple shifts for the special case of uniform shifts of two positions. The method is most frequently employed in this form. However, it is entirely possible to extend the method to uniform shifts of three or more positions. It can be shown that multiplication in groups of three bits requires, for instance, an addition or subtraction by any even multiple of the multiplicand between zero and eight. Only the derivation of 6 times the multiplicand produces any difficulties to speak of. One may add this multiple in two addition cycles (four times the multiplicand plus two times the multiplicand), one may employ logic circuitry to derive it, or one can use special adders with more than three sets of inputs. The larger the group of bits becomes, the more complex becomes the circuitry, but, at least theoretically, the scheme could be expanded to a multiplication simultaneous in all bits of the multiplier.

Use of Carry-Safe Adders: The high-speed multiplication techniques which we have discussed so far, aim at the reduction of the number of addition or shift operations required for a multiplication. However, one can obtain an equally significant improvement by reducing the time required for an addition.

We have already seen techniques to shorten the carry propagation time. However, there is one particular technique which can be applied to repeated additions as in multiplication. The idea is basically simple: It is not necessary to propagate carries beyond a single stage in the adder. One can "save" the carries and add them during the next addition cycle. Of course, after the last cycle, the carries must be propagated in the usual manner. Fig. 8.18 shows the basic arrangement of one stage in a carry-save adder.

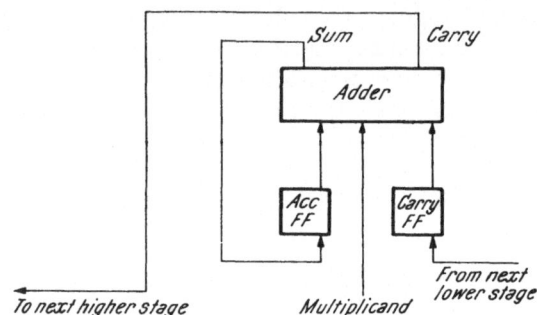

Fig. 8.18. Carry-Save Adder

The multiplicand is added in the usual manner to the contents of the accumulator. However, generated carries are not propagated, but stored in carry flip-flops. The carries are then added to the contents of the accumulator during the next addition cycle (generated carries are again not propagated, but only stored). Only after the last addition cycle, carries are allowed to propagate.

Starting the Multiplication before the Multiplicand is available: There is one final "trick of the trade" to save time for a multiplication: It is possible to start a multiplication before the multiplicand is available.

If the low order bits of a multiplier are such that no operation other than shifts are required (the least significant bits of the multiplier are zeros), the inspection and shifting of the multiplier can commence before the multiplicand is available. Of course, the multiplicand must be put into the proper position, when it becomes available. This problem is easily solved if a shift matrix is employed since the multiplicand has then a fixed position in the multiplicand register and only different sets of outputs are energized.

8.1.3. Division

8.1.3.1. Binary Division: As we have seen, multiplication can be reduced to series a of additions. Similarly, division can be reduced to a number of subtractions. Also, as in multiplication, many different approaches are possible and feasible. Let us again begin with the paper and pencil division of binary numbers. Fig. 8.19 shows the computation for a specific numerical example.

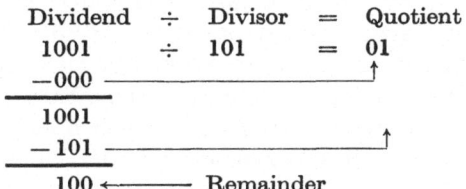

Fig. 8.19. Paper and Pencil Division

Individual digits of the quotient are derived in a series of steps. For the first step, the most significant digits of divisor and dividend are lined up. The divisor and the dividend are then compared. If the divisor is smaller than, or equal to the dividend, a "one" is entered in the quotient and the divisor is subtracted from the dividend. If the divisor is larger than the dividend, the divisor is not subtracted (or zeros are subtracted instead), and a "zero" is entered in the quotient. The divisor is then shifted one place to the right and compared with the modified dividend. This comparison, subtraction and shifting continues with the modified (or partial) dividend replacing the original dividend.

If we try to implement the scheme, we immediately encounter a number of difficulties. The quotient resulting from a division may have any numerical value between 0 and ∞, even when we restrict the size of operands to a certain number of bits (a finite number divided by zero is infinity; a zero divided by any non-zero number is zero). If we provide only a finite number of digital positions in the quotient register, the significant bits of the result may well be outside the range of this register. If we line up the most significant bits of the two operands, as we have done in the paper and pencil calculation, then, even though we may get the significant bits of the quotient into the register, the binary point of the result may be anywhere within or outside of the range.

In practice, either the range of operands is restricted (so that significant results with the binary point in a fixed position may be derived), or the operands are lined up (so that quotients in the order of unity are derived). In the first case, we speak of fixed-point division. The programmer has to "scale" his problem so that operands fall into the usable range and give meaningful results. In the second case, we speak of floating-point division[1]: The computer hardware keeps track of the shifting, during the alignment of operands. The "result" of the division consists not only of

---

[1] See paragraph 8.1.6.3.

the derived quotient, but includes also the number of shifts, or the "scale--factor" of the quotient. In fixed point division, we may distinguish two cases: the division of integers and the division of fractions.

In *integral machines*, the binary point is considered to the right of the least significant bit of a word, and a quotient is derived which also has the binary point to the right of the least significant bit. The operands are restricted such that the quotient is an integer which has a numerical value equal to or smaller than the largest integer which a computer word can hold. In order to find the applicable restriction let us consider the following division:

$$1111. \quad \div \quad 0001. \quad = \quad 1111.$$

We have four-bit operands with the binary point to the right of the least significant bit. The dividend is a large number, and the divisor is a small number so that we obtain a large quotient. In fact, the above computation gives us the largest possible quotient, unless we divide by zero (in which case the quotient would be infinity). We see that a division of integers can give a quotient which is too large only when the divisor is zero. The only "forbidden" operation is, therefore a division by zero. The machine should give an alarm (a divide fault) for this case. If the divisor is larger than the dividend, the integral part of the quotient becomes zero. We see now that, in order to obtain a significant quotient, the divisor must be in the range: $0 < \text{divisor} \leqslant \text{dividend}$. It is the responsibility of the programmer to scale his problem in such a manner that he stays within these limits[1].

In a *fractional machine*, the binary point of operands is considered to the left of the most significant bit, and a quotient is derived which also has the binary point to the left of the most significant bit. Again the operands are restricted, so that the quotient stays in the proper range. Let us consider the following division:

$$.00111 \quad \div \quad .00111 \quad = \quad 1.0$$

The divisor is equal to the dividend and the quotient is one. If the quotient is to be a fraction, the divisor must be larger than the dividend. On the other hand, if the divisor becomes much larger than the dividend, the quotient loses most of its significance as the following example will show:

$$.0001 \quad \div \quad .1111 \quad \approx \quad .0001$$

Again it is the responsibility of the programmer to scale his problem in such a manner that adequate significance is obtained.

---

[1] Even if the integral part of the quotient becomes very small or zero, usually the remainder is available, so that a consecutive division of the remainder with modified scalefactors can produce the fractional part of the quotient.

Division of integers: Let us now try to find the required implementation for a binary division process. Suppose we begin with a fixed-point integer division. First we repeat the computation shown in Fig. 8.19 with four bit operands.

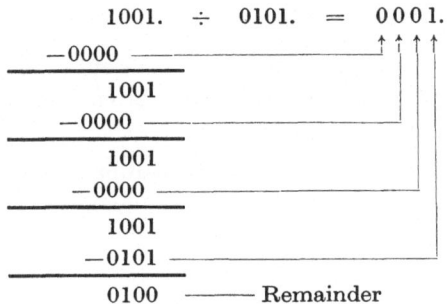

Fig. 8.20. Fixed-Point Division of Integers

We have to perform four "subtractions" in order to derive the four bits of the quotient. The last of the four subtractions determines the quotient bit with the weight one, hence for this step, the bits of the divisor must be lined up with the bits of the same weight in the dividend[1]. Counting backwards from this step, we see that the first subtraction must take place with the divisor shifted three places to the left with respect to the dividend. In general, the divisor must be shifted $n-1$ places with respect to the dividend if we have n-bit operands.

The computation shown in Fig. 8.20 can be directly implemented if we assume for the moment that a double-length accumulator is available.

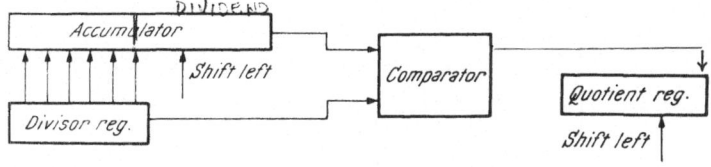

Fig. 8.21. Functional Layout for a Division

The divisor is entered into the divisor register and the dividend into the right half of the accumulator. Instead of shifting the divisor to the right with respect to the dividend, we shift the dividend left with respect to the divisor. The division begins with a left shift of the accumulator. The contents of the left half of the accumulator are then compared with the con-

---

[1] Consider, for example, the division of 0001 by 0001.

tents of the divisor register. If the dividend is larger than the divisor, the divisor is subtracted and a "one" is shifted into the quotient register. For the next step, the contents of the accumulator are shifted one place to the left, and again dividend and divisor are compared. The procedure continues until all bits of the quotient are derived. The remainder is left in the left half of the accumulator.

Problem 42: Show the exact contents of the double-length accumulator, the divisor register, and the quotient register for each step of the computation shown in Fig. 8.20.

The rules governing this division process can be shown in the form of a flow chart:

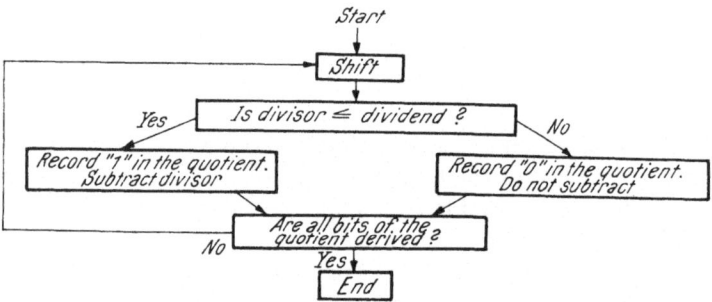

Fig. 8.22. Flow Chart for Division with Test Feature

We notice that three arithmetic registers are sufficient to perform a division. The arrangement of the three registers is very similar to the arrangement of registers required for multiplication in Fig. 8.9 and, in practice, identical

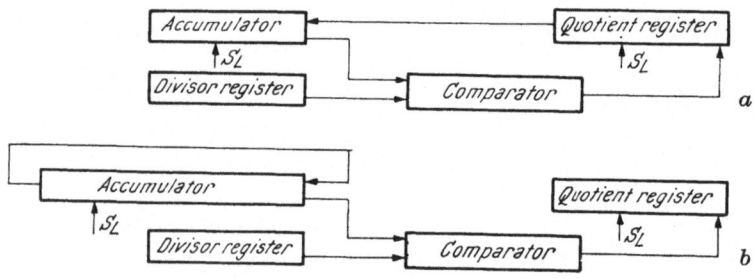

Fig. 8.23. Variations of the Division Scheme, Fig. 8.21

registers are almost always used to perform both, multiplication and division. Also, as in multiplication, several different arrangements of registers may be used to perform divisions.

Fig. 8.23 a shows an arrangement in which only a single-length accumulator is used. The dividend is originally stored in the "quotient" register and shifted bit by bit into the accumulator, while the quotient is being shifted bit by bit into the quotient register. Fig. 8.23 b gives an arrangement where the dividend is originally contained in the left half of the accumulator and shifted bit by bit into the right half by a circular shift.

Problem 43: Show the contents of all registers during the individual steps of the division $1001 \div 0101$. Assume the register configurations given in Fig. 8.23 a and b.

Incidentally, practically all of the indicated schemes can be easily modified for double-length dividends. (We remember that we have either a double-length accumulator, or a double-length storage consisting of a single-length accumulator and the quotient register). In order to avoid a divide fault, the numerical value in the most significant half of the dividend must be smaller than the divisor. The detection of a divide fault is fairly simple. The divisor and the most significant half of the dividend are compared before the initial shift. If the divisor is smaller than, or equal to the dividend, the quotient will exceed the capacity of the quotient register. This test detects any illegal division including a division by zero. We note that it is not required to actually perform a division in order to detect a divide fault.

Problem 44: Assume the layout of registers given in Fig. 8.21. Show the contents of all registers during the following divisions:

a) $0110\ 1010 \div 1001$
b) $1001\ 1010 \div 0110$
c) $0000\ 1001 \div 0000$

Division of Fractions: Let us first consider the computation in Fig. 8.24.

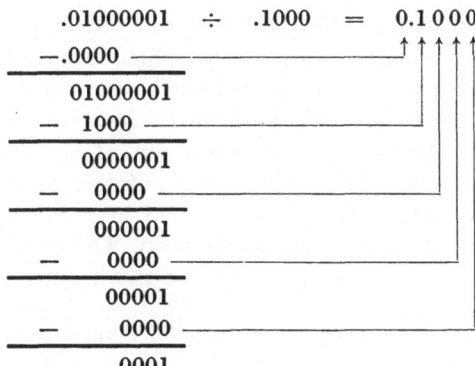

Fig. 8.24. Fixed-Point Division of Fractions

The first comparison in this computation is made where the bits with the same weight in the dividend and the divisor are lined up. The result of this comparison gives us the quotient bit with the weight one. The following comparisons and "subtractions" determine the lower order bits of the quotient.

If we expect only fractions as quotients, the first comparison is not required and the division process can start with the divisor shifted one place to the right. However, the comparison in the original place can be used to detect divide faults in the same manner as for the division of integers. We see that the scheme becomes identical to that for the division of integers if we consider a double-length dividend. The most significant half of the dividend is placed into the left half of the accumulator and the least significant half into the right half. At the beginning, the divisor and the most significant half of the dividend are lined up. A test for divide fault is then performed. For each step of the following division, the divisor is shifted one place to the right with respect to the dividend. The only difference between a division of integers and a division of fractions is thus in the placement of single length dividends. Single-length integer dividends are placed into the right half of the accumulator (considered as the least significant half of a double-length dividend) and single-length fractional dividends are placed into the left half (considered as the most significant half of a double-length dividend). Some computers have both, divide integer and divide fractional instructions. The essential difference in the execution of these instructions is the placement of single-length dividends.

Problem 45: Assume the layout of registers given in Fig. 8.21. Show the contents of all registers during the following divisions:

 a) $.1001 \div .1$
 b) $.1001 \div .0$
 c) $.0111 \div .1$

Restoring Division: In our discussions of the division we have so far assumed that there is a comparator available which is capable of comparing the numerical values of the dividend and the divisor. Such a comparator can be designed but it will be fairly expensive[1]. Many computers perform, therefore, the comparison by a test-subtraction of the divisor from the dividend[2]. If the difference is negative, the divisor is larger than the dividend. The original dividend is then restored by a subsequent addition of the divisor. We speak of a restoring division. The flow chart given in Fig. 8.22 can be modified to reflect the use of this testing procedure:

---

[1] The logic design of such a serial or parallel comparator is offered as a "recreational" exercise.

[2] This type of a "comparator" costs almost nothing if the capability to perform subtractions is already implemented.

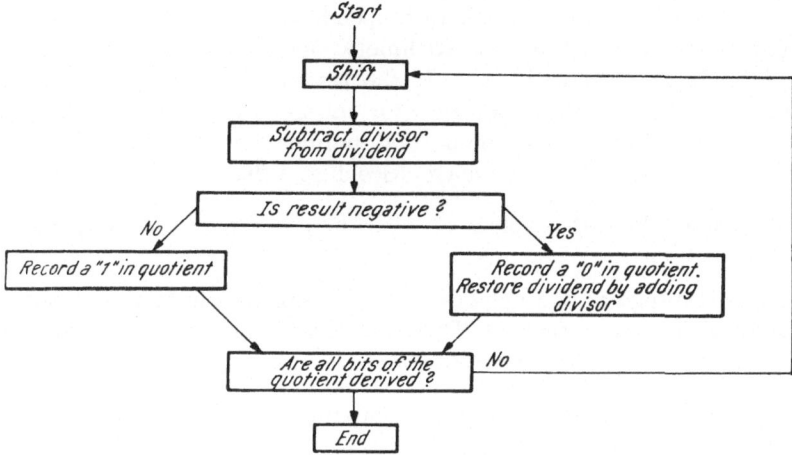

Fig. 8.25. Flow Chart for Restoring Division

Problem 46: Repeat problem 42 with the assumption of a restoring division rather than a division with test feature.

Non-Restoring Division: The restore operations in the previously discussed division method require a certain amount of time for their execution. However, if we inspect the flow chart in Fig. 8.25, we see that each restore operation (i.e., an addition of the divisor) is followed by a subtraction of

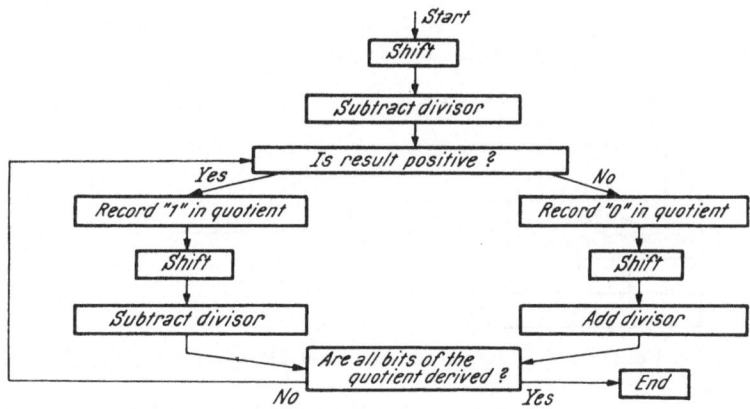

Fig. 8.26. Flow Chart for Non-Restoring Division

the divisor during the next iteration (with the divisor shifted one place to the right, or divided by two). The two operations "add present divisor" and "subtract one half present divisor" can be combined to a single operation

15*

"add one half of present divisor". More specifically, if a test subtraction gives a negative remainder, the next arithmetic operation in the iteration loop should be "add shifted divisor". If the test subtraction gives a positive remainder, the arithmetic operation for the next iteration loop should be "subtract shifted divisor" in the normal manner. Let us indicate these rules again in the form of a flow chart. (See Fig. 8.26.)

Problem 47: Repeat problem 42 with the assumption of a non-restoring division.

A disadvantage of the non-restoring division is the fact that the final "remainder" may be negative. This complicates any program which uses the remainder for further computation.

Problem 48 (Voluntary): Try to compare the average execution times required for a
  a) division with test feature,
  b) restoring division,
  c) non-restoring division.

8.1.3.2. Decimal Division: A decimal division is, of course, more complicated to perform than a binary division. An individual digit in the quotient may have any of the ten values from 0 to 9. The selection of the proper value and the corresponding subtraction or addition of the proper multiple of the divisor from the dividend introduces a number of problems not encountered in binary division. However, at least in principle, we have the same approaches as to binary division, i.e., division with test feature, restoring, or non-restoring division.

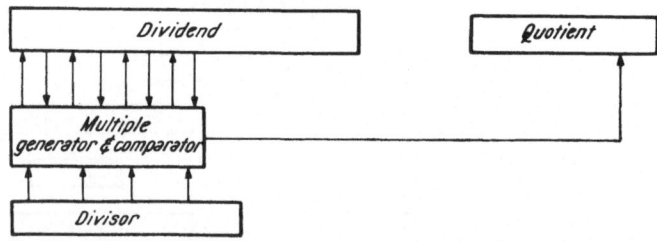

Fig. 8.27. Functional Diagram for Decimal Division with Test Feature

Division with Test Feature: Even though this division method is not very attractive, it may be worthwhile to discuss it briefly in order to see more clearly some of the difficulties encountered in decimal division. Let us, however, restrict the discussion to the derivation of a single digit in the quotient. Fig. 8.27 gives a rough layout for a decimal division with test feature.

The multiple generator develops all multiples of the divisor (from 0 to 9). The ten multiples are compared with the dividend. The largest multiple which is smaller than the dividend is then selected. This multiple is subtracted from the dividend, while the size of the multiple (an integer in the range 0 to 9) is recorded in the quotient. The scheme is simple and straightforward, but the size and the complexity of the multiple generator and comparator is almost prohibitive.

Let us, therefore, see whether we cannot break the function of the multiple generator into a series of less complicated steps. We can, for instance, subtract the single divisor repeatedly, rather than subtracting the appropriate multiple in one step. We keep subtracting as long as the divisor "goes" into the dividend. The number of subtractions which we have to perform determines the desired digit of the quotient.

The comparator only decides whether or not the dividend is larger than or equal to the divisor. If it is, another subtraction takes place; if not, the proper digit in the quotient has been found and the process is terminated.

The following numerical computation may serve as an illustration of the scheme:

$$
\begin{array}{lllll}
64 & \div & 21 & = & 0. \\
-21 & & & & \\
\hline
+43 & \longrightarrow & & & 1. \\
-21 & & & & \\
\hline
+22 & \longrightarrow & & & 2. \\
-21 & & & & \\
\hline
+01 & \longrightarrow & & & 3.
\end{array}
$$

Restoring Division: Again, as in binary division, the comparator is a functional element which is fairly complex. Many computers use, therefore, a scheme in which a test subtraction (or addition) rather than a true comparison is performed. If we use a restoring division, a fairly simple scheme results. The divisor is subtracted from the dividend until a negative remainder results. The proper dividend is then restored by the addition of the divisor[1]. The digit in the quotient reflects the number of subtractions. The numerical computation on top of page 230 may illustrate this procedure.

Non-Restoring Division: As in binary division, the restore operation in the previous scheme can be eliminated. We obtain then a non-restoring division method. Instead of restoring the dividend and subtracting repeatedly one tenth of the present divisor (when we derive the next digit of the quotient) we can omit the restoring and add one tenth of the present

---

[1] Desk calculators use this division method.

dividend		divisor		quotient
64	÷	21	=	0.
−21				
+43			→	1.
−21				
+22			→	2.
−21				
+01			→	3.
−21				
−20			→	(4.)
+21 (Restore)				
+01			→	(3.)

Restoring Division

(1)	268 ÷ 350		
	−350		
(2)	−082	→	(1)
	+350		
(3)	+268	→	0.
	− 35		
(4)	+233	→	0.1
	− 35		
(5)	+198	→	0.2
	− 35		
(6)	+163	→	0.3
	− 35		
(7)	+128	→	0.4
	− 35		
(8)	+093	→	0.5
	− 35		
(9)	+ 58	→	0.6
	− 35		
(10)	+ 23	→	0.7
	− 35		
(11)	− 12	→	(0.8)
	+ 35		
(12)	+ 23	→	(0.7)

Non-Restoring Division

(1)	268 ÷ 350		
	−350		
(2)	−082	→	0.
	+ 35		
(3)	− 47	→	0.9
	+ 35		
(4)	− 12	→	0.8
	+ 35		
(5)	+ 23	→	0.7

divisor repeatedly (during the next iteration). The number of additions required to obtain a positive dividend is then the complement of the desired digit in the quotient. In effect, we remember that we have subtracted the present divisor (or ten times the new divisor) once too much, and we test for the next digit of the quotient in reversed order, i.e., 9, 8, 7, etc. The numerical example on the previous page illustrates this procedure.

Fig. 8.28 shows the dividend in these computations in graphic form. The labeling of the dividend corresponds to the labeling of the individual steps in the previous example.

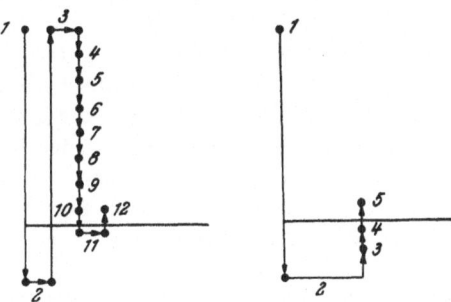

Fig. 8.28. Comparison of Restoring and Non-Restoring Division Methods

Problem 49: Give the computation and a graphic representation of the dividend for the division of $268 \div 350$. Assume a division with test feature.

Problem 50: Derive the quotient of $3806 \div 22$ to three decimal places. Use a:

  a) division with test feature
  b) restoring division
  c) non-restoring division

Problem 51 (Voluntary): Try to compare the execution times for decimal division with test feature, restoring and non-restoring division.

Although non-restoring decimal division is more efficient than restoring division, the degree of improvement is not as great as in binary division. In decimal division, the average number of subtractions before a negative result is obtained is: $(1+2+\ldots+9) \div 9 = 5$. In restoring division, one additional operation is required bringing the average to 6. Thus an improvement of 1 part in 6 or about 20% is obtained. In binary restoring division, the average number of operations per iteration is: $(1+2) \div 2 = 1.5$. In non-restoring binary division, exactly one operation is required during each iteration. Hence an improvement of 0.5 part in 1, or 50% is obtained.

8.1.3.3. High-Speed Division Techniques: *Division of Normalized Operands:* The division methods which we have discussed so far, can be simplified if the operands are "normalized" before the actual division takes place, that is, if both dividend and divisor have a "1" as the most significant bit[1]. Under this condition, there will now be instances during the division process when it is no longer necessary to perform a test subtraction or addition in order to determine whether the divisor is smaller or larger than the dividend. A simple inspection of the most significant bit of the dividend will suffice. Specifically, if the most significant bit of the modified dividend is a "zero", it is not necessary to subtract the divisor (which contains a "one" in the most significant bit position) to find out whether or not the remainder is positive or negative. Under this circumstance, the remainder will always be negative. In other words, if the most significant bit of the dividend contains a "zero", we may enter a "zero" in the quotient and shift without performing any test subtraction. Consequently, if the dividend contains a series of leading "zeros" we may shift across this series entering a corresponding "zero" in the quotient for each place shifted across. In this respect the method becomes equivalent to the multiplication method in which we shift across "zeros" in the multiplier[2].

Problem 52: Compute the integral part of the quotient of $10100110 \div 1011$ to four places. Assume a restoring division, but omit the test substraction if the most significant bit in the dividend is a "zero".

Let us now assume that we perform a non-restoring division. If the modified dividend is negative and contains a zero in the most significant bit, we know immediately that the result of an addition would be positive. In other words, under these circumstances, we can immediately enter a "one" in the quotient. In effect, no arithmetic test operation is required if the dividend is negative or positive *and* contains a high order "zero". Test addition or subtractions are required only if the high order bit in the dividend is a "one".

Fig. 8.29 on the following page shows the flow chart for this type of division.

Problem 53: Compute the quotient of $01000000 \div 1001$ to four binary places. Use a non-restoring division with shifts across zeros in the dividend.

The method allows us to shift across series of zeros in the dividend. We generate thereby strings of "ones" or "zeros" in the quotient. We might say, the method allows us to shift across strings of "ones" or "zeros" in the quotient. In this sense, the method is equivalent to the previously discussed high-speed multiplication method in which we shift across series of ones

---

[1] Floating point numbers, for instance, are usually normalized. The number of shifts required to normalize the operand is reflected in the exponent or characteristic.

[2] See paragraph 8.1.2.3.

or zeros in the multiplier[1]. Again the method becomes much more advantageous to use if the design incorporates a flexible shift feature so that immediate multiple shifts across a number of places are possible.

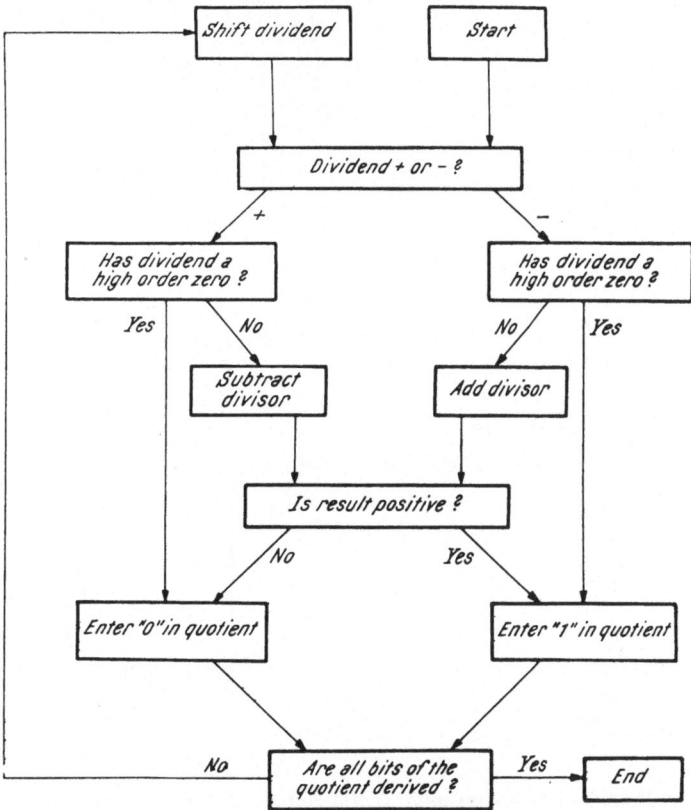

Fig. 8.29. Flow Chart for Non-Restoring-Division With Normalized Operands, and Shifts Across Zeros in Dividend

Problem 54 (Voluntary): Try to estimate the relative average execution times for the division of 36-bit binary numbers, using a:

 a) Division with test feature
 b) Restoring division
 c) Non-restoring division
 d) Non-restoring division with flexible shift feature and shifting across zeros in the dividend.

---

[1] See paragraph 8.1.2.3.

Fig. 8.30 shows the operation, and the dividend in graphical form for three division methods.

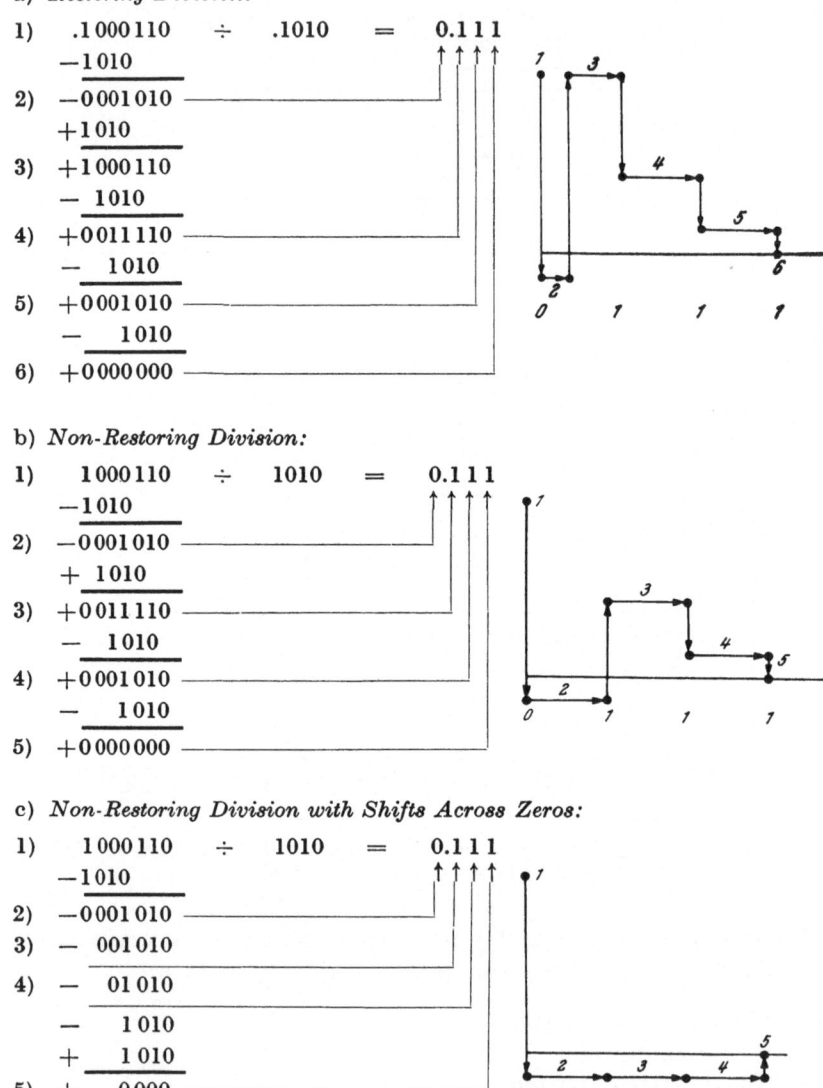

a) *Restoring-Division:*
1)   .1 000 110   ÷   .1010   =   0.1 1 1
      −1 010
      ‾‾‾‾‾‾‾
2)   −0 001 010
      +1 010
      ‾‾‾‾‾‾‾
3)   +1 000 110
      − 1010
      ‾‾‾‾‾‾‾
4)   +0 011 110
      − 1010
      ‾‾‾‾‾‾‾
5)   +0 001 010
      −   1010
      ‾‾‾‾‾‾‾
6)   +0 000 000

b) *Non-Restoring Division:*
1)   1 000 110   ÷   1010   =   0.1 1 1
      −1 010
      ‾‾‾‾‾‾‾
2)   −0 001 010
      + 1010
      ‾‾‾‾‾‾‾
3)   +0 011 110
      − 1010
      ‾‾‾‾‾‾‾
4)   +0 001 010
      −   1010
      ‾‾‾‾‾‾‾
5)   +0 000 000

c) *Non-Restoring Division with Shifts Across Zeros:*
1)   1 000 110   ÷   1010   =   0.1 1 1
      −1 010
      ‾‾‾‾‾‾‾
2)   −0 001 010
3)   −   001 010
4)   −   01 010
      −   1010
      +   1 010
      ‾‾‾‾‾‾‾
5)   +   0 000

Fig. 8.30. Comparison of the Three Division Methods

The actual mathematical operations for the three cases are:

a) dividend $+(-1+1-\frac{1}{2}-\frac{1}{4}-\frac{1}{8})$ divisor
b) dividend $+(-1+\frac{1}{2}-\frac{1}{4}-\frac{1}{8})$ divisor
c) dividend $+(-1+\frac{1}{8})$ divisor

All three formulae yield in effect: dividend $-\frac{7}{8}$ divisor.

Problem 55: Show the computation and the graphic representation of the dividend as in fig. 8.30 for a:

a) division with test feature
b) restoring division with shifts across zeros.

Problem 56: What is the difficulty in adapting the division method with shifts across zeros to un-normalized operands?

8.1.4. Extraction of the Square Root

Most present-day computers have no built-in square root algorithm. Instead, square roots are extracted according to a programmed iterative formula[1]. The fairly long execution time is accepted in favor of savings in hardware. This approach is justified if one considers the extraction of a square root as a relative infrequent operation. On the other hand, the hardware implementation of a binary square root algorithm is not really expensive, once a divide algorithm is implemented.

Let us see what is involved in finding the square root of a binary operand. Suppose we extract the square root of a fraction A. The result will be a binary fraction which we may represent in general form as follows:

$$\sqrt{A} = .f_1 f_2 f_3 \ldots f_n \qquad (8.20)$$

The f's are binary integers; f_1 has the weight $\frac{1}{2}$, f_2 the weight $\frac{1}{4}$, etc. If f_1 is to be a "one", then A must be equal to or greater than $.1^2$. In order to determine f_1, we compare A with $.1^2$. If $A \geq .1^2$, then $f_1 = 1$; if $A < .1^2$, then $f_1 = 0$.

If f_2 is to be a "one", then A must be equal to or greater than $(.f_1 1)^2$. A comparison of A with $(.f_1 1)^2$ will determine f_2. Consecutive digits can be found in an identical manner. Table 8.11 lists the required comparisons.

The extraction of the square root according to this approach requires a repeated comparison of the radicand A with a test value. The first comparison is with a fixed test value $(.1^2)$. The test values used for all further comparisons depend upon the outcome of preceeding tests. Table 8.11

---

[1] The formula $S_i = \frac{1}{2}\left(\frac{x}{S_{i-1}} + S_{i-1}\right)$ describes, for instance, an iterative procedure wherein S_i is a better approximation to $\sqrt{x}$ than S_{i-1}. S_0 may have any value $0 < S_0 < \infty$.

Table 8.11. *List of Comparisons Required for the Extraction of Square Root*

Comparison	Yes	No
$A \geq (.1)^2$	$f_1 = 1$	$f_1 = 0$
$A \geq (.f_1 1)^2$	$f_2 = 1$	$f_2 = 0$
$A \geq (.f_1 f_2 1)^2$	$f_3 = 1$	$f_3 = 0$
$A \geq (.f_1 f_2 f_3 1)^2$	$f_4 = 1$	$f_4 = 0$

describes, in essence, an iterative procedure in which consecutive iterations determine consecutive digits of the result. The following numerical computation may serve to illustrate the procedure:

$$\sqrt{.10\,101\,001} \qquad\qquad \sqrt{} = ?$$

First Iteration:

$$A \geq .1^2 = .01 ?, \text{ Yes} \rightarrow f_1 = 1 \qquad\qquad \sqrt{} = .1$$

Second Iteration:

$$A \geq (.f_1 1)^2 = (.11)^2 = .1001, \text{ Yes} \rightarrow f_2 = 1 \qquad\qquad \sqrt{} = .11$$

Third Iteration:

$$A \geq (.f_1 f_2 1)^2 = (.111)^2 = 110001, \text{ No} \rightarrow f_3 = 0 \qquad\qquad \sqrt{} = .110$$

Fourth Iteration:

$$A \geq (.f_1 f_2 f_3 1)^2 = (.1101)^2 = .10\,101\,001, \text{ Yes} \rightarrow f_4 = 1 \qquad\qquad \sqrt{} = .1101$$

The straightforward derivation of test values as listed in Table 8.11 requires a squaring (multiplication) for each iteration. This is inconvenient and time-consuming. We shall see that it is possible to derive a particular test value from the previous one by simpler operations. Let us first write the consecutive test values in the following form:

$$(.1)^2 = (.1)^2$$

$$(.f_1 1)^2 = (.f_1 + .01)^2 = .f_1^2 + .0f_1 + .0001 = .f_1^2 + .0f_1 01$$

$$(.f_1 f_2 1)^2 = (.f_1 f_2 + .001)^2 = (.f_1 f_2)^2 + .00f_1 f_2 + .000001 = (.f_1 f_2)^2 + .00f_1 f_2 01$$

$$(.f_1 f_2 f_3 1)^2 = (.f_1 f_2 f_3)^2 + .000f_1 f_2 f_3 01 \qquad\qquad (8.21)$$

With these results, we can write an alternate set of comparisons for the extraction of the square root (See Table 8.12.).

The left hand members of the inequalities still contain squares but these can be relatively easily derived. We notice for instance that $.f_1^2$ is either equal to zero (if $f_1 = 0$), or it is equal to .01 (if $f_1 = 1$). Consequently, the term

$A-.f_1^2$ is either equal to A (the term listed above the term $A-.f_1^2$), or equal to $A-.01$ (the term above diminished by the previous test value). In the same manner, $A-(.f_1f_2)^2$ is either equal to $A-.f_1^2$ or equal to $A-.f_1^2$ diminished by $.0f_101$.

Table 8.12. *Alternate Set of Comparisons for the Extraction of a Square Root*

Comparison	Yes	No
$A \geq .01$	$f_1 = 1$	$f_1 = 0$
$A - .f_1^2 \geq .0f_101$	$f_2 = 1$	$f_2 = 0$
$A - (.f_1f_2)^2 \geq .00f_1f_201$	$f_3 = 1$	$f_3 = 0$
$A - (.f_1f_2f_3)^2 \geq .000f_1f_2f_301$	$f_4 = 1$	$f_4 = 0$

We have found a scheme for the extraction of a square root in which no multiplications, but only subtractions are required. A comparison of this technique with the division algorithm reveals many similarities. For the moment, let us think of the left hand members of the inequalities as the "dividend" and of the right-hand members as the "divisor". In each itera-

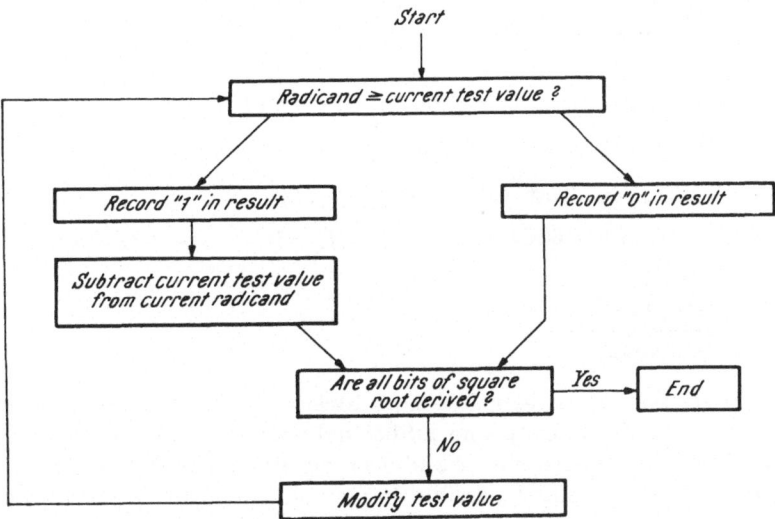

Fig. 8.31. Flow Chart for the Extraction of the Square Root

tion the values of dividend and divisor are compared. If the dividend is larger than or equal to the divisor, a "one" is recorded in the result and the divisor is subtracted from the dividend. If the divisor is larger than the dividend, a "zero" is recorded, and the divisor is not subtracted from the dividend. The only essential difference between a division and a square root algorithm is then the modification of the "divisor" for each iteration.

A flow chart representing this square root algorithm is given in Fig. 8.31.

The term "radicand" is used instead of the term "dividend" and the term "test value" instead of "divisor". The test values for each iteration are listed as the right hand members of the inequalities in Table 8.12.

The following numerical example may serve to illustrate the procedure:

$$\sqrt{.10\,101\,001} = ?$$

First Iteration (Test value .01):

$.10\,101\,001 \geq .01$? Yes $\to f_1 = 1$ $\sqrt{} = .1$

$\quad .10\,101\,001$
$-.01$
$\overline{\quad .01\,101\,001}$

Second Iteration (Test value $.0f_1\,01 = .0101$):

$.01\,101\,001 \geq .0101$? Yes $\to f_2 = 1$ $\sqrt{} = .11$

$\quad .01\,101\,001$
$-.0101$
$\overline{\quad .00\,011\,001}$

Third Iteration (Test value $.00f_1f_2\,01 = .001\,101$):

$.00\,011\,001 \geq .001\,101$? No $\to f_3 = 0$ $\sqrt{} = .110$

Fourth Iteration (Test value $.000f_1f_2f_3\,01$):

$.00\,011\,001 \geq .00\,011\,001$? Yes $\to f_4 = 1$ $\sqrt{} = .1101$

$\quad .00\,011\,001$
$-.00\,011\,001$
$\overline{\quad .00\,000\,000}$

This procedure is almost identical to a division with test feature. Identical circuits can be used. Only one additional step is required: The correct test value is to be placed into the divisor register for each iteration. Let us consider one of the many possible implementations in detail. Suppose a divide algorithm is already implemented which uses a double-length accumulator and a divisor register as shown in Fig. 8.32.

The divisor has a fixed position in the divisor register and the dividend is shifted during the division process from the right half to the left half of the accumulator. For the extraction of the square root, it would be advantageous if we would shift the radicand from the right half of the accumulator to the left half and keep the test value in a fixed position. For the same reason it would be advantageous to enter individual digits of the result into the right-most position of the quotient register, and to shift its contents left for each iteration.

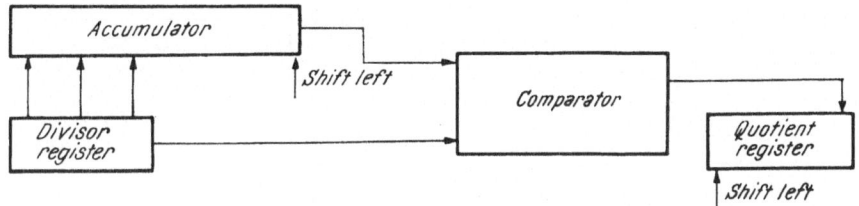

Fig. 8.32. Implementation of the Division Algorithm

Let us now consider the modification of test values. The left column in Table 8.13 lists the consecutive contents of the "quotient" register.

Table 8.13. *Register Contents During the Extraction of a Square Root*

Current Result in "Quotient" Reg.	Test Value for Next Iteration
0000	01
$000f_1$	$f_1 01$
$00f_1f_2$	$f_1f_2 01$
$0f_1f_2f_3$	$f_1f_2f_3 01$
$f_1f_2f_3f_4$	$f_1f_2f_3f_4 01$

The right-hand column lists the test values required for the next iteration (the binary point is not shown). Disregarding the least significant digits "01", we see that the test value for the next iteration is equal to the current result contained in the "quotient" register.

A simple transfer from the quotient register to the dividend register is sufficient to establish the correct test value. We see also that the register for the test value must be longer than the result register, in order to accommodate the additional "01" in the least significant position. This can be easily accomplished by the installation of two dummy positions at the least significant end of the divisor register. These two positions contain the fixed value "01". The implementation of the square root algorithm is then given in Fig. 8.33.

The derivation of the proper test value requires only a simple transfer from the result register to the extended "divisor" register. There remains now only the problem of positioning the radicand correctly. The test value for the first iteration has the binary point just left of the fixed "01". (See Table 8.12.) The radicand has, therefore, to be placed into a position so that its binary

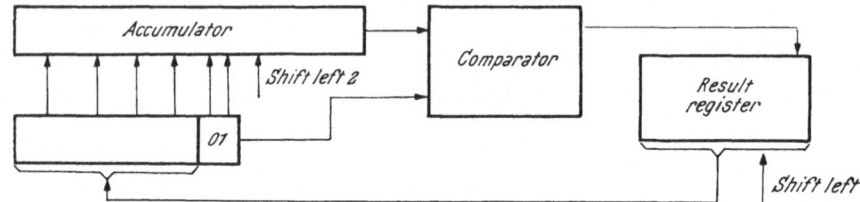

Fig. 8.33. Implementation of the Square Root Algorithm

point lines up with the point of the test value. In other words, the radicand must be placed into the right half of the accumulator for the first iteration. For the second iteration, the test value has the binary point two positions left of the fixed value "01". (See again Table 8.12.) In other words, the radicand must be shifted *two* places left with respect to its position during the first iteration. Consecutive comparisons require a repeated left shift of the radicand by two positions. The computation of a n-bit square root requires a $2n$-bit radicand.

Let us now list the deviations from a straight division algorithm:

1. Extend divisor register by two dummy positions containing "01".

2. Transfer the current result from "quotient" register to the "divisor" register for each iteration.

3. Shift accumulator left by two places instead of one place for each iteration.

None of these modifications should pose any difficult problems or require much hardware.

Problem 57: Show the contents of all three arithmetic registers during the extraction of the square root. Compute $\sqrt{.10101001}$ and assume the previously discussed implementation with double-length accumulator and test feature.

There are, of course, many variations of the previously discussed implementation possible. One may insert the fixed value "01" into the least significant end of the divisor register, rather than into dummy positions. This scheme would produce less than a full word as result and, would require a somewhat different transfer from the result register to the divisor register. One could also think of a scheme in which the contents of the accumulator are shifted one place to the left, while the contents of the divisor register are shifted one place to the right for each iteration. One could keep the radicand in a fixed place and shift test values. One might eliminate the fixed value "01" from the test value by subtracting "01" from the radicand before comparison. The restore operation of adding 01 and then subtracting "01" shifted by two places during the next iteration could be combined to a single subtraction. Which of these alternatives is "best", has to be decided in each individual case. There are, however, a few more basic alternatives possible. So far, we have assumed the equivalent of a division with test feature. We can, however, without difficulty, modify the scheme so that it becomes equivalent to a restoring division. Fig. 8.34 gives the appropriate flow chart.

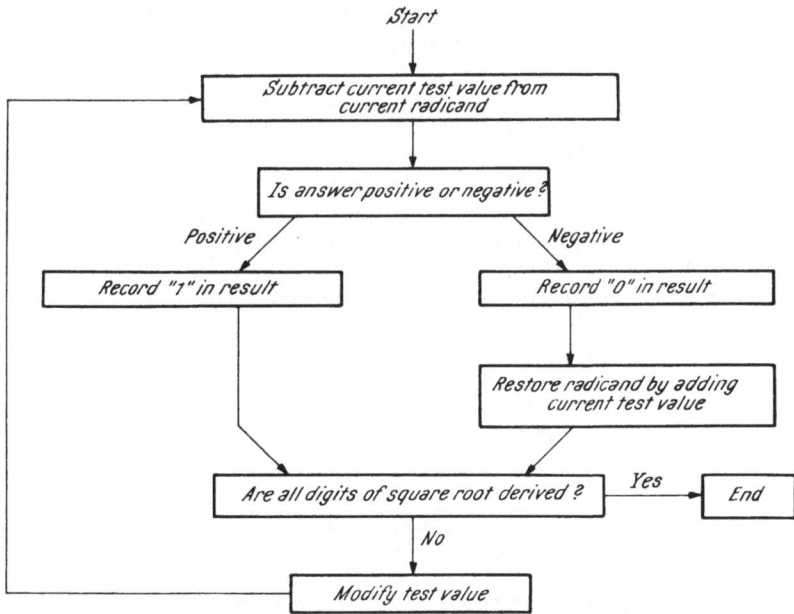

Fig. 8.34. Flow Chart for "Restoring" Extraction of a Square Root

Problem 58: Repeat problem 57. Assume a restoring process rather than a process with test feature.

The two operations of restoring the radicand by adding the test value and then subtracting the new test value during the next iteration can be combined into a single operation as in a non-restoring division.

The leftmost column of Table 8.14 shows the restore operation for consecutive iterations (if there is a restore operation required). The center column gives the subsequent test subtraction during the next iteration.

The two operations can be combined as shown in the right hand column. The correctness of this listing may not be obvious, so let us discuss it in more detail. A restore operation during the first iteration requires the addition of .01. The test subtraction during the second iteration requires the subtraction of $.0f_1$ 01. We know, however, that a restore

Table 8.14. *Non-Restoring Extraction of a Square Root*

Restore Operation	Next Test Subtraction	Combined Operation
$+.01$	$-.0f_1 01$	$+.0011$
$+.0f_1 01$	$-.00f_1f_2 01$	$+.00f_1 011$
$+.00f_1f_2 01$	$-.000f_1f_2f_3 01$	$+.000f_1f_2 011$
$+.000f_1f_2f_3 01$	$-.0000f_1f_2f_3f_4 01$	$+.0000f_1f_2f_3 011$

operation is required, only if f_1 is a "zero". We are, therefore, justified in substituting a value $-.0001$ for the next test subtraction rather than the listed value of $-.0f_1$ 01. The two values $+.01$ and $-.0001$ together give a value of $+.0011$ as listed in the right hand column. In a similar manner, we can combine $+.0f_1$ 01 and $-.00f_1f_2$ 01 to obtain $+.00f_1$ 011.

Problem 59 (Voluntary): Show the correctness of this latter combination.

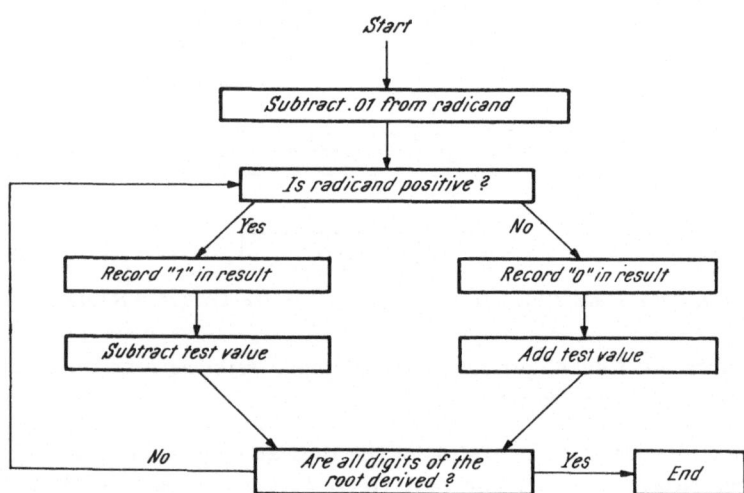

Fig. 8.35. Flow Chart for Non-Restoring Square Root Algorithm

We can now construct the flow chart for a non-restoring square root algorithm.

Table 8.15 lists the appropriate test values for each individual case.

Table 8.15. *List of Test Values for Non-Restoring Square Root Algorithm*

	If radicand is positive		If radicand is negative
$f_1 = 1$	$-.0101$	$f_1 = 0$	$+.0011$
$f_2 = 1$	$-.00f_1 101$	$f_2 = 0$	$+.00f_1 011$
$f_3 = 1$	$-.000f_1 f_2 101$	$f_3 = 0$	$+.000f_1 f_2 011$
$f_4 = 1$	$-.0000f_1 f_2 f_3 101$	$f_4 = 0$	$+.000f_1 f_2 f_3 011$

Problem 60: Extract the square root of .10101001. Use a non-restoring square root algorithm.

Problem 61 (Voluntary): How can the indicated algorithms be adapted to extracting the square root of an integer? Specifically, show a set of test values equivalent to those in Tables 8.11 and 8.12.

8.1.5. Logic Operations

If we were to analyze the purpose of each instruction which is executed by a computer during an "average" computation, we would find a large number of instructions which do not directly contribute to the arithmetic solution of a problem. Some of these are still arithmetic instructions but have to do with "housekeeping". For example, they may keep track of the number of times a program loop is executed, or be used to modify an operand address for each iteration. A large number of other instructions perform operations defined by logic rather than arithmetic rules. They serve to select, sort, re-arrange or re-format information. These latter types of instructions, constitute an important and essential part of a well balanced instruction repertoire of a computer[1]. Their number, variety and flexibility determine, to some extent the convenience and effective speed of the computer operation. The implementation of these instructions is not very difficult. In fact, anyone understanding the principles of digital design can find many alternatives. Consequently, we shall concentrate here more on what rather than on how operations are implemented. Furthermore, we shall consider only those logic operations which employ the arithmetic unit for their execution. Others will be discussed later in connection with the control unit.

8.1.5.1. **Shifting:** Paragraphs 8.1.1 and 8.1.2 have shown that arithmetic registers have provisions for shifting in order to perform multiplications

---

[1] This is reflected in the use of the term "data processor" instead of the term "computer". The latter term implies the performance of purely arithmetic operations, the first also implies operations which have to do with the "handling" of information.

or divisions[1]. These same arithmetic registers can be used to perform the logic operation of shifting. Shift instructions may specify a left or a right shift. Each of these may be a "straight" shift (that is one in which zeros are entered into vacated positions), a "circular" shift (that is one in which bits shifted out of one end of the register are entered into the other), or a "long" shift (that is one in which information contained in one register is shifted into another register). Furthermore, the shifting may affect the entire contents of a register or only part of it. Shifts of the latter kind are frequently used to "scale" numbers[2]. Here, the sign bit is not altered and only the magnitude portion of a number is shifted. If both, positive and negative numbers are represented by sign and magnitude, we may speak of "shift magnitude" instructions. If only positive numbers are represented by sign and magnitude, but negative numbers by sign and complement, we may speak of "sign extended" shifts[3]. Magnitude shifts or sign extended shifts are frequently incorporated into "scale factor" instructions. Here, numbers are shifted left or right until their magnitude (or its complement) falls into a given range (say between 1/2 and 1 for fractional machines). The number of shifts required to get them into this range is reflected in the scale factor of the number which is usually contained in a part of the same word as the number itself or in a separate word.

Problem 62: How is the scale factor of a number to be changed
a) for a left shift by one binary position,
b) for a right shift by one binary position?

Problem 63: How can you achieve a right shift in a machine which has only provisions for a left shift?

8.1.5.2. Testing: Test instructions are basically conditional jumps. Jumps are executed (by the control unit) only if certain conditions are present. Some of the conditions which are frequently tested are concerned with the arithmetic unit. For instance, a computer can ordinarily test for possible overflows resulting from the addition of two numbers, or for positive or negative contents of the accumulator. For such simple tests, the state of an appropriate flip-flop in the arithmetic unit, say the overflow flip-flop or the sign flip-flop of the accumulator is made available to the

---

[1] For the basic design of shift registers, see paragraph 6.1.3.

[2] Each shift of a binary number by one bit position corresponds to a multiplication or a division by 2 and changes the binary "scale factor" of a number by one.

[3] We notice that the right shift of positive numbers requires the entering of zeros at the most significant bit position (disregarding the sign position) but that a right shift of negative numbers requires a shifting in of ones. In other words, the state of the sign bit is to be shifted into the most significant bit position, or the sign is to be "extended" to the right.

control unit. For more complicated test conditions, some operations may have to be performed in the arithmetic unit, before it can be established whether or not a test condition is present. For instance, if the test condition is "$A=0$", we could subtract a "one" in the least significant position of the accumulator. If during this operation, the sign of the accumulator changes from $+$ to $-$, the original contents of the accumulator have been zero. We can subsequently restore the original contents of the accumulator by the addition of "one". We may think of several alternate schemes to implement the same test. For example, if the arithmetic unit has a separate adder network, we could add "-1" to the contents of the accumulator with-

out copying the result of the addition. If the adder shows a negative sum, but the contents of the accumulator are positive, the contents are zero. Alternately, if the accumulator is layed out for subtraction (and addition is performed by the subtraction of the complement), we can introduce an artificial borrow into the least significant position. If the borrow propagates through all stages of the accumulator, its content is "zero". Finally,

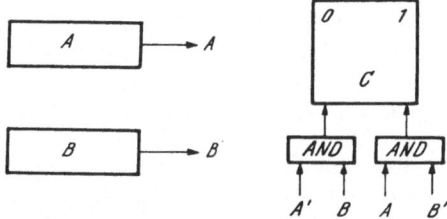

Fig. 8.36. Serial Comparator for $A > B$

there is the most straightforward, but also probably the most expensive implementation of using an AND-circuit with an input connected to the "zero" output of each flip-flop in the accumulator[1].

Up to now, we have considered tests which had to do with the contents of a single register. Frequently, test instructions are implemented which compare the contents of two registers. Examples of conditions for such tests may be $A=Q, A\neq Q, Q\geq A$. Tests of this sort may involve an arithmetic test subtraction and a subsequent restoration of all registers. Again we note that if a separate adder network is available, restore operations are not required.

The contents of two registers can be compared not only arithmetically, but also logically. Logic equality is synonymous with arithmetic equality (all bits are the same), but the logic implication $A\supset Q$ is different from $A>Q$. The condition $A\supset Q$ is true if Q contains "ones" only in those places where there is also a "one" in A (there may be "ones" in A where there are "zeros" in Q). The detection of this or similar conditions can be made by special logic circuitry connected to both registers, or by performing mask or extract operations (see below) and subsequent arithmetic checks.

---

[1] For a zero test by logic operations, see paragraph 8.1.5.3.

Problem 64: Devise a test for $a \supset b$ by performing complement, mask, or similar operations, and a check for equality.

It is worthwhile to note that the arithmetic or logic comparison of the contents of two registers is fairly simple in serial designs. For instance, the comparison $A > B$ requires only one flip-flop and two AND-circuits as indicated in Fig. 8.36.

Assume that the comparator flip-flop C is initially in the "zero" state and that the contents of both, the A- and the B-register are made available in serial form, least significant bit first. If a bit of A is "one", while the corresponding bit of B is a "zero" ($A > B$ for this bit), the comparator flip-flop is set to the "one" state. If A is "zero" and B is "one" ($B > A$, for this bit), the C flip-flop is cleared. A comparison in a higher order bit overrules any comparison of lower order bits. The final state of C, after all bits of A or B have been inspected, indicates whether or not the condition $A > B$ is met. The flip-flop is set if $A > B$, and it is cleared if $A > B$. It remains cleared if $A = B$.

Problem 65: What set and reset inputs to the comparator flip-flop would you use for a comparison:
a) $A = B$,
b) $A \geq B$,
c) $A \supset B$.

Be sure to indicate what the initial state of the comparator flip-flop is and what final state indicates the presence of the test condition.

8.1.5.3. Complement, Mask, Extract, and Related Operations: So far, we used Boolean algebra only as a tool for the design of logic circuits. However, nothing prevents us from giving a computer the capability to apply logic operations to "information" in general.

At this point it may be well to recall a few fundamental findings of Boolean algebra.

The logic operations of AND, OR, NOT are sufficient to represent any logic function or operation. If desired, the operation of AND can be expressed in terms of OR and NOT, and, alternately, the operation of OR can be expressed in terms of AND and NOT. The operations AND and NOT, or OR and NOT are, therefore, sufficient to represent any logic function or operation.

Logic Complementation (the operation of NOT) is probably the most frequently implemented logic operation. In fact, logic complementation is frequently implemented in order to perform strictly arithmetic operations[1]. By logic complementation we mean the recording of a binary "one", where

---

[1] Some number representations require a reversal of all bits in a word if a positive quantity is to be converted into the same negative quantity or vice-versa.

there previously was a "zero" and the recording of a "zero", where there previously was a "one". The operation of NOT, like other logic operations, is applied individually to every bit of a word. Depending upon the type of machine, this may be accomplished serially or in parallel. Fig. 8.37 indicates schematically a few basic approaches.

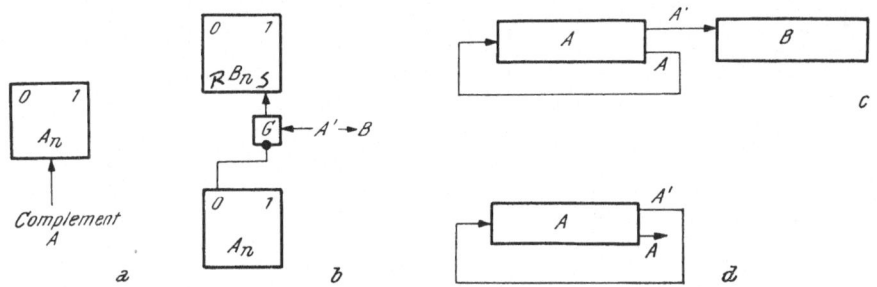

Fig. 8.37. A Few Basic Implementations of Logic Complementation

Fig. 8.37a shows one stage of a parallel register. A complement signal applied simultaneously to all stages of the register, inverts all bits of the word. Fig. 8.37b indicates a circuit in which the complement of the contents of the A-register is transferred in parallel to the B-register by a "set on zero" transfer. Here, it is assumed that the B-register is cleared before the transfer takes place. However, the clearing of B prior to the transfer can be omitted if gates for "zero" and "one" sides are provided, and the two commands "set on zeros" and "clear on ones" are applied simultaneously.

Fig. 8.37c indicates an implementation in which the complement of the contents of the A-register are transferred serially to the B-register. The indicated transfer path is established or enabled for one word time. Fig. 8.37d indicates an approach in which the contents of A are complemented with the result appearing in A itself. Normally, the output A of the register would be used to close the re-circulation loop. If, however, the output A' is used for one word time, all bits of A are complemented.

Logic Summation (the operation of OR), like logic complementation, can be performed in a number of ways. Again, it is probable that the circuitry required for its implementation is a part of the circuitry required for the performance of arithmetic operations.

Fig. 8.38 indicates a few basic possibilities.

Fig. 8.38a shows an arrangement in which the logic sum of the contents of the A- and B-registers is formed by an OR-circuit. The sum can be transferred to the C-register by the command "$(A + B) \rightarrow C$". It is assumed that the C-register is cleared before the transfer.

Problem 66: How is the diagram in Fig. 8.38a to be augmented if the logic sum of the contents of A and B is to be transmitted to C but C is not necessarily cleared before the transfer?

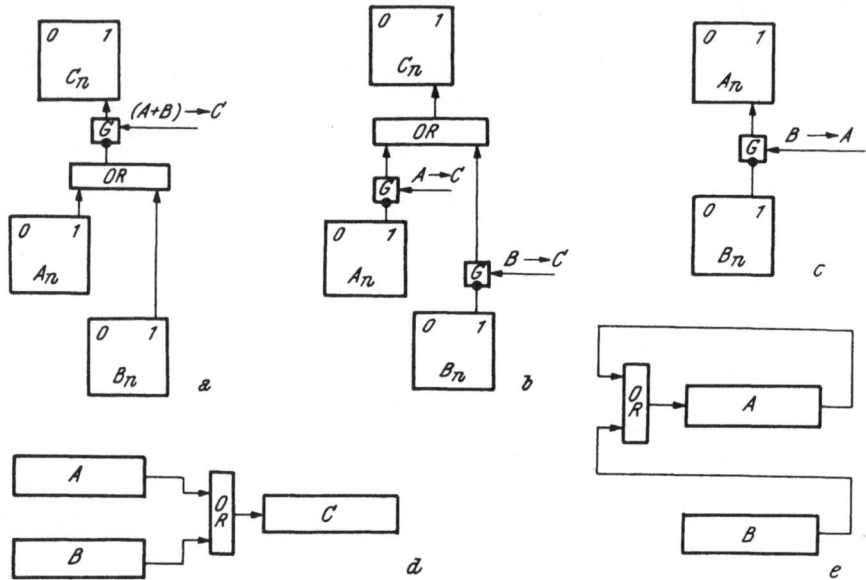

Fig. 8.38. A Few Basic Implementations of Logic Summation

Fig. 8.38 b shows an arrangement in which "ones" from A and B are transmitted to a previously cleared C-register. We note that, as a result, C will contain "ones" where there are "ones" in A, or "ones" in B. We note also that no additional logic circuitry is required, once a basic "one's transfer" from A to C and from B to C is implemented. Fig. 8.38c shows an arrangement in which the logic sum of the contents of A and B is formed in A. After the transfer $B \rightarrow A$, A contains a "one" wherever originally there was a "one" in A, or where a "one" has been transmitted from B, or both.

Figs. 8.38d and e show two self-explanatory schemes for forming the logic sum in serial designs.

Problem 67: How would you modify the circuits in Fig. 8.38 if you had to implement the function $A + B'$?

The *logic product* (the operation of AND), like the previously discussed operations can be implemented in several ways. A few basic approaches are shown in Fig. 8.39.

Fig. 8.39a indicates a scheme in which the logic product of the contents of the *A*- and *B*-register is derived in a straightforward manner by an AND circuit.

Problem 68: How is the circuit shown in Fig. 8.39a to be augmented if the logic product AB is to be transferred to C, but C is not necessarily cleared before transfer?

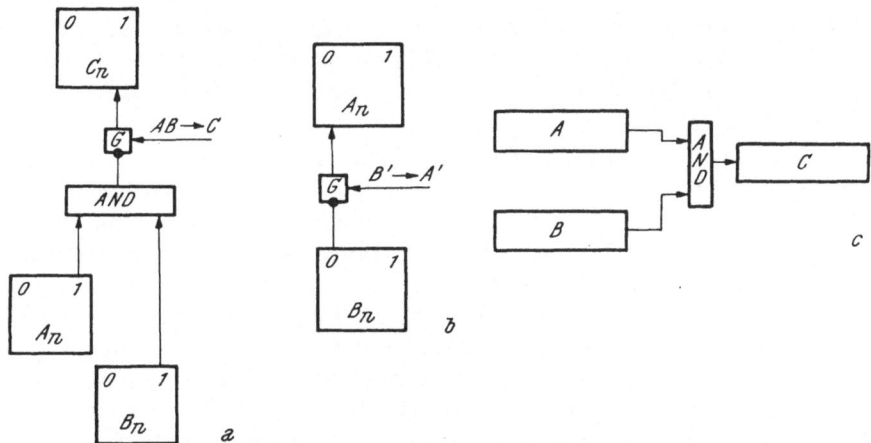

Fig. 8.39. A Few Basic Implementations of the Logic AND

Fig. 8.39b indicates an approach in which the logic product of A and B is formed in A. The zero transfer from B to A clears those positions of A which have a corresponding "zero" in B. As a result, A will contain "ones" only in those positions where originally there was a "one" in both, A and B. We note that no additional logic circuitry is required, once the basic transfer path "clear A on zeros in B" is implemented. Fig. 8.39c shows a straightforward approach for serial designs. We observe that the implementation of logic operations in serial designs requires relatively little hardware. The expense of a single AND or OR circuit is practically negligible, considering the overall circuitry required to implement the computer. However, in parallel designs, AND or OR circuits have to be provided for all bits in a word, and it may well pay to search for the simplest possible implementation. Here then implementations similar to the ones shown in Fig. 8.38a, b and 8.39b which use regular transmission paths become very attractive. In fact, in many cases, the implementation of a particular logic operation may be determined by the kind of transmission paths already available for other purposes. Suppose for example, that the path "clear A on zeros in B" which is indicated in Fig. 8.39b is not required for any purpose, other than the implementation of the logic AND. We do however, have, a path "set A on ones in B" as indicated in Fig. 8.38c. The implementation of the logic OR is, therefore, practically without cost, while the implementation of the logic AND may become relatively expensive. However, we recall that the operation of AND can be replaced by the operation of OR and complementation. If, therefore, the A- and B-registers have provisions for complementation, the operation of AND can be implemented without additional hardware. Of course, a disadvantage of this approach is the slower execution, since a series of operations rather than one is required.

Problem 69: Assume the layout given in fig. 8.38c and the existence of complement inputs to both, A- and B-registers. What sequence of control commands would you apply in order to form the logic product of the contents of A and B.

In many layouts, not only the basic logic operations of NOT, OR, and AND are directly implemented, but also more complicated functions if suitable hardware is already existing. An adder with suppressed carry propagation (modulo 2 adder) is, for instance, a rather straightforward implementation for the EXCLUSIVE OR operation[1].

The implementation of the EXCLUSIVE OR, among other operations, allows the performance of a relatively simple equality test. Suppose the accumulator A contains a certain numerical value. We now want to test whether or not the contents of B are equal to the contents of A. If we add the contents of B to the contents of A modulo 2, the final contents of A will be "zero" only if the two values were equal, whereas A becomes $A \neq 0$ if they are not equal. The operation of the EXCLUSIVE OR, plus a subsequent zero test of the accumulator is, therefore, sufficient to test for equality. We note that this approach to an equality test requires no carry propagation and is, therefore, faster than an algebraic test subtraction. The restore operation, if there is one required, is also faster than the arithmetic restore operation.

Problem 70: Assume that the content of A is 01101110 and the content of B is 00001111. Show the sequence of operations required and the contents of all registers for each step of a logic equality test.

a) if the equality test is performed by a test subtraction and a subsequent restore addition

b) if the equality test is performed by a modulo 2 addition. What restore operation is required in case b?

Assume that $A = 0$ can be tested.

Problem 71: Show that the operation of EXCLUSIVE OR can be used to perform logic complementation.

---

[1] A subtracter with suppressed borrow propagation serves the same purpose. We note also that the first step of the two-step addition scheme discussed in paragraph 6.3.1 implements the EXCLUSIVE OR. The latter may also be considered as a "controlled complement".

Computer instructions which cause the execution of logic operations are frequently referred to as controlled transfer, mask, or extract instructions. The first of these terms reflects the manner in which an instruction is executed, while the latter two reflect more the typical applications of these instructions. Controlled transfers may transfer the contents of one register to another depending upon the contents of a third. The schemes indicated in Fig. 8.39 a and c for instance, may be described as a B-controlled transfer of A to C. (Only where there are "ones" in B, is the content of A transferred to B[1]). The same operation may be described as a mask operation whereby B contains the "mask". Only those positions of B which contain a "one" allow the transfer of information from A to B. Positions of B which contain "zeros" block the transfer. The following specific example may help to clarify this point.

$$
\begin{array}{llll}
0111 & 0101 & 0011 & \text{original information } (A) \\
0000 & 1111 & 0000 & \text{mask} \qquad\qquad\quad (B) \\
0000 & 0000 & 0000 & \text{original contents of } C \\
0000 & 0101 & 0000 & \text{final contents of } C
\end{array}
$$

For clarity, all words are divided into three parts. The bits in the mask are arranged so that the outer parts of the original information are "masked out", and only the center part is transmitted or "extracted".

Information may be extracted not only into a cleared register as in the example above, but also into one which contains some prior information. The following example may serve as an illustration.

$$
\begin{array}{llll}
0111 & 0101 & 0011 & \text{contents of } A \\
0000 & 1111 & 0000 & \text{contents of } B \text{ (mask)} \\
1001 & 1001 & 1001 & \text{original contents of } C \\
1001 & 0101 & 1001 & \text{final contents of } C
\end{array}
$$

As in the previous scheme, information is transferred from A to C where there are "ones" in the mask. However, where there are "zeros" in the mask, the original content of C is not disturbed. Such a scheme is accurately described by the term "controlled substitute" (Bits of A substitute bits of C if the mask contains "ones").

Problem 72: Show the logic circuitry for at least one implementation of the just previously described controlled substitute.

For some instructions, masks may be implied rather than set up specifically. Examples are instructions which transfer or modify only a part of

---

[1] In this case we are, of course, equally well justified to describe the scheme as an A-controlled transfer of B to C.

computer words, e.g. only the least significant half, the most significant half or only the address portion, of an instruction[1]. Here, specific transfer paths may be enabled or disabled by the control unit, rather than by a mask in a register.

Sometimes very elaborate combinations of arithmetic and logic operations are specified by single instructions. A few illustrative examples of such instructions are an instruction which reverses the order of bits in a word (the most significant bit becomes the least significant one, etc.), an instruction which produces the arithmetic sum of all bits in a word (sideways add), or one which adds arithmetically the logic product of the contents of two registers to the contents of the accumulator. Again, the implementation of such instructions is not really difficult. The problem is to implement them in an economical manner in such a way that the most effective use is made of already existing hardware.

8.1.6. Floating-Point Arithmetic Operations

Any numerical quantity can be expressed in the form $C \times B^e$. For instance, the decimal quantity .0007 can be written as 7×10^{-4}, and the quantity 560 as 5.6×10^2. The advantage of such a representation is that a wide range of numerical values can be expressed very conveniently. The three quantities C, B, and e are respectively the coefficient, the base, and the exponent. For general computations (and for decimal computers), it is customary to use the base 10. Binary machines use the base 2. Since the base remains fixed for a specific machine, it does not have to be stated expressly, and two numerical values are sufficient to represent such a quantity: the coefficient (or mantissa), and the exponent (or characteristic). We speak of floating-point numbers. The coefficient and the exponent are generally treated differently when arithmetic operations are performed. The multiplication of two floating-point numbers requires, for instance, a multiplication of the coefficients, but an addition of the exponents. Arithmetic operations with floating-point numbers are, in general, more complicated than arithmetic operations with fixed-point numbers and their execution usually not only takes longer but also requires more complex hardware[2]. On the other hand, the floating-point representa-

---

[1] Such groups of bits within a computer word are frequently referred to as "bytes".

[2] It is to be noted that by no means all computers have the built-in capability to perform floating-point arithmetic operations. However, practically all computers can be programmed to operate with floating-point number representations by, let us say, a dozen individual fixed-point arithmetic or logic, instructions in place of a single floating-point instruction. The individual programmed operations resemble very closely the sequence of steps required for the built-in floating-point capability discussed below.

tion is not only more convenient but allows to perform computations in which the range of operands can not very well be predicted. In such cases, a fixed-point representation might result in overflows or lost significance (too large or too small numbers), but since floating-point numbers are automatically scaled for full significance, the problem does practically not exist.   ↖ I.e LEFT BIT JUSTIFIED.

8.1.6.1. Arithmetic Comparisons: Floating-point numbers are usually represented in a format which retains the validity of tests designed for fixed point numbers. For instance, a floating-point binary number with the value $X \cdot 2^Y$ is usually represented as:

±	Exponent Y	Coefficient X

Fig. 8.40. Representation of a Floating-Point Number

A typical format for 36-bit machines is: 1 sign bit; 8 bits for the exponent; and 27 bits for the coefficient. Typical for 48 bit machines might be: 1 sign bit; 11 bits for the exponent, and 36 bits for the coefficient.[1]

The notation for both, the exponent and the coefficient is such that equal floating-point numbers have identical representations, and so that the larger of two floating-point numbers "appears" to be larger when treated as fixed-point number. In particular, the sign of the coefficient is in the same position as the sign of fixed point numbers. The coefficient (which has less significance than the exponent) occupies the less significant bit positions of the word. Its binary point is considered to the left of its most significant bit, and its magnitude is restricted to values in the range $1 > X \geq 1/2$.[2] The size of the exponent is restricted to values which can be accommodated in the limited space available. If, for instance, eight bits are used to represent exponents, only $2^8 = 256$ different exponents can be accomodated. Usually, half of the 256 available "codes" represents positive exponents, while the other half represents negative exponents. In other words, exponents may be integers in the range $+128 > Y \geq -128$. Their notation is such that the smallest possible exponent (i.e. -128) has the smallest "fixed-point" representation (i.e. eight binary zeroes) and the largest possible exponent (i.e. $+127$) has the largest "fixed-point representation" (i.e. eight ones). In this arrangement, the most significant bit indicates the sign of the

---

[1] This arrangement has been shown to be a good compromise. We note that an increase in bit positions for the exponent would decrease the precision of the coefficient, and an increase in bit positions provided for the coefficient would reduce the range of acceptable exponents.

[2] In other words, the absolute value of the most significant bit of the coefficient is always a "one".

exponent. However, it is customary not to consider the sign as a separate entity, but to speak of "biased" exponents. The bias in the above example is 128, that is, the binary equivalent of $128 + Y$ is recorded instead of the exponent Y. This notation immediately allows the comparison of positive floating-point numbers by the rules applicable to the comparison of fixed-point numbers. This can be easily verified by working out a few numerical examples.

Problem 73: What is the largest and smallest positive quantity which can be represented
   a) by a 36-bit floating-point binary number,
   b) by a 48-bit floating-point binary number?

Problem 74: Why is it necessary to restrict the size of binary coefficients to values $1 > X \geq {}^1/_2$ if fixed-point rules are to be applied to the comparison of floating-point numbers?

Problem 75: Represent the following quantities by 36-bit floating point binary numbers:
   a) $+.007652_8$
   b) $+7363.5_8$
Use 12 octal digits to represent the 36 bits of a floating-point number.

Of course, it is desirable that fixed-point rules remain valid also for the comparison of negative floating-point numbers. This requires that the exponent and the coefficient of negative floating-point numbers be represented in the same manner as negative fixed-point numbers, i.e. either by the "magnitude" (i.e. by the same bit configuration as positive numbers) or by the "1's complement" (i.e. by the inverted bit configuration), or by the 2's complement (i.e. by the inverted bit configuration increased by one).

Problem 76: Represent the following quantities by 36-bit floating-point numbers:
   a) $-.007652_8$
   b) -7363.5_8

Assume that negative fixed-point numbers are represented by their 1's complement. Use 12 octal digits to represent the 36 bits of a floating-point number.

With the given rules, the quantity "zero" can only be approximated. However, since "zero" is so frequently used, an exception is made. It is usually represented by the bit combination $00 \cdots 0$. This violates the rule that $1 > X \geq {}^1/_2$, but otherwise gives a valid representation (e.g. 0×2^{-128} for 36-bit or 0×2^{-1024} for 48-bit floating-point numbers). Furthermore "fixed point" comparisons and arithmetic operations with "zero" produce valid results.

A final remark might be appropriate. The above representation of floating-point numbers may have some inconveniences. The two important advantages are, however, that the word-length is the same as that for fixed-point numbers and that identical comparison procedures can be used for both kinds of number representations.

8.1.6.2. Addition and Subtraction: The addition and subtraction of floating-point numbers is rendered more difficult than that of fixed-point numbers by the requirement for an alignment of coefficients. Before an addition or subtraction can take place, the scale factors (i.e., the exponents) must be examined so that digits of equal weight in the coefficients can be aligned. Let us illustrate this requirement with a simple specific example. Suppose the two decimal quantities 7×10^2 and 6×10^4 are to be added. The coefficient 6 has a different weight than the coefficient 7. We, therefore, cannot immediately perform an addition. The necessary alignment can be acomplished in a number of ways. Two relatively simple possibilities are indicated below:

$$
\begin{array}{rl} 7 & (\times 10^2) \\ +600 & (\times 10^2) \\ \hline 607 & (\times 10^2) \end{array}
\quad \text{or} \quad
\begin{array}{rl} 0.07 & (\times 10^4) \\ +6 & (\times 10^4) \\ \hline 6.07 & (\times 10^4) \end{array}
$$

As we can see, the alignment requires a shift of at least one coefficient. Each shift of a coefficient by one place to the left decreases its associated exponent by one, and each shift to the right increases the exponent. We continue to shift until the exponents of the two quantities are the same. In order to simplify the alignment procedure, a specific computer may either always shift the quantity with the smaller exponent to the right, or always shift the quantity with the larger exponent to the left. Which of these approaches is used depends upon what provisions for the shifting of operands are existing, and also whether or not the machine has a double-length accumulator[1].

So far, we have discussed the initial alignment of operands. We note, however, that in certain cases, a final alignment of a result may be required. A subtraction of nearly equal coefficients results, for instance, in a small difference. A left shift is required to get the coefficient representing the difference into the customary range $1 > X \geq 1/2$. Alternately, an addition of aligned coefficients may result in a sum larger than one. In such a case, a right shift of the coefficient representing the sum is required.

The rounding of coefficients introduces an additional complication. We note that the addition or subtraction of aligned coefficients in their shifted

---

[1] Unless a double-length register is available, the left shift may cause the loss of the most significant digits of the larger operand.

or staggered positions produces a result with more digits than either of the original operands. A truncation and simultaneous rounding is, therefore, required. The truncation and rounding may take place before an addition or subtraction is performed (initial round), or after (final round), or both.

Having stated the basic requirements for floating-point additions and subtractions, let us start to think about possible implementations. Since we have previously seen that the addition and subtraction of fixed-point numbers can be implemented in a number of ways, we should expect an even greater variety of schemes for the equivalent floating-point operations. Let us, therefore, begin with the discussion of a specific, fairly straight-forward, and relatively uncomplicated scheme.

As long as we are aware that we are discussing only one of many possibilities, we are free to make assumptions as they fit our purpose. To begin with, let us assume that separate registers are provided for the handling of coefficients and the handling of exponents. In this manner, our discussion is not unnecessarily complicated by the consideration of time-sharing approaches.

The accumulator is the natural place for the addition or subtraction of coefficients. Most likely, it has provisions for shifting so that it can also be used to perform the alignment. Suppose we have a single-length accumulator, A, with provisions for right and left shifts and a single-length register X whose contents can be added to the contents of the accumulator. As in a fixed-point addition or subtraction, we first bring one coefficient into A and the other into X. This initial placement of coefficients is indicated in Fig. 8.41.

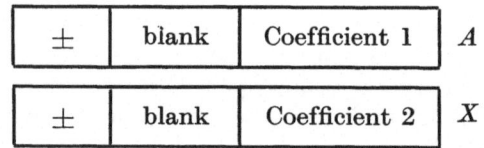

Fig. 8.41. Initial Placement of Coefficients

The coefficients and their signs occupy only a part of the available register space. The positions within the word which are normally filled by the exponent are marked blank[1]. For the initial alignment, it is now necessary to shift one or the other coefficient. Since we previously assumed that the

---

[1] Remember that we agreed to work upon the exponents in separate registers. Note also that the term "blank" means that the absolute value of these digits is zero. Hence, if the computer represents negative numbers by their comple-ment, the space may be filled with ones, i.e. the sign is extended into these positions so that the negative coefficients can be added or subtracted like negative fixed point numbers.

A-register has the provisions for shifting, and since we should shift right rather than left so as not to loose the most significant bits, it is necessary that the coefficient with the smaller exponent be placed into A. In other words if the coefficient initially contained in A is the larger coefficient, we must interchange the contents of A and X[1]. The coefficient contained in A can now be shifted right for alignment[2]. Fig. 8.42 shows the contents of A and X after the initial alignment, assuming that an exchange of operands had been necessary.

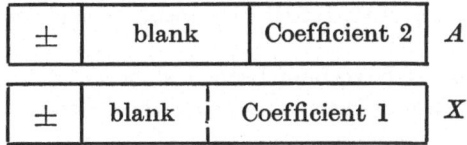

| ± | blank | Coefficient 2 | A |

| ± | blank | Coefficient 1 | X |

Fig. 8.42. Placement of Coefficients after the Initial Alignment

We notice that the sign retains its original position in order to permit the subsequent algebraic addition or subtraction according to fixed-point rules.

Problem 77: What rules apply to the rounding of the shifted coefficient? Assume a binary machine. Can you propose a circuit for the rounding?

Problem 78: Propose a simple arrangement which fills the positions vacated by the coefficient by appropriate "blanks", i.e., zeros if the initially blank space contains zeros, and ones if the original blank space contains ones.

The addition or subtraction of coefficients may follow any one of the already discussed schemes. There remains now only the final alignment or "normalization" of the result. If the accumulator shows a too large result, i.e., a carry into a previously blank position, a right shift of the contents of the accumulator by one place is required. However, if the result is too small and outside the normal range, a left shift by one or more positions is necessary[3].

---

[1] This interchange may pose a problem in some designs, for instance when DC flip-flops are used which cannot be read and set simultaneously, or when the existing transfer paths are not adequate. In such cases it may be necessary to employ a third register, e.g., the Q-register, for the temporary storage of one coefficient, and to perform the exchange in several steps, e.g. $A \rightarrow Q$, $X \rightarrow A$, $Q \rightarrow X$.

[2] The appropriate number of places for this shift is determined by a comparison of the exponents to be discussed below.

[3] If the accumulator has no provisions for a left shift, a right circular shift of A, or a shift from A to Q by an appropriate number of places may be the solution. In this latter case, the result would appear in Q instead of A, and a re-transfer from Q to A may be desirable.

Let us now discuss the treatment of exponents. We assume that they have been brought initially into a pair of registers A' and X'. Fig. 8.43 indicates the initial placement.

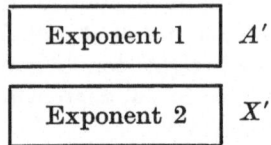

Fig. 8.43. Initial Placement of Exponents

The two registers have here been designated A' and X' to emphasize the close relationship of their contents with the contents of A and X. The first operation should be a comparison of the exponents. Let us assume for the moment that a straightforward comparator for the contents of A' and X' is available. We can then proceed as follows: Compare the two exponents. If exponent 2 is smaller than exponent 1, interchange the contents of A' and X'[1]. Compare the exponents again. If they are not equal, shift the contents of A (the smaller coefficient) one place to the right, and add "1" to the contents of A' (increase the exponent). Repeat this comparison, shift, and addition until the exponents in A' and X' are equal. The coefficients are now aligned. If, after the addition or subtraction of the coefficients, a normalization of the result is necessary, increase the exponent in A' by "1" for a right shift, or, alternately, decrease the exponent by "1" for each left shift of A.

In this scheme, the operations on the contents of the A' and X' registers are not very difficult. However, if no comparator is available, comparisons must be performed by test subtractions[2]. We may, for instance, determine whether or not an initial interchange of operands is required by the sign of the difference of a test subtraction $(A') - (X')$. If the result is negative, that is if exponent 1 is smaller than exponent 2, an interchange is not required. We note that the difference of the exponents is equal to the number of places by which the smaller coefficient in A has to be shifted. We may, therefore, shift the contents of the accumulator while we reduce the difference contained in A' by "1" for each shift until the contents of A' are reduced to "zero"[3]. This procedure is simpler than a continuous test subtraction and restoration of operands. We note also that the procedure

---

[1] This interchange may, again, require the use of an auxiliary register.

[2] This test subtraction in itself may require a complementation, addition, and re-complementation of operands.

[3] This reducing of the negative difference requires the addition of 1 to the contents of A' for every shift of A. The difference has been reduced to zero when the sign in A' changes from — to +. In machines with shift matrices the difference determines immediately the number of places to be shifted.

does not alter the larger exponent contained in X' so that it properly reflects
the exponent of both operands after the initial alignment.

If the test subtraction produces a positive difference, the coefficients
have to be interchanged. In order to be able to follow identical alignment
procedures in both cases, it is desirable to interchange also the exponents.
Since, however, the larger of the two exponents has been replaced by the
difference in A', we first have to perform a restore addition $(A' + X')$,
before we exchange the contents of A' and X'. A subsequent test subtraction produces now a negative difference (or a zero difference) and we
may proceed as explained above. The comparison of exponents in this
manner requires only a sign check of A' in addition to the capability
to subtract the contents of X' from the contents of A'. The whole
procedure has now become fairly complicated so that it may be best to
show all operations in the form of a flow chart. We can then also combine
the handling of exponents with the handling of coefficients. Before we
start, let us assume a specific floating-point format and a layout of registers
and transmission paths as it is indicated in Fig. 8.44.

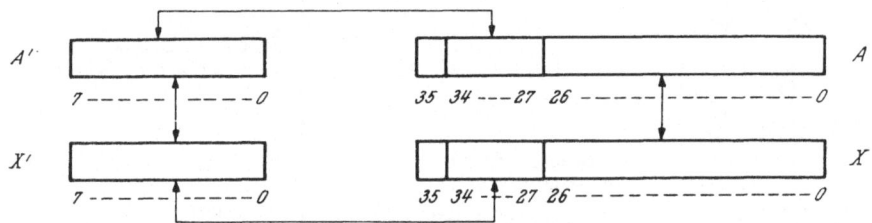

Fig. 8.44. Sample Layout of Registers for Floating-Point Additions or
Subtractions

The sign of a floating-point number is in position 35. Bits 27 through 34
represent the biased exponent, and bits 0 through 26 represent the coefficient.
The contents of X can be added to or subtracted from the contents of A.
Also, the contents of X' can be added to or subtracted from the contents
of A'. The contents of A_{27-34} (the exponent) can be transferred to A' and
the contents of X_{27-34} can be transferred to X'. Negative numbers are
represented by sign and magnitude. The flow chart begins after the two
floating-point numbers have been transferred to A and X, but before the
exponents have been "unpacked". In other words, we assume that the two
floating-point numbers to be added or subtracted are contained in A and X,
and that A' and X' are cleared. We further assume that the final result is
contained in A and "packed" into the normal floating-point format.
Fig. 8.45 shows the applicable flow chart.

The flow chart reflects the previously discussed operations with
both, the coefficients and the exponents. In addition, the unpacking

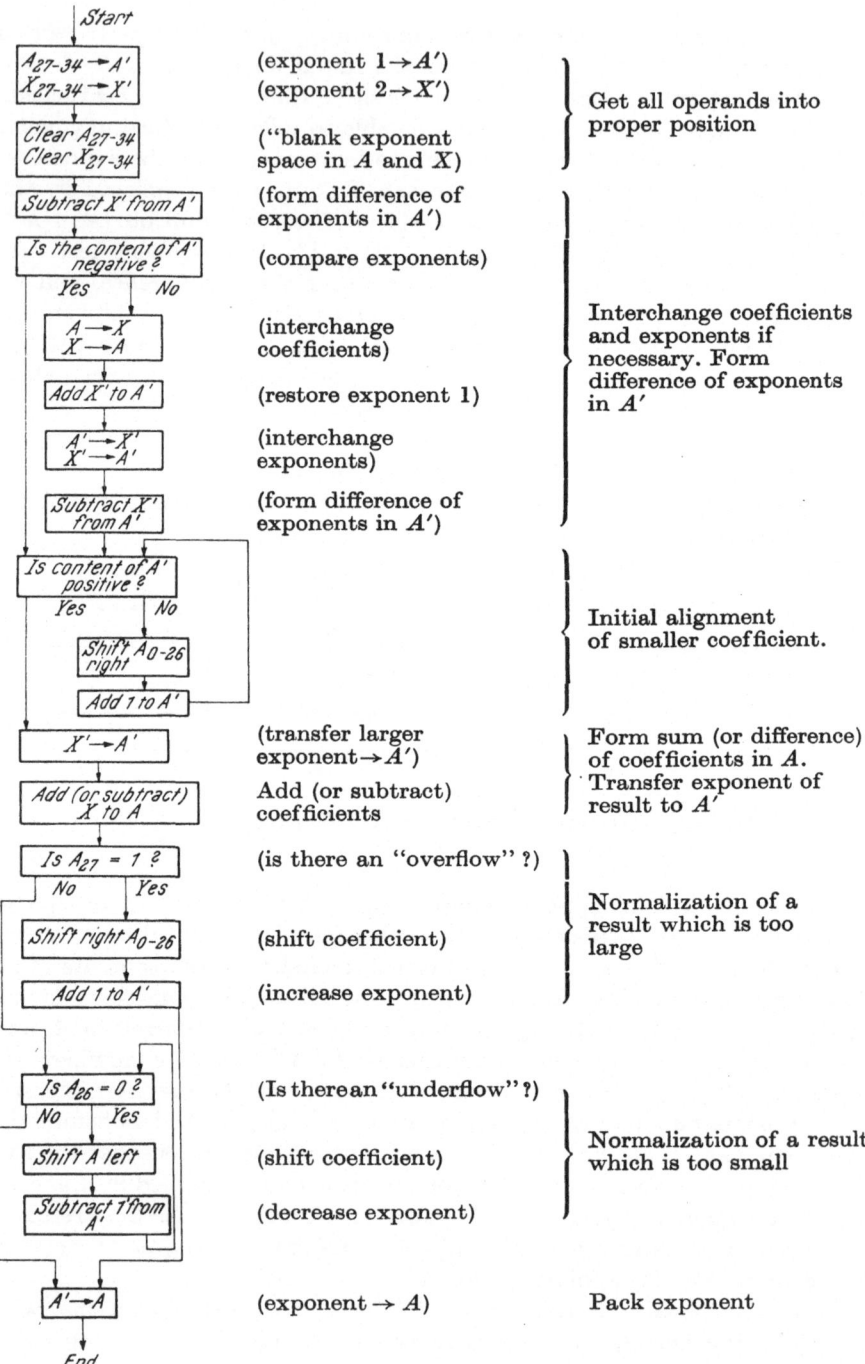

Fig. 8.45. Sample Flow Chart for Floating-Point Addition or Subtraction

and packing of exponents is shown. The parts of the flow chart applicable to these latter operations should be self-explanatory. The flow chart is complete with two exceptions: the initial or final alignment of very small (or zero) coefficients as it is shown in the flow chart may result in a large (or infinite) number of shifts. Computers usually have provisions to terminate the alignment procedure after a limited number of shifts. These provisions have been omitted in Fig. 8.45. A reasonable limit in our example might be 27, since an initial right shift by more than 27 positions, shifts even the most significant bit of a coefficient outside the range of the accumulator and a left shift of a result by more than 27 positions will not bring the coefficient into the range $1 > x \geq {}^1/_2$ if the first 27 shifts have failed to do so. In such a case, both the coefficient and the exponent are usually "forced" to zero. The second omission is a check for an exponent overflow or underflow. In rare cases even the wide range of numbers which can be accommodated by the floating-point representation may be exceeded. If a result is too large, usually an overflow alarm is given. If a result is too small, it is usually forced to zero.

Problem 79: Show the contents of all registers during the addition or subtraction of the following floating point numbers:

$$.56_8 \times 2_8{}^4 + .43_8 \times 2_8{}^6$$

Assume the layout in Fig. 8.44 and follow the flow chart in Fig. 8.45.

We see that this scheme requires a careful sequencing of individual operations, but its implementation poses no unique or particularly difficult design problems. The same can be said for other possible schemes. Rather than discussing a number of variations in detail, let us, therefore, only indicate how the above scheme could be modified. Some of these variations are quite obvious. For instance, a different floating-point format (say 48 instead of 36 bits) requires longer registers and slightly different transfers. If negative numbers are represented by their complements, a re-complementation of exponents may be required[1]. If the accumulator has no provisions for a left shift, a right circular shift in A, or a long right shift of the result into, say, Q may be necessary. The determination of the proper number of shifts becomes then more complicated. Finally, a different layout of registers such as the use of dual-rank registers, or the co-use of already existing registers[2] may require minor or major modifications of the indicated flow chart.

---

[1] These operations may be required in any case if a subtraction of exponents is performed by the addition of the complement.

[2] For example index registers and/or index adders may be used for exponent arithmetic. See paragraph 8.2.5.

Problem 80 (Voluntary): Assume a computer layout of registers as indicated below:

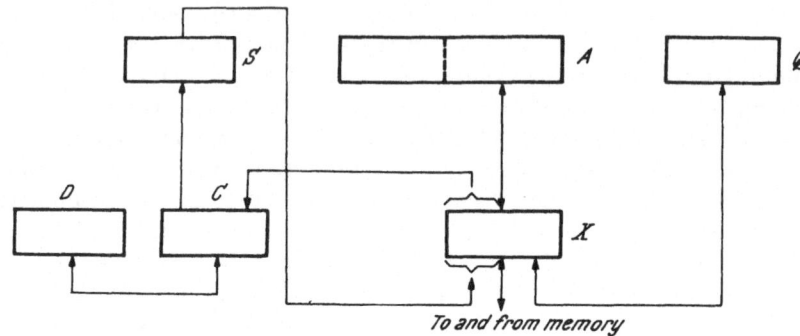

To and from memory

The accumulator A is a 72-bit double-length register. X and Q are 36-bit single-length registers. D, C, and S are 8-bit registers. All operands are entered into the arithmetic unit via X, and all results are transferred to storage via X. The only provision for shifting is a left shift in A (straight or circular). The indicated transfer path from X to A is implemented as "add X to A", and the indicated transfer path from C to S is implemented as "add C to S". Negative numbers are represented by the 1's complement. Design a flow chart for the execution of a floating point add. Assume that all arithmetic registers are cleared at the beginning. Start your flow chart with the transfer of the first operand from the memory to X. Hints: It is advantageous to load the first operand into both A and Q, to simplify an eventually necessary exchange of operands. The initial alignment of operands is here best accomplished by a left shift of the coefficient with the larger exponent.

Again, some final remarks may be appropriate. The indicated methods for the addition and subtraction of floating-point numbers can, of course, be adapted to decimal notation. The proper range of coefficients may then be $1 > x \geq .1$. Shifts by one (decimal) position, dependent upon the BCD code, used, may be equivalent to a shift by 4 to 6 "binary" places. It is also worthwhile to note that some computers have a built-in double-precision floating-point capability. Operands are then represented by two words rather than one. The format of the first word is usually exactly as that for single-precision floating-point numbers. The second word simply contains an extension of the coefficient, that is, those least significant digits of the coefficient which could not be accommodated in the first word. Of course, the flow chart becomes more complex if double precision quantities are handled and more storage registers may be required. However, no problems arise which have not previously encountered in the implementation of single-precision floating-point or double-precision fixed-point operations.

8.1.6.3. Multiplication and Division: Floating-point multiplications or divisions require practically no more complex hardware organization than that required for floating-point additions or subtractions and fixed-point multiplications or divisions. In some respects, floating-point multiplications and divisions might even be considered simpler than the equivalent fixed-point operations[1].

As indicated by the following equation, the multiplication of two floating-point numbers requires a multiplication of the coefficients and an addition of the exponents:

$$(X_1 \cdot 2^{Y_1}) \times (X_2 \cdot 2^{Y_2}) = (X_1 \cdot X_2) \cdot 2^{Y_1 + Y_2} \qquad (8.22)$$

Similarly, a floating-point division requires a division of coefficients and a subtraction of exponents:

$$(X_1 \cdot 2^{Y_1}) \div (X_2 \cdot 2^{Y_2}) = X_1 \div X_2 \cdot 2^{Y_1 - Y_2} \qquad (8.23)$$

No comparison of exponents or initial alignment of coefficients is required, so that the multiplication or division of coefficients can commence immediately after the floating-point numbers have been unpacked. The multiplication or division of the coefficients can be performed by the same mechanism which performs the equivalent fixed-point operations. The addition or subtraction of exponents can be handled in a similar manner as the comparison of exponents in floating-point additions or subtractions.

The normalization of a result is also somewhat simpler than in additions or subtractions, but can be accomplished by the same techniques. We note that the multiplication of two coefficients in the range between 1 and $\frac{1}{2}$ produces a result in the range between 1 and $\frac{1}{4}$. The final alignment of a product requires, therefore, either no shift, or a left shift by one bit.[2] The division of two coefficients in the range between 1 and $\frac{1}{2}$ produces a result in the range between 2 and $\frac{1}{2}$, so that either no shift, or a right shift by one digit is required.

Packing the result, i.e., attaching the proper exponent to a product or quotient may require slightly different procedures than those required for packing of sums or differences. We note that a quotient is usually derived in Q, but a sum or difference in A. Furthermore, a product is of double-length, and if the least significant part of it is to be saved as a floating-point number, a different (smaller) exponent has to be attached. The same is true for saving the remainder of a division. Even though the procedures

---

[1] The fact that coefficients are aligned may, for instance, simplify the division process (see paragraph 8.1.3.). The reduced length of operands may speed up both, multiplications and divisions.

[2] Similarly, in a multiplication of decimal floating-point numbers a shift by at most, one decimal digit is required.

or the transfer paths required for the final packing of a result may be different from those required for the packing of sums or differences, their implementation should be simple and straightforward.

Problem 81: Design a flow chart for the multiplication of two floating-point numbers. Make the same basic assumptions as those on which the flow chart in Fig. 8.45 is based.

8.1.7. The Layout of the Main Arithmetic Registers

As we have seen, arithmetic units require at least two arithmetic registers for addition and subtraction, and at least three registers for multiplication and division. While these registers may be easily identified in most computer designs, many variations in layout are possible. These variations include the number and specific arrangement of arithmetic registers, but, perhaps more important, the manner in which they are connected to the remainder of the computer system.

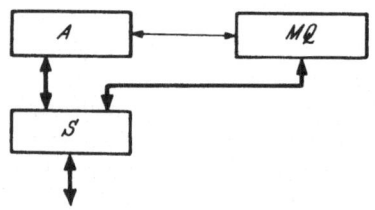

Fig. 8.46. "Typical" Layout of an Arithmetic Unit with Three Single-Length Registers

Let us look here at a few "typical" and some more extreme layouts of arithmetic units. We shall concentrate our attention on the arrangement of main registers and transmission paths and neglect details such as control inputs to registers (e.g. clear; complement; shift-signals), and types of transfers (e.g. transfer of "zeros" and "ones", transfer of "ones" only, complement on "ones" etc.). Also, we shall neglect registers required for the handling of exponents in floating-point operations, since, in many cases functions of these registers are combined with control functions to be discussed later.

Fig. 8.46 shows a rather common arrangement of an arithmetic unit. The three registers, S, A, and MQ are single-length registers. S is the storage register, A the accumulator, and MQ the multiplier/quotient register.

To perform an addition, the first operand is brought from storage into A via S. The second operand is then brought into S. The contents of S and A are added, the sum being recorded in A. The result can be transferred from A back to the memory via S. The sequence for a subtraction is identical to that of an addition, except that the minuend is complemented in S before addition.

To perform a muliplication the multiplicand is placed into S and the multiplier in MQ. Partial products are accumulated in A, starting with the least significant bits. As the no longer needed bits of the multiplier are shifted out of MQ, the least significant bits of the product are shifted from

A to MQ, so that the combination $A\text{-}MQ$ contains a double-length product at the end of the multiplication. Each half of the double-length result may then be transferred to the memory via S.

At the beginning of a division, MQ (or the combination $A\text{-}MQ$) contains the dividend and S contains the divisor. During the division, the dividend is shifted left in the combination $A\text{-}MQ$, and the contents of the accumulator (dividend) are compared with the contents of S (divisor). The digits of the quotient starting with the most significant digit are shifted into MQ from the right. At the end of the division MQ, contains the quotient, and A the remainder.

A number of variations of the basic layout are possible. One may provide a path from either A, or MQ, or both back to the memory which by-passes the S-register. One may also omit the path from S to MQ and load the multiplier or dividend from S into MQ via A. The S-register may be solely an arithmetic register, or it may double as the central exchange register, or even as the memory register itself.

If also the details of the transmission paths are considered, many more variations are possible. For instance, the path from Q to S may not allow a straightforward transfer of "ones" and "zeros" but might be implemented in such a manner that S is first cleared, then complemented (S contains now all "ones"), and finally recomplemented wherever there is a "zero" in the corresponding position of Q. Such a transfer is, of course, rather complicated, but may be preferable where speed is not a consideration and where this specific path must be implemented due to other design considerations. A common example of such a situation is the transmission from S to A. For an addition, S must be transferred to A in such a manner that, after the transfer, A contains the sum of A and S. In other words, the path "Add S to A" must exist. Rather than implementing a second path "transfer S to A", one frequently clears A and then uses the path "Add S to A" to load A with an operand.

One further observation may be appropriate. The heavy lines in Fig. 8.46 are meant to indicate a parallel transfer of information; the thin lines are indicative of serial (1-bit) transfers. This is, of course, applicable only to parallel arithmetic units. Serial arithmetic units with the equivalent layout would employ serial transfers throughout.

Problem 82: Suppose the arithmetic unit shown in Fig. 8.46 has only the following transmission paths and control signals implemented:

Storage $\rightarrow S$ (transfer of "ones"),

Subtract S from A,

$A \rightarrow S$ (transfer of "zeros"),

$S \rightarrow$ Storage (transfer of "zeros" and "ones"),

Clear A,

Clear S,

Complement S.

What sequence of transfer and control signals would you use to add two numbers ? Assume that negative numbers are represented by their B-1 complement, that the operands are stored in the memory, and that the result has to be stored.

An arithmetic unit with a double-length accumulator is shown in Fig. 8.47.

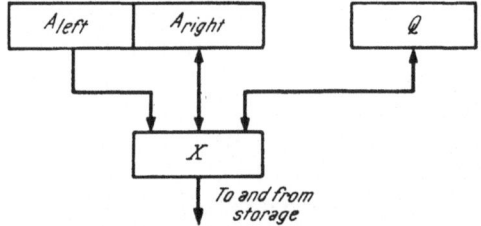

Fig. 8.47. Layout of an Arithmetic Unit with Double-Length Accumulator

The right half of the accumulator and the X-register are used in the normal manner for additions and subtractions. Any carries (or borrows) which may occur in the right half, are propagated into the left half of the accumulator, so that double-length results of additions or subtractions can be derived.

A multiplication is started with the most significant bits of multiplier and product. Both, A and Q are shifted left during the multiplication. The double-length product remains in A. Both halves of the product (A left and A right) can be transferred back to the memory via the X-register. The multiplier may be preserved in Q by a circular shift.

The division starts with the most significant half of the dividend in A_{right} and the least significant half of the dividend in A_{left}[1]. A is shifted left (circular) during the division. The bits of the quotient are recorded in Q.

We notice that A and Q are shifted left for both, multiplication and division. We note also that no paths for a shift from A to Q or from Q to A are required.

Problem 83: What individual steps (arithmetic operations and transfers of information) are required if two double-length numbers are to be added ?

---

[1] The first step in the division algorithm may be an exchange of the contents of A_{left} and A_{right} (e.g. by a left circular shift) so that the programmer may insert the most and least significant halves of the dividend respectively into A_{left} and A_{right}.

Compare these with the steps which are required if the machine has only a single-length accumulator as in Fig. 8.46.

In many cases, arithmetic units incorporate separate adder networks. A "typical" layout may be the following:

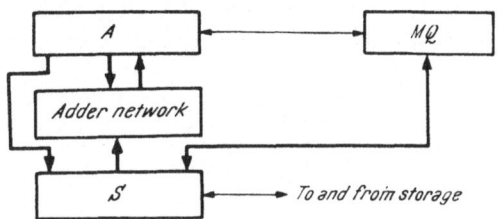

Fig. 8.48. Arithmetic Unit with Adder Network

The operation is essentially the same as that of the layout in Fig. 8.46, except that additions are not performed within the A-register itself (e.g. by the two-step addition method shown in paragraph 6.3.1), but the contents of A and S are used as inputs to the adder network and the sum output of the adder network is copied back into A.

A slight modification of the layout shown in Fig. 8.48 will allow shifts to be performed simultaneously with additions. Thus, the speed of multiplications and divisions is increased. Fig. 8.49 shows one possible layout.

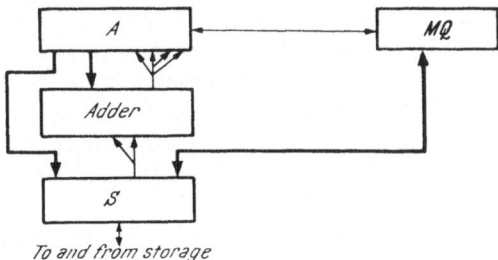

Fig. 8.49. Arithmetic Unit with Provisions for Simultaneous Add and Shift

The S-register can be connected to the adder in a straightforward manner, or shifted one position to the left (contents of S multiplied by two). In a similar manner, the sum output of the adder can be copied into the A-register straight, shifted one position to the left, one position to the right, or two positions to the right. Multiple transfer paths like these are especially valuable in connection with high-speed multiplication or division techniques.

All the layouts shown so far, assume registers constructed with flip-flops which can be set and read-out simultaneously[1]. If this is not the case (if flip-flops with DC coupling are used), duplicate registers have to be provided in order to accumulate, shift, count, etc. The two ranks of a register are usually designated by the same letter, but distinguished by a subscript or superscript e.g. A and A_1, or A and A^*, Fig. 8.50 shows a possible arrangement of the two ranks of the accumulator in such a machine.

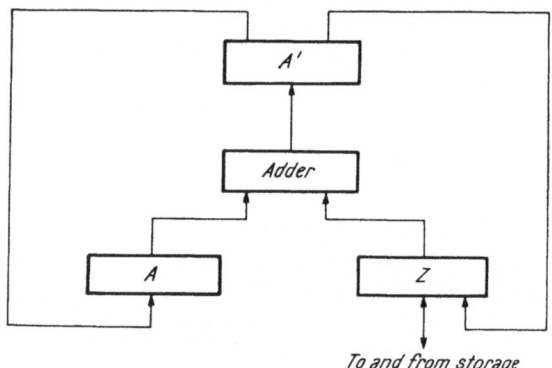

Fig. 8.50. Arrangement of a Dual-Rank Accumulator

The first operand of an addition is loaded into the rank A from storage via Z, adder, and rank A'. The second operand is loaded into Z. The sum of the contents of A and Z is then recorded in A'. If the sum is to be stored, it is first transferred to Z and then to storage. If, however, the sum is to be used further (e.g. in a repeated addition), it can be transferred from A' to A, and then the contents of Z (which may be a previous operand, or a new one) can be added to the previous sum. In essence, two ranks are required for the A-register in order to store the old and the new content of the accumulator.

In practice, not only the paths shown in Fig. 8.50 but also additional paths for the shifting of information are required. Fig. 8.51 shows the more detailed layout.

In this layout the contents of A can be copied into A' straight, shifted left by one or two positions, or shifted right by one position. In other words, the contents of A can be multiplied by 4, 2, 1, or $1/2$. However, the contents of A can be also multiplied by six, that is, the contents of A shifted left by two positions can be added to the contents of A shifted left by one position.

---

[1] For instance, the contents of the A-register are used as input to the adder and the sum is copied into the same A-register. Alternately, it is assumed that a shift can be performed within a single register.

Depending upon which particular combination of transmission paths is enabled, one may obtain any one of the following terms on the output of the adder.

$$
\begin{array}{lll}
{}^{1}/_{2}A & {}^{1}/_{2}A+Z & {}^{1}/_{2}A+2A=2{}^{1}/_{2}A \\
A & A+Z & \\
2A & 2A+Z & A+2A=3A \\
4A & 4A+Z & \\
6A & 2Z &
\end{array}
$$

Certainly, not all the indicated transmission paths have to be implemented to obtain a workable layout, but their provision results in a more flexible and faster computer.

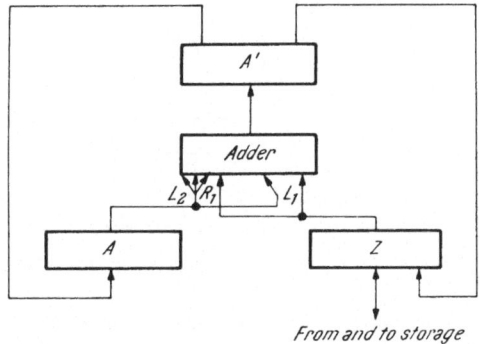

Fig. 8.51. More Detailed Layout

Fig. 8.52 shows the layout of an arithmetic unit which is equivalent to that of Fig. 8.46 except that all registers have dual ranks and that a separate adder network is used.

The accumulator and the Q-register have provisions for left and right shifts. For instance, the contents of A_1 can be transferred to A_2 either straight, or one position to the left, or one position to the right. Provisions have also been made for a transfer from Q to A, or from A to Q, for a left circular shift in A and Q individually, or for a left circular shift in the combination A-Q.

There are also parallel transfer paths from A to Q, and Q to A. Certainly not all the paths are really required, that is to say, one can build a workable computer with less communication paths. However, the diagram may be representative of a moderately complex modern arithmetic unit. The communication network gives the designer a large amount of freedom in the "microprogramming" of computer instructions, i.e. in determining the exact sequence of inter-register transfers required to implement individual computer instructions. Furthermore, a flexible communication network is one essential prerequisite for true high-speed operation.

Let us now, for a moment, examine the layout of the accumulator in machines with dual-rank registers. Many machines have an organization

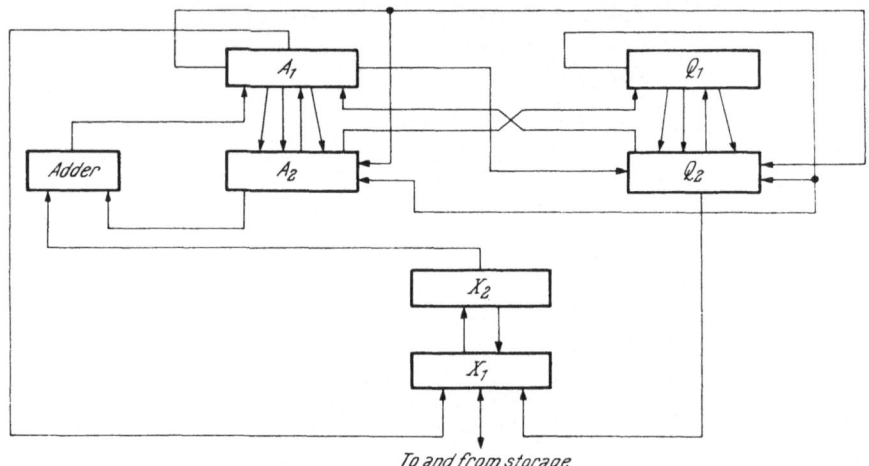

Fig. 8.52. Layout with Dual-Rank Registers and Separate Adders

following closely the arrangement shown in Fig. 8.50. However, for high-speed operation it becomes inconvenient to transfer the result of an addition from A' back to A, before the next accumulative addition can be performed.

Fig. 8.53 shows two basic arrangements in which both halves of a dual-rank accumulator are equivalent as far as their connection to the adder network is concerned.

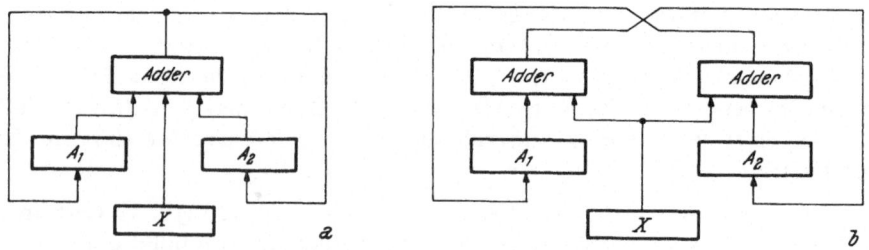

Fig. 8.53. Two Basic Arrangements of Dual-Rank Accumulators

The arrangement in Fig. 8.53a has single adder and can be used in two distinct modes. Either the contents of X can be added to the contents of A_1 and the sum recorded in A_2, or the contents of X can be added to the fontents of A_2 and the sum recorded in A_1. The two modes are alternated during an accumulative addition. We note that no "dummy" transfers crom A_1 to A_2 or vice versa are required.

The arrangement in Fig. 8.53b is very similar to the arrangement shown in Fig. 8.53a except that two adders instead of one are used. The expense for the additional adder can be set off by the less complex gating circuitry at the input of the adder.

In many cases, the output of the adder is connected to a shift matrix in order to perform simultaneously an addition and a shifting of information. Fig. 8.54 shows this arrangement in more detail.

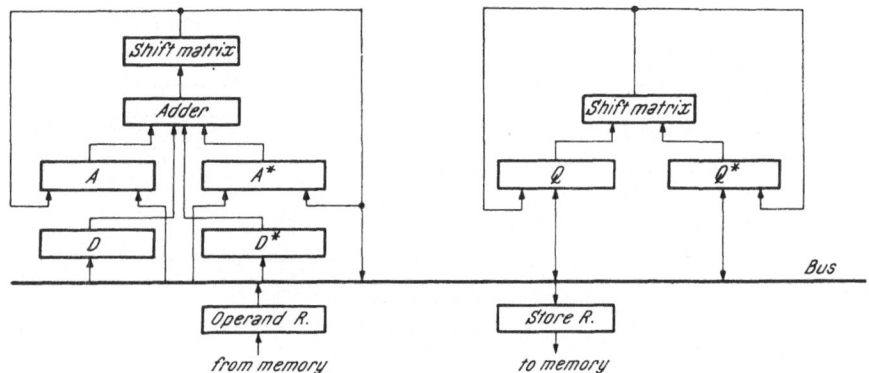

Fig. 8.54. Layout with Dual-Rank Registers and Shift Matrix

All three arithmetic registers (A, Q, D) are of dual rank so that they can be used in an alternate manner. We note that no dummy transfers for addition or shifting are required, in spite of the fact that DC coupled flip-flops are used.

As we have seen, many arithmetic units contain to some extent circuitry and communication paths which are not truly necessary but which are provided for high-speed operation. Figs. 8.55 and 8.56 may be indicative of extremes in this respect. The layout in Fig. 8.55 is that of an accumulator with carry-save feature for high-speed multiplication[1].

The arrangement contains two modulo 2 adders which do not propagate carries but provide separate sum and carry outputs for each stage. For each addition, the modulo 2 sum and the generated carries are recorded in separate registers (A and B, respectively). During one add cycle, the contents of X_1 are added to the contents of A_2 and B_2 and the result is recorded in A_1 and B_1. During the next add cycle, the contents of X_2 are added to the contents of A_1 and B_1 and the result is recorded in A_2 and B_2. Add

---

[1] See paragraph 8.1.2.3.

cycles continue to alternate until a multiplication is practically completed. Only then, the final contents of A and B are added, and the carries propagated. For this purpose, the arrangement contains a full adder which can also be used for normal additions. This adder is omitted in Fig. 8.55. Of course, many other details are omitted. For instance, some transfers paths are arranged so that information is shifted[1].

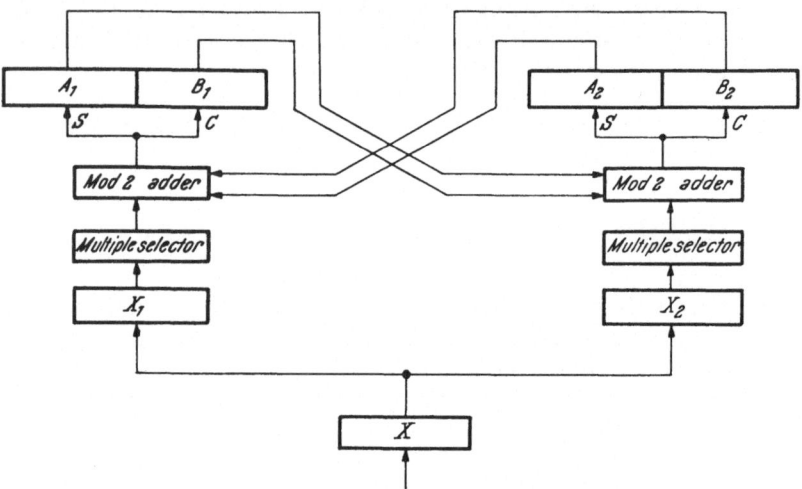

Fig. 8.55. Dual-Rank Accumulator with Carry Save Feature

Provisions can be made to transmit the bits of the least significant half of the product to the omitted Q-register in the order in which they are derived.

The arrangement contains a multiple selector between X registers and the adders, to allow a multiplication by uniform multiple shifts[2].

Problem 84 (Voluntary): Expand the layout of Fig. 8.55 for multiplication by uniform shifts of two places (for each add cycle). Show the multiple selector, and the transmission path to Q for the least significant half of the product. Indicate straight and shifted transmission paths (e.g. S, L_1, R_2, etc.).

Fig. 8.56 finally shows an extremely flexible layout which is well adaptable to double-precision operations.

---

[1] The contents of the accumulator, i.e., the product is shifted right with respect to the contents of X, i.e., the multiplicand. Note also that carry bits contained in B have twice the weight of the sum bits contained in A.

[2] See paragraph 8.1.2.3.

The registers, as shown, are single-length registers. We note, however, that the layout of A- and Q-registers is identical. Since also an A-adder and a Q-adder is available, double-precision quantities can easily be added or

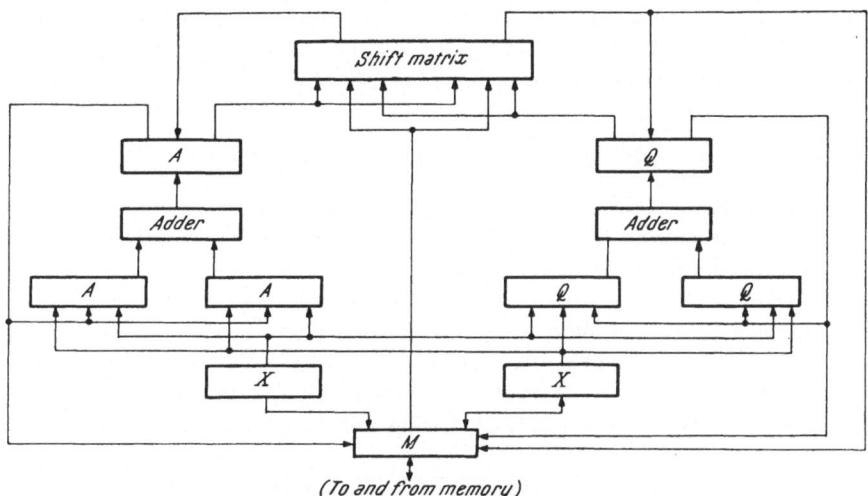

Fig. 8.56. Extremely Flexible Layout, Well Adaptable to Double Precision Operations

subtracted in one operation. We note also that a very universal shift and transfer arrangement is used. The contents of A may be transferred to A while simultaneously the contents of Q are transferred to Q shifted by any number of places. Alternately, Q may be shifted and transferred to A, while A is shifted and transferred to Q. Finally, M may be transferred to A and/or Q straight or shifted. Incidentally, the layout provides three ranks for A and Q. This allows to use the addition scheme shown in Fig. 6.58, in which a part of the addition is performed by transfers.

8.2. The Control Unit

In chapter 7, we defined the over-all function of the control unit as the controlling of all internal operations of the computer. As we have seen, this includes the control of arithmetic or logic operations and that of information transfers. We also identified several individual components and their respective functions:

a *program address register* containing an address which determines the memory location from which a program instruction is fetched;

an *instruction register* which holds an instruction while it is executed;

a *function translator* which decodes the "op-code" of instructions; and a *sequencer* which issues a sequence of control commands for the execution of each instruction.

These four basic components can be found in practically any digital computer. However, the organizations of control units differ widely in many other respects. The range of variations is comparable to that of arithmetic units, where we are usually able to identify the three main arithmetic registers, but where the details of one design can be very unlike those of another.

8.2.1. The Sequencing of Operations

Sequencers are laid out specifically for each computer. It is, therefore, practically impossible to show their organization in a systematic manner. On the other hand, their design is basically simple. Once we have indicated their design in principle and a few possible approaches to their implementation, anyone capable of logic design should be able to implement any combination of desired sequences in a number of ways. The problem becomes difficult only when one attempts to "optimize" the implementation. Let us begin with the rough functional diagram of a sequencer.

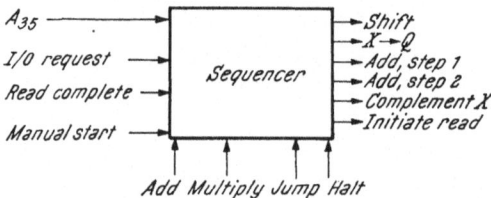

Fig. 8.57. Simplified Functional Diagram of a Sequencer

The sequencer receives inputs and provides outputs on a large number of individual signal lines. Inputs are mainly of two types. One type indicates which instruction is to be executed. Fig. 8.57 shows only a few possibilities: add; multiply; jump; and halt. These inputs come from the function translator which decodes the "op-codes" of the instruction contained in the instruction register. The second type of inputs indicates the status of various units of the computer. Fig. 8.57 again shows only a few diversified examples: "A_{35}" is the sign bit of the accumulator; "I/O request" indicates that the input/output unit is ready for an information transfer; "read complete", indicates that the memory has completed a read operation; and "manual start" is a signal generated when the operator pushes the start button. The outputs of the sequencer represent control commands. Output lines are directly connected to the appropriate circuits in all units. For example, the line labelled "$X \to Q$" may terminate at transfer gates which are connected to the output of the X-register and the input of the Q-register. In a parallel machine, the signal "$X \rightarrow Q$" may, for instance, be a short pulse that strobes all transfer gates simultaneously and, thereby, transfers the contents of the X-register to the Q-register in parallel. In a serial machine, the signal "$X \rightarrow Q$" may be a level prohibiting the re-circulation of the Q-register but enabling the flow of information from X to Q for one word time. The other outputs in Fig. 8.57, similarly, are of appropriate duration. The lines labelled "add, step 1" and "add, step 2", for instance, might terminate at an accumulator using a two-step addition scheme; the line labelled "complement X" might terminate at the complement input of all X-register flip-flops: and the line labelled "initiate read" at the computer memory. Again, only a few diversified examples of command signals are shown in Fig. 8.57, but, by now, it may have become apparent that the exact numbers and kinds of sequencer inputs and outputs depend completely upon the specific organization of all units of the computer.

Let us now try to develop step-by-step one of many possible concepts for the implementation of a sequencer. Suppose we consider for the moment only the sequence of control commands for the execution of a single instruction, say an add. A chain of delays as indicated in Fig. 8.58 could produce such a sequence in a rather straightforward manner.

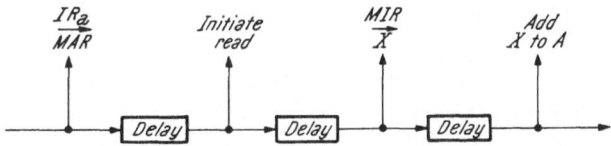

Fig. 8.58. Sample Sequence for the Execution of an Add Instruction

Once started by an input signal, the chain will provide a sequence of individual output signals[1]. The timing is entirely determined by the adjustments of individual delays. Of course, it has to be consistent with the speed of the components which "execute" the commands. The labelling of control commands in Fig. 8.58 is compatible with the model layout of a digital computer discussed in chapter 7. However, it is obvious that the same principle can be used to implement any other desired sequence of control commands.

Let us now expand the concept to provide different sequences for the fetching and the execution of instructions. Fig. 8.59 shows the resulting sequencer.

Again, the indicated sequences are compatible with the model layout in Fig. 7.6. In the right half of Fig. 8.59 we notice that a separate sequence has

---

[1] Dependent upon the type of delays used, these command signals may be either pulses or levels.

been provided for the execution of each individual instruction. The add sequence of Fig. 8.58 is only one of four indicated sequences, the others being jump, multiply, and halt. Since the multiply sequence is quite lengthy, it is abbreviated. The other two are quite short. In fact only one operation is required for each.

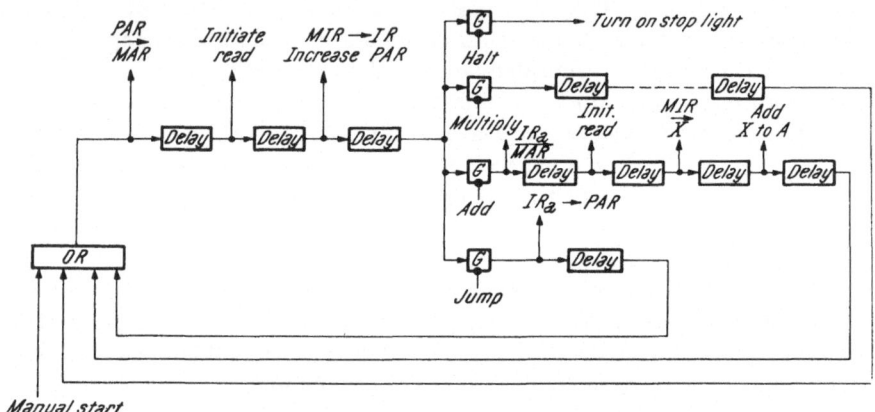

Fig. 8.59. Simplified Asynchronous Sequencer

As soon as the execution of any instruction, other than a halt, is complete, the sequence in the left half of Fig. 8.59 is initiated (note the paths returning from right to left). This sequence ($PAR{\rightarrow}MAR$, initiate read, $MIR{\rightarrow}IR$, increase PAR) fetches a new instruction[1]. After the new instruction has been brought into the instruction register, an instruction "split" takes place. We notice that for each instruction only one of the four indicated gates is enabled and, therefore, only one of the four specific sequences is initiated.

Once started, the sequencer supplies a continuous flow of control commands consistent with the program to be executed. The operation begins with a manual start and ends with the execution of a halt instruction[2]. Commands which fetch instructions alternate with commands which execute instructions. The sequencer, as indicated, is basically complete but, of course, not in all details. Its design can be easily modified to provide different sequences or to take care of a larger instruction repertoire.

Problem 85: Design the sequence for a transmit instruction which loads the accumulator with the contents of the storage location determined

---

[1] This latter sequence is also initiated when the computer is started manually.
[2] Notice that no "return path" for the halt is provided.

by the address part of the instruction. Use the model layout in Fig. 7.6 as a guide.

In many instances it is required that external conditions control the progress of certain sequences. Fig. 8.60 indicates two examples.

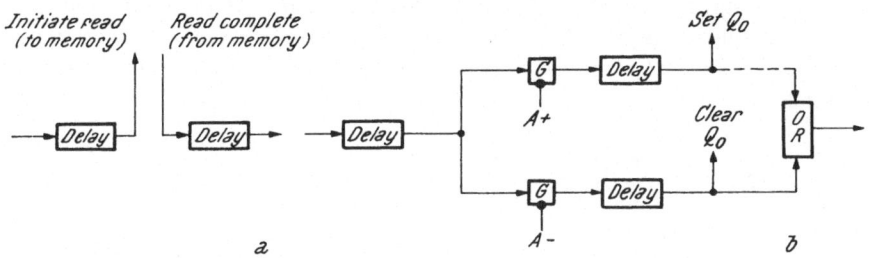

Fig. 8.60. Two Examples Where External Conditions Influence the Progress of a Sequence

Fig. 8.60 a represents a case in which the chain of events internal to the sequencer is interrupted until a resume signal from an external source is received. This approach becomes very attractive if the time required to perform an operation is variable, such as the read or write operation in certain types of memories, or the addition in a completion recognition adder. In such cases, operations can be resumed immediately after the execution is completed. The alternative to this scheme would be the delaying of further operations until sufficient time has elapsed for an execution even under the most unfavorable conditions.

Fig. 8.60b is representative of cases where alternate actions have to be performed dependent upon external conditions. The diagram might implement a part of a divide sequence where, depending upon the sign of the accumulator, a quotient bit has to be set to "one" or "zero". A few other examples of such cases are a multiply sequence, where an "add" is initiated only for a "one" in the multiplier, or an alignment sequence for floating-point numbers, where no further shifts are executed once the coefficient is in the proper numerical range.

Problem 86: Design a sequence for a conditional jump. Take the model layout in Fig. 7.6 as a basis and assume that a jump is executed for positive contents of the accumulator but no jump is executed for negative contents.

One branching of control sequences merits special mentioning: that for "interrupts". Examples for interrupt conditions might be an overflow in an arithmetic register, a faulty read or write operation by a tape unit, or a request for an information transfer by some external equipment. On such occasions it may be desired to interrupt the normal execution of a program temporarily and to execute a short interrupt routine. Interrupts can be handled in many different ways, but a common approach is to store the current program address (contained in PAR) in some predetermined storage location, say 0000, and to perform a jump to an interrupt program starting at some other predetermined storage location, say 0001. At the end of the interrupt program, the original contents of PAR are restored and the execution of the "normal" program resumes exactly where it was interrupted. Fig. 8.61 indicates an implementation of such an interrupt sequence.

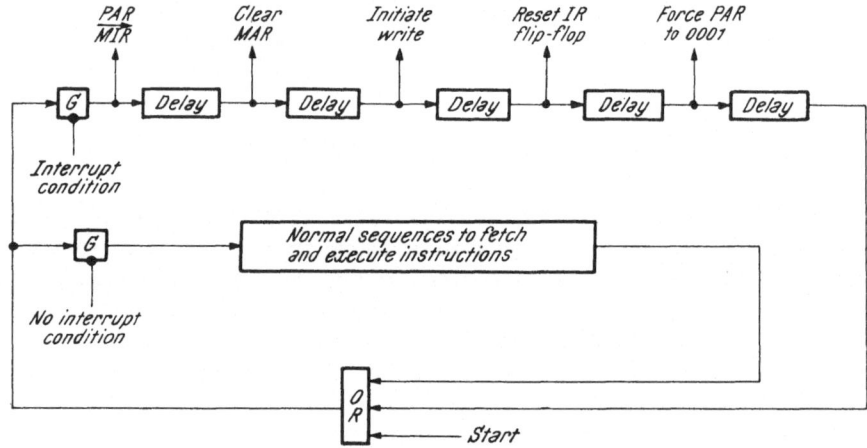

Fig. 8.61. Example of an Interrupt Sequence

If an interrupt occurs, no further "normal" instructions are executed. Instead, the program address is saved in storage location 0000 ($PAR \rightarrow MIR$, clear MAR, initiate write), the interrupt condition is cleared (reset IR flip-flop), and a jump to location 0001 is performed (force PAR to 0001). The interrupt routine can now be executed like any other program. The last instruction of the interrupt routine should be a jump back to the location determined by the address previously saved in storage location 0000, so that the execution of the normal program resumes at exactly the place where it had been interrupted.

The sequence in Fig. 8.61 is, as we have said previously, only an example of the implementation of an interrupt. Its discussion should have confirmed our previous statement that this approach to the implementation of sequencers is very flexible and can be adapted to practically any requirement. Incidentally, the indicated approach is typical for "asynchronous" machines in which the timing of individual operations is practically independent and not controlled by a common timing source.

In many designs, it becomes desired or even mandatory to synchronize various operations. For instance, if recirculating registers are employed, the recirculation and the transfer of information must be synchronized so that a bit of information transferred from one register to another appears at its appropriate place within the train of other bits. This synchronization is usually achieved by timing signals synchronized with a common "clock", e.g. a pulse generator with a constant pulse repetition rate. The same clock can be used to synchronize the control commands issued by the sequencer. Fig. 8.62 shows one possible technique.

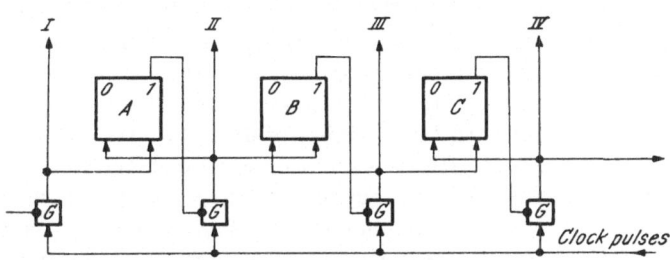

Fig. 8.62. Part of a Ring-Counter as Synchronous Sequencer

The figure shows a series of flip-flops in an arrangement similar to that in "ring-counters". Only one flip-flop at a time is in the "one" state. When flip-flop A has been set by an input pulse, it is cleared by the next clock pulse which also sets flip-flop B. The following clock pulse sets flip-flop C and clears flip-flop B, and so on. Output pulses appear in sequential order on the lines labelled I, II, III, IV. In principle, the arrangement is equivalent to that of Fig. 8.58, except that only delay times equal to the clock period are used[1]. Because of this similarity, the overall approach to the sequencer can be virtually identical to the approach previously described.

Problem 87: Repeat problem 86. Use the basic approach indicated in Fig. 8.62 for the design.

Problem 88: The arrangement in Fig. 8.62 provides pulses as output signals. How could you use the same basic approach to provide level signals with:
a) a duration of one clock period,
b) a duration of several clock periods?

---

[1] Of course, by leaving some outputs unused, individual commands may be separated by any integral multiple of the clock period. In some designs, two clock "phases" are used so that command signals may be spaced in increments equivalent to one half of the basic clock period.

A slight difficulty is encountered in this design when unsynchronized resumes have to be accepted by the sequencer. In such a case, one or two additional flip-flops may be needed to synchronize the resume signal.

Problem 89: Design a circuit equivalent to that given in Fig. 8.60a. Assume that you have to synchronize the "read complete" signal.

The two previously shown approaches to the design of sequencers are rather straightforward, but also relatively uneconomic: A large number of individual delays or flip-flops are needed to implement all sequences required in even a small computer. Let us, therefore, try to indicate a few alternatives.

The arrangement in Fig. 8.63 uses a fixed counter to provide a basic sequence. We imagine the counter progressing through all its possible states for the execution of a single instruction. Some counter states, say the lower ones, can be associated with the fetching of an instruction, others, say the higher ones, are associated with the execution of an instruction. The decoding matrix provides a single, unique output for each counter state. In effect,

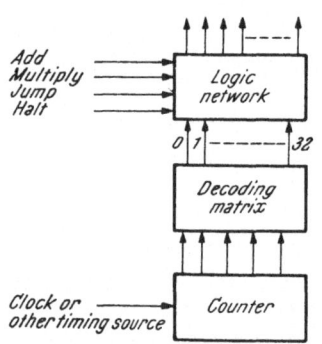

Fig. 8.63. Sequencing with a Fixed Counter

the outputs of the matrix are energized in sequential order during each instruction cycle. The indicated logic network gates now certain of these outputs to produce appropriate control commands. For instance, an output "complement X" may be produced only if the instruction to be executed is a subtract *and* if the, say, fifth output of the decoding matrix is energized. It should be apparent that the number and variety of sequences which can be implemented with this approach is limited only by the size of the counter and the complexity of the logic network. The scheme can be used immediately for synchronous sequencers. For asynchronous sequencers, a slight modification is required. Instead of having clock pulses advance the counter, various resume signals are used. Thus, the counter is advanced to its next sequential state as soon as the current operation is completed.

Compared with previously discussed approaches, a much smaller number of storage elements (flip-flops or delays) is required. This, normally, more than compensates for the larger amount of required logic circuits. However, if faithfully copied, the scheme would have a serious disadvantage. All instructions, no matter how complex or simple, require a full count of the timing counter. For this reason a counter with selectable modulus[1] is fre-

---

[1] See paragraph 6.2.3.

quently substituted for the fixed counter. The counter can then be reset as soon as the execution of an instruction is completed.

In many cases, also flexible counters[1], are used to advantage. Certain counter states can then be skipped or repeated as desired. Such an arrangement becomes the prime example of a sequential mechanism in which a new state is determined by the previous state *and* by a number of other conditions, e.g., the type of instruction being executed, external status conditions, resumes, etc. Fig. 8.64 may be indicative of such an approach.

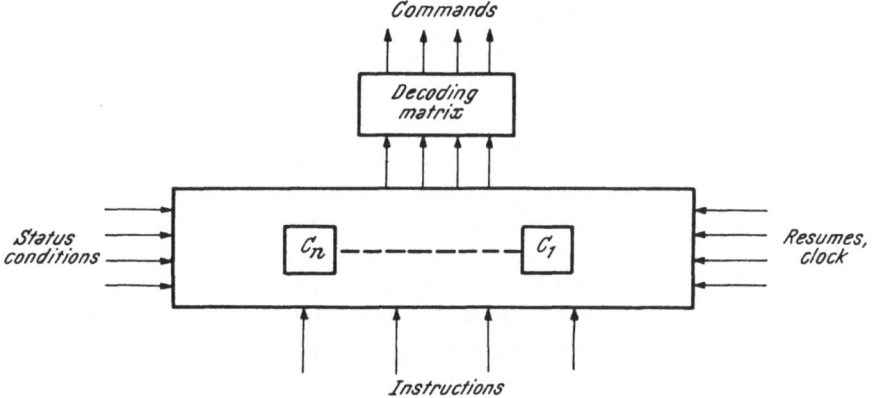

Fig. 8.64. A Flexible Counter as Sequencer

A counter with n flip-flops can be in any of 2^n possible states. These states are decoded by the matrix which produces a control signal associated with each specific state. For example, the state 0—...—00 may produce the signal "$PAR \rightarrow MAR$", the state 0—...—01 the signal "initiate read", and the state 0—...—10 the signals "$MIR \rightarrow IR$" and "increase PAR". If the counter sequentially assumes these three states, a new instruction is fetched. The specific command sequences for the execution of an instruction can be initiated by other sequential or non sequential counts. For example, the state 0—...—11 may follow the state 0—...—10 if an add instruction has been fetched, but the state 01—...—0 follows if the instruction is a multiply, or the state 10—...—10 if it is a jump. Theoretically, counter states can be assigned completely at random, although a nearly consecutive count sequence may simplify the counter circuitry.

Problem 90: Consider the sequences shown in Fig. 8.59. Assume that you have to implement the same sequences with a 5-bit flexible counter. Which control signals would you associate with what counter states? How would you perform the instruction split?

---

[1] See paragraph 6.2.3.

The three or four indicated basic approaches to the implementation of sequencers have here been shown separately. However, it is very probable that we would find a combination of techniques in an actual computer. For instance, sequencers which basically consist of flip-flop or delay chains, may incorporate counters which control the number of times a repetitive sequence has to be executed. Conversely, sequencers using counters as the basic component may contain delays for the purpose of producing very large or very small spacings between some consecutive control signals. The problem of designing a sequencer becomes, therefore, more a problem of selecting the particular combination of techniques and components which gives an economic solution under the circumstances, rather than finding a fundamental solution.

8.2.2. Function Translation

The function translator decodes the "op-code" of the instruction being executed. In parallel machines, the design of a function translator is quite simple. A straightforward approach is indicated in Fig. 8.65.

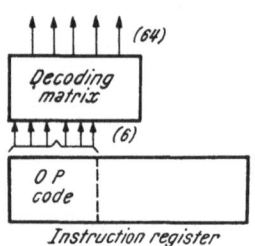

Fig. 8.65. Function Translator in a Parallel Machine

The function translator is merely a decoding matrix. Its inputs are the, say, 6-bits of an instruction designating the type of operation, e.g. multiply, divide, jump, etc. For each of these op-codes, only one of the, say, 64 matrix outputs is energized[1]. The outputs are primarily used by the sequencer, but, of course, can be made available to other components[2]. Serial machines can use the same approach to decode instructions if they contain a static instruction register. However, in many cases, recirculating registers are used to hold the instruction being executed. With such an arrangement it may become impossible to "look" simultaneously at all the bits which define the type of instruction, and the individual bits of the op-code must be decoded one at a time.

Let us show this in a simple, but specific example. Suppose we have only four different instructions: add, multiply, jump, and halt. The binary opcodes are listed below:

Instruction	Op-Code
Add	00
Multiply	01
Jump	10
Halt	11

---

[1] The decoding matrix may also provide special outputs which are energized for certain "groups" of instructions, such as test instructions, input/output instructions and the like, if this should be desirable.

[2] For instance, an operator indicator panel.

Two individual decisions must be made before the type of an instruction is determined and a sequence for its execution can be initiated. A possible implementation is indicated in Fig. 8.66.

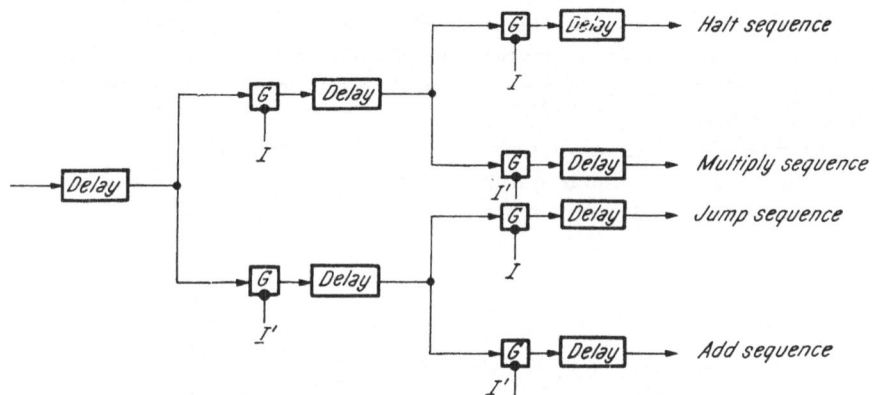

Fig. 8.66. Instruction Split in a Serial Design

The output of the instruction register (I or I') is examined at two different bit times: when the least significant bit of the op-code is available, and when the most significant bit is available[1]. Depending upon the state of I, different sequences are initiated. The whole arrangement resembles very much that of a sequencer, and in fact, the sequencer decodes op-codes in many serial designs. Although an approach employing chains of delays has been illustrated in Fig. 8.66, the scheme can be adapted to other approaches. Fig. 8.67 indicates an approach using a flexible counter.

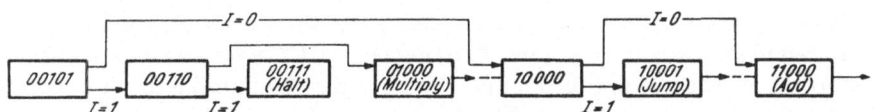

Fig. 8.67. Flow Chart for Sample Instruction Split

The flow chart is labelled with the consecutive states of a 5-bit flexible counter[2]. The counter counts, e.g., from state 00101 to state 00110 if the least significant bit of the op-code (I) is a "one". The counter "skips" from state 00101 to state 10000 if this bit is a "zero". Depending upon the most

---

[1] This can be done at two different bit times within the same word time, or at the same bit times during different word times. In the latter approach the contents of the instruction register are shifted and the same bit position is examined during each word time.

[2] Compare also figure 6.42

significant bit of the op-code, similar skips or counts are performed leaving the states 00110 and 10000. It is assumed that the counter states 01000,10001, and 11000, respectively, are the initial counter states of the sequences which execute, multiply, jump, and add instructions.

We already noted that the instruction split in serial designs usually requires several individual operations. We also note that, even though a decoding matrix can be omitted, the instruction split may require a fairly complex sequencer.

Fig. 8.68 shows an approach to an instruction split in serial designs which, although not much simpler in principle, may in practice allow a reduction of required components at the price of a longer execution time.

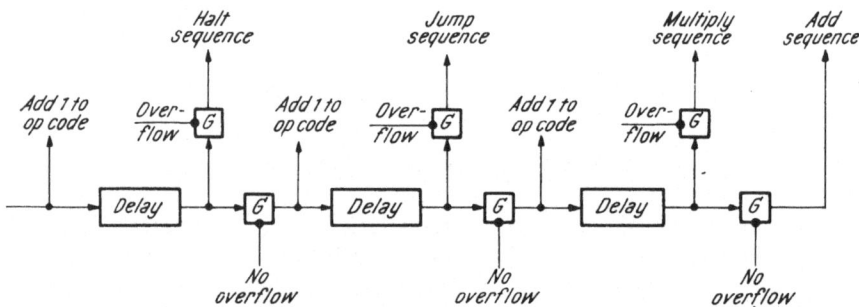

Fig. 8.68. Instruction Split by Counting

The approach is again illustrated with an implementation using delay chains. It also assumes the instruction repertoire shown on page 282. The op-code is repeatedly modified during the instruction split by an addition of "1". If the op-code is "11", the first addition produces an overflow and the halt sequence is initiated. For an op-code "10", two consecutive additions are required to produce an overflow. The op-code "01" requires three additions before an overflow is produced, and the op-code "00" produces no overflow even after three additions. We see that the instruction-split sequence becomes more straightforward than in previous schemes, but, of course, the average time to decode an instruction is increased. Judging from Fig. 8.68, the approach may seem to require even an increase in hardware. However, if the arithmetic adder can be used to perform the addition and overflow test, and if the sequencing is performed by a counter, rather than a chain of delays, a rather economic solution may result.

The two basic approaches indicated in Fig. 8.68 and 8.66 can again be combined if this should be desired. For instance the high-order bit (or bits) of the op-code can be used to split "groups" of instructions in the manner indicated in Fig. 8.66, while the selection of individual instructions within a group follows the approach indicated in Fig. 8.68.

Problem 91 (Voluntary): Compare the **expense** for the decoding of op-codes in a serial design

a) assuming a static instruction register and a decoding matrix;

b) assuming a recirculating instruction register and a sequencer with flexible counter.

8.2.3. Addressing

An important function of the control unit is the addressing of the computer storage[1]. Before we begin to discuss sophisticated addressing schemes, it may be well to clarify and summarize some basic concepts concerned with addressing.

An address is a bit combination which identifies a storage location. It is normally assumed that a storage location has the capacity of one word, that each storage location has a unique address, that each address has a unique storage location associated with it, and that the address is not affected by the information which is stored in the associated location and vice versa. As we shall see, there are important exceptions to all these assumptions but, for the time being, let us accept them as valid.

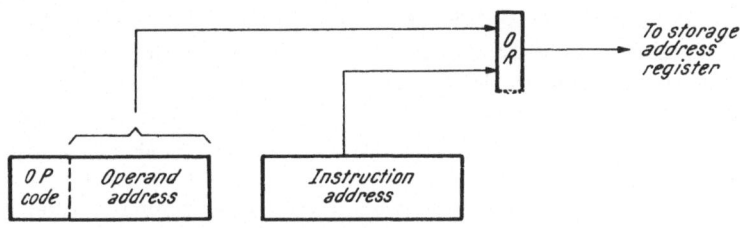

Fig. 8.69. Basic Addressing

The computer finds or selects information by addressing. Addresses facilitate the "access" to the stored information. An individual computer instruction is associated with addresses in two respects: For one, the instruction itself is normally contained in a storage location. An access to this location must be made before the instruction can be brought into the instruction register, and then executed by the machine. We speak of instruction addresses. The fetching of instructions is a straightforward process which we will not consider in further detail. Secondly, the instruction in turn usually refers to one or more addresses. These addresses specify storage locations, whose content is used by the computer while executing the instruction. We speak of operand addresses. Operand addresses are usually contained in the instruction word, together with the operation code.

---

[1] In some designs not only the memory, but also registers are addressed.

In the most basic design, only two registers and a few associated transfer paths accomplish the addressing:

The operand address is normally specified by the instruction being executed, and is, therefore, contained in the instruction register. The instruction address is contained in the program address register. For the fetching of an instruction, the contents of the program address register are sent to the storage address register, while the address portion of the instruction is sent to the storage address register for the reading or storing of an operand. This approach is basically simple and straightforward. However, instructions may contain more than one address. There are one, two, three and four address instructions in practical use.

A typical one-address instruction might be the following: Add the contents of location U to the contents of the accumulator. Abbreviated, the instruction would appear as: $AD\ U$. Of course, the actual computer representation is by a binary code, say 6 bits for the operation "AD" and 15 bits for the address "U". A corresponding two-address instruction may be $AD\ U\ V$. It could be interpreted as: Add the two operands at locations U and V, leave the result in the accumulator. Using three addresses, we could have an instruction: $AD\ U\ V\ W$, to be interpreted as: Add the operands at locations U and V, store the result in location W. It may be hard to imagine the usefulness of a fourth address in an instruction. However, some computers use it, for instance, to specify the location of the next instruction to be executed. $AD\ U\ V\ W\ X$ would then be interpreted as: Add the contents of U and V, store the result in W, and go to X for the next instruction[1].

When the instructions of a particular computer normally refer to one, two, three or four addresses, we speak of one, two, three and four-address machines. There have been extended arguments concerning the relative merits of these machines. Unfortunately, the problem eludes an unbiased and objective analysis; it is largely a matter of personal preference. Considering the two extremes, we might say that a single four-address instruction can accomplish more than a single one-address instruction, but with several one-address instructions, we can accomplish the same and possibly in the same time. Recently, the trend seems to be towards flexible formats which allow the mixing of instructions with a differing number of addresses.

Immediate Addressing: There are a few instances in which the bits contained in the "address" field of an instruction word do not specify the address of an operand, but the operand itself. This is practical only if the required operand is rather short (since the address part is only a fraction of a normal word), but may be useful to specify, for instance, the number of places

---

[1] This is especially useful in computers with sequential access storages like drums, delay lines, etc. See paragraph 8.3.4.

in a shift instruction or to specify a (small) increment by which the contents of a register is to be modified. Since no additional memory access is required for the fetching of the operand, we speak of "immediate addressing". In the case of one-address machines, we may also speak of "no-address" instructions. The advantages of immediate addressing include: shorter execution times because an additional memory access to obtain the operand is not required, and conservation of memory locations.

Implied or Implicit Addressing: In many instances it is not necessary to expressly specify an address because it can be implied from the type of operation to be performed. For example, an add instruction implicitly involves the accumulator. The accumulator itself may or may not have an address in a specific machine. If it has one, we can make use of it for the transferring of information. However, an add command does not have to specify the address of the accumulator since it is obvious that any addition has to be performed there, and, therefore, any quantity to be added has to be transferred to the accumulator. Similar considerations may be applied to other registers in the machine. Some machines address all internal registers implicitly. Others use either implied or direct addressing as it fits the particular design considerations. In some isolated cases, even main storage addresses may be implied.[1]

Block Addressing: So far, we have assumed that an address always refers to a single storage location. There are several common exceptions to this rule. Frequently, an address specifies a block of information. The address usually gives the location of the first word within the block. Blocks may be of various lengths, e.g., all the, say 128, words stored on a channel of a magnetic drum or disk storage may be considered a block, or all the, say 8, words read from a punched card. Blocks of information can be handled conveniently by block transfer commands. The programmer, normally, specifies only the address of the first storage location within the block and the computer itself effects the transfer of the entire block. When the block-length within a machine is variable, it is further necessary to specify either the last address within the block or the total number of words. In any event, block addressing simplifies the handling of larger amounts of information. The design of some tape, drum, or disk storage does not even allow the addressing of single words. An address is provided only for blocks of information. In such cases, we speak of a "block access". A whole block has to be read or recorded, even though only one word within the block may be of interest.

In many instances, computers use blocks consisting only of two words. We may consider this to be the case when a double-length accumulator

---

[1] For instance, in case of program interrupts, the program address may always be saved in location 0000, and the next instruction fetched from location 0001.

has only one address. However, more clearly in this category belong schemes for double-precision operations, wherein the most and least significant halves of an operand are contained in two consecutive, individual storage locations. The address portion of double-precision instructions usually refers to the most-significant half and it is understood that the least-significant half is stored or has to be stored in the next consecutive location. One may, however, argue whether this scheme should be considered as block addressing or implied addressing.

Addressing by Tags: Some — especially larger — computers may have several internal registers performing equivalent functions. In such cases, it may be implied from the type of the instruction that a register of a particular type is to be used (e.g., an accumulator, a specific input/output or index register), but it cannot be implied which one of the several. A specific register may then be designated by a "tag", i.e., by individual bits, or short bit combinations in predetermined positions within the instruction word. A short tag of this type can usually be accommodated in addition to the operation code and the regular address or addresses, while it would be prohibitive or impossible to accommodate an additional full-length address within the limited length of a word.

Indirect Addressing: Up to now we have assumed (and this is normally true) that the address designates the physical location which contains the desired information. However, there are several important exceptions to this rule. For one, the address may specify the location which contains the address of the desired information. In this case, we speak of indirect addressing. The storage location which is specified in the program acts then similar to a "catalog" which refers the computer to the location of the proper item. There are several advantages to indirect addressing. For instance, the operands needed for a particular program may be relocated without making any modification to the program itself. It is only necessary to correct the "catalog" of the operands to which the program refers. If we imagine a program in which a large number of instructions is concerned with a specific operand, we can easily see that it is much simpler to change one entry in the operand catalog, than to change all instructions referring to the particular operand. Of course, with indirect addressing there is no requirement to have the operands in any particular order, as long as the catalog is reasonably ordered. In general, indirect addressing is advantageous to use if information (programs or operands) has to be relocated.

Relative Addressing: Relative addresses do not indicate a storage location itself, but a location relative to a reference address. The reference address might be, for example, the address of the storage location which contains the first word in a block of information. The program can then simply refer to word 3 and the computer will interpret this as word 3 in relation to the reference address. Programming can be simplified by relative addressing (the programmer does not have to anticipate in all detail where the operands are located as long as they are in the proper order) but the required hardware will be slightly more complex (the computer control unit must have a register in which the reference address can be stored). Some computers use the program address as the reference address. (This address is stored in the control unit in any case, so that no additional register is required). In this event, we speak of "forward" (or "backward") addressing. This scheme is perhaps not as powerful as the one previously mentioned but quite useful in many instances[1]. Relative addressing is sometimes used to advantage if the word-length of a computer is relatively short or the memory relatively large. It is then either not feasible, or simply impossible to accommodate all the bits of a complete address within an instruction word. The reference address, i.e., the designation of the part of the memory which the program refers to, is then stored in the control unit[2] and only those bits of the address which designate individual locations within this part are carried in instructions. The content of the register containing the relative address can be changed by the computer program if desired. Since the operand addresses are shorter than full addresses, we also speak of "abbreviated" or "truncated" addressing.

Indexing: In the concept of relative addressing, the instruction word contains a modifier which together with the reference address (stored in a register of the control unit) makes up the effective address of the desired operand. The concept of indexing is concerned with exactly the opposite procedure: the instruction word contains a base address which together with the contents of an index register[3] (modifier or simply index) determines the actual address to be used. The results of both methods are nearly equivalent. However, those computers which are capable of indexing usually have several individual index registers, so that there is considerably more flexibility in indexing than in relative addressing. Moreover, the instruction repertoire of such machines usually contains a variety of instructions which modify, test, or store the contents of index registers (e.g., load index, store index, decrement or increment index, jump on index$=0$.) Thus, the programming becomes very convenient. The convenience of indexing can be demonstrated by a simple example in which the same calculation is made upon a sequence of operands x_i $(0 \leq i < n)$. The program contains instructions which perform the desired calculation on x_0. Indexing is then used to make the identical instructions operate upon x_1, x_2, etc. Sometimes it may be desired to refer to operands, e.g., matrix elements by two indices

---

[1] A program employing this addressing scheme is, for instance, completely "relocatable".

[2] For instance in "bank indicators".

[3] Frequently also referred to as B-register or B-box. This is usually a separate register, but some designs use a conventional storage location instead.

(x_{ij}). The two indices can be stored in separate index registers and modified separately as it fits the calculation. Of course, it is slightly more difficult (and requires additional hardware) to modify a base address by two indices, rather than one. Not all computers capable of indexing can perform such multiple indexing, even though programmers accustomed to indexing certainly wish they would.

Computers, to varying degrees, can perform what might be called an "automatic index modification". As an example, let us consider a scheme in which the contents of the index register are increased by "one", every time the index is used. This scheme could be very convenient to obtain consecutive values of x_i as in the above example. The index does not have to be modified by the program, if the incrementation is performed automatically by the hardware. Since no additional commands need to be executed, the modification is performed much faster than by programming. Let us consider another example: An instruction, "jump on index $\neq 0$", decrements the index automatically by "one". If the index register contains originally, say the value 10, such a jump can be performed ten times before the index is decremented to zero and the jump is omitted. The scheme is rather convenient for leaving a program loop after a specified number of executions. Indices may be incremented or decremented by predetermined values or by values which are specified for each individual instruction (e.g., by the contents of an increment or decrement field, that is a bit combination contained at certain positions within the instruction word.

Address Modification by Repeat Instructions: A capability similar to indexing can be achieved by repeat instructions. While indexing requires a fair amount of specific hardware, the incorporation of repeat instructions requires very little additional hardware[1]. The repeat instruction causes another instruction to be executed more than once. In the simplest approach the instruction following the repeat instruction is executed a number of times, while its address is incremented (or decremented) after each execution. The number of times which the instruction is to be executed is specified by the repeat instruction.

A typical example of an application may be the case where the sum of numbers in consecutive storage locations has to be computed. A single instruction "add to accumulator" is sufficient to do this if it is preceded by an appropriate repeat instruction.

In more sophisticated schemes, the address part of the repeated instruction can be modified by a specified amount. It is also possible to repeat a sequence of several instructions. In this case, however, usually instruction registers for several instructions are provided.

---

[1] In general, the repeat instruction does not offer the same flexibility as indexing. However in some isolated but entirely practical examples, the repeat may be faster.

Problem 92: How would you arrange the normal instruction format (i.e. how many bits would you assign to the operations code, to the address or addresses, to tags, to increment or decrement fields) for a computer with a word length of 36 bits and 32,768 storage locations?

a) Show the format for a two-address machine without any particularly advanced addressing techniques.

b) Show the format for a one-address machine with provisions for indirect and relative addressing, indexing, and automatic index incrementation and decrementation.

Problem 93: Can you find any addressing technique or combination of addressing techniques which would allow you to construct a computer with a word-length of 12 bits and 4,096 storage locations? How could you specify a jump instruction to an arbitrary location with a 12-bit instruction word?

Addressing by Content: Many computational procedures would be simplified if storage locations could be accessed not by their address but by the information they contain. Analogous to such a procedure is the use of a dictionary in which we usually do not find the desired information by the physical location, i.e. by the number of page and line, but by part of the stored information itself. Various techniques are being used or have been proposed allowing computers to use data as addresses or instead of addresses.

Data Used as Addresses: The use of this technique[1] is within the capabilities of any normal stored program computer; no special hardware is required. Let us take a simple example. Suppose it is required to translate decimal digits given in the 8421 code to the 2-out-of-5 code[2]. We could take the bit-combination representing a digit in the 8421 code directly as an address. The binary address 0000 would then correspond to the decimal digit 0, the address 0111 to the digit 7 and 1001 to the digit 9. If we store the bit combination which corresponds to the 2-out-of-5 code in the appropriate location, we have a simple translation table. For example, looking in cell 0000, we would find the bit combination 11000, i.e. the 2-out-of-5 code representing the decimal digit 0. Storage location 0111 would contain the number 10001, location 1001 contains 10100 and so forth. Simply by taking the given data as an address, we can immediately look up its translated equivalent. This technique is quite practical for many applications, such as conversion from octal to decimal digits, print code translation, mathematical tables and the like. However, the scheme is not as universally applicable as one might think. The major drawback is that tables

---

[1] Programmers frequently refer to this techniques "Table look-up".
[2] See, for instance, Table 2.6

frequently cannot be brought into a continuous and consecutive order so that appreciable storage space is wasted. We can see this easily when we consider the translation from the 2-out-of-5 code to the 8421 code. We obviously need 5 bit addresses since the 2-out-of-5 code uses five bits. 32 physical storage locations correspond to 5-bit addresses but we use only a total of 10 out of the 32 for the storage of the entries in our table. The used addresses are not consecutive but are scattered throughout the 32 physical locations.

Another example which illustrates the restrictions of the technique even better may be the following: A computer is set up to handle a supplier's stock record. The file consists of part numbers, minimum and actual stock levels, unit prices, discount rates, shipping weights, etc. All activities such as sales, receivings, price changes have to be reflected in the file. The problem would be relatively simple if one could always use the part number as an address. The access would then be by part number. Stock levels, prices, etc. would be easily found. However, the scheme will not be practical in many instances since the part numbers may not be consecutive, and since they may contain letters, symbols, prefixes and affixes. Even a fixed association between part numbers and addresses (such as a mathematical formula) may be impractical since it should be possible to add or delete items of various lengths. If such changes are frequent and if tables are re-arranged after each change, it may be very difficult or even impossible for the programmer (or the computer program) to anticipate the exact present physical location of information in the computer memory.

The best solution, under such circumstances, may be the storage of the part number together with the associated information. The desired item can then be located by "searching" through the stored information.

Programmed Memory Search: The advantages of a memory search can be seen immediately. The information does not have to be stored in a particular sequence; additions and deletions are relatively simple processes. Searches can be performed not only for part numbers, but also for other details. For instance, one could easily search for all items weighing more than 100 lbs, for items to which special excise taxes apply, or for items for which the stock level is below the given minimum. In all these cases, a search is performed for a part of the stored information. We can speak of an access by association, rather than by addressing. One of the disadvantages of a search is, of course, that it takes longer than an access to a known location.

Any computer can be programmed to perform a search. A program loop consisting of, say, three or four instructions can make the computer obtain the content of a storage location, compare it with the desired pattern, terminate the search if a match is found, but continue to look in consecutive locations if there is a mis-match. For the search of long, but ordered[1] tables one may even propose certain search strategies. A "binary" search, for instance, would make the first access to the middle of the applicable storage region. The information which is found there would determine whether the desired item is stored in the first, or the last half of the table (unless a match is found, in which case the search can be terminated). The second access would be made to the middle of the appropriate half of the table. This determines then the appropriate fourth of the table. The following accesses determine then the applicable $1/8$, $1/16$, $1/32$ of the table, so that the location of the desired item is rapidly approached. A search strategy of this type requires, of course, a search program which is more elaborate than one for a straightforward search through consecutive locations, but its execution will take less time if tables are long.

Table Search Instruction: A table search by program is a rather lengthly process. Many of the larger computers incorporate, therefore, special table search instructions which initiate all the internal operations of a sequential search. Such an instruction may specify the pattern to be located and the addresses of the first and the last entries in the table[2] (or the first address and the number of words to be searched). Any address modifications required to pick up the contents of intermediate addresses and the determination of matches or mis-matches are performed by logic circuits in the computer control unit. Computers may incorporate table search instructions which look for the equality between the items to be located and the given pattern, or search for items which are numerically or logically larger or smaller than the given information.

Associative Memories: A table search technique which operates even more automatically than table search instructions has been proposed in the form of "associative" or "content addressed" memories[3]. Here, the search is performed entirely by logic circuitry which is part of the memory itself. The control unit of the computer is not involved other than in initiating the searching process. A search can be performed much more rapidly in this manner, although the additional complexity makes the cost of an associative memory higher than that of a conventionally addressed memory.

It may be appropriate to pause here for a moment and consider the relative advantages of the addressing schemes we have discussed.

In general, we can state that a computer acquires no capability by using sophisticated addressing schemes which could not be achieved also with

---

[1] An ordered table is one in which the elements are arranged according to a known rule, e.g.: The larger of two numbers will always stored in the location with the larger address.

[2] Perhaps by giving the address of a "block" of information which, in turn, contains all the required specifications.

[3] See paragraph 8.3.5.

simple and straightforward addressing. On the other hand, such schemes certainly do increase the efficiency, speed, and convenience of the computer operation.

Some addressing modes—and this applies particularly to implicit addressing, addressing by tags and relative addressing—make rather efficient use of the available bit positions in the instruction format. Fewer bits are required to specify a particular register or storage location, than by straightforward addressing. This economy permits the construction of machines with rather short word-lengths. It also allows to pack more than one instruction into a normal computer word, or to make room within an instruction word for the designation of additional operations such as indirect addressing, indexing, index modifications and the like.

Addressing modes which modify the base address contained in the instruction perform operations which otherwise would have to be accomplished in one way or another by additional instructions. Programs, consequently, become shorter. When only the internal registers of the control unit are involved in such modifications, the operations also become faster. An additional benefit of this technique is the relative convenience for the programmer. The "housekeeping" of a program is simplified, that is, those parts of a program which are concerned with repeating loops a specified number of times, getting operands in the proper order, setting up exits, and the like, require less detailed attention than otherwise.

An important consideration is also the restoring of programs. Any normal program which modifies itself during execution must be "restored", before it can be run again. Additional instructions are required to "initialize", (the program i.e. to set to its initial state). In contrast, address modification schemes such as indexing and indirect addressing do not change a program as it is located in the memory. Only the individual instructions are modified before execution. Programs are, therefore, left intact and no (or very few) restore operations are required, before a repeated run. This property is not only a convenience but also shortens programs and execution times.

The capabilities of individual addressing schemes overlap to a certain extent[1]. One scheme accomplishes what could possibly be accomplished by another scheme in a slightly different manner. Some schemes really can be considered special cases of others. Probably no existing computer employs all possible schemes, but each computer uses them to some extent. As with many more advanced concepts in computer design, it is not a simple matter to judge the values or relative merits of the various schemes. The problem is complicated because the value of the different methods depends upon the application of the computer. Only a very detailed analysis

---

[1] It is also entirely possible to specify combinations of techniques for a single instruction, e.g. relative addressing *and* indexing.

of both, the computer and its application will disclose the benefits in a particular instance.

8.2.4. The Instruction Format and Repertoire

The instruction format defines the use or the "meaning" of bits within an instruction word. Instruction words are usually divided into "fields" which have their own specific functions. Separate fields are provided for the op-code, for the address or addresses, and for tags, flags and the like. Generally, a number of compromises have to be made before the instruction format of a machine is finalized. For instance, the size of individual fields is influenced by the following considerations: the word-length restricts the number and lengths of fields; the number of instructions in the instruction repertoire determines the minimum length of the field which represents the op-code; and the length of address fields depends upon the size of the memory[1].

A few practical instruction formats are indicated below. The numbers underneath each format indicate the length (in bits) of individual fields:

42-bit, 3-address instructions:

OP	X-Address	Y-Address	Z-Address
6	12	12	12

The machine has an instruction repertoire of at most $2^6 = 64$ different instructions. $2^{12} = 4,096$ storage locations can be directly addressed. For most instructions, the three addresses determine the location of two operands, and the location where the result of an arithmetic or logic operation is to be stored. For jump instructions, the Z-address field contains the jump address, and for shift instructions, the Y-address field contains the shift count.

36-bit, 2-address instructions:

OP	X-Address	Y-Address
6	15	15

The two addresses determine the location of two operands, or the location of one operand and the location of a result. $2^{15} = 32,768$ storage locations

---

[1] A memory with a storage capacity of 4,096 requires, for instance, 12-bit addresses.

can be directly addressed. The Y-address field may contain a shift count. For some specific instructions, the X-address field may contain tags, indicating address modifications and the like.

36-bit, 1-address instructions:

OP	Tags	Address
12	9	15

This machine has provisions for a rather large instruction repertoire ($2^{12} = 4,096$ different instructions). However, in practice only a fraction of it is implemented. The actual op-codes are selected so that the function translation is simplified. The address field contains an operand address, a jump address, or a shift count. For some instructions, like index load or unload instructions, the address field may contain a short operand. Tags indicate the use of indirect addressing, relative addressing, indexing and the like.

Alternate 36-bit, 1-address format:

OP	Minor Function	Tags	Address
6	4	10	16

The op-code field has a basic length of six bits. However, for some instructions it may be expanded to 10 bits. The additional four bits may then specify the "minor function"; e.g., the particular type of a jump, or the byte used for a partial transmission instruction. The address field comprises 16-bits so that a larger memory can be addressed, but in other respects the format is very similar to the previous one.

48-bit, 1, 2 or 3-address instructions:

OP	Tag	Address	OP	Tag	Address
6	3	15	6	3	15

OP	Minor Function	X-Address	Tags	Y-Address
6	3	15	9	15

OP	Minor Function	X-Address	Y-Address	Z-Address
6	3	5	5	5

The machine uses three different instruction formats. Normally two separate instructions (or a "double instruction") are contained in a single instruction word. However, for more complex instructions, a single "double-length" instruction occupies the same space. This allows the extension of the op-code by a minor function code and the use of a larger tag field, in addition to the use of a second address. The third format is used for "inter-register transmissions". A half-word of 24 bits contains an op-code, a minor function code and three (short) register addresses. This arrangement combines the advantages of one, two and three-address machines very effectively.

Alternate 48-bit format:

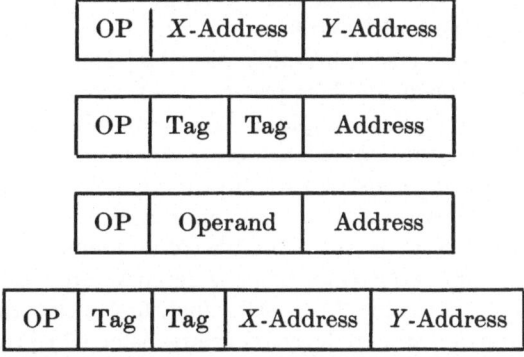

Instructions may have a length of 16, 32, or 48 bits. An instruction with 16 bits specifies two short addresses and is used for inter-register transfers. Instructions with 32 bits specify one full address and have either two additional tag fields of four bits, or an additional operand field of eight bits. 48-bit instructions contain two full addresses in addition to two 4-bit tag fields.

12-bit, 1-address format:

OP	Modifier	Address
6	6	12

The relatively small word length of 12 bits is not sufficient to accomodate an op-code and a full address. For this reason, normally two words are used to represent a single instruction. The two words are fetched subsequently and from consecutive locations. Dependent upon the op-code, the contents of the modifier field may be interpreted as a minor function code, as tags for indirect or relative addressing, as a short operand, or even as a truncated address. Similarly, the contents of the address "field" may be interpreted as operand address, as jump address, or as an operand.

The indicated instruction formats are to be taken only as illustrative examples representing an almost unlimited number of conceivable formats. In practice, each existing computer has its own particular instruction format dictated by the particular design considerations.

Problem 94: Propose an instruction format for a 24-bit machine. Provide for indexing, relative, and indirect addressing. Assume that the machine has several memory "banks" with a storage capacity of 4,096 words each. Make the instruction format as effective as you can.

The instruction repertoire of an "average" computer may comprise somewhere between 50 and 100 different instructions. Out of this number, only perhaps a dozen or so are really essential. That is, with this dozen instructions, any problem could be programmed, whose solution is within the capabilities of a digital computer. The remainder of the instruction repertoire is provided for convenience and, more important, for increased operational speed. A single instruction, if properly selected, may accomplish the same as several instructions of the "minimum" set. For example, an add instruction which adds the contents of a memory location to the contents of the accumulator could be considered as an "essential" instruction. A second instruction or "Load A" which first clears the accumulator before the contents of a storage location are transferred to the (now empty) accumulator may not be essential, but is very convenient. If only a minimum repertoire were provided, it might be necessary to store the previous contents of the accumulator, then subtract this value from A, before the first operand can be loaded by the add instruction. Similarly, floating-point arithmetic operations are not absolutely necessary. They could be performed by programming a number of fixed-point operations. Even multiply and divide instructions could be replaced by a sequence of individual add, subtract and shift instructions.

It is interesting to note that programming can be considered as an extension of the built-in sequencing of operations. The following table reveals this aspect more clearly.

The table lists operations of increasing complexity. A small computer may provide hardware sequences only for the most elementary operations of addition and subtraction. Consequently, more complex operations including multiplications and divisions have to be programmed. Another computer may have the capability to sequence internally floating-point operations in addition to the four basic fixed-point arithmetic operations. Identical operations are built-in in one case, and programmed in the other. In this

Table 8.16. *Continuum of Hardware and Software*

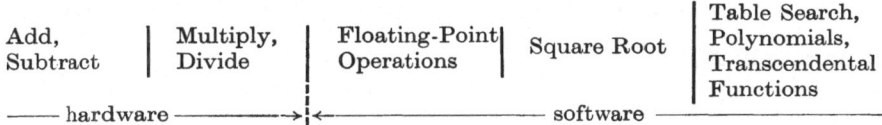

Add, Subtract	Multiply, Divide	Floating-Point Operations	Square Root	Table Search, Polynomials, Transcendental Functions
—— hardware ——→	←—————————— software ————————

respect, there exists a continuum of hardware and software. The software (programming) takes over where the hardware (the internal sequencing) leaves off. The borderline between hardware and software is determined by the implemented instruction repertoire and may coincide with any one of the dividing lines indicated in the above table. Incidentally, in this respect, the term "microprogramming" becomes very descriptive for the sequencing of complex operations[1].

The provision of "non-essential" instructions can considerably increase the operational speed of a computer. It may give the intrinsically slower of two machines, the higher "effective" speed for many applications. The selection of a well-balanced instruction repertoire is, therefore, extremely important. Unfortunately, no general rules can be given. Details depend upon specific design characteristics, and upon the anticipated application. In the following, we shall, therefore, only indicate the possible variety of instructions within the instruction repertoire. All of the shown examples have been implemented in one or the other machine, but no single design incorporates all of them. In general, we shall restrict examples to one-address instructions. Instructions with more than one address are indicated only when they permit interesting or particularly useful variations of the basic operation. The letters X and Y denote the bit combination contained in the address fields of instruction (normally an address). The letters A and Q denote respectively the accumulator and the Q-register. Parentheses are used to indicate "the contents of". The expression (A) denotes for instance "the contents of the accumulator", and (Y) denotes "the contents of storage location Y". An arrow indicates a transfer of information to a given location.

---

[1] See paragraph 9.5.

Fixed-Point Add Instructions

Add Y	$(A)+(Y) \rightarrow A$
Clear Add Y	$0+(Y) \rightarrow A$
Add Magnitude Y	$(A)+|(Y)| \rightarrow A$
Replace Add Y	$(A)+(Y) \rightarrow A, Y$
Replace Add One Y	$(Y)+1 \rightarrow A, Y$

Add instuctions with two or more addresses allow the combination of add operations with information transmission operations. For example:

Replace Add XY	$(X)+(Y) \rightarrow A, X$
Add and Transmit XY	$(A)+(X) \rightarrow A, Y.$

Floating-Point Add Instructions

Floating Add Y	$(A)+(Y) \rightarrow A$
Floating Add Magnitude Y	$(A)+|(Y)| \rightarrow A$
Unnormalized Floating Add Y	$(A)+(Y) \rightarrow A$ (result unnormalized)
Add to Exponent Y	$(A)+(Y) \rightarrow A$ (the addition is performed only in the exponent field.)

Double-Precision Add Instructions

All previously shown fixed and floating point add instructions may be expanded for double-length operands. For instance, a double-precision add may be performed as:

Double Precision Add Y \qquad $(A, Q)+(Y, Y+1) \rightarrow A, Q$

The expansion to double-length operands may be indicated by different op-codes, minor function codes or tags.

Subtract Instructions

Subtract instructions have the same variety as add instructions. In general, they have different op-codes but their execution is identical to that of add instructions, except that additions $(+)$ are replaced by subtractions $(-)$.

Multiply Instructions

Multiply Y \qquad\qquad $(Q) \times (Y) \rightarrow A, Q$

Note: In general the binary point is considered to the right of the sign bit (fractional machine) or to the right of the least significant bit (integral machine). However, some machines have the capability to treat numbers either as fractions or integers and have consequently two different multiply instructions: "Multiply Integer" and "Multiply Fraction".

Multiply Round Y $(Q) \times (Y) \to A$
Variable Length Multiply Y, C $(Q) \times (Y) \to A, Q$

Note: Only the C most significant bits of the product are derived; the numerical value of C is contained in a designator field of the instruction.

Floating Multiply Y $(Y) \times (Q) \to A, Q$
Unnormalized Floating $(Y) \times (Q) \to A, Q$ (result is not normal-
  Multiply Y ized)

Note: Different op-codes or tags may be used for all multiply instructions to specify single-length or double-length operands, rounding, normalizing, and the use of magnitude or complement of operands. Multiply instructions with more than one address allow the combination of multiply operations with information transfer operations. For example:

Multiply Add XY $(A) + (X)\,(Y) \to A$
Polynomial Multiply XY $(Q)\,(X) + (Y) \to Q$

Divide Instructions

Divide instructions have the same variety as multiply instructions.

Logical Instructions

Selective Set Y	Set bits in A where there are corresponding "ones" in (Y). Do not affect bit positions which have a "zero" in (Y)
Selective Clear Y	Clear bits in A where there are corresponding "ones" in (Y)
Selective Complement Y	Complement bits in A where there are corresponding "ones" in (Y)
Selective Substitute Y	Transfer bits of (Y) into A where there are corresponding "ones" in (Q)
Q Controlled Add Y	Form the logical product of (Q), (Y), and add it arithmetically to (A)
Q Controlled Subtract Y	Form the logical product of (Q), (Y), and subtract it from (A)

Note: More complex logical operations can be performed with two or three address instructions. For example:

Controlled Complement XY	$(X) \oplus (Y) \to X$
Q Controlled Transmit XY	Form the logic product of (Q), (X). Store result in Y
Q Controlled Transmit XYZ	Form the logic product of (X), (Y). Store result in Z

Extract $X Y Z$ Replace bits in (Z) by bits of (X) where there are corresponding "ones" in (Y). Do not change bits in (Z) where there are corresponding "zeros" in (Y).

Shift Instructions

Shift A Right Y	Shift (A) by Y places
Shift Q Right Y	Shift (Q) by Y places
Long Right Shift Y	Shift the combination $(A—Q)$ by Y places

Note: Shifts may be straight, circular, or sign-extended. Equivalent left shift instructions may exist. The various modes may be specified by different op-codes, minor function codes, or tags.

Scale $A Y$ Shift (A) left until the absolute value of the most significant bit is "one". Subtract the number of required shifts from (Y)

Scale $A, Q \ Y$ Same as above except that the double-length quantity $(A—Q)$ is shifted

Note: Scale instructions can be used conveniently in the process of converting fixed-point to floating-point numbers. Shift instructions with more than one address allow the combination of shift and add or transmit operations. For example:

Add and Shift $X \ Y$	$(A)+(X) \to A$; shift (A) by Y places
Load A and Shift $X \ Y$	$(X) \to A$; shift by Y places
Scale $X \ Y$	$(X) \to A$; scale (A); number of shifts $\to Y$
Normalize and Pack $X \ Y$	Pack the number (coefficient) contained in (X) and the scale factor (exponent) contained in (Y) into a normalized floating-point number. Store result in X.

Full Word Transmit Instructions

Load $A \ Y$	$(Y) \to A$
Load A Complement Y	$-(Y) \to A$
Load A Magnitude Y	$|(Y)| \to A$
Load $Q \ Y$	$(Y) \to Q$
Load Q Complement Y	$-(Y) \to Q$
Load Q Magnitude Y	$|(Y)| \to Q$
Store $A \ Y$	$(A) \to Y$
Store $Q \ Y$	$(Q) \to Y$
Exchange $A \ Q$	$(A) \to Q$; $(Q) \to A$

Note: Similar instructions may exist for transmissions to and from other registers in the machine. Source or destination registers may be designated by tags or different op-codes.

Enter Keys Y	The state of operator control switches is copied into Y
Set Indicators Y	The contents of location Y is copied into a display register
Store Zero Y	$0 \to Y$
Block Transfer Y	Transfer n words in consecutive order from locations X, $X+1$, $X+2$, etc. to locations Z, $Z+1$, $Z+2$, etc.

Note: The numerical values for n, X, Z are contained in the control word with the address Y. Some designs allow also for address increments ± 1 for both sources and destinations. In this case, the same or another control word contains address decrements or increments.

Note: Some arithmetic and logic instructions simultaneously transmit information. Such instructions are listed here under the heading reflecting their primary purpose. Some examples of 2- or 3-address transmit instructions are shown below:

Transmit $X\ Y$	$(X) \to Y$
Transmit Negative $X\ Y$	$-(X) \to Y$
Transmit Magnitude $X\ Y$	$\lvert(X)\rvert \to Y$
Transmit Negative Magnitude $X\ Y$	$-\lvert(X)\rvert \to Y$

Partial Word Transmit Instructions

Store Left Half Y ⎫	The appropriate field of the word contained in Y is replaced by the contents of A. Bits outside this field are not affected.
Store Right Half Y ⎪	
Store Address Y ⎬	
Store Tag Field Y ⎪	
etc. ⎭	

Note: Similar instructions may use Q, PAR or other registers as source or destination. Source and/or destination registers may be designated by op-codes or tags.

Transmit Byte	A byte of information is transmitted from one location to another.

Note: Bytes may be flexible as far as size and location within a word is concerned. They may be specified by codes, tags, etc. Sources or destination may be registers (specified by tags) or storage locations (specified by addresses). Some designs allow to specify bytes also for arithmetic and logic operations.

Jump Instructions

(Frequently also called test, examine, or transfer instructions)

Unconditional Jump Y	Go to Y for next instruction
Selective Jump $Y\ C$	Go to Y for next instruction if operator switch C is "on"

Note: Switch C is usually designated in a tag field of the jump instruction.

Jump on Overflow Y	Go to Y for next instruction if the accumulator contains an overflow
A Jump Y	Go to Y for next instruction if condition is met

Note: The condition may be $A = 0$; $A \neq 0$; $A +$; $A-$ etc. These different A Jumps may be specified by different op-codes or by tags.

Q Jump Y	See A Jump
Index Jump	See indexing instructions
Examine Status Register Y	Go to Y for next instruction if the contents of the status register $\neq 0$

Note: In some designs, individual bits rather than the entire contents of the status register can be examined.

Register Jump Y	Jump if a specified register meets a specified condition (see also A Jump)

Note: Jump instructions with more than one address may specify two-way jumps, e.g. go to Y if the jump condition is present, but go to X if the condition is absent.

Return Jump Y	Go to Y for next instruction. Record the current program address (PAR) in a predetermined storage location

Note: The "return address" (PAR) may be recorded, for instance, in storage location 00000, a designated index register, or in the memory location to which the jump is made. Two-address return jump instructions may also specify an arbitrary storage location where the return address is to be stored. In some designs, each jump instruction may be designated as return jump by the use of tags.

Storage Test Y	Skip next instruction if the contents of storage location Y meet certain conditions

Note: Conditions may be specified by different op-codes, minor function codes, or tags. A few examples of test conditions are: $(X) = 0$; $(X) \neq 0$; $(X) +$; $(X) \leq (A)$; $(X) < (Q)$; $(X) \subset (Q)$. An example of a two-address storage test is:

Storage Test X Y	Go to Y for next instruction if (X) meets certain conditions
Compare Algebraically Y	Compare contents of A and Q. If $(A) \leq (Q)$, go to Y for next instruction
Compare Logically Y	Compare contents of A and Q. If $(A) \subset (Q)$, go to Y for next instruction

Halt or Stop Instructions

Stop Y	Stop. After manual start, go to Y for first instruction

Selective Stop $Y\,C$	Stop if operator switch C is set. If switch C is not set, go to Y for next instruction
Final Stop	Stop. Ignore subsequent manual starts until a master clear is given

Indexing Instructions

Load Index Y	Replace the contents of a specified index register with (Y)
Store Index Y	Replace (Y) by the contents of a specified index register

Note: Index registers are usually shorter than the normal word-length. A specific index register may be designated by tags. Op-codes, minor function codes, or tags may also be used to specify certain fields of Y, e.g. the address field, the increment field, etc., or the use of complemented operands, or the use of A as source or destination.

Index Jump Y	If the contents of the specified index register $\neq 0$, subtract one from index and go to Y for next instruction. If contents $= 0$, go to next instruction
Index Jump $Y\,C$	Subtract the numerical value of C from the contents of a specified index register. Go to Y for next instruction
Jump on High Index YC	If contents of index register $> C$ go to Y for next instruction. If not, take next instruction

Note: A number of variations of the shown index jumps are in practical use. They differ with respect to jump conditions, and index modifications. Some designs also allow to use normal storage locations as "index registers". For example:

Index Jump $X\,Y$	$(X) - 1 \to X$; if (X) final ≥ 0, go to Y for next instruction.

Input/Output Instructions

Many I/O instructions perform operations which are specifically designed for the particular layout of the I/O unit and the specific external equipment connected to it. Following are a few of the more typical instructions.

External Function Y	Select and connect one of several external devices to the I/O unit and makes the equipment perform a certain operation (e.g. read, move forward, print, rewind, etc.). This is accomplished by transmitting the external function code from Y to an I/O control register

External Read Y	Transfer the contents of the I/O information register to Y
External Write Y	Transfer (Y) to the I/O information register
Connect Y	Perform the equivalent of the external function instruction, except that also one of several channels may be specified by the control word located in Y
Copy Status Y	Transfer the contents of the I/O or channel control register to location Y
External Read Y	Transfers the contents of the I/O information register to location Y
External Write Y	Transfer (Y) to the I/O information register
Read Y C	Transfer C words from a channel information register into consecutive storage locations starting with location Y

Note: An information transfer of this type is usually executed independently, that is, the transfer goes on simultaneously with the execution of further program instructions. An alternate version of the read instruction is:

Read Y	Read the control word located in Y to determine the number of words to be transferred and the first storage location
Write Y	Transfer information from a storage location to a channel information register. See also notes for read instruction.

Table Search Instructions

Equality Search Y C	Inspect C words in storage beginning with address Y. If $(Y)=A$, go to next instruction. If no $(Y)=A$, skip next instruction. Record last address, Y, in a designated index register
Threshold Search Y C	Similar to equality search, except that $(Y)>A$ terminates the search
Masked Equality Search Y C	Similar to equality search, except that the logic product of (Y), (Q) is compared with (A) for equality

| Masked Threshold Search | Similar to previous threshold search instructions except that the logic product of (Y), (Q) is compared with (A). for (Y), $(Q) \geqslant A$ |

Note: Similar searches may be implemented with conditions:

$$(A) \geq (Y) > (Q), \text{ or } (A) \geq |(Y)| > Q \text{ etc.}$$

| Convert Y | $Y+(A)$ forms an address. The contents of this address replaces the contents of the accumulator |

Note: This instruction converts or translates a bit combination contained in A into another bit combination, according to a stored translation table. The translation of individual bytes is performed individually and consecutively. (Note that the full word contained in A is too long to be used as an address.) The capability to use different translation tables for the different bytes may exist. In this case, the instruction contains only the origin address for the first table Y_1. The origin address for the second table Y_2 is stored within the body of the first table. Y_3 is stored within the body of the second table and so on.

| Locate List Element Y_n | Locates the nth entry of a list with the origin address Y |

Note: The list may have entries with one or more words and entries may be scattered throughout storage. The only requirement is that the first word of each entry contains the address of the next "consecutive" entry. The result of the instruction is the starting address of the nth entry (appearing in a designated index register). (See also paragraph 9.7.)

Miscellaneous Instructions

Round	Round the quantity contained in (A), (Q). Leave result in A
Normalize	Normalize the floating-point number contained in (A)
Pass	Do nothing
Execute Y	Execute the instruction in storage location Y
Perform Algorithm	Perform a sequence of operations specified by a "microprogram"

Note: This microprogram may be wired on custom designed patchboards or it may be determined by external control circuitry. This instruction gives the computer user the possibility to perform frequently used algorithms (e.g. square root, parity checks, code conversions, etc.) which are normally not implemented. See also paragraph 9.5.

Unpack Y Transmit the coefficient of the floating-
point number contained in A to (Y).
Transmit the exponent to $Y+1$.

8.2.5. The Registers of the Control Unit

As we have said at the beginning of paragraph 8.2, the organizations
of control units differ widely. As far as registers are concerned, only the
instruction register and the program address register are common to most
designs and readily identifiable. A number of other registers may be added
in varying configurations, as required for implementing the specific capabilities of a machine.

We will show here shortly only two representative arrangements, but
indicate some possible variations. The layout of a relatively simple control
unit is given in Fig. 8.70.

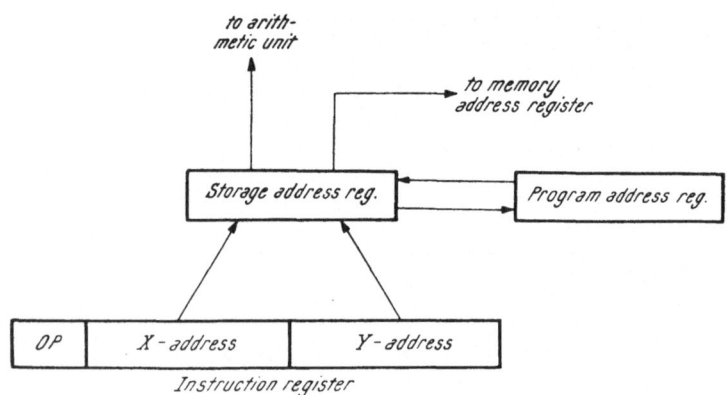

Fig. 8.70. Layout of Registers in a Relatively Simple Control Unit

The machine has a two-address instruction format, reflected by the
X- and Y-address fields in the instruction register. Depending upon the
purpose of a memory access, either the X-address, the Y-address, or the
program address is sent to the memory address register. The three corresponding paths join in the storage address register. In this respect, the
storage address register acts as an exchange register in the network of
information paths. It also acts as an exchange register for the transmission
of an address to the arithmetic unit. An operand "address" might be sent
there, because it really is no address but an immediate operand. The program address might be sent through the arithmetic unit to the memory
for storage during the execution of a "return jump". The remaining path

from the storage address register to the program address register serves to transfer the new program address during the execution of a "jump" instruction[1].

The use of the storage address register as exchange register simplifies the gating arrangement at the input of the destination registers. The usefulness of the storage address register may be increased by additional capabilities. It may, for instance, have the properties of a counter, so that it can be used to keep track of the number of shifts during the execution of a "shift" instruction.

Problem 95: Propose a sequence of operations and information transfers for the execution of a computer interrupt. Assume the layout of registers shown in Fig. 8.70. Store the current program address in location 0000 of the memory and fetch the next instruction from location 0001.

Problem 96 (Voluntary): Propose a sequence of operations and information transfers for the execution of a "repeat" instruction, i.e. of an instruction which causes the next consecutive instruction to be executed n times. Assume that the X-address field of the repeat instruction specifies the number n. Hints: Store the current program address in location 0000 of the memory, and use the program address register to count the number of times the repeated instruction is executed.

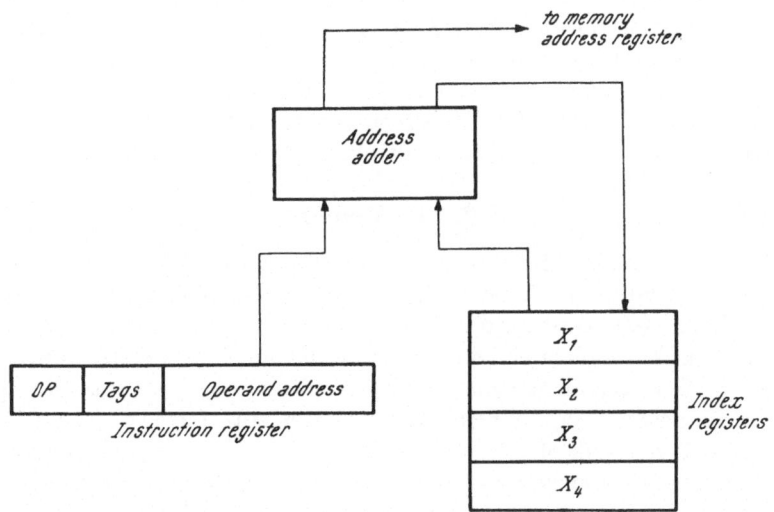

Fig. 8.71. Design Detail of a Control Unit with Indexing Capability

---

[1] For an explanation of the mentioned types of instructions, see paragraph 8.2.4.

The arrangement shown in Fig. 8.71 is a "typical" detail of a machine with indexing capability.

The address adder serves to augment the operand address by the contents of a selected index register, before it is transmitted to the memory[1]. Alternately, an index register may be loaded with the output of the address adder. This may serve to execute a "load index" instruction in which the contents of the operand address field are transferred to an index register.

The same path can be used for the execution of other indexing operations in which the contents of an index register are incremented or decremented by modifiers contained in the instruction format. Incidentally, the address adder in many designs is used as exponent adder during floating-point operations[2].

A number of additional registers may be a part of the control unit. A few examples of such registers are:

Bound registers containing the upper and lower addresses of memory regions which are to be protected against unintentional accesses.

Address monitor registers containing a specific address, so that an alarm (error, stop, or interrupt) can be given whenever the corresponding memory location is accessed.

Status and *interrupt registers* for the communication with external devices[3].

Look-ahead registers which serve as buffer for several instructions prior to their execution. This increases the effective speed since it is possible to fetch instructions ahead of time, whenever a memory cycle is available. Moreover, operand addresses can be modified (e.g. by indexing) prior to the entering of an instruction into the instruction register.

Counters which serve for the execution of instructions which require repetitive cycles, such as shift, multiply and add instructions.

8.3. The Memory

The basic function of the memory is the storage of information. As we have seen in chapter 7, this includes the storage of instructions, operands, and results, as well as any other information required for the execution of programs, such as addresses, indices, patterns, etc. The transfer of information to and from the memory is accomplished by only a few internal operations: read, write, select. When information is transferred to the memory to be stored, we speak of a write operation. When information is retrieved from the memory and made available to other units of the computer, we speak of a read operation. Information is normally transferred in words,

---

[1] See paragraph 8.2.3.
[2] See paragraph 8.1.6.
[3] See paragraph 8.4.2.

that is, in relatively small groups of bits as compared to the over-all size of the memory. An information transfer affects, therefore, normally only a small portion of the memory and an appropriate section of the memory, the "storage location" must be selected. Let us use the rough functional diagram shown in Fig. 8.72 to discuss these operations in general terms.

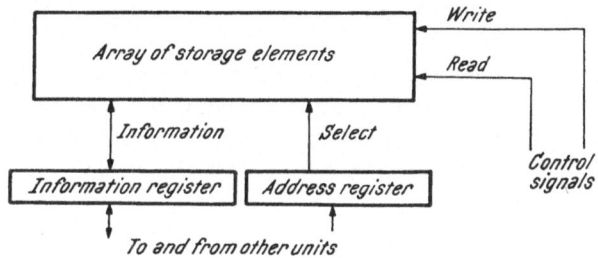

Fig. 8.72. Rough Functional Diagram of a Computer Memory

The memory consists of an array of individual storage elements organized into a number of storage locations, each having the capacity to store one word. The address contained in the memory address register is used to select an individual storage location. An appropriate address is, therefore, placed into the address register for each read or write operation. The information register is used to hold the word of information to be transferred. During a write operation, the word contained in the information register is stored at the selected location. During a read operation, a word is retrieved from the selected storage location and placed into the information register. A typical write sequence might be the following: address → address register; information → information register; initiate write signal; information from information register → selected location. A typical read sequence might be: address → address register; initiate read signal; information from selected storage location → information register.

The above general functional description applies to most memory designs. However, many of the finer details depend upon the specific design. Memories are different with respect to the basic storage elements they use. We speak then of core, drum, or delay line memories, as the case may be. Memories may be designed for parallel or serial information. We then speak of parallel or serial memories. They may also be designed for random access or sequential access. We speak of random access or sequential access memories. In the first case, the access to any storage location requires the same amount of time while in the latter case, storage locations are basically accessible only in a sequential order, and read or write operations are delayed until the desired storage location becomes accessible. Some memories are destructive, that is, the reading process destroys the information stored in the selected storage location, while others are non-destructive. Not all combinations of these characteristics are of practical interest, and we shall restrict our discussion of memories to the few most commonly used types[1].

8.3.1. Core Memories with Linear Selection

The properties of magnetic cores[2] make them very attractive as storage elements for the construction of computer memories. One of the simplest — and easiest to understand — organizations of core memories is that of a memory with linear selection. Fig. 8.73 shows the basic layout of such a memory.

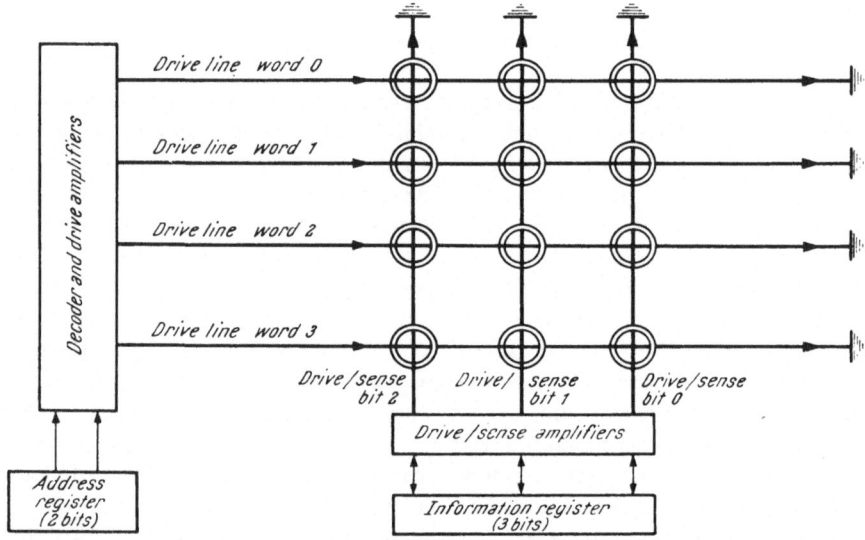

Fig. 8.73. Basic Layout of a Four-Word, Three-Bit Core Memory with Linear Selection

Twelve cores give the memory a total storage capacity of twelve bits. Although this memory would be too small to be of any practical value, its size is adequate to illustrate the principle, and its layout can easily be expanded. Cores are threaded by a matrix of word and bit-lines. Three cores are associated with each word-line and four cores are associated with each bit-line, resulting in a memory organization of four words with three bits

---

[1] Some unusual memory elements are discussed in chapter 11.

[2] See paragraph 5.1.2.

each. The memory has a 2-bit address register, allowing for four unique addresses for the four storage locations. It has a 3-bit information register, corresponding to a word-length of three bits.

For the discussion of the *write operation*, let us assume that all cores are initially cleared, this is, they are magnetized in a counterclockwise direction. In order to write into a specific storage location, the address contained in the address register is decoded and the appropriate word drive line is energized by a drive amplifier. The current through the selected word drive line is smaller than the critical current[1], say $1/2I_C$. This current tends to magnetize all the cores in the selected storage location in a clockwise direction, but is too small to cause the actual switching of the flux. A core switches only if there is some additional current applied to its specific bit line. This current, like the word drive current is smaller than the critical current, again say $1/2I_C$. It is applied by the bit drive amplifier only if there is a "1" in the corresponding bit position of the information register. We see, that the magnetization of cores is switched to a clockwise direction only where word and bit-drive currents intersect, that is, only in the selected storage location, and only in those bit positions into which a "1" has to be written.

Table 8.17 summarizes the possible combinations of drive currents which the various cores in the memory experience during a write operation.

Table 8.17. *Summary of Drive Currents for Write Operation*

	Selected Core (Word-Drive Current $= 1/2I_C$)	Unselected Core (Word-Drive Current $= 0$)
Write "1" (Bit-Drive Current $= 1/2I_C$)	$+1/2I_C + 1/2I_C = I_C$	$+1/2I_C$
Write "0" (Bit-Drive Current $= 0$)	$+1/2I_C$	0

Selected cores, that is cores of the selected storage location, experience a word drive current of magnitude $1/2I_C$; cores threaded by energized bit lines experience a bit drive current of magnitude $1/2I_C$. Cores which are to be switched experience a combined drive-current I_C, while all other cores experience at the most a drive current of the magnitude $1/2I_C$. This 2:1 current ratio is sufficient for reliable operation. However, a wider

---

[1] See paragraph 5.1.2.

operational range can be achieved fairly easily. Suppose we apply a word drive current $2/3I_C$ to the selected storage location. We further apply a bit drive current $1/3I_C$ to the bit drive line for writing a "1" and a bit drive current $1/3I_C$, but in opposite direction, for writing a "0". Table 8.18 summarizes the drive conditions for this alternate arrangement.

Table 8.18. *Summary for Alternate Arrangement of Drive Currents*

	Selected Core (Word-Drive Current $=+2/3I_C$)	Unselected Core (Word-Drive Current $=0$)
Write "1" (Bit-Drive Current $=+1/2I_C$)	$+2/3I_C+1/3I_C=I_C$	$+1/3I_C$
Write "0" (Bit-Drive Current $=-1/3I_C$)	$+2/3I_C-1/3I_C=1/3I_C$	$-1/3I_C$

As we can see, the arrangement provides a $3:1$ ratio of combined drive currents for cores which are to be switched, versus cores which are not to be switched. This wider range allows a more reliable operation[1]. The additional cost is insignificant: essentially, only slightly more complex bit drive amplifiers have to be provided.

The *read operation* is rather straightforward. A word-drive current is applied to the selected storage location. This current is directed opposite to the direction of the word-drive for writing and exceeds the critical value I_C. All cores of the selected storage location are, therefore, set to their zero state independent of their previous magnetization. Cores which previously contained a "1" switch their flux, and induce a signal into the bit line[2]. The flux of cores which already contained "zeros" is not changed and no signal is induced. The presence of a signal on a bit line signifies, therefore, a previously stored "1", while the absence of a signal signifies a previously stored "0" in the selected location. The read signals are amplified by the sense amplifier and used to set corresponding bit positions in the information register to the "1" state.

The memory, as it stands, has two operational characteristics which, in many cases, are undesired: The information stored in the selected location is destroyed by the reading process; and the writing works properly only if the storage location is initially cleared[3]. For these reasons, a read cycle is

---

[1] Or a faster operation, since the cores can be driven "harder".

[2] Some memory designs use separate bit drive and sense lines.

[3] We note that the "writing" of "0" does not clear a previously stored "1".

normally followed by a "restore" cycle, and a write cycle is normally preceded by a "clear" cycle.

The restore operation is equivalent to a write operation which, in effect, rewrites the previously read information from the information register[1] back into the storage location. The clear operation is equivalent to a dummy read operation, destroying the stored information. Restore and clear cycles are initiated by the control circuitry internal to the memory, so that the memory appears to the outside world as having a non-destructive read and full writing capabilities.

Although clear/write and read/restore are the normal modes of operation for computer memories, it is interesting to note that there are applications for memories in which read and write operations for each storage location strictly alternate. We speak of buffer memories. Here, clear and restore cycles can be eliminated, resulting in a faster overall operation.

8.3.2. Core Memories with Coincident Selection

While linear selection is normally used for small, fast memories, most of the larger core memories use "coincident" selection. Let us again use the sample layout of a relatively small memory to illustrate the principle.

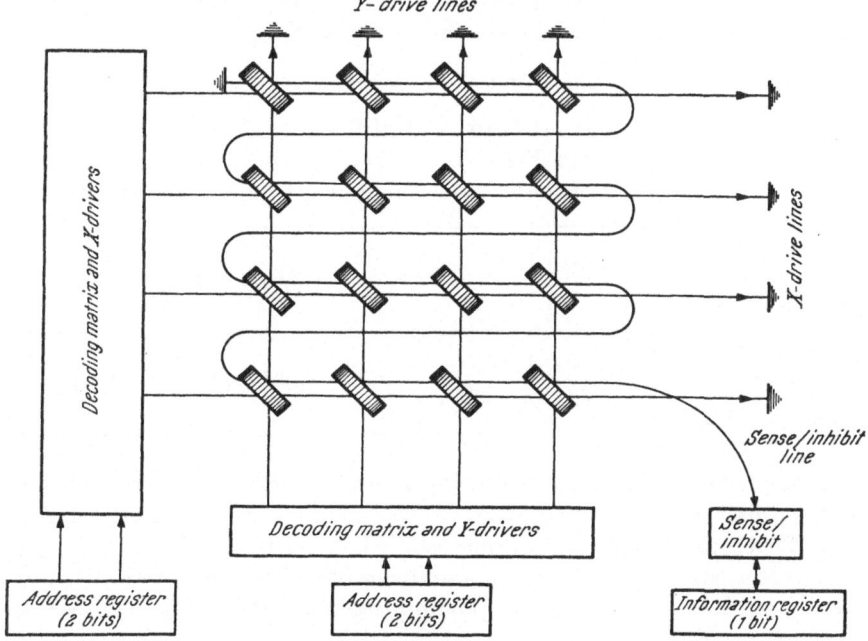

Fig. 8.74. Core Plane for Memory with Coincident Selection

---

[1] In this mode of operation it is also called the restore register.

The sixteen cores in Fig. 8.74 are arranged into a "core plane".[1] As we shall see later, there are as many core planes to the memory as there are bits in a word, but for the moment let us consider Fig. 8.74 as representing a complete memory with sixteen storage locations of one bit each. The 4-bit address required to address the sixteen locations is accommodated by two 2-bit address registers. This split in the address register has been made to show the selection mechanism more clearly: two address bits are decoded and used to select one of the four X-drive lines; the other two address bits are decoded and used to select one of the four Y-drive lines. The intent of this arrangement is to select a storage location by the appropriate combination of X- and Y-drives[2].

In considering the *write* operation, let us again assume that all cores are initially cleared. In order to select a location (a single core in Fig. 8.74, one X- and one Y-drive line is energized by a current $1/2 I_C$. Only the core at the intersection of the two energized drive lines experiences a drive current sufficient to switch its magnetization. Whether or not the core actually switches depends upon the current through the inhibit line. If there is no inhibit current, the core switches and a "1" has become stored. If there is an inhibit current (of magnitude $1/2 I_C$ and in a direction opposite to the drive currents), the core remains in the "0" state. The various possible drive conditions are shown in the following table.

Table 8.19. *Summary of Drive Currents for Write Operation*

	Selected Core (X- and Y-Drive Current $=1/2 I_C$)	Half-Selected Core (X- or Y-Drive Current $=1/2 I_C$)	Unselected Core (X- and Y-Drive Current $=0$)
Write "1" (Inhibit Current $=0$)	$1/2 I_C + 1/2 I_C = = +I_C$	$+1/2 I_C$	0
Write "0" (Inhibit Current $=-1/2 I_C$	$1/2 I_C + 1/2 I_C - -1/2 I_C = +1/2 I_C$	$1/2 I_C - 1/2 I_C = 0$	$-1/2 I_C$

A core is "selected" when both, its X- and Y-drive lines are energized; a core is "half-selected" when either its X-, or its Y-drive line is energized (but not both); and a core is "not selected" when neither its X-nor its Y-drive line is energized. Considering the possibilities of inhibit current or no inhibit current, we have the six different possibilities indicated in Fig. 8.79. We see

---

[1] Also called "bit plane" or "memory plane".

[2] This is in contrast to a memory with linear selection in which each storage location has its own word drive line.

that the combined drive current is sufficient to switch the core under only one of the six possible conditions[1].

In order to *read*, both the appropriate X-drive line and the appropriate Y-drive are energized again by a current of magnitude $1/2\,I_C$, but now in the direction opposite to that for a write operation. If the selected core is in the "1" state, it becomes reset and a signal is induced into the sense line. If the selected core is in the "0" state, its flux is not changed and no signal is induced into the sense line. The presence or absence of a signal on the sense line is detected by the read amplifier, and used to set the information register to its proper state in the same manner, as in memories with linear selection.

The arrangement, as it stands, has very little, if any advantage over a memory with linear selection. However, the advantage becomes quite apparent when we consider memories with word-lengths of more than one bit. Since the current in both, X- and Y-drive lines is independent of any information and dependent only upon the address, we can use the drive amplifiers of one bit-plane to drive others. We have only to connect X- and Y-drive lines serially as indicated in Fig. 8.75.

We see that a particular X-drive line intersects a particular Y-drive line in each of the indicated core planes. By energizing one combination of X- and Y-drives, we select, therefore, N cores if there are N bit planes. In other words, we immediately have a memory with N bit words. The N-cores in the same position in each of the N planes are used to store one word of information. Conversely, one plane is used to store bit 0 of all words, another bit 1, and so on. Of course, each bit plane must have its own individual sense/inhibit line and amplifier to control the writing of zeros and ones, and to sense the output of the bit plane during a read operation.

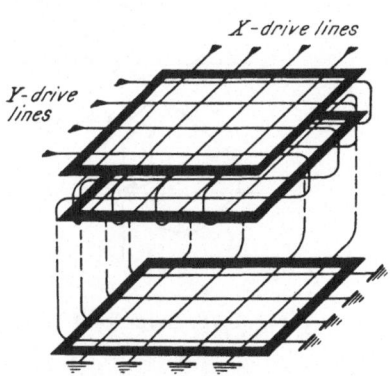

Fig. 8.75. The "Stacking" of Bit Planes

Problem 97: Determine the total number of drive and sense amplifiers required for a 4,096-word, 36-bit memory.

a) With linear selection;

b) With coincident selection.

---

[1] It is interesting to note that cores with a more rectangular hysteresis curve would allow a three-dimensional selection. A single core would be selected by the coincidence of three drive currents and the bit plane would become a "bit cube".

We notice that the memory with coincident selection has a "destructive" read, like a linear selection memory. Similarly, the writing of "zeros" does clear previously stored "ones". Both memory types are, therefore, normally operated in the clear/write, read/restore mode.

Certain aspects make the construction of very large memories with coincident selection difficult. For instance, the wiring of a bit plane as shown in Fig. 8.74 is no longer adequate. If the sense line and the X-drive line are routed in parallel, there will be considerable cross-talk. Consequently it becomes difficult to distinguish between "zeros" and "ones". The problem is further aggravated by the fact that half-selected cores change their flux slightly[1]. They will therefore induce a small but additive signal into the sense line during a read operation. To overcome these difficulties in practice, there exist a number of wiring techniques used to minimize these effects. Fig. 8.76 shows an example of such a technique.

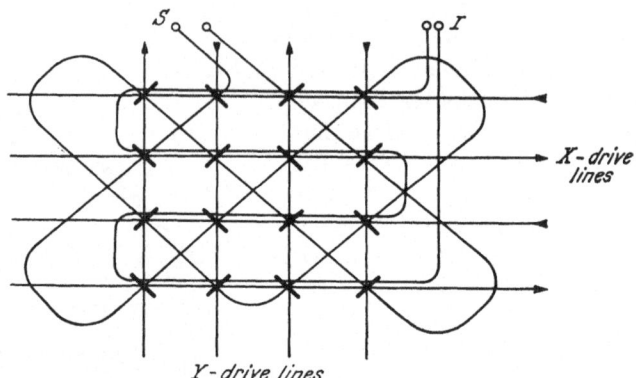

Fig. 8.76. Sample of Wiring Technique to Minimize Cross-Talk

The arrangement uses separate sense (S) and inhibit (I) lines. We notice that the sense line threads half the cores in the same direction as a particular drive line and the other half of the cores in the opposite direction. The undesired signals generated by half-selected cores tend, therefore, to compensate each other. Moreover, the sense line crosses drive lines with an angle of 45° and winds back and forth over each drive line so that the cross-talk between drive and sense lines is minimized. A disadvantage of this and similar arrangements is that the reading of a stored "1" may provide a signal of either polarity, depending on the particular core selected. Fig. 8.77 shows a few representative outputs of the sense amplifier for the reading of "ones" and "zeros".

---

[1] Compare Fig. 5.14.

The reading of a "zero" will provide a small undesired signal as indicated in Fig. 8.77 a. This is due to imperfections in the wiring, and the non-ideal and non-uniform characteristics of individual cores. The reading of a "one" produces a larger desired signal as indicated in Fig. 8.77 b. Both the "zero" and the "one" signal may have either polarity. In practice, the

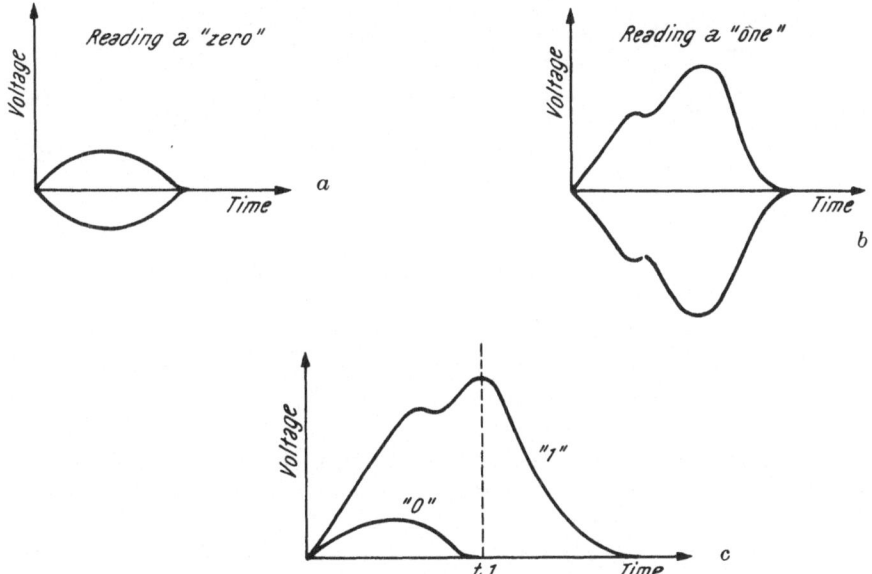

Fig. 8.77. Typical Output Waveforms for Reading Operation

signals are, therefore, rectified. Fig. 8.77 c shows the rectified "zero" and "one" signals superimposed. This output waveform is strobed at a time t_1 which gives the best voltage discrimination between "zeros" and "ones". Depending upon the result of the strobe, the information register is either set or cleared.

Let us add here a few practical observations. "Full-cycle times" (this is the time for a complete clear/write cycle, or a complete read/restore cycle) are in the range between 1 and 10 microseconds. "Access times" (that is the time it takes the outside world to communicate with the memory) have approximately 1/2 or 3/4 of these values[1]. One of the restrictions in achieving higher operating speed is the physical core size. Smaller cores operate generally faster than larger cores, but there is a limit on the "smallness" of cores which can be handled.

---

[1] Part of each full clear/write or read/restore cycle is controlled by the internal circuitry of the memory. Equipment external to the memory may be released during these times.

8.3.3. Delay Line Memories

Delay lines are well-suited as basic components for the construction of memories for small, serial computers[1]. Fig. 8.78 shows a model layout of such a memory.

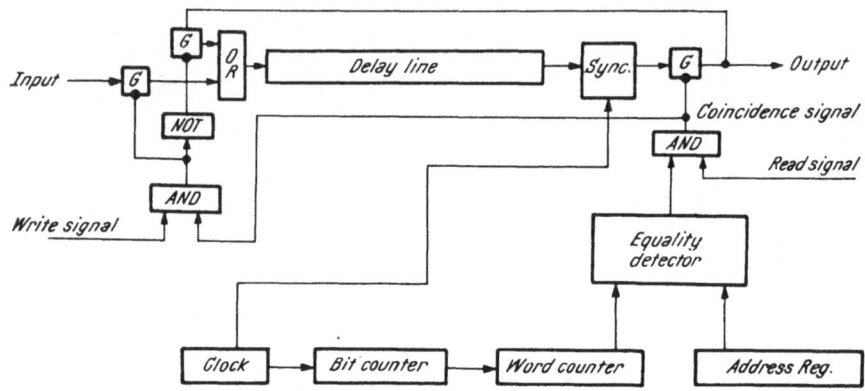

Fig. 8.78. Model Layout of a Delay Line Memory

The stored information recirculates through the delay line bit by bit, one word following another. Individual bits or words have, therefore, no fixed physical location and "storage locations" are defined only as time slots within the recirculation cycle. The "Addresses" are determined by two counters: a bit and a word counter. The modulus of the bit counter equals the number of bits in a word. The bit counter re-cycles, therefore, for each word time and advances the word count by one. The modulus of the word counter equals the number of words stored in the delay line. The word counter re-cycles, therefore, for each recirculation of the delay line and assumes always the same state during each specific word-time.

In effect, the state of the word counter indicates the address of storage locations. In order to make an access to a specific location, it is only necessary to compare the contents of the address register, i.e., the desired address, with the contents of the word counter, i.e., the actual address. If equality is found, the reading or writing can commence. In a read operation, the coincidence signal enables a read gate and makes the stored word available on the output of the memory. In case of a write operation, the recirculation of the delay line is interrupted, and the serial information at the input of the memory is allowed to enter the delay line. The coincidence exists for exactly one word time, since the word counter stays in any one

---

[1] For a description of delay lines, see paragraph 5.2.2.

state for only one word-time. Read and write gates are disabled after this time and the normal recirculation continues.

We notice that storage locations are basically accessible in sequential order. We speak of a sequential access in contrast to the random access of core memories. The actual access time to a specific storage location depends upon the time it takes for a desired address to appear.

The basic layout shown in Fig. 8.78 may be varied in a number of details. For instance, the memory may have an information register for the temporary storage of the word to be stored or retrieved. However, many of the smaller computers, to which the memory design is particularly applicable co-use one of the arithmetic registers for this purpose. The word counter and the address register may present their contents as parallel or as serial information. This, of course, changes the design of the equality detector[1]. The memory may comprise several delay lines in order to increase its storage capacity. The high-order bits of an address might then be used to select a specific line, while the low-order bits determine the specific location in the previously explained manner. It also should be noted that it is possible to build parallel memories with delay lines. Such a memory might have as many delay lines as there are bits in a word. The coincidence detector would provide a coincidence signal for only one bit-time and information would be transferred in parallel to or from the delay lines at a single bit-time. Finally, addresses can be "interlaced" in order to reduce the average access time or the memory[2].

It may be appropriate to add one final observation. Delay memories are volatile memories, that is, the stored information is lost when the power is turned off.

Problem 98: Assume that a serial memory uses a magnetostrictive delay line with a storage capacity of 1,024 bits and a bit rate of 200 kc as a basic component.

a) How many 32-bit words can the delay line store?

b) How many bits are required for addressing?

c) What is the average access time of the memory?

8.3.4. Drum or Disk Memories

Modern large computers employ disk and drum memories only as peripheral or auxiliary mass storage devices. However, drums or disks are still used to construct economic memories for small or special purpose computers. Information is stored by recording it magnetically on a rotating

---

[1] Compare also paragraph 8.1.5.2.

[2] Interlacing is an addressing technique which is frequently used for disk or drum memories and is discussed in paragraph 8.3.4.

surface in the manner described in paragraph 5.1.3. The physical location of a piece of recorded information is determined by the channel, i.e., the recording track, and the sector, i.e., location on the circumference of the recording medium.

Two techniques for addressing are in practical use: recorded addresses, and addressing by counting. Fig. 8.79 shows a sample layout of a parallel drum memory with addressing by counting.

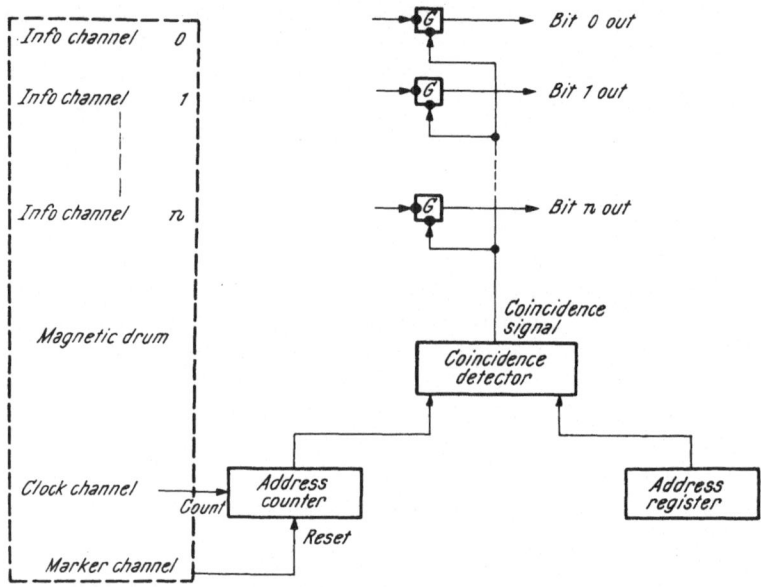

Fig. 8.79. Sample Layout of a Parallel Drum Memory

Information is recorded in parallel, that is in such a manner that all bits of a word appear simultaneously at the output of several drum channels (information channels 0 to n in Fig. 8.79). In order to accomplish the addressing, the drum carries two permanently recorded channels, designated as clock and marker channels. The marker channel produces an output signal once per drum revolution. The clock channel provides signals signifying each rotation of the drum by one sector or bit space. The clock and marker signals are used to advance and reset an address counter. With this arrangement, the contents of the address counter indicate the physical position of the drum (i.e., the accessible storage location) similarly to the manner in which the bit and word counter indicate the accessible location of a delay line memory[1]. Also, like in delay line memories, a coincidence detector

---

[1] See paragraph 8.3.3.

detects the equality between the desired address contained in the address register and the actual address contained in the address counter. For read operations, the coincidence signal strobes the output of the information channels and makes the recorded information available at the output of the memory. Similarly, but not shown in Fig. 8.79, the coincidence signal is used during a write operation to strobe information at the input of the memory and cause it to be recorded.

Large parallel memories may contain several sets of information channels. In effect, they contain several logically separate drums (or disks) combined into a single physical drum. An individual logical drum may be selected by the high-order bits of an address while a particular sector is determined by the low-order bits.

Clock and marker signals are recorded (in contrast to the external clock of delay line memories) because it is practically not feasible to synchronize the drum rotation with some external timing. Instead, the clock is synchronized with the physical movement of the drum and, incidentally, also serves as a timing signal for the remainder of the computer system. We observe that this arrangement makes drum or disk memories non-volatile storages. The address counter is re-synchronized by clock and marker pulses after power is turned off and re-applied.

Problem 99: a) What drum diameter is required for a 4,096 word parallel memory if the recording density is set at 400 bits per inch?

b) What is the average access time of this memory if the drum rotates with 1,800 rpm?

c) What is the clock rate?

The drum memory, like the delay line memory, is a sequential access storage. Read and write operations are delayed until the desired storage location becomes accessible. The minimum access time approaches zero; the maximum access time equals the time for one complete drum-revolution. If one assumes accesses to random addresses, the average access time becomes equal to the time for a half drum-revolution. However, if accesses are not completely random, other average values for the access time can result. Suppose, for instance, that the execution of a computer program requires many consecutive accesses to consecutive addresses. Let us also assume that the time between such accesses is fairly short, but slightly longer than the time required for the drum to rotate by one sector. In this case, the desired sector has normally just passed the read head when an access is attempted and it becomes necessary to wait for almost a complete drum-revolution until the sector becomes accessible again. The average access time may approach the maximum access time.

In such cases, a re-programming of the problem may improve the situation. For instance, a different choice of operand addresses (which for many types of programs can be selected as desired) may reduce the average access time and, therefore, also computer running time. We speak of "minimum access programming".

Another approach is the provision of several read or write heads around the circumference, and a selection of the head most suitable located for each memory access. A third possibility is the assignment of non-sequential addresses to sequential drum sectors or vice versa. This can be accomplished by the memory hardware and is known as "interlacing". Fig. 8.80 shows, as an example, an interlace for four-bit addresses.

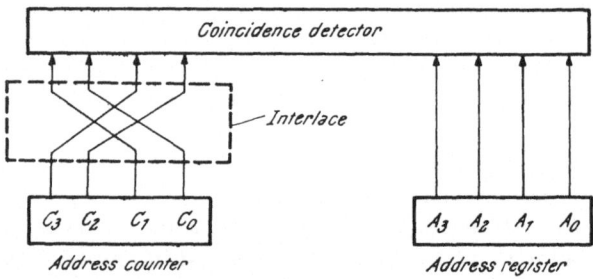

Fig. 8.80. Example of an Address Interlace

The output lines of the address counter are transposed, so that the coincidence detector compares bit positions with different weights in the address register and the address counter. In effect, the coincidence detector "sees" the address counter count in a somewhat irregular manner. The consecutive apparent counter states are listed in the right-hand column of Table 8.20.

Table 8.20. *Effect of Sample Address Interlace*

Sector	Address
0000	0000
0001	0100
0010	1000
0011	1100
0100	0001
0101	0101
0110	1001
0111	1101
1000	0010
etc.	etc.

We notice that the interlacing does not change the number of unique bit-combinations. Each sector has, therefore, still its individual address. However, consecutive addresses no longer designate consecutive sectors

In the example above, consecutive addresses are assigned to storage locations which are in general four sectors apart. Interlaces can, of course, be designed to provide other spacings and to fit longer addresses.

Problem 100: Design an interlace for a drum memory with 4,096 storage locations which assigns consecutive addresses to sectors which are in general one half of a drum-revolution apart.

Interlaces are sometimes designed so that they can be exchanged or rewired. Specific programs are then run on the computer with the most efficient interlaces inserted.

So far we have discussed the addressing by *counting*. An alternate technique is addressing by *recorded addresses*. This technique is most frequently found in serial memories. Fig. 8.81 shows the basic arrangement.

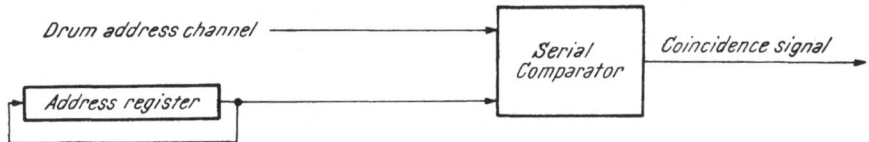

Fig. 8.81. Addressing by Recorded Addresses

The drum carries an address channel in addition to information, clock, and marker channels. The address channel contains pre-recorded addresses, that is serial recordings of bit combinations which are unique for each sector. The pre-recorded address is compared during each word time with the desired address, recirculating in the serial address register. When equality is found, a coincidence signal allows the writing or reading of information to, or from one of the information channels.

Addresses may be recorded in a strictly sequential manner. However, for the same reasons for which interlaces are found in connection with addressing by counting, addresses are frequently recorded in an interlaced manner. We then speak of minimum access address channels. Access times can be minimized by this technique better than with interlaces. We notice that there is complete freedom in the sequence in which addresses may be recorded. At least in special purpose computers[1], addresses may be assigned in the exact sequence which produces the minimum average access time. The same efficiency cannot be achieved with interlacing as described previously, unless a complex flexible counter is used instead of the simple address counter. General purpose computers employing recorded addresses, frequently have several address channels recorded with differently arranged

---

[1] For instance, the sole purpose of some airborne computers is the solution of guidance equations.

sequences of addresses. The most effective of these may then be used for each run of a computer program.

Let us conclude this paragraph by mentioning a few sufficiently interesting technical details. Usually only the low-order bits of an address are used to select a storage location in the previously discussed manner, while the higher order bits of the address are used to select one of several information channels. The address channel may have identical addresses recorded more than once within a word. This is usually done so that these addresses line up with the several address fields of the instruction format[1]. Addresses can then easily be compared with the desired address in either address field without shifting.

Drum or disk memories, in general, may have only block access, that is, each read or write operation transfers more than one word. In some instances they have "dead spaces", i.e., portions which are normally not used for storage[2]. They usually have some spare channels which can be used if one of the normal channels, becomes inoperative.

8.3.5. Associative Memories

Associative memories are presently used only in special purpose computers[3]. However, disregarding their cost, their properties make them very attractive for general purpose use. It has been shortly mentioned in paragraph 8.2.3. that these memories are designed to retrieve information not by conventional addressing, but by searching for the information itself. For this purpose, the memory is normally given a "descriptor", i.e., a bit pattern which characterizes the desired information. The memory tries then to locate that information in its storage which contains this descriptor. It responds either with the desired information, or, at least, with its address. Let us begin by considering the sample layout of an associative memory shown in Fig. 8.82.

The diagram indicates the individual storage elements of the memory, organized by bit and word. In some respects, the organization is that of a parallel memory. A word can be selected, and all bits of this word can be transferred in parallel to the information register. If the memory were used in the conventional manner, one word at a time would be read. The word would then be compared to the desired pattern and a match or mis-match would be determined. If the memory had. say, 32,768 words, 32,768 read

---

[1] See paragraph 8.2.4.

[2] It is rather difficult to record clock signals with equal spacing over the entire circumference of the drum. In some cases, therefore, clock signals are recorded over the major part with equal spacing. A small part of the circumference is left blank.

[3] For a discussion of associative computers, see paragraph 9.8.

operations would be necessary for a complete scan of the memory. However, if the memory is used in an associative manner, the search is by bit rather than by word. Each storage location (i.e. each word) has a match indicator. When the search operation is started, all match indicators are

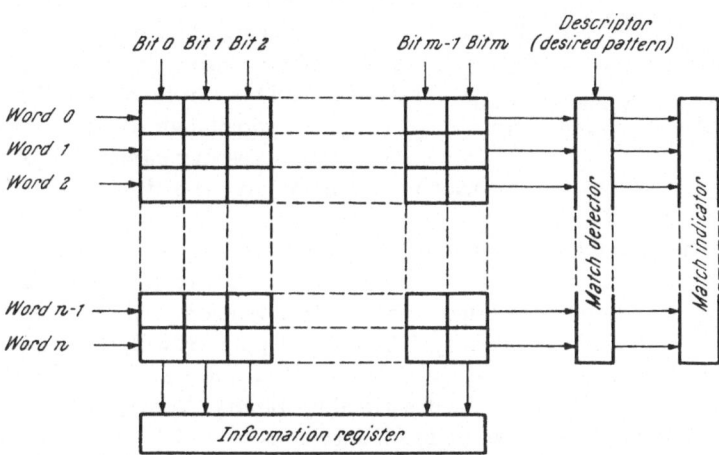

Fig. 8.82. Organization of an Associative Memory with Serial-by-Bit and Parallel-by-Word Search

in the "set" condition. The first access is to bit "0" in all words simultaneously, that is, bit "0" is read simultaneously in all words. The contents of this bit "0" location are transferred in parallel to the match detector. The match detector compares then the outputs of all words with bit "0" of the desired pattern or the "descriptor". The match indicator of those words in which bit "0" equals the descriptor are left set. However, the match indicators of those words which mis-match the desired pattern are cleared. The status of the match indicators after the first read operation indicates then whether or not the corresponding words match the first bit in the desired pattern. The second read operation is concerned with bit "1". The match indicators of all words mis-matching the second bit in the desired patterns are cleared if they have not already been cleared previously. Consecutive read operations read the remaining bits, with the final result that only the match indicators of those words are left set which match the descriptor in all bits.

We note that reading operations are serial-by-bit and parallel-by-word compared to a conventional parallel-by-bit and serial-by-word reading operation. If we assume again a memory with 32,768 words of, say, 36 bits each, only 36 read operations are necessary to scan the entire content of the memory, as compared to the 32,768 read operations which are necessary with a conventional search.

The diagram given in Fig. 8.82 omits one essential detail. The scheme, as it stands, indicates basically only whether or not a desired pattern is stored in the memory. A translation of the position of the remaining "set" match indicator into an address, and an actual access to this memory address (by word) has to be made, before the desired information is externally available. This can be accomplished with straightforward logic design. Supposing that the necessary circuitry is available, we may represent an associative memory by the functional diagram shown in Fig. 8.83[1]:

Fig. 8.83. Functional Diagram of an Associative Memory

In general the descriptor will be not a complete word. As a matter of fact, the retrieval of information associated with a certain pattern is practically more important than the determination whether or not a pattern is stored in the memory. The desired pattern will, therefore, almost always be concerned with selected bits within a word. The desired status of bits in these positions can be described by binary "zeros" or "ones". However, provisions have to be made for locations in which we do not want to prescribe a fixed state. The information to be searched for could be best

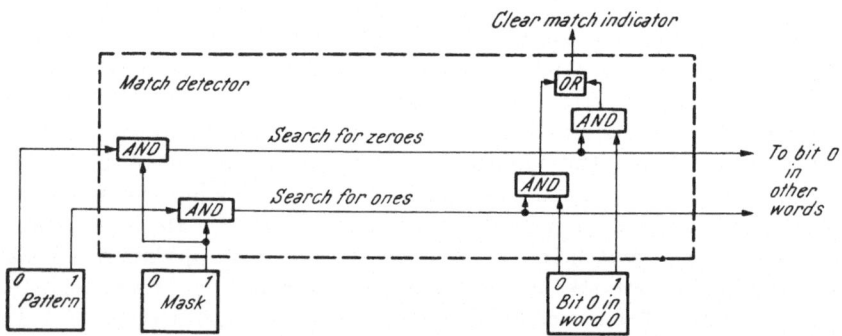

Fig. 8.84. Partial Logic Diagram for the Specification of the Descriptor and the Match Detector

described by a ternary configuration (0, 1, don't care) for each bit-position. However, ternary information is inconvenient to handle, so that most designs of associative memories specify the search pattern by an information word and a mask word (see Fig. 8.84).

---

[1] An identical functional diagram also describes the capabilities of table search instructions. The difference is only in the manner in which the capability is implemented.

The diagram shows the logic circuitry for only one bit-position. A "1" in the mask indicates that we want to specify a pattern in this bit-position. The information pattern determines then the wanted state of the bit. A "0" in the mask register indicates that we "don't care", i.e. that we do not want to prescribe a required state. This arrangement is completely flexible, i. e. any bit-position or combinations of bit-positions can serve for the selection of information. Alternate schemes use fixed bit-positions for the selection. A separate mask register is therefore not required, but the flexibility of the operation is reduced[1].

According to the diagram given in Fig. 8.83, the desired response to the select pattern or the descriptor is the information which is associated with it. Practically, there have been three slightly different responses implemented. In the simplest scheme, the response is in form of the address of the desired information. For all practical purposes, this is a valid response. It assures that the desired information is stored in the memory (this assurance may be all which is required in some cases). If it is required to retrieve the information itself, the computer can make an access to the given address in the conventional manner. In the most elaborate scheme, the associative memory responds with the full word containing the descriptor and its associated information. In an alternate scheme, the associative memory responds with the word, less the descriptor. The latter is a less flexible arrangement, but adequate when the descriptors are always placed in a fixed portion of all words (say e.g. the first 15 bits of a word).

There is no question about the usefulness and desirability of associative memories. Also, there are, in principle, no technical difficulties in building them. However, there are many open questions concerned with their design. Many technical details can be solved only in connection with the proper choice of components, that is such components which make associative memories economically feasable. Let us indicate a few of the problems and open questions in order to give an idea of the broadness of the area in which decisions have to be made.

Let us take the diagram in Fig. 8.83 as a starting point. It shows the basic logic function to be performed by the selection mechanism. However, there is considerable latitude in implementing the logic function by the memory hardware. Even if we take the diagram to represent a parallel-by-word and serial-by-bit search (as e.g. indicated in Fig. 8.82), one might consider memories with linear selection or coincidence selection of individual cells. One might also implement the masking by the match detector not in the manner shown in Fig. 8.84 but simply by skipping the reading operation for certain bits if the data mask specifies that they are to be

---

[1] In the previously cited example it may be possible to search for a certain part number, but not for, say, items supplied by a specific manufacturer.

ignored in the selection process. But really one is not limited to a parallel-by-word and serial-by-bit search. At least theoretically, one can extend the selection scheme given in Fig. 8.84 to a simultaneous read in all bits of all words[1]. One can also go to the opposite extreme and investigate individually each bit stored in the memory in serial manner. This solution may be of interest if a serial storage is used[2].

In either case it is desirable, if not a necessity, to construct the storage of elements which allow a non-destructive readout. Memory cells will, therefore, be of different design than those of contemporary computer memories. The requirement for a nondestructive read-out in itself will make associative memories basically slower and more expensive.

So far we have indicated only various approaches to a search for an exact match. However, one may think of several possibilities to retrieve information with the best match, even though it is not perfect. Two possible criteria for the "best" match would be the highest number of matches in single bit positions, or the longest match in consecutive positions[3]. Another approach might be a search in which certain matches are prescribed as mandatory and others as desired. If we extend the basic concept further, we may want to design an associative memory which searches for a descriptor which is numerically larger or smaller than a given value. In all cases we essentially require a match detector which is more complex than Fig. 8.84 implies. There is no fundamental theoretical or practical problem other than the question of just how much complexity one can economically incorporate.

.   Let us finally pose some questions which can be answered intelligently only in connection with the application of the associative memory or the computer system of which it is a part. What provisions have to be made for no matches or multiple matches? Should only associative reading or also associative writing be incorporated? What should the length of one word in the memory be? How should long items of information be handled in a computer with relative short word-length? Should the descriptor and the associated information be logically separated? If yes, by an "end of descriptor" character, or simply by the position within a word? Is a word consisting of all "zeros" legal information? (In other words do we have to distinguish between an empty cell and a cell containing zeros?)

---

[1] Fig 8.84 would then be overlayed with identical circuits concerned with the remaining bits of the memory. A mis-match detected in any layer would clear the match indicator.

[2] Actually, language translation dictionaries have been implemented in this manner.

[3] The first technique might enable a computer to find that document in a library easily which best fits a particular combination of subject headings. The second scheme could be useful (and has already been used) for language translation by machines.

Problem 101: Assume that you have a serial storage device (e.g. magneto-strictive delay line) which has a storage capacity of 512 words with 24 bits each. Show the logic diagram of a selection mechanism which makes the storage device an associative memory.

Problem 102 (Voluntary): Try to estimate the number of instructions required and the relative program execution times for locating an item of information in a memory with 32,768 words of 48 bits each

a) with normal programming,

b) with programming, but assuming that the machine is capable of indexing, indirect and relative addressing, etc.,

c) with a table look-up instruction,

d) with an associative memory.

8.4. The Input/Output Unit

The purpose of the input/output unit, or *I/O* unit for short, is to serve as a communication link between the computer and its attached peripheral or external devices. Such devices include magnetic tape units, card readers and punches, printers, disk storages, on-line data transmitters and receivers.

The term "communication" is used here in a rather broad sense. It is meant to include not only the transfer of information itself, but also the interchange of signals necessary to control this transfer. In particular, the *I/O* unit causes the selection of one of several external units and its connection to the computer; it requests a specific operation or function to be performed, such as read, punch, print, or rewind; and it coordinates the actual information exchange between the computer and the external device. More complex *I/O* units may also perform the additional tasks of multiplexing several independent information transfers and of addressing the memory.

I/O units, perhaps, exhibit more variations in their design details and operational aspects than other computer units. On the other hand, they pose very few difficult design problems, once their operation is understood. Let us begin by describing the operation of a relatively simple, hypothetical unit. Alternative techniques and layouts may then be discussed in relation to this model.

8.4.1. A Model Input/Output Unit

Let us assume a model *I/O* unit similar to the one shown in chapter 7. It consists essentially of two registers as indicated in Fig. 8.85, the *I/O* information register, and the *I/O* control register.

The *information register* is used exclusively for the exchange of information. One word at a time is transferred between the computer and the external device. During an input operation, the selected external device places a word into the information register whenever the information becomes available. The execution of a programmed external read instruction[1]

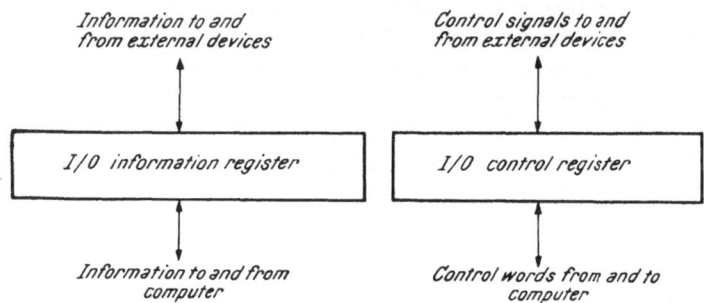

Fig. 8.85. Model Layout of a Simple *I/O* Unit

causes then the transfer of this word to the memory or an arithmetic register. During an output operation, a programmed external write instruction transfers a word of information from the memory or an arithmetic register to the information register. The word is then transmitted to the selected external unit whenever the device is ready to accept the information. Thus, the information register acts as a buffer, i.e. as a temporary storage device for one word during this information exchange[2].

The *control register* is used solely for the interchange of control words. Individual bits or bit-combinations within these control words are assigned certain meanings. Specific codes are used to designate the various external units, various operations or functions, and various conditions. Both, the computer and the external device may put words into the control register. The computer places codes in the *I/O* control register during the execution of *I/O* select or *I/O* function instructions[3]. The code, i.e. the states of individual flip-flops in the control register, is then sensed and interpreted by the attached external devices and causes one of the units to perform a specific function. Conversely, the external device may transfer codes to the *I/O* control register which signify certain external status conditions, such as "ready", "not ready", "error", "rewound", and "function completed". These status conditions can be read and interpreted by the computer program.

---

[1] See paragraph 8.2.4.
[2] Paragraph 8.5. shows the implications of this buffering in more detail.
[3] See paragraph 8.2.4.

In practice, there exists no convention for the assignment of external equipment codes. Each machine has its own particularities. The assignment depends upon the complement of peripheral devices manufactured for the specific computer, the word-length, the ease in decoding, and so on. Many computer centers employ codes which are unique for their installation and are designed for the control of special purpose equipment. In this connection, it is important to note that the addition or reassignment of such codes requires no change in the computer hardware. Codes are embedded in the program like other operands. A change in their assignment affects only the program[1].

So far, we have disregarded one very important operational aspect: the interaction between the operation of the external device and the program execution by the computer. The transfer of information and control words between the I/O unit and other parts of the computer is under program control. The transfer of words between the I/O unit and external devices may be in response to a computer action, but is under control of the external device. Since the computer has no direct control over the timing of transfers, it must receive some indication whenever such transfer is made or requested by the external device. Several different techniques for the coordination of external requests with the execution of the computer program are in practical use, each having its own particular advantages and disadvantages. Let us assume that our model incorporates only one of these, the "delay execution" scheme[2]. A sequence of I/O instructions appropriate for the particular I/O operation is embedded in the program. The execution of each of these instructions is delayed until the external device has initiated a request or has otherwise indicated that it is ready.

Let us use an example to illustrate this technique. Suppose a block of information recorded on magnetic tape contains 120 words. To read this information, it is necessary to execute one external function and 120 external read instructions in sequence. The purpose of the external read instruction is to transfer a control word from the memory into the I/O control register which then causes the selection of a magnetic tape unit and requests it to read one record. The purpose of the external read instructions it to transfer, consecutively, the 120 words from the I/O information register to the computer memory. The program may schedule the execution of these instructions one after another, but their actual execution is delayed. Specifically the execution of the external function instruction is delayed until the tape unit indicates that it is ready, that is, until it has completed

---

[1] A change of codes does affect the hardware of peripheral devices, especially the decoding circuits.

[2] Alternate schemes will be discussed in paragraph 8.4.2.

any previous operation, until the power is turned on, etc. The execution of each external read instruction is delayed until the tape unit has placed a new word into the information register. This delaying of the instruction execution can be accomplished without difficulty by the computer control unit. It is only necessary that the selected tape unit provides to the control unit appropriate "go ahead" signals[1].

Problem 103: a) At what time should the external unit provide a "go ahead" signal during a tape write operation ?

b) Propose a simple scheme for delaying the execution of external write instructions.

Problem 104: When should the I/O control register and the I/O information register be cleared:

a) during an input operation,

b) an output operation ?

Propose a simple scheme to accomplish this clearing.

Problem 105 (Voluntary): Design a sequencer for the execution of an external function instruction. Assume a synchronous computer and a control unit implemented with flip-flops (see Fig. 8.62).

If the program schedules I/O instructions one after another, as indicated in the above example, the computer may waste considerable time waiting for the external unit[2]. This situation can be improved by interspersing I/O instructions and other instructions in the computer program, that is, by performing other meaningful operations between consecutive I/O transfers. This, however, has to be done with care: if the execution of instructions not concerned with I/O operations takes relatively long, the computer may respond too late to an external request. The timing of operations is not a simple task since many execution times are variable and cannot be exactly anticipated[3].

Problem 106: What information is received by either the computer or the external device when the computer program responds too late to an external request

a) during an input operation,

b) during an output operation ?

---

[1] Fig. 8.60.a shows, for example, the design detail of a control unit in which the execution of an instruction is delayed until an external signal is received.

[2] The word transfer rate of external devices is usually much lower than the rate at which instructions can be executed.

[3] For instance, the execution time of a multiply command may depend upon the distribution of "ones" and "zeros" in the multiplier; the rewind time for a tape unit depends upon the position of the tape; and the access time to a sequential memory depends upon the address specified.

Assume that the data transmission rate of the external equipment is fixed (e.g. determined by the mechanical movement of the unit).

Problem 107: Assume that a peripheral unit transfers every millisecond one word of information to the I/O information register. A programmer wants to execute as many program steps as possible between two consecutive external read instructions.

Approximately how much time is available for this purpose

a) in the average

b) at the maximum ?

8.4.2. The Monitoring of External Operations

The delay execution scheme discussed in the last paragraph provides a rather close coordination between the execution of the computer program and the operation of the external equipment. No specific monitoring of the external equipment is required: the program execution proceeds step-in-step with the external operations. However, such a close tie has its disadvantages, and is not always desired. As we have already seen, the computer may be slowed down in its program execution by the external equipment if the programmer is not careful in the timing of operations. Even more important, in this respect, is that the programmer is forced to anticipate the exact sequence of events. This is not always possible. Suppose that a computer receives inputs from two different data sources.[1] If the two devices transmit information at unsynchronized rates, individual words will arrive in a sequence which is not exactly predictable. It becomes, therefore, impossible to write a predetermined sequence of appropriate input/output instructions. The delay execution scheme, in effect, is able to synchronize the computer operation with only one external device at a time.

For similar reasons, the scheme does not allow the writing of programs which recover from unexpected fault conditions. Let us again take an example. Suppose a computer executes a program designed to read the, say, 120 words contained in a block of information recorded on magnetic tape. The actual record, however, erroneously contains only 119 words. The program will "hang up" on the execution of the 120th external read instruction. The computer waits for the transfer of the 120th word and ceases all further operation. Of course, the computer would also hang up if the record actually had the correct length but the tape unit made a sprocket error, that is if it accidentally skipped a line during the reading of the tape.

---

[1] Perhaps via two separate I/O information registers.

These operational difficulties can be resolved when the computer program is able to monitor the operation of the external device.

8.4.2.1. The Status Scheme: In the status scheme, the execution of an I/O instruction is attempted only when the execution can be successfully completed. The request from the external unit sets a status indicator[1]. The state of this indicator can be tested by programmed test instructions. Programs are then written in such a manner that the status indicator is tested at fairly regular intervals throughout the execution of the main program. If the status condition is not present when a test instruction is executed, the test instruction has no effect, and the execution of the main program continues without interruption. If, however, the status condition is present, the test instruction becomes in effect a jump instruction. Typically, a subroutine is entered which responds to the external request by appropriate I/O instructions.

One advantage of the status scheme in comparison with the delay execution scheme is that the execution of an I/O instruction is attempted only when the external unit has already made a request for a corresponding action. No unnecessary waiting periods are, therefore, encountered. A disadvantage for some applications, is perhaps the fact that the programmer cannot predict exactly where in the instruction sequence of the main program an I/O operation will take place and that the status must be checked periodically.

It is important to note that the status scheme can be designed so that it becomes possible to write programs which are able to recover from unexpected conditions. Let us again take the example of reading a magnetic tape record which is shorter than expected. Suppose that two different status indicators can be set by the external unit, one, S_i, for requests concerning the I/O information register and another, S_c, for requests concerning the I/O control register. The two different status conditions can be distinguished by computer test instructions. Suppose further that the tape unit always signifies the end-of-record by putting an "end-of-record" code into the I/O control register. The reading operation may now proceed as follows: The main program contains test instructions which test periodically for the presence of either status condition, S_i or S_c. When either status condition is found, an I/O subroutine is entered. The subroutine determines now whether the status is S_i or S_c. If the status is S_i, an external read instruction is executed. If the status is S_c, the contents of the control register are read and investigated. If an end-of-record code is found, before the full number of external reads has been executed, the record has been

---

[1] Perhaps in the form of a flip-flop in the control unit.

shorter than expected, but the computer does not hang up and the program is free to take whatever corrective action is appropriate[1].

Problem 108: How can you adapt the status scheme to handle the information exchange with two independent, unsynchronized external devices?

Problem 109: When or by what action should the status indicator be reset?

8.4.2.2. The Program Interrupt: In an alternate scheme, the external request causes a "program interrupt". The computer control unit ceases the execution of the normal computer program and enters an interrupt routine instead. The interrupt routine responds to the external request with an appropriate I/O instruction or a sequence of instructions. The "jump" is effected by the computer control unit and not by a programmed jump instruction[2].

The interrupt scheme, like the status scheme, allows the writing of programs which recover from unexpected conditions. However, the main advantage of the interrupt scheme is that it is not necessary to intersperse status test instructions in the main program. It appears to the programmer as if the main program and the I/O interrupt routine are executed by the computer essentially independent of each other[3]. In order to create this effect, care must be taken that the execution of the interrupt routine does not destroy information which was left in arithmetic and control registers when the interrupt occured. This usually infers that the initial program steps of an interrupt routine store the contents of these registers in a reserved region of the computer memory and that the final steps retrieve and restore this information.

Problem 110: How can you adapt the interrupt scheme to handle the concurrent communication with two separate external devices at unsynchronized data rates? Specifically, how do you propose to handle interrupts during interrupts?

Although the interrupt scheme is rather flexible, there may be occasions when its use is disadvantageous. For example, during a real-time or on-line computation, there may be time periods during which no interruption

---

[1] Many possible actions are conceivable, for instance: back up tape and attempt to re-read; print the fact that there has been an error and stop; skip the computations concerned with this particular record of information and go on with the remainder of the problem.

[2] The handling of interrupts by the computer control unit is discussed in paragraph 8.2.1.

[3] Only the execution time of the main program seems to increase if interrupts occur.

of any kind can be tolerated. An interruption by one external device may also be undesired while the computer is in the process of responding to a previous interrupt by another unit. In these and similar circumstances, the status scheme may be preferable[1].

8.4.2.3. Combination of Techniques: So far, we have treated the delay execution scheme, the status scheme, and the interrupt scheme as if they always would be employed separately. This was done to show the characteristics of the different techniques more distinctly. In practice, designs frequently use a combination of techniques. For instance, a particular computer might always employ the delay execution scheme for the exchange of information via the I/O information register, while program interrupts are used to signify that an external device has placed a status code into the I/O control register. In many designs, however, the selection of a specific scheme is at the option of the programmer. He can then select the scheme best suited for the particular application. In this connection, we should note that the use of the status scheme itself is optional. The

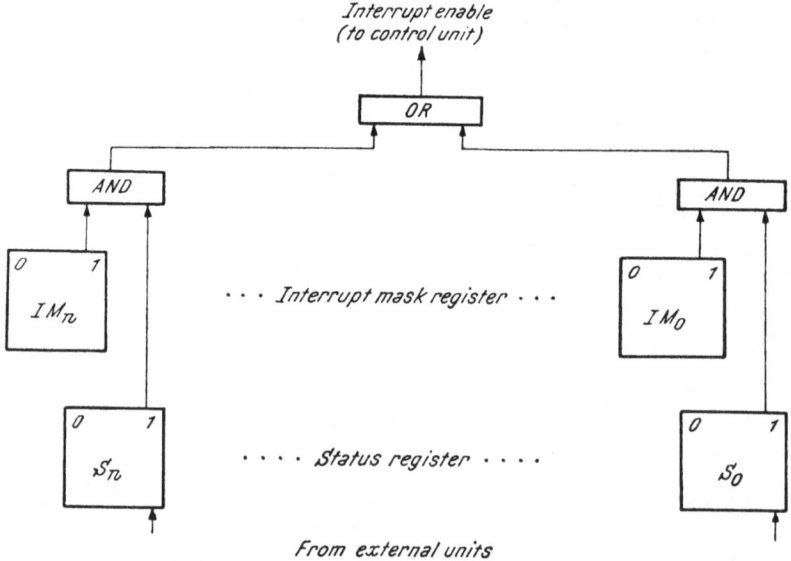

Fig. 8.86. Optional Interrupt System

programmer may test for status conditions, but he does not have to do so. The use of the program interrupt can be made optional if interrupts can be enabled or disabled under program control. Fig. 8.86 indicates one possible approach.

---

[1] In paragraph 8.4.2.3 we shall see an approach by which undesired interrupts can be locked out under program control.

The arrangement contains two registers, the status register and the interrupt mask register. Both are accessible by program instructions.

External devices place their status codes into the status register. The status conditions can then be monitored, i.e. read out and tested by program whenever desired. The presence of an individual status bit generates a program interrupt, *provided* the corresponding bit in the interrupt mask register contains a "1". Since the contents of the mask register can be modified by the program, it is completely at the discretion of the programmer to determine which particular status conditions shall generate an interrupt. These status conditions may include internal conditions such as overflows, divide faults, illegal instructions, etc., in addition to external status conditions.

The selection of interrupting conditions may change repeatedly during the execution of a program. The programmer may, for example, want to block all possible further interrupts, once an interrupt has occurred. Alternately, he may allow a higher priority interrupt during the response to a lower priority interrupt[1]. In order to accomplish this, he makes the program change the contents of the interrupt mask register according to the needs.

Several basic variations of the indicated approach are feasible. For instance, the status register, or at least some of its flip-flops, may be physically located in the external unit rather than the *I/O* unit. Alternately, a combination of interrupt and status registers may perform the function of the single status register indicated in Fig. 8.86. In this event, the contents of the interrupt register would indicate only which external device generated the interrupt and an access to the appropriate status register would be required to determine the specific reason for this interrupt. With the incorporation of an interrupt register, the status register may be omitted altogether. Status information may then be obtained from external devices upon execution of a specific "request status" instruction. This latter approach has the advantage that detailed status information for many external devices can be made available with a rather limited number of registers.

The more detailed possible variations are too numerous to be considered here. In all of them we find the essential distinction: The status scheme leaves the initiative, but also the burden of controlling the communication to the computer. (The program tests for external requests and determines

---

[1] The highest priority might be assigned to a power fault. An interrupt routine may then terminate the execution of a program in an orderly manner in the fraction of a second which is available between the failure of the prime power and the complete shutdown of the computer. This, normally, involves the saving of all the information contained in volatile storages, such as flip-flop registers.

when an input or output operation is appropriate.) The program interrupt gives the initiative for communication to the external equipment. Which one of the two approaches is preferrable depends upon the specific circumstances.

8.4.2.4. Fault Detection: The communication between the computer and external devices is prone to error. Such errors may be caused by external devices (when, for instance, a card reader or punch is mechanically jammed, or a defect in the magnetic tape causes the skipping of a line during a read operation), or by the computer program (when, for instance, an unassigned select code is used, when an insufficient number of external read instructions is programmed, or when the transfer of information into the I/O register comes too late). Since such errors are relatively frequent, it is customary to monitor the exchange of information and to provide fault indications.

External fault conditions are normally detected in the external device itself. The presence of a fault causes the transmission of a fault status code or a program interrupt. Errors in the sequence of the information exchange are usually detected in the I/O unit. The two circuits indicated in Fig. 8.87 may be representative of such fault detectors.

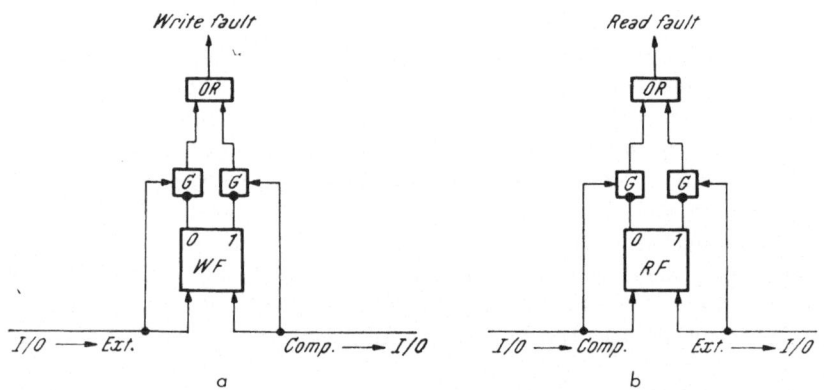

Fig. 8.87. Examples of I/O Fault Detectors

The circuit in Fig. 8.87a detects "write faults". The monitoring flip-flop is set for each information transfer from the computer to the I/O register, and it is cleared for each information transfer from the I/O register to an external device. A write fault is generated whenever the computer transfers information to the I/O register twice in succession without the external device taking information out of the register (information pile-up), or whenever the external device reads information twice in succession without the computer bringing in new information (no-information fault). The circuit in Fig. 8.87b, similarly, detects information pile-ups or no-information faults during a read operation.

The states of the monitoring flip-flops can be made available to the computer control unit and used as go ahead signals for the delay execution scheme. Alternately, they can be used as status indications in a status scheme.

Although the fault detection has been discussed here only for the transfer of information, similar fault detectors can be used to monitor the transmission of control codes. The presence of a fault condition causes a fault stop in some designs; in others it produces a program interrupt.

8.4.3. Variations in the Layout of Input/Output Units

Having discussed several approaches to the monitoring of input/output operations, we are now in a position to consider the variations in the layout of I/O units. Let us again take the model layout in Fig. 8.85 as the starting point of our discussion.

8.4.3.1. Omission of the I/O Control Register: The I/O unit of some smaller computers does not contain an I/O control register. Instead, the external device is controlled in exactly the same manner as the internal units of the computer. The sequencer in the control unit issues an appropriate sequence of control commands for each operation of the external device. The instruction repertoire contains I/O instructions concerned with specific functions of specific devices, such as read one record from magnetic tape, print one character on the typewriter, and so on. A disadvantage of this approach is that — depending on the number of external devices and their various functions — the instruction repertoire must contain a relatively large number of specific I/O instructions. Moreover, the addition of a new external device to the computer system may require a re-design of the computer control unit.

Some designs omit even the I/O information register. Information is then directly transferred to and from the external unit in the same manner as between the internal units of the computer. This approach eliminates the one-word buffer in the form of the I/O register, and allows no independence whatsoever between the operation or the timing of the computer and the external equipment.

8.4.3.2. Use of a Single I/O Register: Some computer designs use a single register for the transfer of both, information and control words. One or more flip-flops associated with the I/O unit are set or cleared depending upon whether a control word or an information word is placed into the I/O register. The state of this (or these) flip-flop(s) is interpreted by both,

the computer and the external device together with the contents of the
I/O register. Fig. 8.88 shows an example of such a layout.

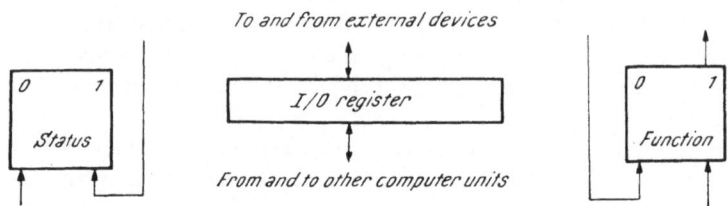

Fig. 8.88. Partial Layout of an I/O Unit with Single Register

The layout contains two control flip-flops, labelled respectively "function" and "status". The state of the function flip-flop indicates to the
external unit whether the computer has placed a function code or a word
of information into the I/O register. The flip-flop remains cleared when an
external write instruction transfers an information word, but is set when
a select instruction places a function code into the register. The flip-flop
is usually cleared, together with the code word contained in the I/O register,
by the external unit when it has recognized and accepted the code.

The state of the status flip-flop indicates to the computer whether the
external device has placed a word of information or a status code into the
I/O register. The flip-flop is set when a status code is transmitted but left
cleared when an information word is transmitted.

Problem 111: At what time should the status flip-flop be cleared?

8.4.3.3. The Use of Multiple I/O Registers: Several I/O registers, or combinations of I/O information and control registers, may be provided for the
concurrent communication with several external devices. A specific register or set of registers is then usually designated by tags contained in the
format of the I/O instructions[1]. The execution of I/O instructions serving
individual registers is interlaced by the program as appropriate. A prerequisite for the concurrent communication with several devices is the
implementation of at least one of the monitoring techniques discussed in
paragraph 8.4.2.

8.4.3.4. I/O Channels: The layouts of I/O units which we have discussed
so far, are typical for small or medium-size computers. Practically all
larger modern machines have I/O units in the form of I/O channels. The
essential characteristic of a channel is the capability to address the computer memory and to control the transfer of information to and from it.

---

[1] See paragraph 8.2.4.

Since this transfer no longer requires the execution of programmed *I/O* instructions, the channel is able to provide truly program independent input and output[1].

Let us initially assume the layout indicated in Fig. 8.89 when we discuss the operation of a channel.

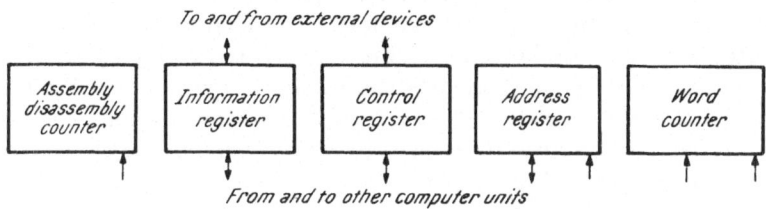

Fig. 8.89. Representative Layout of an *I/O* Channel

The *I/O information register* is used for the information exchange between the external device and the computer in much the same manner as in previously shown layouts. The remaining registers serve to control the channel operation.

In preparation for an input or output operation, several pieces of information are transmitted to the channel by an *I/O* instruction: The designation of an individual external unit; the specific function to be performed by this equipment; the memory address to or from which information is to be transferred; and the total number of words to be transmitted. This control information is stored in the control register, the address register and the word counter respectively.

The *control register* retains the selection code and the function code throughout the entire *I/O* operation and makes these codes available to the external equipment. In addition, the control register may contain some channel or equipment status information and make it available to the computer.

The *address register* contains the address of a memory location. The address is sent to the computer memory for each access by the channel. It is initially set by the computer program and incremented for each transfer of a word. Consecutive words are, therefore, stored in consecutive memory locations during an input operation, and consecutive words are obtained from consecutive locations during an output operation.

The *word counter* is used to count the number of words transmitted. Its contents are decreased for each transfer. When the contents have decreased

---

[1] The advantages of this approach will be shown in more detail in paragraph 8.5.1.

to zero, the channel terminates the input/output operation and ceases the transfer of further words[1].

The incorporation of an *assembly-disassembly counter* gives the channel a capability which we have not yet mentioned: it is able to communicate with the external equipment in bytes. A channel may, for example, transmit bytes of six bits each to a magnetic tape unit, while it in turn receives 36-bit words from the computer memory. The channel disassembles each computer word into six bytes which are transmitted to the external device sequentially. For this purpose, the information register usually has the properties of a shift register. The assembly-disassembly counter is used to count the number of bytes in each word and to control the shifting operation. Similarly, for an input operation, the channel may assemble 36-bit words from the 6-bit bytes which it receives from the external device.

Even though the transfer of information is independent of any program execution, the program can usually monitor the operation of the channel and that of the external device, either by testing for status conditions or by receiving program interrupts. Status information concerning the channel contained in the channel control register can be obtained by "copy channel status" instructions. The execution of such an instruction may cause the transfer of this information to an arithmetic register, where it can be investigated by test instructions. Channel status codes may be provided for such conditions as: "channel busy"; "function rejected"; "function in progress"; "function completed"; and "external error".

More detailed status information concerning the external unit can be obtained by "copy external status" instructions. The execution of such an instruction selects an external unit and causes it to transmit a status word via the information register. This transfer is accomplished in the same manner as the normal transfer of information words. External status codes may be provided for such detailed and varied conditions as: "not ready"; "rewind in progress"; "lateral parity error"; "card jammed"; and "paper supply exhausted". Since the transmission of external status conditions involves the *I/O* information register, copy external status instructions should not be executed while an input or output operation is in progress.

In addition to providing status information, both the channel and the connected external equipment may produce program interrupts. The channel may, for instance, interrupt when it erroneously receives a new function code while some other function is still in progress. The external device may produce a program interrupt upon the detection of errors in the external device, such as a parity error. Either the channel, or the external device,

---

[1] Simultaneously, a status bit may be set into the control register, or a program interrupt may be produced.

may produce an interrupt when a function has been completed. These interrupts may be selectable under program option[1].

Problem 112: a) What do you consider the essential advantage of a channel in comparison with simpler I/O units?

b) What do you think is the reason that I/O channels are incorporated only in larger machines?

Problem 113: How can you arrange for the transmission of all control information with one program instruction if the necessary control information contains more bits than a computer word? Propose a sample format for such an instruction.

Although the layout and the operation which we have discussed so far are representative for most channels, there exist variations in many respects.

A relatively simple variation is the replacement of the word counter indicated in Fig. 8.89 by a "last address" register. In this approach, words are not counted during the input or output operation, but the current memory address in the address register is compared with the contents of the last address register. When equality is found, the I/O operation is terminated in much the same manner as if the word counter had counted down to zero. In an alternate scheme—possibly selectable under program option—the determination of words to be transferred is left to the external device. The I/O instruction then sets up the channel not for the transfer of a specified number of words, but to read or write a record. Whenever the external device determines that a record has been completed, the I/O operation is terminated. This approach can be quite advantageous in situations where the block-length is subject to change, for instance, in reading data tapes with records of variable lengths.

The manner in which program instructions set up the operation of the channel and of the external device may differ also in other respects. In one approach, one I/O instruction causes the selection of an external equipment and, simultaneously, specifies its function. This usually implies a relatively large channel control register to hold both, the equipment and the function code. In an alternate approach, first one instruction causes the connection of a specific equipment. A second instruction specifies subsequently the particular function for the connected equipment. This usually requires only a short control register in the channel (holding only the equipment code), but individual function registers in each of the external devices. In this latter approach, the function code is usually transmitted to the external device via the information register.

Two other programming features should be mentioned. One of them is the "clear channel" instruction. This instruction is contained in the instruc-

---

[1] See paragraph 8.4.2.3.

tion repertoire of practically all computers incorporating channels. The execution of this instruction clears the control registers of the channel and those of the connected external device. This causes the disconnection of all external devices and halts their operation. The other programming feature is the "linking" of I/O operations. The channels of some computers contain an additional register which is used to specify the address of a control word. Upon termination of one input or output operation, the channel makes an access to the memory location specified by this address and obtains a new control word specifying another operation. The channel can, therefore, link a series of input/output operations without program assistance.

Rather significant is also the manner in which the channel control is implemented. One approach provides sufficient control circuits in the channel itself to control its own operation. The control circuits sequence the operation of the channel in a similar manner as the computer control unit sequences the operation of the remainder of the computer. Fig. 8.90 shows the consequences of this arrangement schematically.

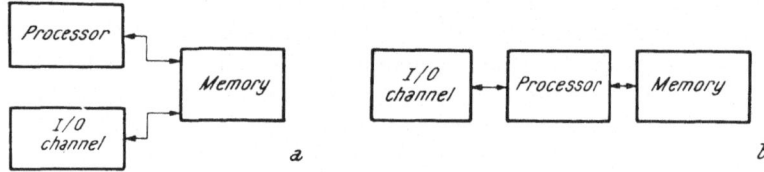

Fig. 8.90. Two Approaches to Channel Control

Both, the I/O unit and the "processor" (i.e. the combination of control and arithmetic unit) operate independently. Their operation interfers only as far as the memory is concerned: one device may have to wait until an access by the other device is completed. We say colloquially: the I/O unit "steals" memory cycles from the processor.

Problem 114: Show the sequence of actions required to input an individual word within a relatively long input operation. Assume that six bytes have to be assembled into one computer word, that the memory is addressed by the channel and that the contents of the channel registers have to be kept current.

Problem 115 (Voluntary): Design the control circuitry required for a channel input/output operation. Assume the channel layout given in Fig. 8.89.

In an alternate approach, the computer control unit sequences the channel operations. Whenever required, the control unit ceases the execution of programmed instructions and issues instead a sequence of control signals for the channel. This approach is similar to a program interrupt, but there is an essential difference: No program is executed for this control; no jump is involved; and it is not necessary to save the current program address. The control unit simply defers the program execution while it exercises the control over the channel. Colloquially, we say: the channel "steals" compute cycles.

Problem 116: What do you consider the main advantages and disadvantages of each of the two approaches to channel control?

8.5. Communications in Digital Computer Systems

During the initial discussion of the layout of digital computers in chapter 7, we found a requirement for communication paths between the various functional units of a computer. But at that time, communication was a topic only incidental to the more basic purpose of explaining how computers work. In fact, even experts failed to look at computers seriously from a communication point of view for a surprisingly long time. On the other hand, the manner in which communications between the units of a computer are provided, can make the essential difference between a computer which just works and one which is well laid out, organized, and effective. To consider the interaction of computer units as a communications problem not only shows a new aspect of computer operations, but it is also one of the prerequisites to a thorough understanding of the effectiveness of the computer as a system.

Our main objective here will not so much be the description of specific solutions to communication problems, but we shall try to construct a frame of reference so that we may see the significance of individual arrangements with respect to the overall problem.

Speaking very generally, we have four main units in a computer which have to communicate with each other. This is schematically indicated in Fig. 8.91.

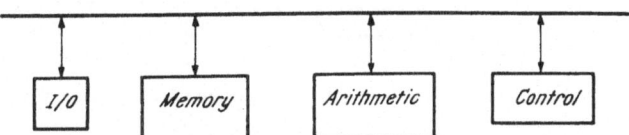

Fig. 8.91. General Communication Requirements in a Digital Computer

The diagram is meant to indicate communication requirements of the computer units in general. If taken literally, the diagram is representative for computers with an exchange register or a common information "bus". A single path is used for the transfer of information between any two computer units, and only one transfer of information can take place at any one time.

With some freedom of interpretation, we may consider the diagram to be representative also of computer layouts which contain several independent communication paths. The heavy line in Fig. 8.91 is then representative of a communication network between the computer units in which probably not all the possible individual paths are implemented, but only a certain part of them. Fig. 8.92 shows an example.

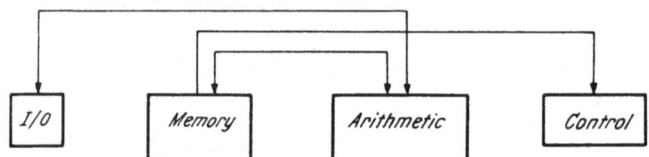

Fig. 8.92. Communication Paths of a "Standard" Computer

A similar figure could be drawn for any other layout. The advantage of Fig. 8.91 is that it is a valid representation for a large number of individual and different computer layouts. Of course, it has the disadvantage that it does not show the details of the communication network itself. The number of possible different arrangements of communication paths in digital computers is rather large. In order to get a feeling for the variety of different possibilities let us try to show the simplest and the most elaborate possible layout in two diagrams.

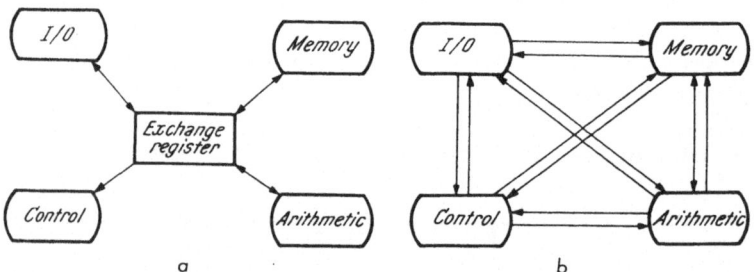

Fig. 8.93. Two Extremes in Computer Communication

Fig. 8.93a shows a computer system in which only one word of information can be transferred at any one time between the central exchange register and one of the computer units. In a sense, the layout is equivalent to that of a telephone exchange in which, at any one time, only one of the subscribers can be connected to another one and can only either talk or listen.

Conversely, Fig. 8.93b represents a communication system in which every possible path is implemented. Each "subscriber" has his own private line to any of the others, and each can simultaneously talk and listen to each other. Moreover, "conference" calls may be possible.

The two diagrams show only the communication paths between the main internal computer units. The layout could be expanded to show the individual registers and external devices like tape units, card punches, etc. In one extreme, only the bare minimum of communication paths is provided and only one transfer can take place at any one time. In the other extreme, all units have direct communication paths to all other units and all units can communicate simultaneously.

In computer systems (as in telephone systems) both approaches are impractical. The art of an effective system design consists in both cases to a large extent in the selection of a reasonable amount of communication paths[1]. The detailed provisions have to depend upon the usage. Computer units which communicate very frequently or constantly should have a direct communication path between each other. On the other hand, infrequent communications between specific units, or communications at a slow data rate can be handled by selection and time-sharing techniques. Undoubtedly, a configuration desirable for a specific application may handle a different application rather badly. Conversely, the existence or non-existence of a particular communication path in a computer can make a large difference in the speed, and, therefore, the economy with which a particular problem can be handled.

Let us now turn to one of the most serious individual communication problems, and one which was recognized as such probably earlier than any other, the input/output of information. Before we discuss individual techniques let us indicate the problem which exists in computer layouts similar to the one shown in chapter 7.

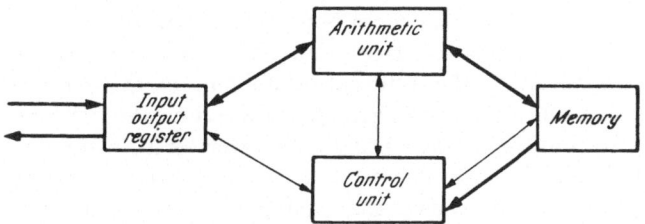

Fig. 8.94. Communication Paths in a Simple Digital Computer

The heavy lines represent communication paths for the transfer of information. The thin lines show control signals which we shall neglect in

---

[1] In a telephone system it would be obviously unreasonable and uneconomic to provide each subscriber in Los Angeles with a direct line to each subscriber in New York.

our present considerations. The diagram, correctly, should include the connections to the "external equipment", such as tape units, card reader and punches, printers, etc. But even that we will neglect for the moment.

According to the diagram, any input/output operation uses the arithmetic unit[1]. Such an arrangement may have been logical and desired for externally programmed computers, where practically all inputs were to the arithmetic unit, and where practically all outputs came from the arithmetic unit. But with the event of stored program machines, the arrangement is far from ideal. Practically, all inputs and output information is now transferred to and from the memory. Here then, the arithmetic unit serves only as an additional link between the input/output unit and the memory. This complicates the communication unnecessarily. The use of the arithmetic unit for all input/output operations has several additional disadvantages: The arithmetic unit cannot be used for arithmetic operations while I/O operations are in progress. The lock-out time comprises not only the transfer time itself, but includes unavoidable waiting periods when the arithmetic unit is ready for a transfer but the I/O circuitry or the external equipment is not yet ready. In many instances, the transfer rate will be such that the time between consecutive transfers is too short for any arithmetic operation. The arithmetic unit must then be ready for a transfer at the earliest possible time while the actual transfer may take place considerably later. Even though actual transfers take only a fraction of the total time, the arithmetic unit may not be able to perform any really useful operation. In effect, a complicated and expensive unit is wasted by periods of idleness and by being forced to perform operations which could be performed by equipment of much less complexity.

Over the years, a number of arrangements have been found which improve the situation. A certain help is the input/output register within the I/O unit. It usually accomodates one word and can be used to hold information until either the external equipment or the arithmetic unit is ready for a transfer. This arrangement may allow the arithmetic unit to finish an operation, before accepting a transfer, or to transfer information to the I/O circuitry, before it can be accepted by the external equipment. The arrangement does reduce the timing problem which the computer programmer has to face, but it is a measure of rather limited effectiveness.

Its effectiveness can be improved to a certain extent by "program interrupts". In this arrangement, the external equipment, when ready, sends an interrupt signal to the computer control unit requesting a transfer. The

---

[1] Either the main arithmetic register (the accumulator), or the exchange register (which is also used in arithmetic operations) may be used for the transfer of information.

control unit interrupts then the normal program execution, performs the *I/O* operation, and resumes operations where it has left off.

The burden on the programmer is greatly reduced by this arrangement. He does not have to time his program critically and intersperse *I/O* operations at appropriate places. It will appear to him like input/output operations are performed independent of the program. On the other hand, additional circuitry is required in the computer control unit to allow the interruption of a program, to "remember" the exact point where the operation was interrupted, to perform the *I/O* transfer, and to resume the computation. But really, even though the scheme is rather convenient and unnecessary waiting periods may be greatly decreased in many cases, the fact remains that the arrangement does not help at all if the input/output rate is so high that no operations can be performed between consecutive transfers[1].

A solution to this latter problem is a truly "buffered" input/output as indicated in Fig. 8.95.

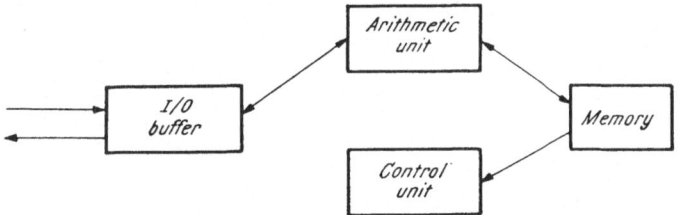

Fig. 8.95. Block Diagram of Computer with *I/O* Buffer

The *I/O* register with a one-word capacity is replaced by a buffer i.e. a storage device for a number of words. One of the main advantages of this arrangement is the possibility to load and unload the buffer at unsynchronized data rates. For instance, information from the external equipment can be transferred to the buffer at the rate dictated by the speed of the external device; the buffer can then be unloaded through the arithmetic unit to the memory at the highest possible rate. No waiting periods are encountered and the total time during which the arithmetic unit is used for the input operation is decreased to a minimum. For an output operation, the buffer can be loaded by the computer as it is convenient for the program being executed. The information is then transferred from the buffer to the external equipment at the proper rate and independent of the computer operation.

The incorporation of the input/output buffer has the following essential advantages over previously discussed input/output techniques:

---

[1] Also: interrupting and resuming operation may (and probably will) take longer than the execution of the majority of computer instructions so that the time gained by the elimination of waiting periods is partly lost again.

a) There are fewer interactions between input/output operations and the computation. That is, the number of times when the computer program switches from arithmetic to input/output operations and vice versa is drastically reduced.

b) The waiting periods for individual I/O operations are eliminated.

c) it is no longer necessary for the programmer to time individual input/output instructions so that the data rate dictated by the external device is matched.

d) The transfers of information are lumped in time so that a maximum of useful instructions can be performed by the computer between I/O operations.

Even though the use of an I/O buffer is very effective as far as the computer operation is concerned, it does not avoid the already mentioned illogical use of the complex arithmetic unit for such simple tasks as information transfers. Most larger modern computers are, therefore, equipped with "compute-independent" inputs/outputs. Fig. 8.96 shows its arrangement in principle.

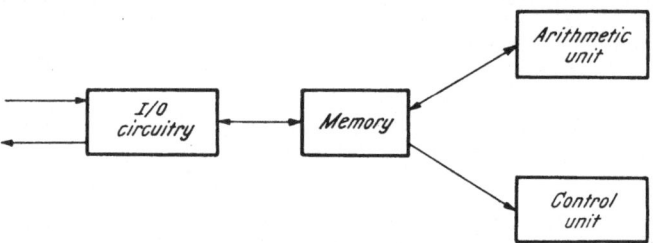

Fig. 8.96. Computer with Inputs and Outputs Independent of Arithmetic and Control Units

The diagram is meant to indicate that transfers between the memory and the I/O unit are not routed through the arithmetic unit, and can take place independent of the execution of the computer program. Of course, even in this scheme it may happen that there is a conflict between the execution of the program and I/O operations. Both may require an access to the memory at the same time, and one of them, obviously, has to wait until the other request is taken care of. But now we are concerned with memory lockouts rather than with the lockout of the arithmetic unit. A memory cycle is shorter than the execution time of a typical instruction and, furthermore, a simple priority circuit in the memory can take care of the situation while in previously discussed schemes the program had to be terminated at the appropriate time and the I/O transfer effected.

In principle, the arrangement is very simple and effective. Having recognized all the disadvantages of using the arithmetic unit for the transfer of inputs and outputs, one cannot help but wonder why computer designers have not used such a layout for a long time. There may have been several reasons why this layout has not been incorporated in even the earliest stored program computers. First of all, it may have been too radical a change from the historic externally programmed or plug-board programmed computer. The second reason may also be more or less psychological in nature. It was felt entirely adequate, and even desired, to have a computer which performs one and only one operation at a time: the program has absolute control over the sequence of instructions to be performed. There is no question of where in the sequence an input/output takes place. The increase in speed or the additional flexibility obtained by performing input/output operations concurrently with arithmetic operations was not appreciated. The third reason may have been one of cost. An additional unit for the control of I/O operations, as simple as it may be, has to be provided.

Schemes of various sophistication following the basic arrangement of Fig. 8.96 have been built. It may be useful to show some of them in more detail.

A relatively simple arrangement for real-time inputs is shown in Fig. 8.97.

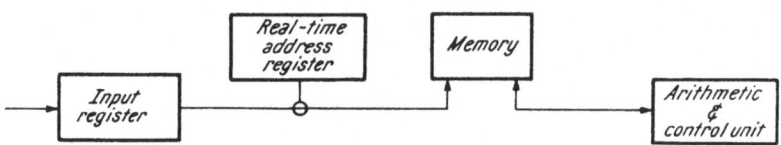

Fig. 8.97. Simple Mechanization of Real-Time Inputs

The information enters the input register (at a rate which is not under control of the computer). The transfer from the input register to the memory is then effected without disturbing the computation in process. The only control is provided by the real-time address register which contains the address of the location to which the information is to be transferred. The content of this register is incremented with each transfer so that the individual words of the incoming information are automatically stored in consecutive locations of the memory. If it should be desired, the real-time address can be altered at any time under computer program control.

A more elaborate arrangement for real-time inputs is shown in Fig. 8.98. A large number of different information sources feed into the computer complex. Here we have the additional problem of keeping the information of the various sources apart even though the individual data rates may vary widely.

The information coming from the various sources is temporarily stored in terminal registers. As soon as a memory cycle is available, the information is transferred to a storage location. A fairly large part of the memory is

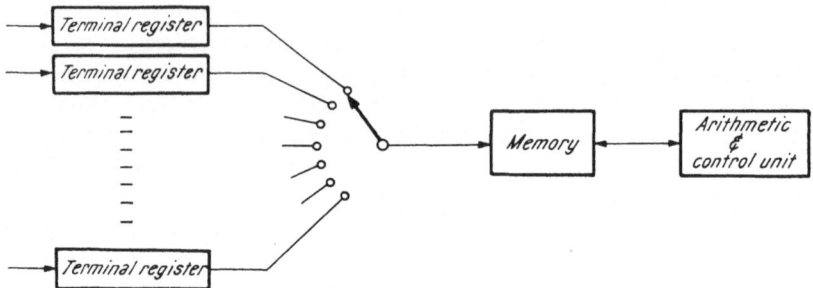

Fig. 8.98. More Complex Mechanization of Real-Time Inputs

reserved for the storage of incoming information and addressing is performed in such a manner that the information seems to "flow" through this part as indicated in Fig. 8.99.

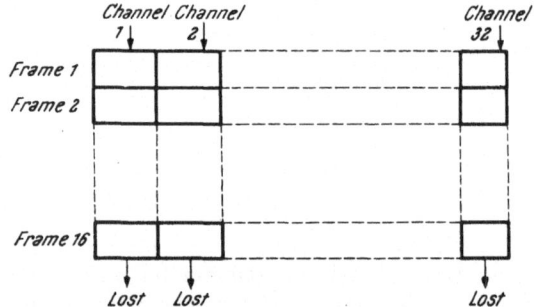

Fig. 8.99. The Flow of Information through the Storage Region

The flow is mechanized so that the computer can address a certain "channel" by the most significant bits in the address. The least significant bits of the address determine the relative "age" of a sample within a channel. The arrangement is so that the computer always has access to 16 "frames" of information[1].

The two indicated implementations of compute-independent inputs represent solutions to a specific communication problem, the real-time

---

[1] What seems to be a shifting of information through the storage region is really performed by an automatic address modification.

input to digital computers. The schemes have to be modified in some manner before they are applicable to the input/output of general purpose computers. Fig. 8.100 shows the arrangement in principle.

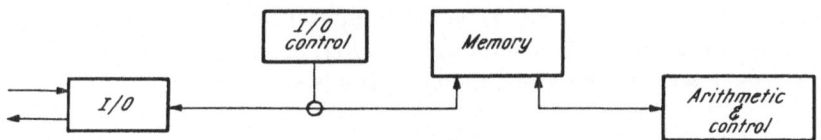

Fig. 8.100. Block Diagram of a Computer with Inputs/Outputs Independent of the Main Processor

A separate *I/O* control has to be provided if the main control unit is used exclusively for the control of the actual computer program and if *I/O* operations are to be performed concurrently with the main program. This *I/O* control is indicated schematically in Fig. 8.100. Actually, there are various ways in which it can be implemented.

Fig. 8.101 shows the lay-out of a computer which incorporates a *I/O* processor or data synchronizer.

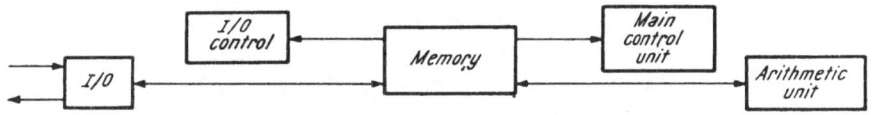

Fig. 8.101. Computer with *I/O* Processor

The main control, together with the memory and the arithmetic unit constitute what we may call the main computer. The *I/O* control, together with the memory and the *I/O* circuitry, can be considered as an auxiliary computer which — although it has essentially no arithmetic capabilities — arranges the data and addresses the memory. It controls the quantity of data to be transferred and their destination. Furthermore, it may perform some testing and editing of information. The auxiliary computer, like the main computer, uses the memory for its program storage. In principle, we have the concept of two independent stored program computers sharing the same memory. It is to be noted that each computer can affect the operation of the other computer. Particularly, the main computer can change the program to be executed by the auxiliary computer if this is so desired. All in all, we have an extremely flexible scheme of operation. The sequence of internal operations of the *I/O* control as indicated in Fig. 8.100 is not predetermined, but can be set up as required by the program. If we compare the scheme to a "standard" layout as indicated, for instance, in Fig. 8.94, we see that additional hardware (i.e. the *I/O* control) is required, but that the arithmetic and main control unit is completely free to perform meaningful operations. Rather than forcing the expensive main control and arithmetic units to participate in relative simple I/O operations, a separate control unit is provided. Several computers have been built using the layout shown in Fig. 8.101. In some computer systems, several I/O synchronizers serve the main computer (see Fig. 8.102a). In others, a single I/O processor serves several main computers (see Fig. 8.102b).

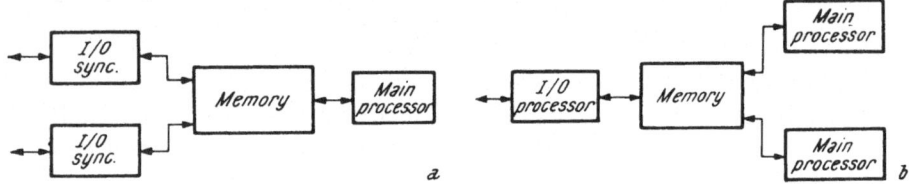

Fig. 8.102. Extensions of the Concept Shown in Fig. 8.101

In many computer systems, the I/O control is not as elaborate or flexible as an auxiliary stored program computer. Frequently, "data channels" are used for the transfer and control of input/output information. Fig. 8.103 shows the basic arrangement.

Fig. 8.103. Computer with Single Data Channel

The operations to be performed by the data channel are necessarily more limited in scope than the operations of an I/O processor. No actual "program" is executed, but the sequence and the type of operations is largely predetermined by the channel design[1].

Computers usually have provisions for more than one channel. A multiplexer arrangement as shown in Fig. 8.104 allows to operate several channels independently, that is on a time-sharing basis.

The layout shown in Fig. 8.104 can be considered representative for many modern computers although their detailed design may vary. A detail which has been overlooked is, for instance, the exact manner in which the data channels together with the arithmetic and control units time-share the memory. For instance, the multiplexer may have additional "contacts" to which the arithmetic and control units are connected.

---

[1] See paragraph 8.4.3.4.

Let us show here some practical layouts which can be considered as representative of modern computer systems. The differences in these layouts may serve to indicate the extent to which computer designers of today are

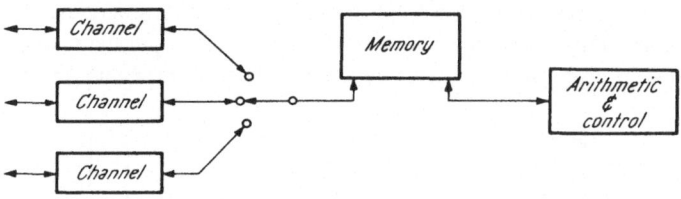

Fig. 8.104. Multiplexing of Several I/O Channels

communication minded. In addition, the excursion will show that the layout of computers is by no means fixed, but that there is a large number of different ways in which a system designer can rearrange the "black boxes" and their interconnections until a system is achieved which he considers optimum.

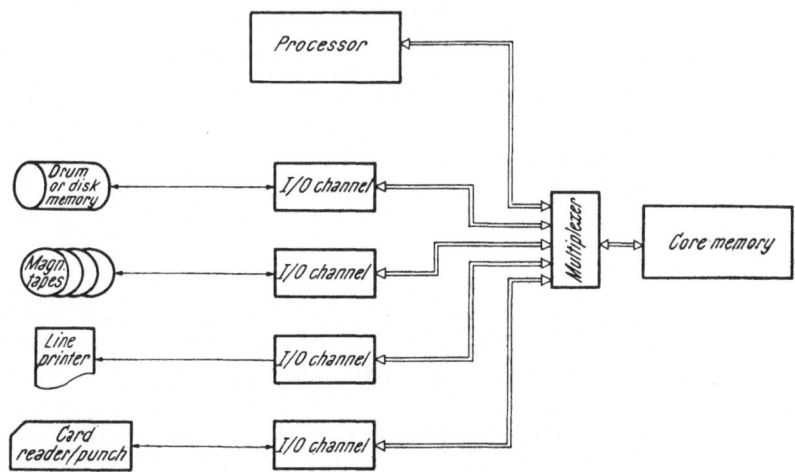

Fig. 8.105. A Minimum Computer System

Fig. 8.105 indicates what one might consider a "minimum" system, that is one which consists of more than simply a computer and attached peripheral devices. The diagram shows functionally (and possibly physically) separate units, and the information paths between them. Control lines are omitted. The processor represents a combination of the arithmetic and control unit. It has the capability to read instructions and operands from the memory and to write results into the memory. A multiplexer is inserted between the memory and the processor, so that also the I/O channels can communicate with the memory. Only one read or write operation can take place at any one time, and memory cycles are distributed by the multiplexer to the processor and the I/O channels as required. The I/O channels have the capability to address the memory and to transfer information to and from it. This provides the capability for independent operations in the processor and all four channels; only the memory is "time-shared".

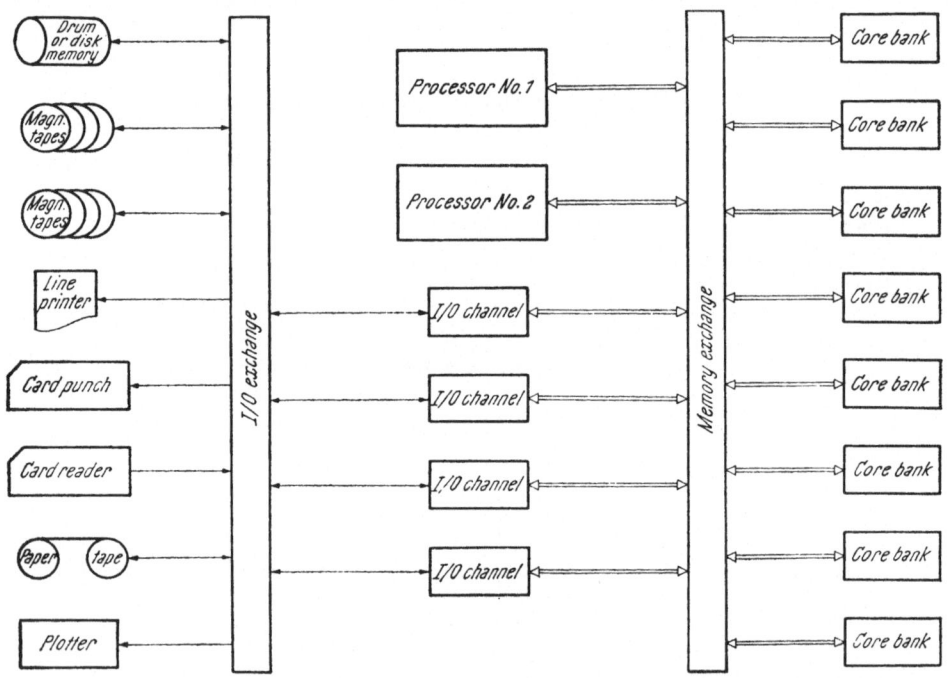

Fig. 8.106. A More Flexible System

I/O channels are designed to interface with the specific type of peripheral device attached to them. In the sample system, there is a drum or disk memory, several magnetic tape units, a card reader/punch, and a line printer. A single channel communicates with only one device at a time, and performs either an input, or an output operation. The single lines between I/O channels and peripheral devices indicate an information transfer which is parallel by byte; the double lines in the remainder of the system represent parallel-by-word communication.

The system configuration may be varied within limits. In particular, it is possible for an installation to start initially with a complement of channels and peripheral devices smaller than the one shown, and to expand to a larger configuration containing additional channels and devices, such as typewriters, plotters, displays, and document readers. One of the practical limitations on system size is the maximum number of channels which can be accommodated.

A more flexible system is shown in Fig. 8.106. The memory consists of several operationally independent memory banks, which are connected to the processors and the I/O channels through a "memory exchange". This exchange is functionally equivalent to a crossbar switch which allows the establishment of independent paths between multiple devices, so that each of the processors and each of the I/O channels can communicate concurrently with a (different) memory bank.

The system also includes an I/O exchange, again a crossbar-like switching device which permits the simultaneous connection of up to four peripheral devices to the four I/O channels. The channels are identical in their design so that each may communicate with any peripheral device.

We should note here that the two processors can exchange information since they share the same memory. If desired, they can also exchange information through peripheral storage devices. One processor would then transfer information to a magnetic tape or disk unit, and the other would read it.

The range of possible expansion is larger with this system than with that of the previously indicated system. In particular, a small system may perhaps consist of only one processor, two memory banks, and one I/O channel. If such a system becomes restricted as far as arithmetic operations are concerned, a second processor can be attached. If the memory should become too small, additional memory banks can be attached, and if the I/O capabilities should restrict the usefulness, more channels can be added. The system can be made to "grow" in exactly that respect which poses the actual restriction. We note that a small system, having only one I/O channel, can operate all types of peripheral devices, while a system similar to the one shown in Fig. 8.105 requires one channel for each type of equipment. A similar advantage shows up in larger systems containing several channels: any momentarily available channel can be designated to communicate with any idle peripheral equipment.

Fig. 8.107 represents a modular computer system. The term "modular" refers here to the possibility to arrange individual modules or building blocks into systems of various configurations, sizes and capabilities. As we shall see, this is facilitated by the arrangement of "spare connectors" indicated in Fig. 8.107.

Communication lines are multiplexed at several levels: in the memory bank, the I/O multiplexer, the I/O channels, the magnetic tape controller, and, finally, the satellite computer.

The multiplexer associated with the memory bank permits the bank to communicate in a time-shared manner with more than one other device. Only two such devices are shown in the sample system: the processor, and the *I/O* multiplexer. However, if desired, more processors and/or multiplexers can be attached. Conversely, a processor or an *I/O* multiplexer can be connected to more than the indicated number of memory banks.

The *I/O* multiplexer permits the time-sharing of a single input of a memory bank by several *I/O* channels. We notice that this arrangement, at least in systems with many channels, reduces the number of separate terminations required at each bank.

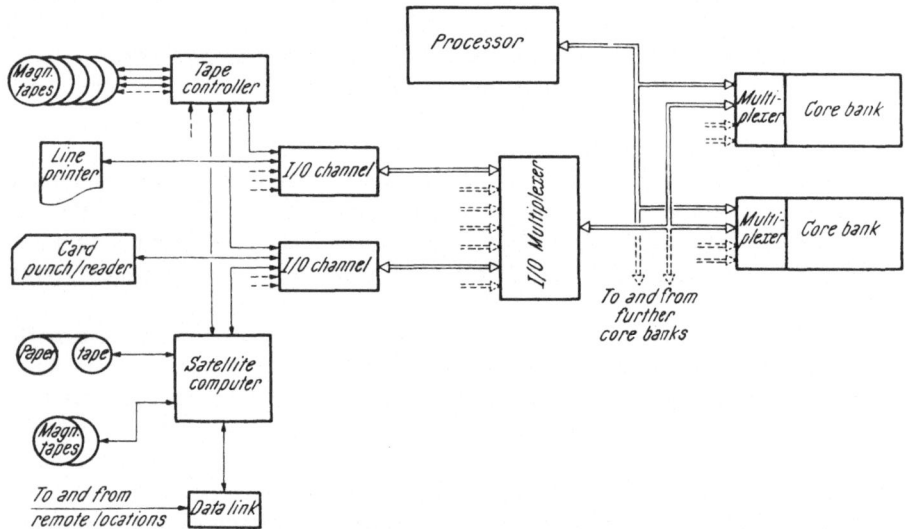

· Fig. 8.107. Sample Configuration of a Modular Computer System

Channels multiplex at a third level: they have the capability to select one of several attached peripheral devices. However, a channel remains connected to a peripheral device until the entire *I/O* operation is completed, while the previously discussed multiplexers connect only momentarily during the transfer of a single word. In contrast to the channels shown in Fig. 8.105, a channel is here able to communicate with all types of devices. The arrangement is perhaps not quite as flexible as the one shown in Fig. 8.106, where each peripheral device can be connected through any channel, but, at least, the number of devices attached to each channel can be selected according to the frequency of the use. Moreover, the tape controller is, at least in some respects, equivalent to the crossbar-like *I/O* exchange in Fig. 8.106. Here, however, the freedom to connect any peripheral device to any channel is restricted to the normally most frequently used peripheral devices, the magnetic tape units.

The satellite computer frees the main computer from the burden of communicating with selected peripheral devices. In many installations, satellite computers control the operation of printers and card reader/punches since these devices normally require extensive re-formatting. In the sample system, there are four devices shown: a paper tape punch/reader, two magnetic tape units, and a data link to a remote location. The satellite computer is able to transfer information between any of these. In addition, the satellite computer can select and communicate with one of the tape units of the main computer, and it can transfer information to and from the main memory through one of the I/O channels.

The satellite computer can communicate with the main computer either off-line, on-line or through a shared peripheral device. An example of an off-line operation is the recording of information received on the data

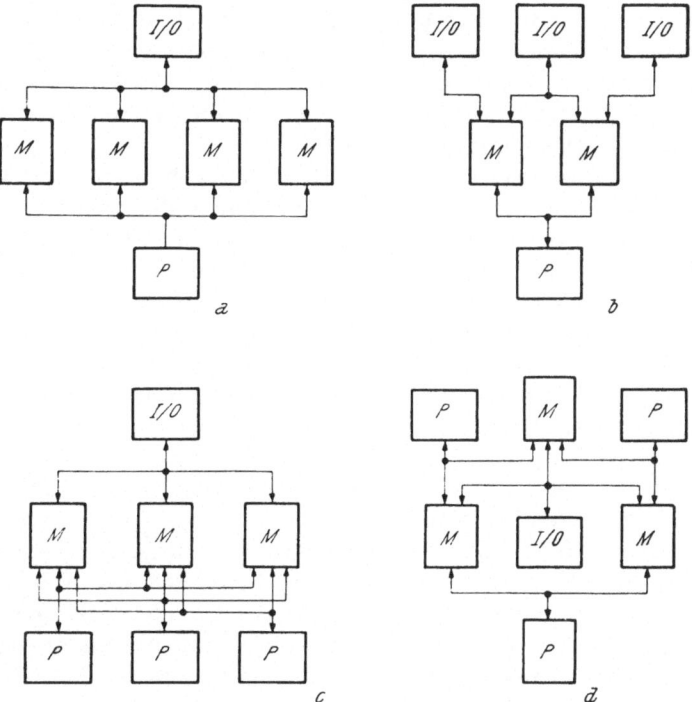

Fig. 8.108. Expanded System Configurations

link on a magnetic tape unit. The tape reel can subsequently be brought to one of the tape units of the main computer. No direct communication between the satellite and the main computer is involved. In a possible on-line mode of operation, the satellite might transfer the received information directly to the memory of the main computer. In a "semi on-line" mode, the satellite computer could record the received information on one of the tape units accessible to the main computer. The main computer can then later read the information without physical transportation of the tape reel.

Let us now consider the growth potential of the indicated system. Obviously, the initial system could be smaller than the one indicated. The satellite computer could be omitted, and the number of memory banks and channels could be reduced. On the other hand, the indicated system can be expanded in several respects. A system with expanded memory is shown in simplified form in Fig. 8.108a. The diagram shows only the processor, the memory banks, and the I/O multiplexer. The details concerning the peripheral devices are omitted. If the sample system should become limited as far as the processing power is concerned, it can be expanded by the addition of more processors. Such a system is shown in Fig. 8.108c. A system with more I/O multiplexers is shown in Fig. 8.108b. Finally, to illustrate the extreme flexibility of the modular system, Fig. 8.108d shows a hypothetical layout in which three computers and their memories are arranged in the form of a triangle.

Fig. 8.108 shows only a few expansions of the basic system. With the possible variations in the complement of peripheral devices and, especially, with different arrangements of satellite computers and data links, the modular system allows to configure a practically unlimited number of different layouts. The flexibility of the modular system is even further increased with the possibility to provide interchangeable modules which perform identical functions, but with different speeds. A system can then "grow" in speed, without change in the basic configuration. This is desirable since a change in configuration usually requires also a change in the computer programs.

As a final representative example of a modern system, let us consider the layout shown in Fig. 8.109.

Here, a fairly drastic step away from the conventional concept has been taken. Several computers have been integrated into one system in such a manner that an individual computer by itself is no longer operationally meaningful. The two basic motives for such a configuration are the desire to perform parallel operations in several processors so that the overall speed is increased, and the assignment of specialized tasks to each type of processor so that the most efficient design can be realized.

The *main processor* concentrates on arithmetic operations and has practically no capability to control data transfers in the system, other than its own accesses to the main memory .The processor is able to overlap the execution of several instructions and has the capability to access the memory through several independent paths. Of course, the implementation of such multiple paths is advantageous only if overlapping accesses are to different memory banks[1]. This condition is difficult to meet satisfactorily by programming alone, and addresses are assigned so that the least significant bits determine the memory bank. Consecutive addresses are thus distributed over several or all of the banks. Although this approach does not guarantee that concurrent accesses are always possible, it increases the probability that this will be the case.

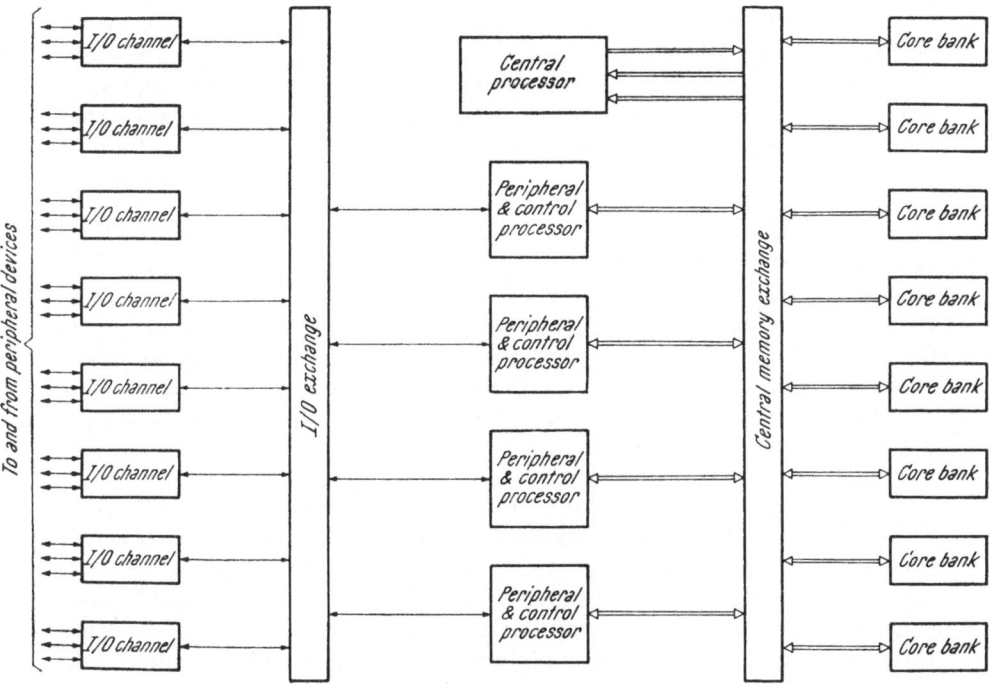

Fig. 8.109. Multiprocessor System

The *peripheral processors* perform housekeeping and control functions. In particular, one of the peripheral processors may be assigned the role of the systems "executive". It schedules jobs according to their priority, reserves memory regions, and assigns peripheral equipment to each job. The other peripheral processors input programs and operands, unload results, re-format information, and control the operation of peripheral devices.

We have seen here only a few of many possible system organizations. The tacit intent was to show important features in at least one configuration. In reality, configurations may, of course, combine features which were

---

[1] A single memory bank can perform only one operation at a time.

attributed here to different systems. For instance, a certain peripheral device may conceivably be connected to the main computer via a specifically designed channel, via a "selector channel", a multiplexer, or via a satellite computer in almost any system configuration. A device may also be "shared" by different computers either through a manual switching arrangement, or through a multiplexer, similar to the tape control in Fig. 8.107. In this connection, it is important to note that even in systems where the processor has only one path to the memory, banks may be assigned to the low-order address bits. This increases the chance that memory cycles (not instructions) can be overlapped. One bank can then still finish a memory cycle, while the processor already accesses another bank. The approach, therefore, decreases the effective cycle time of the memory. On the other hand, the assignment of banks to the high-order bits has the advantage that banks can be added or removed without re-assignment of addresses.

Problem 117: Assume that the complete cycle time of a core memory bank is 4 μs, but that equipment is released by the bank after 2 μs. What are the maximum information transfer rates for the memories in each of the three layouts given in Figs. 8.105, 8.106, and 8.107 under the most favorable conditions? What is the corresponding effective cycle time? What is the effective cycle time under the worst possible conditions? Rate the three systems according to their effective cycle time under "average" conditions.

Problem 118 (Voluntary): Following the layouts in Figs. 8.105, 8.106, and 8.107, "design" three computer systems which have the capability to operate concurrently any of the following combinations of peripheral devices: four magnetic tape units; three magnetic tape units and a line printer; two magnetic tape units, a line printer, and a card reader. Discuss the resulting three systems in terms of the minimum number of channels required, and the maximum number of concurrent operations possible under the most favorable conditions. How would the three systems have to be modified if the requirement were for the concurrent operation of *any* four magnetic tape units out of a total of eight?

The on-line communication over data links has been shortly mentioned incidental to the discussion of Fig. 8.107, but so far, the system aspects have been ignored. Let us indicate the range of possible configurations and applications by three illustrative, but highly hypothetical examples.

The first system is located at the administrative headquarters of a large business concern. It has remote card readers and printers at the branch offices in various cities. Each day, at the close of business, the daily transactions recorded on punched cards are transmitted over telephone lines to the central facility. The data are analyzed by the computer during the night, and summary reports are printed at the remote locations in the morning.

The second system is owned by a transcontinental airline. Computers are installed at several strategic locations. The computers are linked to each other and to input/output typewriters at the desks of all airline ticket agents. The agents can type inquiries concerning available space on all flights, and they can request reservations and confirmations. A typed message is received in response to each inquiry. In order to reduce the number of lines required, the communication net includes computer controlled switches and buffers which route, store, and forward individual messages.

The third system is operated by a research laboratory. It consists of a central computer and input/output consoles at the desks of individual scientists. The users enter their problems on a key-board, and receive solutions on the screen of a cathode ray display. They time-share the central computer, but are not necessarily aware of this fact. Each may feel that the console connects him to his own computer. The advantage of such a layout is that it is economic, but still provides the individual scientist with immediate access to a large computer.

The three indicated systems have altogether different configurations, operational requirements, and communication equipments. In fact, they should prove the point that in certain cases the communication aspects can be as important as the capabilities of the computer itself.

Bibliography

WILLIAMS, KILBURN, and TOOTILL: Universal High-Speed Digital Computers, Proceedings IEE, vol. 98, part II, pp. 13–28. Feb. 1951.

KILBURN, TOOTILL, EDWARDS, and POLLARD: Digital Computers at Manchester University, Proceedings, IEE, vol. 100, part II, pp. 487–500. Oct. 1953.

McCRACKEN D. D.: Digital Computer Programming. New York: John Wiley and Sons. 1957.

BLAAUW G. A.: Indexing and Control-Word Technique, IBM Journal of Research and Development, vol. 3, No. 3, pp. 288–301. July 1959.

BECKMANN, BROOKS, and LAWLESS: Developments in the Logical Organization of Computer Arithmetic and Control Units, Proceedings IRE, vol. 49, No. 1, pp. 53–66. Jan. 1961.

McSORLEY: High-Speed Arithmetic in Binary Computers, Proceedings IRE, vol. 49, No. 1, pp. 67—91. Jan. 1961.

RAJCHMAN J. A.: Computer Memories: A Survey of the State of the Art, Proceedings IRE, vol. 49, No. 1, pp. 104—127. Jan. 1961.

ADAMS C. W.: Design Trends for Large Computer Systems, Datamation, pp. 20—22. May 1961.

WILSON, and LEDLEY: An Algorithm for Rapid Binary Division, Transactions IRE, vol. EC-10, No. 4. Dec. 1961.

GRASELLI A.: Control Units for Sequencing Complex Asynchronous Operations, Transactions IRE, vol. EC — 11, No. 4, pp. 483—493. Aug. 1962.

BURKS, GOLDSTINE, and VON NEUMANN: Preliminary Discussion of the Logica Design of an Electronic Computing Instrument, 1946. Reprinted in Datamation, vol. 8, No. 9, pp. 24–31, Sept. 1962, and vol. 8. No. 10, pp. 36–41. Oct. 1962.
WALLACE C. S.: A Suggestion for a Fast Multiplier, Transactions IEEE, vol. EC — 13, No. 1, pp. 14—17. Feb. 1964.
LANDSVERK O.: A Fast Coincident Current Magnetic Core Memory, Transactions IEEE, vol. EC — 13, No. 5, pp. 580—585. Oct. 1964.
BARTEE, and CHAPMAN: Design of an Accumulator for a General Purpose Computer, Transactions IEEE, vol. EC — 14, No. 4, pp. 570—574. Aug. 1965.
RAJCHMAN J. A.: Memories in Present and Future Generations of Computers, IEEE Spectrum. Nov. 1965.

Following are references on associative memories:

SLADE, and SMALLMAN: Thin Film Cryotron Catalog Memory System, Proceedings Eastern Joint Computer Conference, pp. 115–120. Dec. 1956.
TAUBE, and HEILPRIN: Automatic Dictionaries for Machine Translation, Proceedings IRE, vol. 45, No. 7, pp. 1020–1021. July 1957.
McDERMID, and PETERSEN: A Magnetic Associative Memory System, IBM Journal of Research and Development, vol. 5, No. 1, pp. 59–62. Jan. 1961.
KISEDA, PETERSEN, SEELBACH, and TEIG: A Magnetic Associative Memory, IBM Journal of Research and Development, vol. 5, No. 2, pp. 106–121. Apr. 1961.
CROFT, GOLDMAN, and STROHM: A Table Look-up Machine for Processing of Natural Languages, IBM Journal of Research and Development, vol. 5, No. 3, pp. 192–203. July 1961.
SEEBER, and LINDQUIST: Associative Memory with Ordered Retrieval, IBM Journal of Research and Development, pp. 126—136. Jan. 1962.
LUSSIER, and SCHNEIDER: All-Magnetic Content Addressed Memory, Electronic Industries, pp. 92–98. March 1963.
LEE, and PAULI: A Content Addressable Distributed Logic Memory with Applications to Information Retrieval, Proceedings IEEE, vol. 51, No. 6, pp. 924—932. June 1963.
JOHNSON, and MC ANDREW: On Ordered Retrieval from an Associative Memory, IBM Journal of Research and Development, vol. 8, pp. 189—193. Apr. 1964.
CHU Y.: A Destructive Readout Associative Memory, Transactions IEEE, vol. EC — 14, No. 4, pp. 600—605. Aug. 1965.
McKEEVER B. T.: The Associative Memory Structure, AFIPS Conference Proceedings, vol. 27, part 1, pp. 371—388. Fall Joint Computer Conference, 1965.

9. Unorthodox Concepts

In prior chapters we were concerned with conventional digital computers. We have seen their building blocks and their concepts, and a number of variations in their implementation. Many development efforts aim at an improvement in their characteristics, such as reliability, cost, speed, size, and convenience of operation. Mostly they do not attempt to change the basic structure of the stored program computer. On the other hand, the layout of present day digital computers is not ideal, or at least not ideal for all applications. The structure of digital computers is perhaps still more determined by history than by a superior knowledge of conceivable structures. It would be presumptious to predict the structure of future computer systems. However, we can show a number of ideas concerned with unorthodox concepts. Whether or not these will find a wide practical application remains to be seen. In any event, a discussion of novel ideas should give us a more thorough understanding of present layouts and enable us to re-consider presently accepted concepts.

9.1. Polymorphic Computer Systems

Polymorphic (i.e. many-shaped) computer systems[1] are characterized by having a relatively large number of individual units like compute modules, memory modules, input/output modules etc. These modules can be connected with each other as the particular application requires. In one instance, the full complement may be connected together to handle a single large computational problem. At other times, the modules are connected so that several independent "computers" handle independent problems.

Even more conventional computer systems[2] are modular to some extent: units of specific capabilities like memory banks or I/O channels may be added or removed from a system. However, the essential characteristic of polymorphic computers is the flexibility and speed with which changes in the system can be made.

Connections and disconnections in a polymorphic system are made under program control and at electronic speeds by a switching center. The layout

---

[1] Also called restructurable or variable structure computers.
[2] See paragraph 8.5.

or shape of individual "computers" may, therefore, conceivably change continuously during the operation.

Fig. 9.1 shows a simplified diagram of a polymorphic computer system.

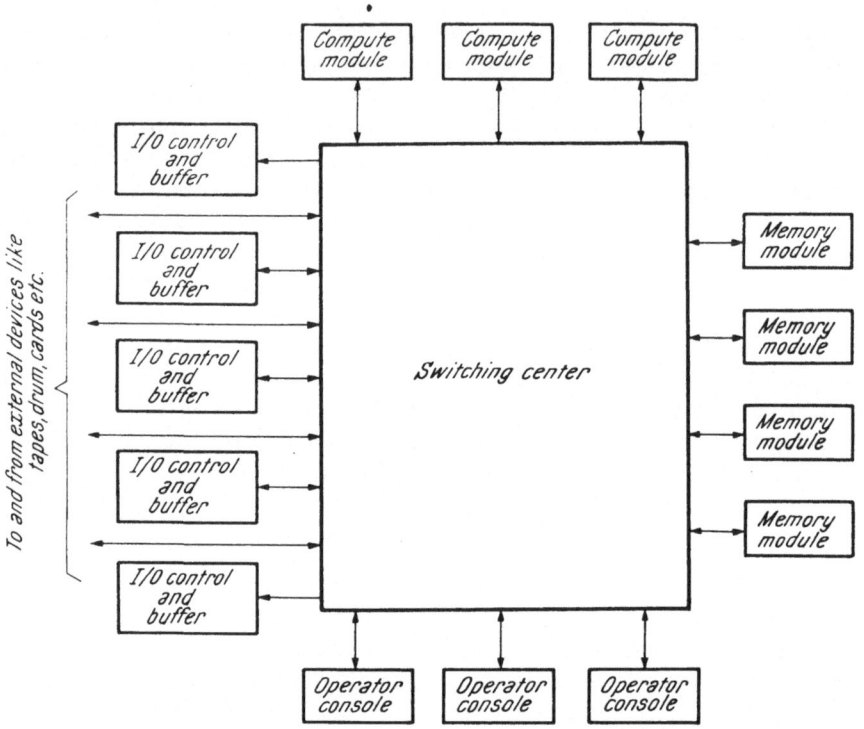

Fig. 9.1. Simplified Layout of a Polymorphic Computer System

A system like this has several desired features: First of all, taylormade computer configurations can be used for specific problems. "Large scale" and "small scale" computers are available when there is a specific need for them. Also, the right amount of external equipment can be connected for each job. In spite of the freedom to select the appropriate equipment, the system can be economic. Units not required in a particular setup are not necessarily idle as in conventional systems, but can be used for another concurrent setup if individual jobs are properly scheduled. Another advantage is the property that a failure may mean that an individual processor or an individual memory module is out of order, but at least a certain amount of computational capability remains. We may speak here of a system with "graceful" degradation.

Problem 1: State some disadvantages of a polymorphic computer system.

9.2. Arithmetic Units with Problem-Dependent Interconnections

Those circuits of a conventional digital computer which perform arithmetic operations are concentrated in the arithmetic unit. Normally, there is only one arithmetic unit and all arithmetic operations are performed sequentially The branches or various requirements of a problem, in a sense, time-share the arithmetic unit. In contrast to such an arrangement is one in which arithmetic operations are performed concurrently in independent arithmetic units with problem-dependent interconnections. Fig. 9.2 may serve to convey the basic idea.

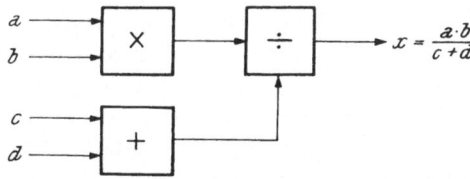

Fig. 9.2. Computation with Distributed Arithmetic Units

The arrangement of distributed arithmetic units has several advantages over a normal digital computer: there is no requirement for a large internal memory[1]; the control circuitry can be much simpler; and its operational speed is potentially higher due to the parallel operation. Of course, there are also some disadvantages: The system is not as flexible as a stored program computer. In fact, the concept is not practical at all when a complicated computation is performed on only one set of input data. Moreover, the amount of hardware available limits the complexity of the job which can be handled. In spite of the severe limitations, the concept has been successfully applied to some computer designs. The analog computer follows the indicated principle, the digital differential analyzer[2] is almost a direct implementation of the scheme, and the principle can be found in highly parallel machines[3]. In all these cases, however, a supply of general purpose arithmetic units is provided and the functions of individual units can be specified as required.

Let us add here some further observations. The manner of looking at the individual circuits of a computer as indicated in Fig. 9.2 will probably become increasingly important. The speed of future computers is undoubtedly as severely restricted by the speed of light, that is the speed

---

[1] Operations are determined by the wiring.
[2] See paragraph 9.4.
[3] See paragraph 9.6.2.

with which signals from one circuit to the other can be transmitted, as by the speed of individual components. Here then, it simply may no longer be feasible to transfer instructions and operands back and forth between concentrated memories, concentrated arithmetic units and concentrated control units. Transmission lines (e.g. wave guides) may have to act as storage devices, while logic elements placed in strategic positions will perform the necessary logic or arithmetic operations on the transmitted information. In any event, the computer designer will be forced to pay utmost attention to the physical location of the individual circuits and incorporate parallel operations as much as possible if he wants to achieve a significant improvement in speed. Even though he may not be designing a computer with distributed arithmetic units, he will at least be concerned with parallel operations in a system of distributed logic.

Following the essence of the argument, it is interesting to note that the scheme of distributed units and parallel operations is seemingly the only promising approach to extreme high speeds[1]. One might consider the step from truly sequential operations to truly parallel operations as significant as the step from serial-by-bit operations to parallel-by-bit operations as far as the speed of computers with comparable components is concerned.

9.3. Hybrid Computation

Both, digital and analog computers have their own merits[2]. Table 9.1 may serve as a basis of comparison for the more salient features of both types of computers for typical applications.

We see that the features of the two types of computers complement each other very well. It is, therefore, not surprising that there has been a multitude of proposals attempting to combine the advantages of analog and digital techniques.

9.3.1. Concurrent Operations of Analog and Digital Computers

One of the most obvious approaches is indicated in Fig. 9.3.

The two types of computers work simultaneously on the solution of a single problem. Individual parts of the overall problem are delegated to

---

[1] The miniaturization of components and the consequent shortening of transmission lines still can go a long way. But pursuing this avenue of approach one is constantly up against the current state of the art, whereas the design of a computer with multiple operations in a system of distributed logic is well within the present state of the art as far as the hardware is concerned.

[2] Presently, there seems to be a tendency to consider digital computers as the cure-all to any computational problem. Many problems are run on digital computers for which quite obviously an analog computer would be a much better choice.

either one or the other machine. Computations which require a high accuracy or the handling of slowly changing variables are best performed by the digital computer, while the analog computer may handle a variety of

Table 9.1. *Short Comparison of Analog and Digital Computers*

Feature	Analog Computer	Digital Computer
Speed for specific application	×	×
Accuracy		×
Handling of high frequency variables in real-time	×	
Handling of low frequency variables in real-time		×
Repeatability of solutions		×
Concurrent arithmetic operations	×	
Modular design	×	

variables with a higher frequency content at lower accuracies. The computers exchange information through digital-to-analog and analog-to-digital converters.

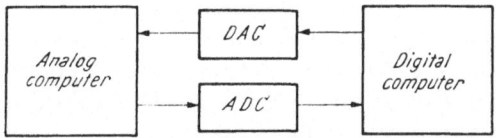

Fig. 9.3. Digital and Analog Computers Cooperating on the Solution of a Single Problem

The indicated arrangement puts equal emphasis on both types of computers. Other combinations have been proposed in which either the digital or the analog part is dominant. Fig. 9.4 should be taken as representative for the range of these proposals.

The digital components incorporated into an analog computer (as indicated in Fig. 9.4a) might perform functions which are awkward to implement with true analog components. Examples may be switching functions required for the representation of backlash or for the change of parameters during a computation. Examples of analog components in-

24*

corporated into a digital computer (as indicated in Fig. 9.4b) might be
analog integrators which perform integration of variables very economically
if not too high an accuracy is required. Other, and perhaps more typical,
examples might be analog curve followers or plotters attached to digital
computers and digital function generators for analog computers.

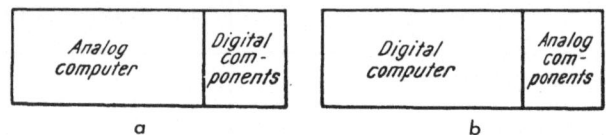

Fig. 9.4. a) Analog Computer, Including Digital Components
b) Digital Computer, Including Analog Components

In any event, the above remarks are intended to illustrate that there
is almost a continuum of possible combinations of analog and digital
computers with the emphasis shifting from purely digital to purely analog
techniques.

In some isolated instances, even those properties of analog computers
which are basically undesired are proposed for incorporation into digital
systems. An example of such a case is the introduction of "noise" into
digital computations.

Noise can be helpful in the detection of lost significance. It frequently
happens that a digital computation shows a result with a precision of, say
36 bits, but only a few of these bits are really significant. The significance
may have been lost e.g. when two almost equally large numbers were subtracted. A small error in either operand produces then a relative large
error in the difference. It is not a simple task to detect or avoid such situations, especially since subsequent shifting or multiplication can give the
impression of numbers with full significance. The equivalent problem is less
serious, or at least troubles are more easily detected, in analog computers.
Small signals are subject to noise and any subsequent amplification increases the random components. A solution with lost significance will therefore look "noisy" and evaluators will automatically place less confidence on
the results. An equivalent noise does normally not appear in digital computations. Only a detailed (and complex) analysis of the problem and its manner
of solution will give an indication of lost significance. However, artificial
noise can be introduced into a digital computation. One possible way of
doing so is to have an additional bit in each word (neighboring the least
significant bit) the state of which is not determined by arithmetic rules but
simply by a random bit generator. As soon as intermediate results are
"amplified" (i.e. shifted to the left or multiplied by values 1), the functional
values will contain random components in higher-order bits. The "noise

content" of the final results can therefore be an indication of their significance. Problems which have not continuous functions, but individual numerical values as solutions have to be run repeatedly before the superimposed "noise" can be detected.

9.3.2. Combinations of Analog and Digital Representations

It is characteristic for most proposals combining analog and digital computations to use digital techniques for requirements with high accuracy and analog techniques for requirements with lower accuracy. The division into digital and analog representations cannot only be made as far as parts of an overall problem or individual variables within this problem are concerned, but can also be applied to the more and less significant parts of a single variable. Fig. 9.5 indicates such a technique which is sometimes used in connection with computations of mainly analog nature.

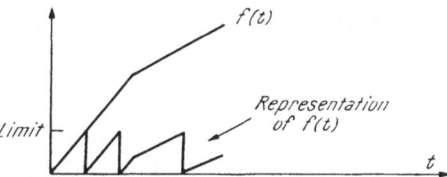

Fig. 9.5. Example of a Variable Represented by a Combination of Analog and Digital Techniques

The representation of $f(t)$ is only partially of an analog nature. The analog value is reset to zero as soon as a given limit is reached. This representation has the advantage that details of the variable can be displayed which would be entirely lost if one would compress the functional values into the operating range (say ± 100 volts) and use the true analog representation (indicated by the steadily increasing value in Fig. 9.5). The actual value of the variable in this representation is not given by its

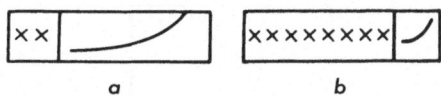

Fig. 9.6. Combination of Analog and Digital Representations for a Single Variable

analog value alone, but by its analog value *plus* the number of resets. In effect, it is given by an analog value plus a digital value. In some cases, the number of resets is counted and numerically displayed. We then have a representation which is schematically indicated in Fig. 9.6a.

The most significant part of a computational value is represented in digital form (usually only a few bits or one decimal digit) and the least significant part is represented in analog form.

We have here a representation in which the emphasis is on analog techniques. However, it is also conceivable to build essentially a digital computer with a relatively large number of bits representing the most significant part of computational values in the usual manner and, in addition, an analog value representing the more detailed structure within the finite number of digital values possible. We would then have a representation in which the emphasis is on digital techniques as schematically indicated in Fig. 9.6b. Again it is interesting to note that there is conceivably a continuum of combinations ranging from purely analog to purely digital representations.

9.3.3. Computers Using Hybrid Elements

Several computing systems have been proposed in which the basic hardware components are analog-to-digital and digital-to-analog converters.

A digital-to-analog converter can be used to derive the product of an analog and a digital value:

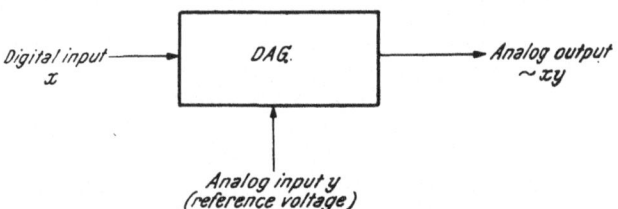

Fig. 9.7. Digital-to-Analog Converter as Multiplier

Conversely, an analog-to-digital converter may be used to derive the quotient of two analog values:

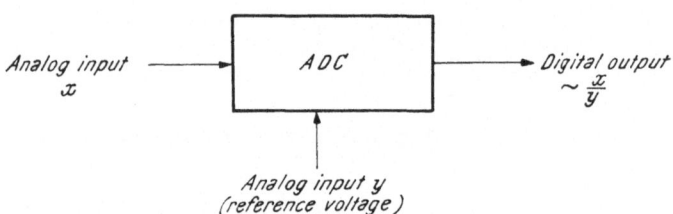

Fig. 9.8. Analog-to-Digital Converter as Divider

Of course, these converters can also be used for the conversion of analog to digital or digital to analog values when, for instance, an operand is available only in the "wrong" representation or when an analog display of a digital quantity (or vice versa) is desired.

Operations other than multiplication or division (addition, subtraction, integration) can be performed either by analog or digital components as it is more advantageous under the circumstances.

The indicated approach can be adapted to the solutions of problems which require the repetition of identical computations on different operands. In this case, a distributed hardware system[1] results except that the basic component in itself is a hybrid.

The indicated approach can also be combined with digital memories or digital control units. In this case, a computer results which has many salient features of the digital computer but has a simple "arithmetic unit" for multiplications and one divisions and which may also allow the operation of several "arithmetic units" in parallel.

9.4. Digital Differential Analyzers

Digital differential analyzers, or DDA's for short, are often regarded as a separate class of computers, on equal standing with digital or analog computers. On the other hand, there is justification to consider these machines as hybrids, having a digital design, but many of the operational characteristics of an analog computer.

The basic functional components of digital differential analyzers are integrators. In principle, an integrator combines two input variables, x and y, to produce an output variable z according to Equation (9.1).

$$z = z_0 + \int_{x_0}^{x} y \, dx \qquad (9.1)$$

However, since the operation of a digital integrator is easier understood in terms of changes, it is advantageous to show Equation (9.1) in its differential form.

$$dz = y \, dx \qquad (9.2)$$

This equation states that the change in the output variable, dz, is proportional to the change of the input variable, dx, and also proportional to the current value of the "integrand", y. We may indicate this relationship in the functional diagram shown in Fig. 9.9.

If y has the value of unity, the change at the output dz is equal to the change at the input dx. If y has a value different from unity, changes at the input dx produce proportionally smaller or larger changes at the output. If y has the value 0, there will be no change at the output, no

---

[1] See paragraph 9.2.

matter how large the change dx is, and if y has a negative value, the change at the output will have the opposite direction or polarity of the change on the input. It is to be noted that the value of y may vary during a computation. The capability to change the value of y is reflected in Fig. 9.9 by the arrow labelled dy.

So far, we have discussed integrators in general. Many electrical or mechanical implementations of the device are known. The integrators of digital differential analyzers use numerical values to represent computational values. Fig. 9.10 shows a representative integrator block diagram.

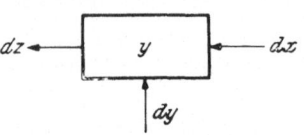

The arrangement contains two registers. The Y-register holds the value of the integrand y. Increments (or decrements) dy can be added to its contents in order to keep the value of the integrand current. In other words, the Y-register has the properties of an up-down counter. The R-register has the properties of an accumulator: The contents of Y can be added to the contents of R. An increment dz is generated wherever there is an overflow of the R-register.

Fig. 9.9. Functional Diagram of an Integrator

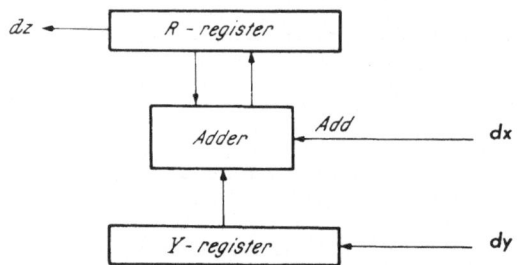

Fig. 9.10. Block Diagram of a Digital Integrator

Let us now consider the operation of the integrator in more detail. Suppose for the moment that the integrand y has the numerical value 1. Increments dx arrive and are interpreted as add commands: the contents of Y are added to the contents of R. Each addition produces an overflow and, therefore, an increment dz. If the integrand is equal to zero, no overflow and no dz increments are generated. If the integrand has the value .5, there will be a dz increment for every second dx increment. We see, the output rate dz is equal to the input rate dx multiplied by the numerical value of Y. Negative increments (i.e., decrements) are handled in the following manner: Decrements dx are interpreted as subtract commands, rather than add commands: the contents of Y are subtracted from the contents of R. If there is a borrow by the most-significant position, a decrement dz is generated. Again, the rate of dz decrements is proportional to the rate of dx decrements and the value of y. Negative integrands are handled without difficulty: If y is negative and dx is positive, the negative value of y is added to the contents of the R-register, that is, in effect, a subtraction is performed. Borrows which are produced are interpreted as decrements dz. When dx is negative while the integrand is negative, the negative value of y is subtracted from the contents of R, that is an actual addition is performed. Overflows are interpreted as increments dz.

The indicated method of integration is closely related to the "rectangular", graphic integration shown in Fig. 9.11.

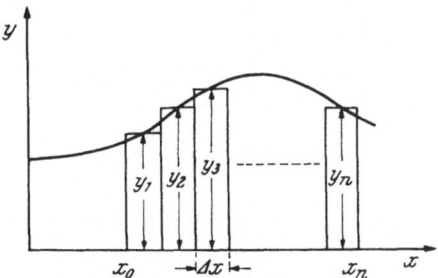

Fig. 9.11. Rectangular Integration

The area between the curve y and the abscissa in the range between x_0 and x_n is approximated by the area of the rectangles of width Δx and height y_r. The area of an individual rectangle is $y_r\, \Delta x$. If Δx has the value of unity[1], the area of an individual rectangle becomes equal to y_r, and the total area is given by $y_1 + y_2 + \ldots y_n$.

This summation of integrand values is exactly what is performed by the digital integrator of Fig. 9.10. For each step in x-direction, the current value of y is added to the already accumulated sum. That not all digits of the sum are retained in the accumulator is peculiar for the digital integrator, but not really a deviation from the indicated scheme. Increments and decrements can be accumulated elsewhere, for example, in the integrand register of another integrator.

Fig. 9.11 gives the impression that rectangular integration produces only a "rough" approximation to the true value of the integral. In practice, any desired accuracy can be obtained simply by using sufficiently small increments. However, we should note that any decrease of the size of

---

[1] It is arbitrary what we call a unit. The "scaling" of the problem can make other values of Δx appear as unity.

increments requires a corresponding increase in their number and therefore, ultimately, in the computing time.

Problem 2 (Voluntary): Design the logic diagram of a digital integrator for "trapezoidal" integration. Hint: The trapezodial integrator does not add the current value of y when a step in x-direction is taken, but adds the mean value between the current value and the value at the time when the last previous step was taken (linear interpolation).

A computer which essentially consists of a number of individual integrators is particularly suited for the solution of differential equations. In this respect, the capabilities of the digital differential analyzer correspond to those of an analog computer. The potentially higher accuracy, the freedom in variables[1], and the fact that solutions are repeatable consistute distinct advantages of the digital differential analyzer. Compared to conventional digital computers, the digital differential analyzer can solve differential equations more economically, simply because it is a special purpose computer for this application.

To describe the application of digital differential analyzers in any detail would exceed the present scope. However, it may be appropriate to show one illustrative example.

Suppose we are given the following differential equation:

$$dy = y \, dx \tag{9.3}$$

This equation describes the specific relation between a dependent variable y, and an independent variable x. The indicated relationship can be "portrayed" by a single integrator according to Fig. 9.12.

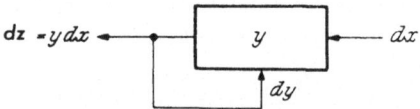

Fig. 9.12. Integrator Setup for the Solution of the Equation (9.3)

The integrator produces the output $dz = y \, dx$ in the previously explained manner.[2] Since Equation (9.3) states that the value $y \, dx$ is equal to the value dy, we simply connect the output of the integrator and its dy input. This arrangement "forces" the variable y to assume the values postulated by Equation (9.3). In effect, the integrator "computes" the appropriate value of y for each value of x. More complex differential equations require, of course, more complicated networks of integrators for their representation.

---

[1] The integrators in an analog computer integrate only with respect to time.
[2] See Equation (9.2) and Fig. 9.9.

Problem 3 (Voluntary): The exponential function $y = e^x$ is, by definition, the solution of Equation (9.3). The integrator setup of Fig. 9.12 may, therefore, be considered a "generator" for the function e^x. Label the diagram in terms of x and e^x, rather than in terms of x and y.

Problem 4 (Voluntary): Find the integrator setup for the solution of the following differential equation:

$$\frac{d^2 y}{dx^2} = -y$$

Hint: Two integrators are required to relate a variable to its second derivative.

What well-known trigonometric function is generated by the setup?

9.5. Micro-Programmed Computers

The term "micro-program" is used to denote the sequence of individual actions or operations which accomplish the execution of a program instruction. As an example, the micro-program for an "Add X to A" instruction may comprise the steps: PAR $\rightarrow$ MAR; initiate read: MIR $\rightarrow$ IR; increase PAR; IR$_a$ $\rightarrow$ MAR; initiate read; MIR $\rightarrow$ X; add X to A[1]. Conventional digital computers are micro-programmed by the computer designer. The micro-program is represented by the sequence of control commands issued by the control unit for each of the instructions contained in the instruction repertoire. We may speak of a fixed, or wired-in micro-program.

In contrast, micro-programmed computers have a programmable sequencer. The micro-program can be designed and altered by the computer user. We may compare the micro-programmed computer to a conventional digital computer in which the control unit has been replaced by a programmable control computer. The "main computer" breaks the program into a sequence of standard instructions. The "control computer" executes a sequence of micro-steps for each of these program instructions. For brevity, we shall refer in the following to a program instruction of the main computer as macro-instruction[2] and to a program instruction of the control computer as micro-order or micro-instruction.

The concept of micro-programming has a number of desired characteristics:

The computer designer may make last minute design changes with relative ease. For example, a change in the arithmetic unit, more than

---

[1] See paragraphs 7.5 and 8.2.

[2] Programmers use the term "macro-instruction" to denote a symbolic instruction which requires for its execution a series of program instructions. In the present context, a macro-instruction may or may not be a symbolic instruction.

likely, affects only the rules of micro-programming, but not the hardware of the control unit.

The maintenance of the computer is simplified, since only sequences of rather trivial operations are implemented in the hardware. Consequently, the structure of micro-programmed computers is relatively uncomplicated, and very effective test programs can be devised.

Since the "meaning" of macro-instructions is defined only by the micro-program, each computer user can design his own optimum instruction repertoire. The instruction set may be changed according to the specific application.

In addition, the micro-programmed computer is well suited to simulate other computers. Since the instruction repertoire can be designed as desired, it can be made to reflect that of another computer. The simulation may be very detailed and not only concern the number and types of instructions but also the exact instruction format, including length, number of addresses, tag fields, op-codes, etc. Such a simulation may be used to advantage by a computer designer who would like to have a working model of the machine which he is designing, or by programmers whose task it is to check out programs for a not yet existing machine. However, it is important to realize that such a simulation will not be realistic as far as program execution times are concerned. Care must be exercised before meaningful conclusions about the speed of the simulated computer can be drawn.

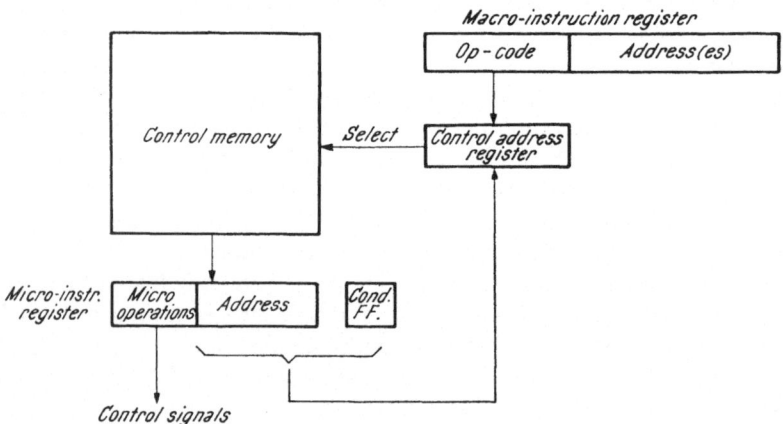

Fig. 9.13. Representative Layout of a Micro-Programmed Control Unit

Before we show several variations in the structure of micro-programmed control units, let us first indicate their basic mode of operation in a representative layout.

The basic components of the layout are the control memory, the control address register, and the micro-instruction register. The *control memory* stores the micro-program. Micro-instructions are retrieved individually by read accesses. The *control address register* serves as the address register for such memory accesses. Its contents determine the location from which a micro-instruction is retrieved. The *micro-instruction register* is the read register of the memory and holds one micro-instruction during execution. The *macro-instruction register* holds the macro-instruction to be executed by the micro-program. As such, it performs the same function as the instruction register of a conventional digital computer. It is indicated in Fig. 9.13 to show the important interconnection of the main computer and the micro-programmed control unit.

In the assumed layout, a micro-instruction specifies a certain micro-operation, but also the address of the control memory location which contains the micro-instruction next in sequence. This address can be transferred from the micro-instruction register to the control address register. The normal sequence of events is then straightforward: A micro-instruction is retrieved and executed. As a by-product, the contents of the control address register are up-dated. The next micro-instruction is then retrieved and executed, and the cycle repeats.

The normal sequence of events is altered when the execution of a macro-instruction is completed. The op-code of the new macro-instruction (and not the address specified by the previously executed micro-instruction) is then transferred from the macro-instruction register to the control address register. In effect, the op-code of the macro-instruction acts as jump address for the execution of the micro-program. We see, that the meaning of the macro op-code is defined only by the micro-program which begins at the control memory location determined by the numerical value of the op-code. This feature justifies the use of the terms "variable instruction computer" and "stored logic computer" synonymously with "micro-programmed computer".

A second deviation from the normal sequence of events is concerned with the program branching. This capability is provided in order to execute conditional jumps of both, the micro- and the macro-program. The layout in Fig. 9.13 provides for this purpose a conditional flip-flop. Its state can be altered by micro-operations, depending upon a selected and restricted number of conditions external to the control unit. An example of such a condition is "accumulator positive". The state of the conditional flip-flop, together with the contents of the control address register, determines the address of the next micro-instruction to be executed.

In summary, there are basically three possibilities for up-dating the address contained in the control address register:

1. It is specified by the address contained in the previously executed micro-instruction.

2. It is determined by the op-code of the macro-instruction.

3. It is determined by the address contained in the micro-instruction as modified by the conditional flip-flop.

Problem 5: Is there complete freedom in assigning normal jump and op-code addresses? If not, what are some of the restrictions?

Let us now look at some actual implementations of micro-programmed control units in more detail.

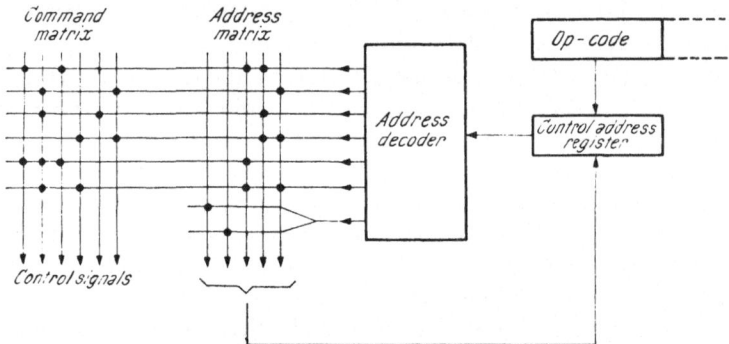

Fig. 9.14. Wilkes' Scheme

Fig. 9.14 shows a layout which is known as Wilkes' scheme. The control memory is a "read only" memory. It consists essentially of an address decoder, a command, and an address matrix. The address decoder decodes the address contained in the control address register. It selects a single matrix row and energizes one of the horizontal matrix lines. This, in turn, energizes selected columns in the command and address matrices. The outputs of the command matrix serve as control commands for the execution of the micro-order. The outputs of the address matrix specify the next control address.

The arrangement is equivalent to a conventional memory which provides a word in response to a read access. Here, however, the "contents" of the memory location are determined by the wiring of the command and address matrix. A change of the contents requires a re-wiring of these matrices or an exchange of matrix boards.

The specification of micro-operations in this scheme is simple and straightforward. Individual bit positions of the "output word" control directly the specific commands to be issued. No micro-instruction register is provided. There are as many bit-positions in a word as there are control lines. A "1" in a certain bit-position may, for example, specify that an "add" signal be sent to the accumulator. A "0" in this position indicates that such a signal should not be generated during the execution of this micro-instruction.

The branching of one of the horizontal matrix lines is meant to indicate the capability to execute conditional jumps. The state of the conditional flip-flop serves to select one or the other row of the address matrix. It determines, therefore, which of two addresses serves as next control address.

Problem 6: How can you accomplish that the last instruction of the micro-program which executes a macro-instruction causes the transfer of the new op-code from the macro-instruction register to the control address register, but that all other micro-instructions transfer the address from the address matrix. Assume the basic layout given in Fig. 9.14 and show the appropriate details.

Problem 7 (Voluntary): Assume that the decoder and sequencer in the layout shown in Fig. 7.6 is to be replaced by a micro-programmed control unit according to Fig. 9.14. Design a command and address matrix required for a repertoire containing the following three macro-instructions:

> load X,   op-code 01;
> add X to A, op-code 10;
> store A,   op-code 11.

Label all command lines. Remember that the execution of a macro-instruction must include the updating of the (macro-) program address register and the fetching of the next macro-instruction.

Let us now discuss a few possible variations in the operational details, still assuming the basic layout of Fig. 9.14.

The first of these is concerned with the design of the address matrix. Fig. 9.14 indicates only one possibility for branching. In reality, of course, multiple branching dependent upon the states of several conditional flip-flops is possible.

Another variation concerns the command matrix. In practice, it will be designed so that several control commands are issued concurrently, when this is appropriate for the micro-instruction to be executed. In addition, it may be layed out so that the selected output columns are energized in a predetermined time sequence. In this event, the execution of a micro-instruction is accomplished in several distinct time phases. During each time period, either a single control command is issued, or several control commands are issued concurrently. When the execution of a micro-instruction always results in a single simultaneous issue of control signals, we speak of *vertical* micro-programming. In this approach, micro-instructions consist of relatively short "words", and a series of micro-instructions is necessary to execute even the simplest macro-instruction. If, however, a single micro-instruction specifies micro-operations independently for several time-phases, we speak of *horizontal* micro-programming. The extreme in this direction would be a micro-instruction format sufficiently long to specify all micro-operations for even the most complex macro-instruction. Fig. 9.15 indicates such a hypothetical format.

Fig. 9.15. Hypothetical Instruction Format for Horizontal
Micro-Programming

In practice, most micro-programmed computers employ a combination of vertical and horizontal micro-programming.

Problem 8: Why is neither pure vertical nor pure horizontal micro-programming entirely satisfactory? Hint: Consider the number of read accesses to the control memory and its storage capacity.

A third variation of Wilkes' scheme concerns the manner in which the control address is updated. So far, we have assumed that a new address is placed into the control address register during the execution of each micro-instruction. This is not necessary if one uses the approach indicated in Fig. 9.16.

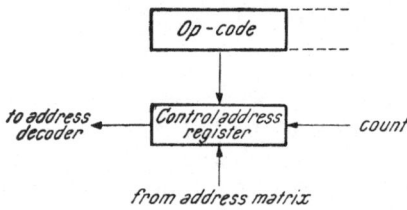

Fig. 9.16. Alternate Scheme for the Updating of the Control Address

The control address register (similar to the program address register of a conventional computer) has the capability to count. During the execution of a micro-instruction, the count is increased by "one". A new address is placed into the control address register only when a jump is executed or when a new macro op-code must be obtained.

Control memories in the form of wired-in matrices have the disadvantage that it is relatively complicated to alter their contents. In addition, there are practical limitations to their size. Most of the more recent control memory designs incorporate, therefore, some other mechanically alterable read-only memory[1] or more conventional read-write memories. The latter

---

[1] See paragraph 11.1.5.

frequently have a non-destructive read-out, and are designed so that the read cycle is as short as possible, even at the price of a relatively long write cycle. This is justifiable since read accesses are much more frequent than write accesses.

The use of a read-write control memory has the important operational advantage that the micro-program can be set up and modified under program control. This allows to load automatically the optimum micro-program for each computer run; it allows to "pull in" applicable sections of the micro-program if the size of the micro-program exceeds the storage capacity of the control memory; and it permits "house keeping" functions to be performed within the control memory. The latter possibility is of particular importance in applications, where there is no macro-program, and all operations

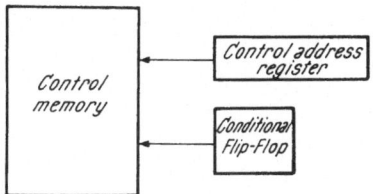

Fig. 9.17. Alternate Implementation of the Jump Capability

are solely controlled by the micro-program. Such may be the case in some on-line process-control or information-handling systems where relatively short sequences of simple operations are repeated over and over. In some cases, it may even be possible to use incoming information instead of a macro-program. That is, this information, rather than a macro-program would be interpreted by the micro-program.

The use of a read/write memory instead of read-only matrices requires a slight change of the previously discussed layout. For the execution of conditional jumps, it is no longer practical to alter the control address as it is contained in the memory. Instead, it is simpler to modify this address prior to the look-up of the micro-instruction following the conditional jump. Fig. 9.17 indicates one possible implementation.

The conditional flip-flop specifies one bit of the control address. Dependent upon its state, micro-instructions are obtained from one or another location. Of course, the scheme can be expanded to more than one condition if this should be desired.

Another relatively simple implementation of conditional jumps with read/write memories is analogous to Fig. 9.16. The jump address contained in the jump instruction is transferred to the control address register only when a jump is to be executed. In case of "no jump" conditions, the contents of the control address register are increased by counting.

Problem 9: Would you say that the use of a regular memory restricts or increases the flexibility in the programming of conditional jumps?

The use of a conventional memory makes another modification desirable: In order to make more efficient use of the available memory word-length, micro-operations should be specified by micro op-codes rather than by individual bit positions within the instruction format. It is rather common to find two or three op-fields within the format of one micro-instruction.

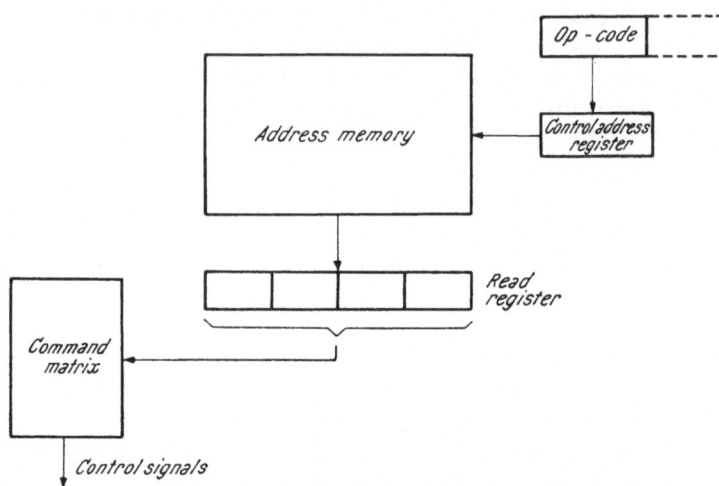

Fig. 9.18. Simplified Layout of a Micro-Programmed Control Unit with Address Memory and Command Matrix

Let us conclude this paragraph by indicating a few of the more interesting variations of micro-programming.

Fig. 9.18 shows the simplified layout of a micro-programmed control unit which contains an address *memory* but a command *matrix*.

The op-code of the macro-instruction serves as address for an access to the address memory. This access makes available a word containing a string of addresses which, in turn, are used for consecutive accesses to the command matrix. Even though the command matrix is permanently wired, any sequence of control signals can be issued for a macro-instruction, simply by storing appropriate strings of addresses in the address memory.

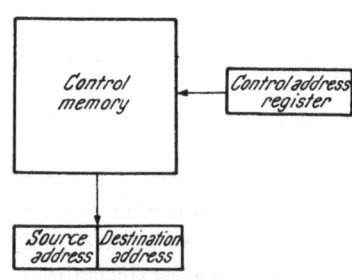

Fig. 9.19. Simplified Layout of a Micro-Programmed Control Unit for Computer with Inter-Register Transfers

Fig. 9.19 is meant to represent the micro-programmed control unit of a computer organization in which all operations of the "main computer" are performed solely by inter-register transfers. Consequently, the execution of all instructions is controlled by identical sequences of control signals, and micro-instructions need to specify only the addresses of source and destination registers.

Fig. 9.20 may help to convince us that it is possible to execute all operations by transfers.

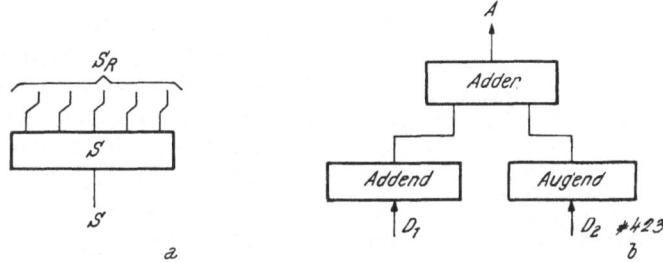

Fig. 9.20. Shift and Add Operations Performed by Transfers

Fig. 9.20a indicates the implementation of a shift operation by transfers. The register S is "loaded" by a transfer using "S" as destination. The shifted contents can be transferred to another register (or back to S) by using "S_R" as a source. Fig. 9.20b indicates the possibility to perform additions by transfers. Augend and addend registers are loaded by transfers using "D_1" and "D_2" as destination addresses. The sum is available at the source "A". Operations such as main memory reads or writes, can be executed by loading a memory control register with appropriate codes. Alternately, the sole fact that a memory register is used as a source or a destination may initiate a read or write cycle.

As a final variation let us mention that some conventional computers incorporate a limited capability for "micro-programming". The normally fixed (macro-) instruction repertoire contains an instruction "perform algorithm". The sequence of micro-operations executed for this macro-instruction can be selected by the user and is determined by the wiring of a patchboard or custom-designed control circuits.

9.6. Machines with Cellular Organization

In this paragraph, we shall shortly survey concepts for computers with structures characterized by arrays of identical cells. Such structures can be attractive because they are either well suited for modern mass fabrication techniques, or because they have an inherent capability to perform parallel operations. Proposals range from relatively simple cells containing only a single logic element, to very complex cells containing their own arithmetic unit, memory and control. Designs within this continuum are usually referred to by terms such as: cellular logic; array processor; iterative circuit computer; highly parallel machine.

9.6.1. Cellular Logic

Let us initially consider the functional diagram of an arbitrary logic unit as indicated in Fig. 9.21.

In general, such a unit receives a number of inputs, and produces a number of outputs. The outputs are functions of the inputs, the specific

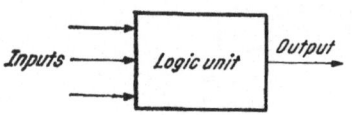

Fig. 9.21. Generalized Functional Diagram of an Arbitrary Logic Unit

function depending upon the specific purpose of the circuit. In conventional designs, the actual composition of such units will be quite different for, say, a decoding matrix, an adder, or perhaps a shift register. However, it is possible to find "universal" structures which can be adapted to specific requirements by simple means, such as the particular connection of input and output lines, the application of external "steering signals", the bridging of terminals, the removal of excess connections, etc. For ease in manufacturing, two-dimensional arrangements are preferable.

A number of such structures has been found. Fig. 9.22 shows one of the simplest, known as a majority array.

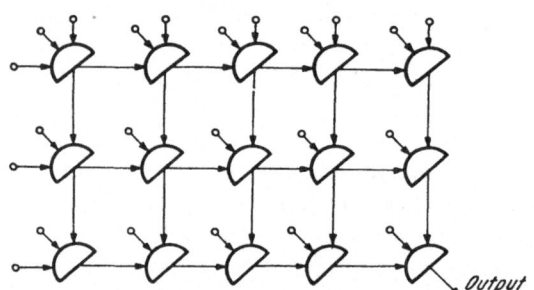

Fig. 9.22. Array of Three-Input Majority Elements

The arrangement consists of an array of identical cells. Each cell consists of a three-input majority element which provides a "1" output if two or more of its inputs are in the "1" state. The outputs of all cells are permanently connected to neighboring cells as indicated, with the exception of the cell at the lower right, which provides the final output of the array. Provided the network contains a sufficient number of cells, any logic function can be implemented, simply by connecting the inputs in an appropriate manner. As an example, it can be shown that a 4×6 array can implement any four-variable function. Many four-variable functions can be implemented by smaller arrays [1].

An approach which is different in several respects is indicated in Fig. 9.23.

This array is able to produce more than one function of the input variables. Moreover, connections to the array are made only along the edges. We speak of "edge-feeding" the array. These desirable characteristics tend to be offset by the increased complexity of the individual cell.

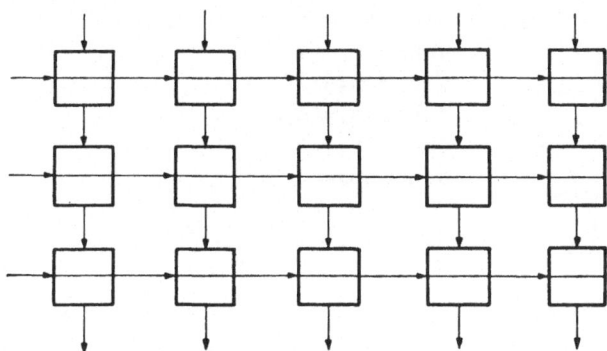

Fig. 9.23. Array of Cutpoint Cells

Each cell consists of a two-input, two-output logic element. The horizontal output af a cell simply duplicates the horizontal input. The array has, therefore, in effect, signal buses from left to right. In vertical direction, the cell produces an output which is a function of both, the horizontal and the vertical input. The specific function which a cell implements is determined by relatively simple alterations within the cell itself. If we assume for the moment that a cell has the potential capability to perform any one of the 16 different functions of two binary variables, then four bits are required to specify the particular function to be implemented. It has been proposed to provide these specification bits in the form of cuts in the internal wiring of the identically fabricated cells. The name "cutpoint cell" is descriptive for this approach.

The study of cutpoint arrays has shown that it can perform any logic function. In fact, this is so even if the single cell cannot perform all, but only a limited set of the 16 functions of two binary variables [2]. Such a set is listed in Fig. 9.24.

A cell can be instructed to perform any one of the indicated nine functions. The listing contains two trivial functions, providing a constant "0" or "1" output, six more complex logic functions, and the function of a set/reset flip-flop. This latter possibility provides the cutpoint array with the capability to implement not only arbitrary logic circuits, but also more complex units requiring the function of storage.

The capabilities of cutpoint arrays have been studied in detail, and a number of rules have been established which are concerned with the number of cells required to implement logic functions, with minimizing procedures, and with methods of repair, that is the re-routing of signals, so that they by-pass faulty cells. It has been shown that any function of four variables can be implemented by a 5×4 array and that many four-variable functions can be implemented by smaller arrays [2].

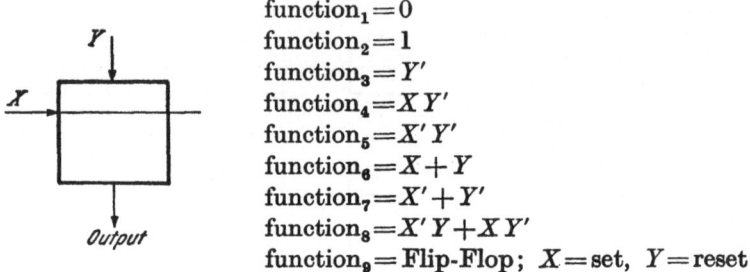

$$\text{function}_1 = 0$$
$$\text{function}_2 = 1$$
$$\text{function}_3 = Y'$$
$$\text{function}_4 = X Y'$$
$$\text{function}_5 = X' Y'$$
$$\text{function}_6 = X + Y$$
$$\text{function}_7 = X' + Y'$$
$$\text{function}_8 = X' Y + X Y'$$
$$\text{function}_9 = \text{Flip-Flop}; \quad X = \text{set}, \quad Y = \text{reset}$$

Fig. 9.24. Cutpoint Cell and Sample Set of Functions

Fig. 9.25 shows a cellular structure which has been developed from the cutpoint array and which avoids some of its disadvantages. Because of the particular manner in which cells are interconnected, the arrangement is referred to as cobweb array [3].

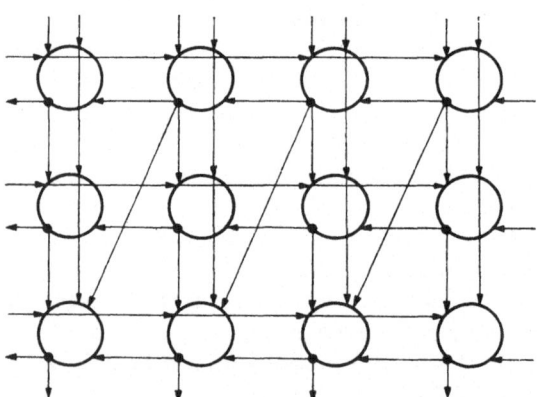

Fig. 9.25. Array of Cobweb Cells

The cobweb cell has one output and five potential inputs. Two of the potential inputs are from the horizontal and vertical signal buses respectively. The other three come from the cell on the right, the cell above, and the cell two rows above and one column to the right. Only two of these five potential inputs are selected by cutpoints and actually used.

The logic capabilities of the cell are the same as those of the cutpoint cell and are listed in Fig. 9.24. The essential difference between the cutpoint array and the cobweb array lies in the increased freedom of inter-connection. In particular, less cells are wasted for the propagation of signals.

Fig. 9.26 shows, as a final example of cellular logic, a cascade of two-rail cells [4].

Each individual cell has three inputs and two outputs. The cell can be instructed to provide any function of three variables on either of its two outputs. This capability requires, of course, a rather complex cell. However, the arrangement has the—at least theoretically interesting—property that any logic function of an arbitrary number of variables can be implemented by a single cascade, that is by a one-dimensional array. All of the previously discussed structures require, in general, a number of cascades to implement arbitrary functions.

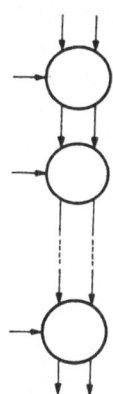

Fig. 9.26. Cascade of Two-Rail Cells

Some of the previously indicated cellular structures allow certain freedom to specify the function of a cell, but practically no freedom to alter the interconnection of cells. Others allow certain freedom in the interconnection, but do not allow to alter the function of the individual cell. In fact, these structures suggest a continuum into which every digital design can be fitted. On one extreme, we have an arrangement of cells which are interconnected in a fixed, predetermined manner, but are performing different functions. On the other extreme, there is the arrangement of cells inter-connected in a flexible manner, but performing identical functions[1]. This continuum contains, of course, a region in which there is reasonable freedom to specify both, the function of an element, and the manner in which elements are interconnected. This is the region into which almost all present designs fall.

Interestingly enough, one can also achieve meaningful designs with arrangements in which all cells perform identical functions and are inter-connected in an identical manner. This is possible if the capability exists to alter the specifications for interconnections and/or functions by program. The next paragraph will show several such structures incidental to another topic.

Problem 10 (Voluntary): Implement the function $X = (A + B) C$ by:
a) a majority array;
b) a cutpoint array;
c) a cobweb array;
d) a two-rail cascade.
Indicate the logic function performed by each individual cell.

---

[1] Representative of this extreme might be a design employing only NAND modules, but having complete freedom in the interconnection of these modules.

9.6.2. Highly Parallel Machines

It is characteristic for machines with highly parallel organization to perform a number of operations concurrently, where conventional computers would execute them sequentially. In this respect, these machines represent a further step beyond the sequence: serial machine; parallel machine. The serial machine operates upon one bit at a time, the parallel machine operates on several bits or one word, and the highly parallel machine operates on a number of words simultaneously. Potentially, the highly parallel machine has a multiple of the speed of the parallel machine, like the parallel machine has potentially a multiple of the speed of the serial machine.

A number of differing concepts for machines with highly parallel organizations have been developped. One of the earliest, and theoretically most interesting, is the *Holland organization,* later also referred to as *iterative circuit computer* [5]. The machine contains no central control, central memory, or central arithmetic unit. The functions of these units are performed by a large number of identical cells. The cells are arranged into a two-dimensional array as indicated in Fig. 9.27. Each cell is able to communicate with its immediate neighbors, or through these with more distant cells.

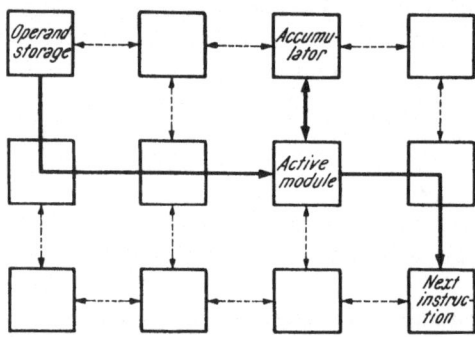

Fig. 9.27. Simplified Holland Organization

A cell has the capability to store a one-word operand, or a one-word instruction. We may compare this arrangement to a conventional computer in which the contents of the centralized memory (instructions, operands, working space, etc.) have been distributed word by word to individual cells.

Let us for the moment consider only those cells which contain the *instructions* of a single program branch. Only one of these cells is active, that is currently controlling the program execution. When the execution of the instruction contained in the active cell is completed, a path is established to the cell containing the next instruction. The latter cell is activated, and the former de-activated. The new cell governs now the execution of the next program step. We see that the individual steps of a program branch are executed sequentially as in a conventional computer. However it is not difficult to imagine that a number of cells are activated at any one moment, if several branches of a program can be executed simultaneously, or if several unrelated programs are being executed concurrently.

In order to activate the proper cell in sequence, each instruction specifies the address of the cell containing the next instruction. This address determines the path from one cell to the next. A similar path-building technique is used for the execution of instructions: the active cell establishes one path to a cell serving as accumulator, and another to a cell serving as operand storage. The three paths from the presently active cell to the cell containing the next instruction, the accumulator and the operand storage are indicated in Fig. 9.27.

Only a minimum instruction set is proposed for the machine. It contains simple arithmetic and transfer instructions such as:

$$\text{operand} + A \rightarrow A \quad \text{(add)},$$
$$A \rightarrow \text{operand} \quad \text{(store)},$$

and a few control instructions such as:

activate operand module if A (—) (conditional jump),
$A \rightarrow$ operand auxiliary register (modify control information),
operand auxiliary register $\rightarrow A$ (copy control information).

In spite of its simplicity, the instruction set is intended to be complete. Moreover, we notice that the same cell, in one instance, may serve as operand storage, in another as accumulator, and in a third as instruction register. This gives the machine the potential capability to alter its program and, therefore, capabilities equivalent to a conventional stored program computer.

A number of attempts have been made to improve the basic Holland machine. The *Comfort organization* [6] aims at simplifying the programming task and at the reduction of required hardware. The basic change is the incorporation of specialized arithmetic units as indicated in Fig. 9.28.

The arithmetic units have more arithmetic capabilities than Holland cells. This permits a more powerful instruction repertoire. On the other hand, the arithmetic capabilities, and therefore, the complexity of the remaining cells can be reduced.

Another study [7] aims at increasing the flexibility of communication between cells. For this purpose, a hypothetical n-dimensional arrangement of cells is proposed in which each cell has neighbors in 2n directions. Moreover cells are assumed to have the capability to establish communication paths in more than one direction, so that a crossing of the paths pertaining to different but concurrently executed programs is possible.

The *Gonzalez organization* attempts an improvement of the Holland organization by providing for the execution of identical operations upon multiple operands [8]. The machine contains three functionally specialized arrays of cells as indicated in Fig. 9.29.

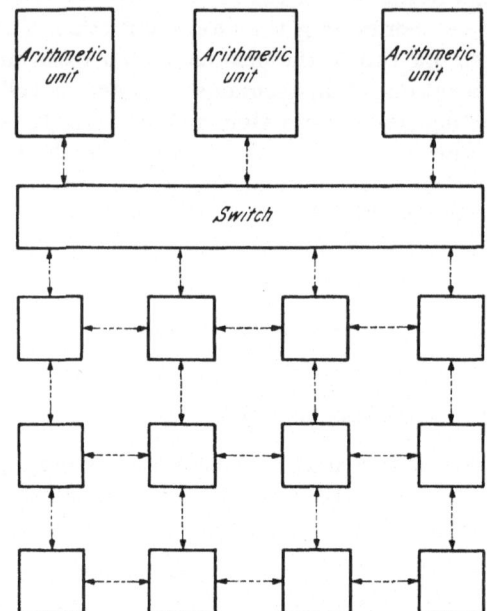

Fig. 9.28. Simplified Comfort Organization

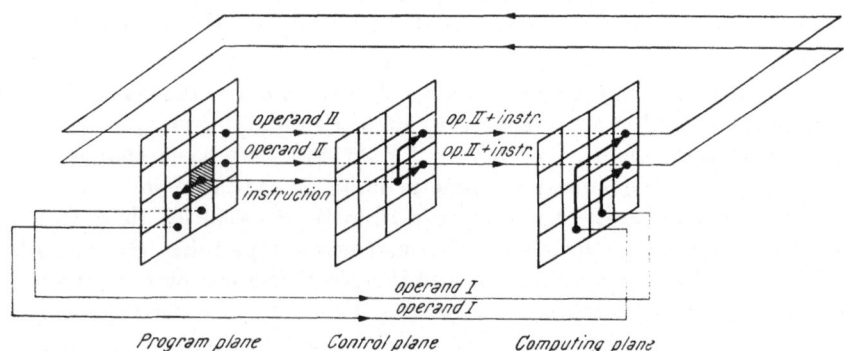

Fig. 9.29. Simplified Gonzalez Organization

The three arrays are referred to as program plane, control plane and computing plane. Communication lines exist between all corresponding cells of the three planes.

The program plane contains the complete program, including the instructions, the operands, the working space, etc. In this respect, the function of the program plane is identical to the Holland array. Identical is also the manner in which the sequential execution of instructions is controlled: the cell containing the instruction currently to be executed is active (indicated by the shaded area in Fig. 9.29); a path to the cell containing the next instruction is established (indicated by the arrow in the program plane); the new cell is activated; and the presently active cell is de-activated. Quite different is, however, the manner in which an instruction is executed.

All instructions are basically three-address instructions. One of the three addresses specifies the cell containing the next instruction in sequence. This address is used only in the program plane. The other two addresses specify operands, designated as operands I and operands II. The instruction, including operand I and operand II addresses, is transferred from the active cell in the program plane to the corresponding cell in the control plane. Here, depending upon the operand II address or addresses, paths are established to one or more cells. The instruction is now transferred to these selected cells. The cells obtain operands II from their corresponding cells in the program plane. The operands II and the instructions including operand I addresses are now transferred to the computing plane. Here, depending upon operand I addresses, paths are established to one or more additional cells. These cells obtain operands I from corresponding cells in the program plane, and transfer them to the cells already containing the instructions and the operands II. The instructions are now executed and the results are transferred back to the program plane into the locations from which the operands II were obtained.

The Gonzalez organization, like the Holland organization allows to execute independent programs concurrently. In addition, the Gonzalez organization permits one instruction to operate upon multiple operands. At the first glance, the price to be paid for this additional feature seems rather high: it requires approximately a triplication of hardware. However, the Gonzalez organization compensates for this increase in cost by a gain in speed also by approximately a factor of three: the three planes operate in an overlapping manner. While an instruction is executed in the computing plane, the control plane already propagates the next instruction and obtains operands II from the program plane, and the program plane establishes a path to the cell containing the then following instruction.

So far, we have discussed concepts for highly parallel machines which are characterized by the absence of a central control unit. However, several proposed machines contain centrally controlled arrays of arithmetic and memory cells. Generally, the aim of such organizations is the concurrent application of identical arithmetic or logic operations upon a large number of operands. The central control specifies a single operation which is then executed simultaneously by a large number of identical processing modules. Of course, the complexity of such machines is justified only if the problems to be attacked have a high degree of "parallelism". A few examples of such problems are: computations involving matrices; calibrations of data samples; correlation of individual weather observations; and step-by-step integration of complex differential equations.

Perhaps the most thoroughly investigated machine of this type is the SOLOMON computer [9][1]. Its current design is based upon several evolutionary steps. Moreover, a simplified model of SOLOMON has been built [10]. A simplified block diagram of SOLOMON II is shown in Fig. 9.30.

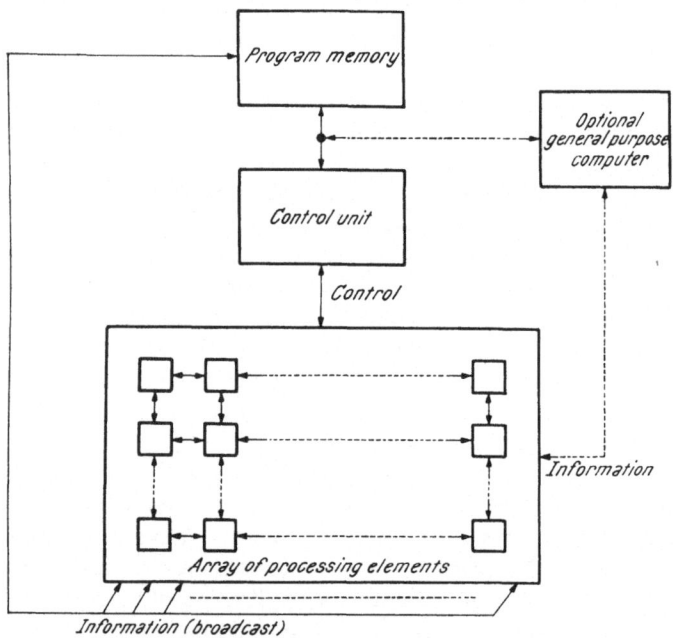

Fig. 9.30. Simplified SOLOMON Organization

The computer consists of three major components: the program memory, the control unit, and the array of processing elements. Provisions are made for an optional fourth component, a conventional general purpose computer.

The program memory contains the program instructions. The control unit, much like the control unit of a conventional computer, fetches individual instructions from the program memory and issues sequences of control

---

[1] "SOLOMON" is short for Simultaneous Operation Linked Ordinal Network.

signals for their execution. Instructions are executed concurrently in some or all of the 1024 elements of the processing array.

Each processing element contains its own arithmetic unit, capable of performing arithmetic and logic operations, and its own 4096-word operand memory. In their capabilities, processing elements are comparable to conventional computers without control units. The common control unit provides detailed control sequences with appropriate timing for all processing elements.

Instructions are executed only by those processing elements which are selected by both, mode and geometric control. The *mode control* is concerned with internal conditions of the processing element. In particular, a processing element may be in one of four possible modes, determined by the states of its mode flip-flops. These flip-flops can be set or cleared under various internal conditions. The processing element executes an instruction only if its mode corresponds to that specified by the instruction. The mode control can, therefore, be used to create a similar effect as the conditional jump of conventional computers: a program branch is executed only if certain conditions are present. The *geometric control* is used to select rows or columns of processing elements for instruction execution. The selection is effected by the contents of row and column registers, which can be modified by program.

Information can be entered into the processing array through the "broadcast" input. This input makes a word of information stored in the program memory available to all processing elements. Again, only the selected processing elements actually use this information. Information can be exchanged between neighboring processing elements through the communication paths indicated in Fig. 9.30. This transfer is under program control. Several geometrical arrangements of processing elements can thereby be specified. One of them is the *rectangular array* as shown in Fig. 9.30. Another configuration is the *vertical cylinder* in which there are additional communication paths established between the leftmost and the rightmost columns of processing elements. A third configuration is the *horizontal cylinder* in which additional paths between the top and bottom rows exist. Finally, there is the *torus* configuration with additional paths between the end rows and also between the end columns.

Let us conclude the description of the machine with an observation on the usefulness of the optional general purpose computer. Its function might be compared to that of an overall system controller: it can load, initiate, monitor, and change programs for the array processor. In addition, its arithmetic capabilities complement those of the basic computer very well, especially for housekeeping tasks, and those aspects of a problem which do not require the concurrent processing of multiple sets of operands. Finally, its existing capability to communicate with peripheral devices such as disk storages, tape units, and card equipment is quite useful. Its usefulness in this respect is enhanced by the (in Fig. 9.30, indicated) possibility to transfer information directly to and from the processing array. The actual implementation of this communication path includes an information buffer which allows the parallel transfer of words either to or from all rows, or to and from all columns of processing elements.

The layout of an array processing computer with an organization closer to that of a conventional computer, is shown in Fig. 9.31.

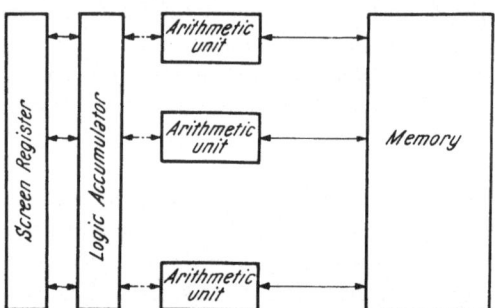

Fig. 9.31. Simplified VAMP Organization

The machine is called by its designer the Vector Arithmetic Multi-Processor, or VAMP for short [11]. It has one central control (not shown in Fig. 9.31), one memory, but 16 arithmetic units. As in SOLOMON, the arithmetic units are able to execute concurrently the same instruction. Operands are obtained from the memory which is common to all arithmetic units. An instruction normally specifies a basic memory address, a, and a modifier, d, so that the various arithmetic units access the locations: a, $a+d$, $a+2d$, $\ldots$, $a+15d$. However, the design has also provisions for indirect addressing, by which more randomly distributed locations can be accessed.

Since each arithmetic unit, potentially, has access to the entire memory, each has access to the information of all other arithmetic units. The information exchange between these units requires, therefore, no special communication paths as in SOLOMON.

The screen register and the logic accumulator shown in Fig. 9.31 perform similar functions as the geometric and mode control in SOLOMON: individual bit positions of the screen register enable or inhibit the operation of individual arithmetic units. Individual bit positions of the logic accumulator are set or cleared dependent upon various internal conditions of the arithmetic units. Since the contents of the screen register and the logic accumulator can be interchanged, internal conditions can be used to enable or inhibit the operation of individual arithmetic units.

The highly parallel machines which we have seen so far, employ arrays of arithmetic units, in order to execute concurrent operations. There exists, however, a number of proposals for machines with a different kind of cellularity. These machines contain, what could be considered as cellular "arithmetic units". These units usually have only limited arithmetic capabilities, but more the logic capabilities required for the processing of pictorial data. The processing of a single picture, and particularly the recognition of patterns, may require the handling of thousands or even millions of data points. Conventional computers seem entirely inadequate for such tasks.

Let us indicate here two representative organizations. Fig. 9.32 shows a structure proposed by Hawkins and Munsey [12].

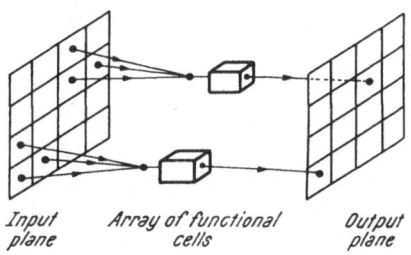

Input plane   *Array of functional cells*   *Output plane*

Fig. 9.32. Planar Organizations of Hawkins and Munsey

The machine has an array of functional cells[1] located between an input plane and an output plane. In its overall function, the arrangement acts like a translator: the input pattern is translated into a different output pattern. This is not unlike the function of a conventional arithmetic unit in which the inputs (operands) are operated upon, in order to provide an output (result). Here, the specific transformation taking place depends upon the particular function specified for the logic cells, and on the particular connections established between functional and input cells. Both are under program control, but the repertoire of functions and connection patterns is rather limited. The same function is specified for all cells, and the same geometrical pattern of connections applies to all functional cells. With such severe limitations, it is obviously not possible to implement all, or even all desired operations. The design compensates for this by providing for the repeated or "iterative" processing of data.

Incidentally, what may here seem to be a complex arrangement of electronic circuitry, can possibly be implemented by relatively simple optical masking and scanning techniques.

Fig. 9.33 shows, as a second example of cellular "arithmetic" units, the layout of a "pattern articulation unit" [13].

---

[1] For clarity, only two of them are shown.

The unit is a special purpose design for the interpretation of particle tracks in bubble chamber pictures. It consists of a two-dimensional arrangement of functional cells, the "stalactites". Grossly simplified, the stalactites are arranged into a 32×32 array, corresponding to the 1024 black or white raster points of a bubble chamber picture. The latter also correspond to the 1024 bits in a word of the machine.

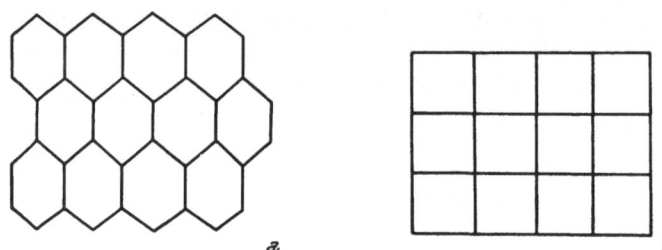

Fig. 9.33. Geometric Arrangement of Cells in PAU

Optionally, one of two arrangements of cells can be selected by program: a rhombic array in which each cell has six neighbors, or a rectangular array in which each cell has eight neighbors (vertically, horizontally, and diagonally). Each cell can sense the binary output state of its neighbors, it can receive one bit from the memory, it can transmit one bit of information to its neighbors, and it can transmit one bit to the memory. The cells perform program selectable logic operations upon their internally stored information (10 bits), and on the input from up to eight neighbor cells. These operations serve to reduce or eliminate optical noise, to reduce particle tracks to lines of uniform thickness, to fill gaps in particle tracks, and to determine track nodes and end points.

Problem 11: Make a short list of what you consider basically new ideas in paragraph 9.6.

9.7. List Processors and List Processing Features

List processing is more concerned with "information manipulation" than with computing. Typical list processing operations are: delete items from a list; add items to a list; locate specific entries; combine lists; and extract items from a list. Of course, these operations can be performed by conventional computers, but, being designed primarily for numerical computations, their operating characteristics are less than ideal for list processing applications. The use of symbolic programming languages and of compilers can help considerably to ease the programming task. In fact, they may give the programmer the impression of having a convenient system for the solution of list processing problems. In reality, however, the computer will probably perform long sequences of "awkward" operations to execute a single symbolic list processing statement. In order to see some of the problems involved, let us discuss here initially two basic concepts, the "list", and the "stack".

A *list* is an ordered assembly of items. Using a conventional approach in a conventional computer, we would simply set aside a certain storage region and store the consecutive items of the list in consecutive locations. This approach is entirely satisfactory if the configuration of the list stays reasonably constant. Several difficulties arise, however, when the list is subject to change. For one, the reserved storage region must be large enough to hold the maximum number of items. This requirement may seem simple, but, in many

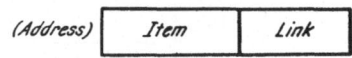

Fig. 9.34. Entry Corresponding to a Single Item

cases, the number of items in the list will depend upon the particular job to be handled, and cannot be determined at the time a program is written. A second difficulty concerns additions and deletions of items during computation. If the listed items are to remain ordered and in contiguous locations, the addition or deletion of a single item may require the relocation of a large number of other items. Basically simple operations may, therefore, require many time-consuming steps.

The indicated problems do not exist with a list structure in which the order of items is determined by entries in the list itself[1]. Fig. 9.34 shows a possible format for a single entry.

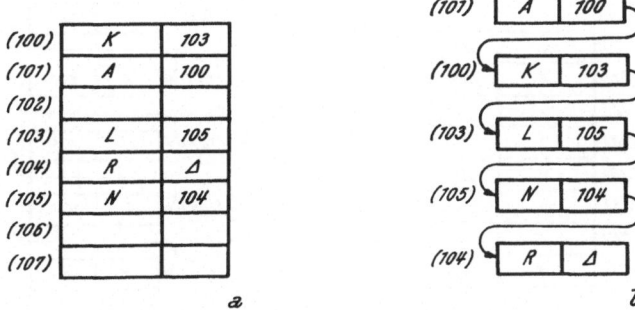

Fig. 9.35. Sample List Structure

The stored information consists of the item and a "link". The link specifies the address of the next entry in the list. Having found one item, we can find the next item in sequence, no matter where it is located physically.

---

[1] Many of these problems can also be eliminated by the use of content-addressed memories. (See paragraphs 8.3.5. and 9.8.)

With such an arrangement, there is no need to keep consecutive items in consecutive locations. The example shown in Fig. 9.35 may help to illustrate this point.

Items *A*, *K*, *L*, *N*, and *R* constitute the list. The items are stored in arbitrary locations of the memory as shown in Fig. 9.35a, but are logically

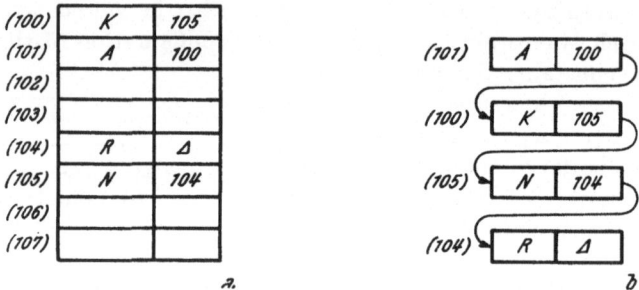

Fig. 9.36. Sample List with Item *L* Deleted

ordered as shown in Fig. 9.35b. We should note here that the link field of the last entry in the list contains a special symbol, Δ, signifying "end of list". With this arrangement, the knowledge of the location of the first item in the list is sufficient to access the entire list. We should also note that the deletion or insertion of items requires no re-organization of the list. Fig. 9.36 shows the sample list with item *L* deleted.

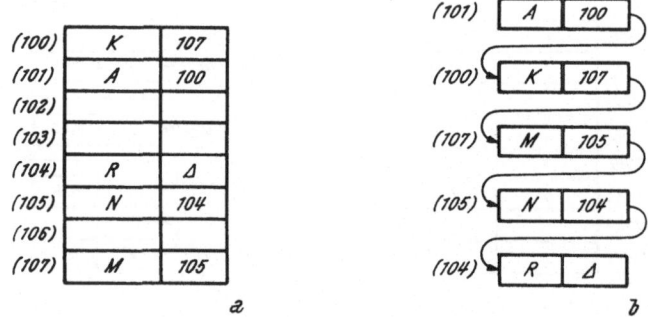

Fig. 9.37. Sample List with Item *M* Inserted

Comparing Figs. 9.35 and 9.36, we find that the only change in the remaining list is the modification of the link associated with item *K*. It reflects now the address of item *N*, instead of the address of the deleted item *L*.

Similarly, a new item may be inserted into any available storage location, but appear logically at the appropriate place within the list. Fig. 9.37 reflects the list of Fig. 9.36 with the item M inserted between items K, and N. Again, the change requires the modification of only one existing entry.

Two advantages of the indicated list structure are apparent: additions and deletions are simple operations; and memory space can be allocated as needed and available. Perhaps not so obvious is that lists of this type can be used to implement many computer housekeeping and control functions. For instance, a list may be set up to keep track of "empty" locations. Needed locations are removed from the list, and no longer needed locations are returned to the list.

More unconventional is the idea to employ a list structure for the computer program. Instructions constitute then the individual items in a program list as indicated in Fig. 9.38.

The entry contains the op-code and an operand address (or operand addresses), but also a link specifying the next instruction in sequence. With this instruction format, there is no need for a program address counter in the computer, the links assure the sequential execution of the program instructions.

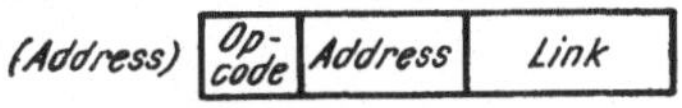

(Address) | Op-code | Address | Link

Fig. 9.38. Sample Instruction Format for Programs with List Structure

Quite important is it to note that the indicated list structure is not limited to one dimension, but can be adapted to more complex configurations. Fig. 9.39 shows, for example, a hypothetical two-dimensional list in which each entry has two links. The first link facilitates the search of the list in vertical direction, the second link allows a search in horizontal direction.

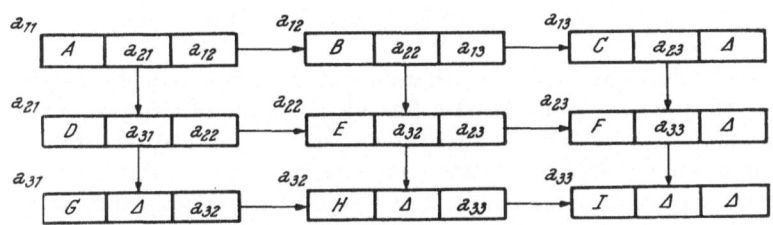

Fig. 9.39. Two-Dimensional List Structure

Even if the format of entries allows only one link, nodes and branches in the list can be realized. Fig. 9.40 indicates a possible approach.

The links associated with items I_1 and I_2 specify the same address, c. Consequently, item I_3 is a part of the list containing item I_1, but also a part of the list containing item I_2. In contrast, either item I_4 or item I_5

may become a part of the list containing item I_3. The entry in location d contains no item in the usual sense, but instead a second link marked by a special code or symbol. A search of the list may leave location d either in the normal manner, proceeding to item I_5 in location f, or, in the presence of certain conditions, proceed to item I_4 in location e. If the indicated list were a program list, the entry in location d would be a conditional jump instruction.

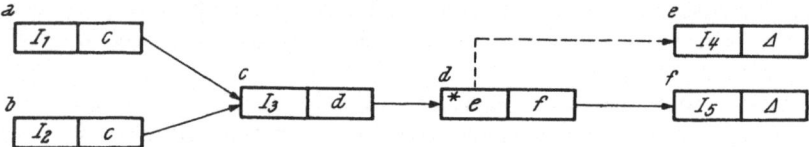

Fig. 9.40. Branching of a List

List arrangements may not only have several dimensions, but also several "levels". Fig. 9.41 shows, for example, a two-level structure.

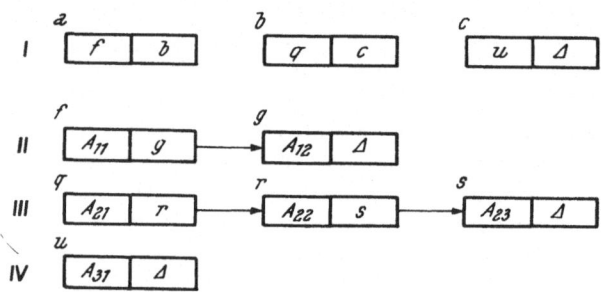

Fig. 9.41. Two-Level List Structure

List I contains as items the beginning locations of lists II, III, and IV. It may, therefore, be considered a "catalog" of the other lower-level lists. If the list were representing programs, list I might play the role of the main program, specifying only the particular sequence in which the various sub-programs are to be executed. List I could then contain the same items repeatedly, that is, specify repeatedly the execution of the same sub-program.

The second concept to be discussed is that of the *stack*. A stack is a storage device which stores information in such a manner that the item stored last, is the first item retrieved. The operation of a stack can be compared to that of the mechanical spring-loaded hopper shown in Fig. 9.42a.

Items are inserted and removed from the top. We speak of a "push down, pop up" storage. Fig. 9.42b shows a functional equivalent. The access to the individual locations of a memory is controlled by a stepping switch. Each write operation advances the switch by one position. Conversely, the switch is stepped back by one position for each read operation.

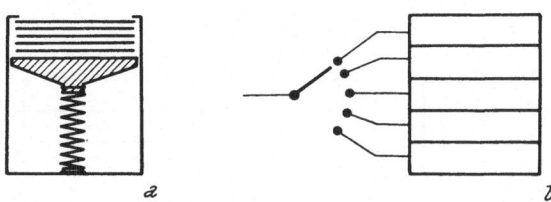

Fig. 9.42. A Mechanical Stack and Its Equivalent

The effect is that of a "last in, first out" buffer. Of course, the actual implementation may employ a memory addressed by the contents of an up-down counter, rather than accessed by a stepping switch. With either arrangement, no external addressing of the stack is required; addresses are completely determined by the sequence of read and write operations.

To conclude the discussion of the stack, let us show here one practical application of a stack in a conventional computer. The stack serves here as temporary storage for intermediate results.

Suppose the arithmetic unit of a computer has a layout as shown in Fig. 9.43. The stack comprises a series of registers, A, B, C, D, etc. When a word is obtained from the memory, it is always entered into register A, while the previous contents of A are pushed down into register B, the contents of the latter are pushed down into C, and so on. When a word is remov-

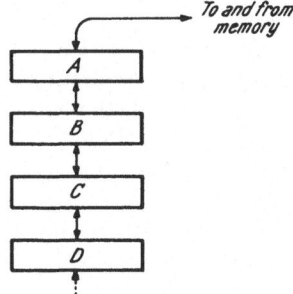

Fig. 9.43. Arithmetic Unit with the Structure of a Stack

ed from register A, the contents of the remaining registers pop up. The two top registers of the stack, A and B, serve as arithmetic registers. More specifically, the execution of an add instruction causes the contents of A and B to be added, and the result to be recorded in A: $A + B \rightarrow A$. A subtraction is performed as: $A - B \rightarrow A$; a multiplication as: $A \times B \rightarrow A$; and a division as: $A \div B \rightarrow A$. Whenever the execution of an arithmetic instruction vacates the B-register, the contents of the lower registers pop up.

Suppose now that the computer is programmed to calculate the value of the expression: $U \times V + X/(Y - Z)$. Fig. 9.44 shows an applicable sequence of program instructions and the contents of the stack after the execution of each instruction.

We see that intermediate results are pushed down into the stack so that they pop up later when they are needed. The stack eliminates many program steps (and the time to execute them) to store intermediate results and to retrieve them later.

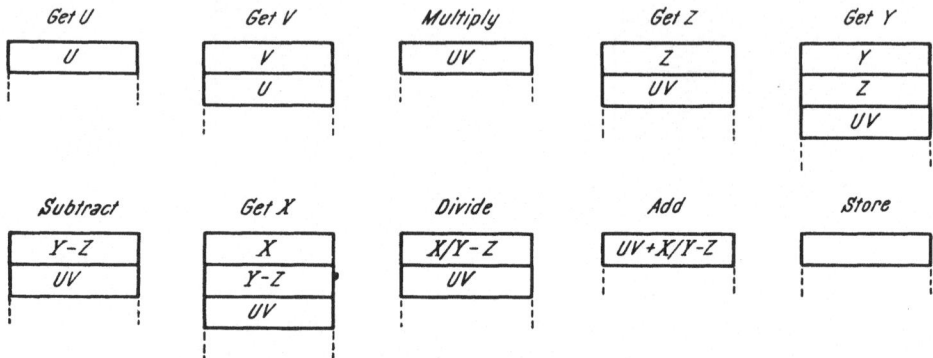

Fig. 9.44. Use of the Stack

If we express the fetching of an operand merely by the letter of the operand, and an arithmetic operation by its symbol, we may represent our sample computation as: $UV \times ZY - X \div +$. This notation is known as *Polish notation* or as *Polish string*. Its main advantage is that algebraic or logic expressions can be stated without the use of parentheses.

Problem 12: Show the normal algebraic expressions for the following Polish strings:

a) $XYZ++$　　　　　　　c) $XYZ+\times$

b) $XY+Z+$　　　　　　　d) $XY+Z\times$

Problem 13: Translate the following algebraic expression into a Polish string:

$$1 + U \frac{X+Y}{Z}$$

Arithmetic instructions, as they are shown, are "no-address" instructions. This may allow to accomodate several instructions in one instruction word. A prerequisite to this approach is that the stack can be loaded with all required operands before the execution of a multiple arithmetic instruction is attempted.

Although Fig. 9.43 shows individual registers for the stack, it is entirely practical to assign locations in a "scratch-pad" memory, or even the main memory to the registers C, D, E, etc. We would then follow the basic approach indicated in Fig. 9.42, but "top" the stack with the two arithmetic registers A, and B.

It is important to realize that stacks may have list structures. Fig. 9.45 shows the basic approach.

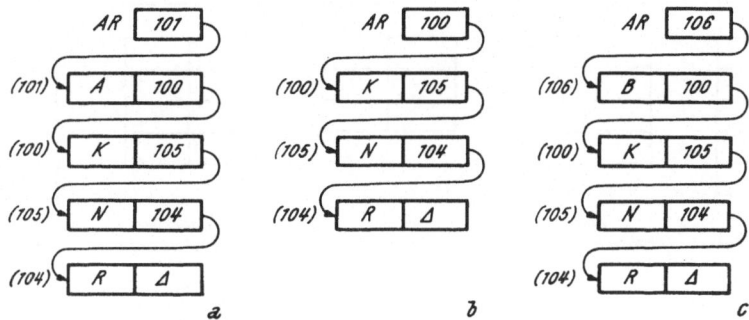

Fig. 9.45. Stack with List Structure

The items A, K, N, and R are stored in arbitrary locations of the memory, but belong logically to a list which begins at location 101. With a proper mode of operation, the items may also be considered to be stored in a stack, in which "consecutive" locations are determined by the links. Item A is in the top location, and item R is in the bottom location. The address of the top item is retained in the stack address register AR. When the top item A is removed from the stack, its link, 100, replaces the previous contents of the stack address register, and item K "pops up" to the top position. Fig. 9.45 b shows the logic organization of the stack after this step. Alternately when a new item is inserted into the stack, the address contained in the stack address register is attached as link to the new item, and both are stored in a previously empty storage location[1]. The address of the new location is inserted into the stack address register. Fig. 9.45 c shows the stack in this position.

In our discussion of list processing we have, so far, considered certain individual concepts and implementations. Many conventional computers incorporate hardware features which simplify or speed up list processing operations. Examples are: stacks, list manipulation instructions, and content-addressable memories[2]. In addition, there exist a number of studies which aim at computer layouts specifically designed for list processing. Fig. 9.46 shows a simplified, but representative sample.

---

[1] The address of a previously unused location might be obtained from a list or stack containing all empty locations.

[2] See paragraph 8.3.5.

At the top of the figure, we see several special purpose and general utility registers. The remainder of the layout is meant to represent one physical memory which is logically allocated to various separate functions[1].

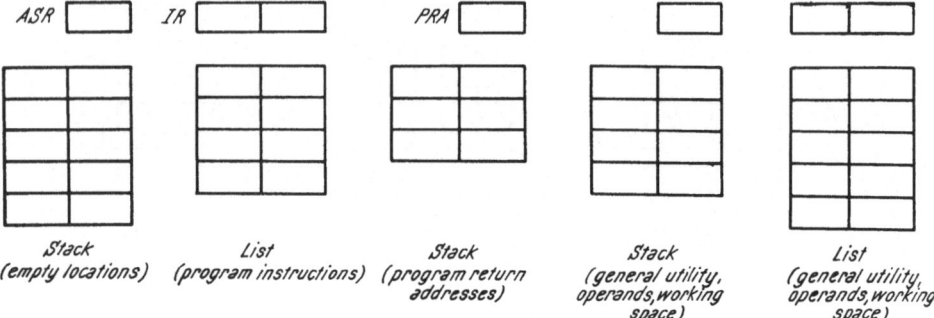

Fig. 9.46. Simplified Layout of a List Processing Computer

Each storage location provides space for an item and a link, permitting a list structure for both, operands and program instructions. Functionally, the machine consists of a number of stacks and lists, having the basic organization indicated in figs. 9.35 and 9.45.

The two structures on the right are representative of several lists and stacks which the program may use as required for the storage of operands or as working space.

On the left is a stack containing all empty locations. Its stack address register is designated as the available space register, ASR. Whenever a memory location is assigned to operand storage, program storage, etc., the top location contained in the ASR stack is given to one of the other structures. Conversely, when a location is no longer used in one of the other stacks or lists, it is returned to the ASR stack. This arrangement permits complete flexibility in the assignment of memory space: individual lists or stacks extend as far "down into the memory" as required. The operation of the ASR stack does not require the execution of housekeeping instructions, but is controlled by the sequencer in the control unit or by the micro-program during the execution of regular program instructions. (See also problem 14 below.)

The second structure from the left represents the instruction list. It comprises as many storage locations as needed to accommodate the applicable program or programs. Associated with the program list is the instruction

---

[1] A corresponding logic allocation of memory space is made in conventional computers: Depending upon its contents, a storage location may be considered to be part of the operand storage, part of the program storage, or part of the working space.

register IR. It holds instructions during their execution and provides space for both, the instruction proper, and its link. Compared with the registers of a conventional computer, the instruction register performs the functions of both, the instruction register and the program address register.

In general, the link determines the location of the next instruction to be executed, but there are two exceptions. One of them is concerned with conditional jumps. When the jump condition is present, the "operand" address, rather than the link determines the next instruction. The second exception concerns return jumps. A return jump[1] serves to jump from one program to another, while simultaneously storing pertinent information so that the original program can be re-entered at a later time and its execution resumed at exactly the place it was left. With conventional computers, the program to be entered must be modified so that its last executable instruction becomes a jump back to the appropriate location of the original program[2]. The model list processor uses a more convenient approach: programs are not modified, but the information required for the return is stored in a program return address stack. The PRA stack is represented in Fig. 9.46 by the third structure from the left.

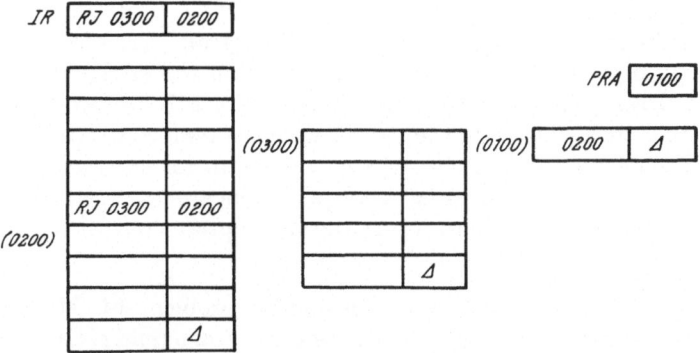

Fig. 9.47. Illustrative Example for the Execution of a Return Jump

Let us use the specific program example in Fig. 9.47 to discuss this approach in more detail. The list on the left represents the main program. One of its entries is shown in detail, the return jump. The intent of this particular return jump could be stated as: "Go to the sub-program beginning at location 0300, but, upon its completion, return to location 0200 of the present program." The return jump is executed in the following manner: the return address 0200 is stored as *item* in the PRA stack as indicated on the right

---

[1] Also referred to as "execute instruction".
[2] This can be accomplished either by the hardware or by the program.

of Fig. 9.47; the next instruction is then taken from location 0300. When the execution of the sub-program is completed, and the Δ symbol is encountered, the return address is retrieved from the PRA stack and used to obtain the next instruction. The main program is, therefore, re-entered at location 0200.

Problem 14: Assume that the registers IR, ASR, and PRA are functionally separate entities, but that all entries of the IR list and the ASR and PRA stacks are stored in a single central memory. The memory has an address register AR and a read/write register RWR. Show a sequence of micro-steps which execute a return jump. When necessary, designate the item and link fields of registers by the indices i and l. Note that three tasks must be accomplished: obtaining an empty location from the ASR stack for use in the PRA stack; storing the return address in the PRA stack; and fetching the next instruction from the program list of the subprogram.

Problem 15: Show the sequence of micro-steps required for the return to the main program after completion of the sub-program.

The indicated approach is not limited to two levels of programs. If there should be a sub-program to the sub-program, the applicable return address is simply pushed into the PRA stack on top of the already stored return address. Whenever the execution of the sub-sub-program is completed, the return address to the sub-program is removed from the top and the return address to the main program pops up for later use. In fact, no matter how the return jumps between programs are arranged, the top item in the PRA stack always contains the appropriate return address in its top position.

Problem 16: a) Review your solution to problem 14. Make sure the indicated sequence of micro-steps allows the storage (and retrieval) of more than one return address in the PRA stack.

b) What condition, would you say, indicates that the PRA stack is empty?

An interesting operational aspect is revealed when one considers the possibility to load the PRA stack by transfer instructions rather than return jumps. The program at the highest level might then simply consist of instructions which load the beginning addresses of lower level programs into the PRA stack. As soon as the Δ symbol in the high level program is encountered, the computer begins to execute the specified sequence of lower level programs[1].

---

[1] Compare also Fig. 9.41.

Perhaps the most interesting aspect of the PRA stack is that a program may call upon itself as sub-program. We speak then of a recursive program. The call may be accomplished either directly by a return jump, or through one or more levels of other programs. If recursive programs require their own working space, the applicable storage locations can be pushed down into one of the general utility stacks in the same manner as the return address is pushed into the PRA stack. The operands applicable to one "pass" of the recursive program may then be covered temporarily by the operands of other programs or by the operands of different passes of the same program, but pop up toward the top positions, while the returns to the various higher level programs are made. A comparable capability could be achieved with conventional computers only by rather complex programs, but is, here, accomplished by the organization of the computer hardware[1].

We have considered here only a few operational aspects of one model list processor and have omitted many details such as the instruction repertoire, the layout of information handling registers, and so on. However, the discussion should have been detailed enough to convey a feeling for the rather unique capabilities of computers with list processing organizations.

Problem 17 (Voluntary): Design an appropriate sequence of micro-steps to execute a program instruction which

a) inserts an item into an existing list,

b) deletes an item from an existing list,

c) compares items in a list with a given pattern and retrieves the first matching item into a working register.

Start with the basic layout given in Fig. 9.46, but assume that working registers are available as needed.

9.8. Associative Computers

The term "associative" is attributed to computers which contain an associative memory[2], or, perhaps more precisely, to computers in which the *functions* performed by such a memory are an integral part of the internal operation Associative computers are particularly suited for non-numeric computations. In this respect, and also in their mode of operation, they are related to list processors. On the other hand, some of their operational aspects justify their classification as highly parallel machines. Let us discuss here shortly only those ideas and principles which are unique for associative

---

[1] With careful programming, it is conceivable to use only one stack for the temporary storage of both, operands and return addresses. Each program must then "un-cover" the return address by removing its operands from the stack, before another program is re-entered.

[2] See paragraph 8.3.5

computers, and which have not previously been brought out in our discussions of either list processors, or highly parallel machines.

One basic difference between associative computers and other machines lies in the accessing of stored information. Fig. 9.48 shows representative formats for instructions and for operands.

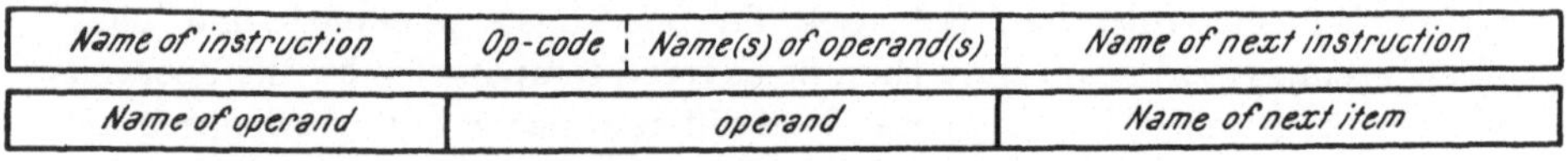

Name of instruction	Op-code	Name(s) of operand(s)	Name of next instruction
Name of operand	operand		Name of next item

Fig. 9.48. Representative Instruction and Operand Formats

The computer accesses instructions and operands by their names. Thus in certain respects, the name of an item has a similar purpose as a conventional address. Names may even consist of address-like numeric values. However, names are associated with items of information, rather than with physical locations of the memory, and are stored together with their associated information. This arrangement has several consequences: Any combination of characters or symbols which is meaningful for the application can be selected as a name. In fact, the symbolic names used in compiler languages are acceptable without change. Moreover, the arrangement eliminates the need to store items in an orderly fashion, a requirement which can become quite cumbersome with conventional computers. Even when compared to list processing organizations, the associative access eliminates or simplifies many address or link manipulations.

The use of associative memories has another important aspect: parallel-by-word operations are not restricted to memory reads and writes, but may include other arithmetic and logic operations. The associative computer

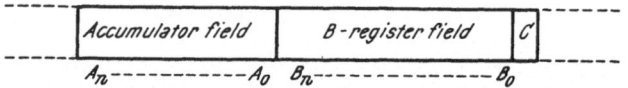

Fig. 9.49. Sample Assignment of Fields Within a Word

acquires then the potential capabilities of a highly parallel machine: the operation specified by a single program instruction may be applied concurrently to multiple operands. Let us illustrate this approach with a specific example.

Assume that the computer memory has the capability for parallel-by-word associative writes, that is, the capability to search for words containing certain specified bit-combinations and to replace the contents of selected fields within these words by other specified bits. Assume further that certain fields within the word format are assigned specific functions as indicated in Fig. 9.49.

Each word contains two numerical values acting as operands. Their respective fields are designated as accumulator field and as B-register field, with the respective bit positions $A_0 \ldots A_n$, and $B_0 \ldots B_n$. In addition, there is a one-bit field designated as carry field, C. The contents of B shall be added to the contents of A, and the sum recorded in A.

Let us begin with the least significant bit. We may find any one of the four possible conditions indicated in the first column of Table 9.2.

Table 9.2. *Truth Table for the Addition of the Least-Significant Bits*

Condition A_0	B_0	Desired result A_0	C	Required Action
0	0	0	0	none
0	1	1	0	set A
1	0	1	0	none
1	1	0	1	clear A; set C

The second column lists the desired results, i.e. the state of A_0 representing the sum and the existence of carries. Only two of the four cases requires a change of the contents of A. The modification of the contents of A can be accomplished by two consecutive associative write operations: the first of these writes a "1" into bit-positions A_0 of those words which contain the bit-combination $A_0'B_0$; the second writes a "0" into bit positions A_0 of those words which contain the bit combination $A_0 B_0$.

Let us now consider the carries. Whenever a carry is generated, we wish to record it in field C so that we can add it to the next significant bit-position in another operational cycle. In case of the least-significant bit, there is only one condition requiring the recording of a carry (see Table 9.2), and this condition coincides with the one already identified as requiring a change of A. We can, therefore, combine the two actions, and accomplish the recording of both, the least-significant bit of the sum and of the carry with two associative write operations as shown in Table 9.3.

Table 9.3.
Summary of Operations for the Addition of the Least-Significant Bit-Fields

	Descriptor	Replacement Pattern
1st associative write	$A_0 = 0$; $B_0 = 1$	$A_0 = 1$
2nd associative write	$A_0 = 1$; $B_0 = 1$	$A_0 = 0$; $C = 1$

The descriptor specifies the words to be selected for each operation; the replacement pattern indicates the bit-configuration to be written into the selected words. The two indicated associative write operations are sufficient for the addition of the least-significant bits in *all* words contained in the associative memory. We have accomplished the first step of an addition which is serial-by-bit, but parallel-by-word.

For the more significant bit-positions, the sum depends upon the state of C, in addition to the state of A and B. The applicable truth table has, therefore, eight entries rather than four. Actions are required for only four of the eight conditions. These actions can be accomplished by four associative write operations as shown in Table 9.4.

Table 9.4. *Summary of Operations for the Addition of two Arbitrary Bit-Fields*

	Descriptor	Replacement Pattern
1st associative write	$A_x = 0$; $B_x = 0$; $C = 1$	$A_x = 1$; $C = 0$
2nd associative write	$A_x = 0$; $B_x = 1$; $C = 0$	$A_x = 1$
3rd associative write	$A_x = 1$; $B_x = 0$; $C = 1$	$A_x = 0$
4th associative write	$A_x = 1$; $B_x = 1$; $C = 0$	$A_x = 0$; $C = 1$

Problem 18: Verify the correctness of Table 9.4 by comparing its entries with those of an applicable truth table, similar to the one given in Table 9.2.

The sequence of four associative write operations indicated in Table 9.4 accomplishes the addition of two arbitrary bit-fields. As many sequences must be performed as there are bits in the respective operand-fields. We should note here that, if the addition is to be performed only in selected words, the descriptor must contain an additional bit or bit-combination characteristic for the applicable words. This requirement may lengthen the associative search, but does not increase the number of write operations required for the addition.

Problem 19: a) Estimate the execution time for the addition of 24-bit operands and compare it to that of a parallel-by-bit addition in a computer with conventional organization.

b) How many concurrent additions must be performed in order to make the serial-by-bit, parallel-by-word operation faster?

Assume an organization of the associative memory as given in Fig. 8.82, and assume equal cycle times for the conventional and the associative memory. Note, however, that several memory cycles are required for an associative write.

Even though we have discussed here only a parallel-by-word addition, it should be apparent that other arithmetic and logic operations can be performed in a similar manner. In fact, the execution of logic operations is, in general, simpler than that of arithmetic operations.

Problem 20: Find an appropriate sequence of associative write operations which

 a) subtract the contents of the B-field from the contents of the A-field,

 b) replace the contents of the A-field by the contents of the B-field,

 c) replace the contents of the A-field by the logic product of the contents of A and B.

We have seen here the potential capability of the associative computer for the operation with multiple operands. This capability is equivalent to that of certain computers with cellular organization: each storage location of the associative memory acts as an "arithmetic" module. However, in order for the scheme to be practical, there must be some means of communication between modules. This communication might be accomplished by conventional parallel-by-bit accesses, and transfers of individual words under program control. With this approach, the concept looses much of its potential power: only arithmetic or logic operations are performed in the manner of a highly parallel machine, while the communication between modules is sequential by word.

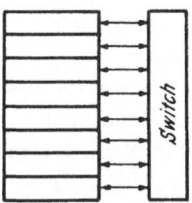

An improvement in some respects can be achieved by having transfer instructions operate upon entire list structures, rather than on single operands. A transfer instruction might then specify only the names of the first items in the "to" and "from" lists, while some hardware control schedules the sequential transfers of all items in the list. Such an approach, undoubtedly, simplifies the programming task (a single instruction causes the transfer of many items), but it would probably not significantly increase the speed of communication.

Fig. 9.50. Associative Memory with Hypothetical Switch

One hypothetical approach to a highly parallel transfer of information is indicated in Fig. 9.50.

The figure is meant to represent an associative memory with the capability for read and write operations which are serial-by-bit and parallel-by-word. Individual storage locations are connected to a switch in a similar manner as they are connected to the match detector[1]. One bit is read from all selected words, transferred in parallel to the switch, re-routed, and written back into different locations of the memory. Of course, the actual implementation of the switch may be rather complex, and it remains to be seen wether the approach finds practical application.

A possible alternative is the use of a cellular machine layout, basically

---

[1] See, for example, Fig. 8.84.

corresponding to the Holland organization[1]. Here, however, we would have the equivalent of an *associative* memory with its storage locations distributed to functional cells. Paths between cells would be established not by matching operand addresses with the physical addresses of other modules, but by matching the operand names contained in the instruction with the names of items stored in other cells. A similar procedure would be used to assure the sequential execution of program instructions: When the execution of the instruction contained in the active module is completed, a path is established to the module storing an instruction whose associated name matches the name of the next instruction specified in the active module. The new module is activated, and the previously active module is de-activated. The cellular organization would enable the associative computer not only to execute one instruction with multiple operands, but also to execute several instructions concurrently.

9.9. Learning Structures, Adaptive and Self-Organizing Systems

The subject which the above title suggests, but only loosely defines, receives close attention from both, engineers and physiologists. Physiologists are interested in physical structures which may provide clues to the better understanding of the functioning of biological nervous systems; engineers are interested in biological systems because their principle may provide suggestions for the design of novel machines. So far, neither group has achieved complete success. No existing model exhibits all the characteristics of even a single nerve cell, and no "computer" has been designed which has even roughly the structure of a biological system. On the other hand, some interesting progress has been made.

Much of this progress is in the area of heuristic programming. Here, a conventional computer is programmed to duplicate some of the capabilities of biological systems, without necessarily simulating their structure or mode of operation. More specifically, heuristic programming is concerned with strategic searches for solutions or even methods of solution, automatic recognition of classes of situations or patterns requiring a different treatment, learning from previous successes or failures, and selection of promising approaches to new situations. Interesting as these studies and their results are, they are only remotely related to our present undertaking, the survey of physical structures which may conceivably serve as models for computer-like machines.

A second area in which progress has been made, is the investigation of mathematical and physical models of nervous systems. We shall show here in some detail one specific design, but indicate modifications which lead to a large number of alternate, implemented or proposed structures.

---

[1] See paragraph 9.6.

9.9.1. A Sample Perceptron

The design of the perceptron has been prompted by the, admittedly incomplete, understanding of the functioning of the cerebral cortex. It is a rough analog to a neuron of which the human nervous system contains in the order of 10^{10}. The perceptron learns to recognize and classify input patterns. The term "pattern" has here a very general meaning: it can conceivably comprise binary patterns, visual patterns, audio patterns, or any other pattern which can be translated into electrical signals. The circuit responds with an output only if it "considers" the input pattern to be a member of the class of patterns which it has "learned" to recognize. Learning is accomplished by the self-adjusting of circuit parameters during a training period in which the perceptron is "taught" the correct responses for a number of trials. Let us use the sample circuit shown in Fig. 9.51 to explain this principle of operation.

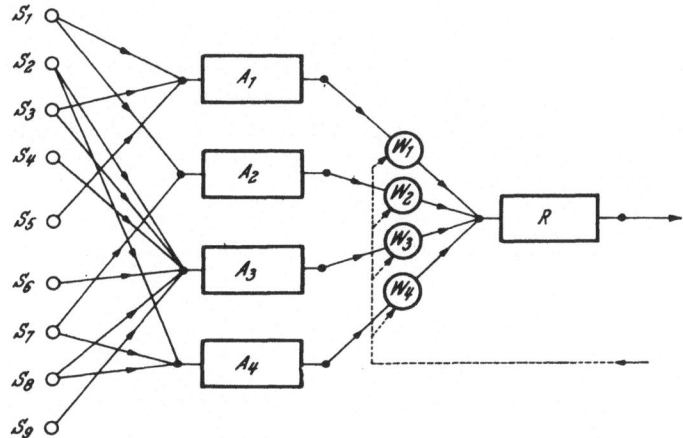

Fig. 9.51. Layout of Sample Perceptron

The arrangement consists of three "layers" of units: sensory units, S; associative units, A; and a response unit, R. The numbers of individual units in Fig. 9.51 have been selected arbitrarily. Also arbitrary is the manner in which the S-units are connected to the A-units[1].

The *S-units* sense the input pattern. They are threshold units providing a "1" output if the input stimulus exceeds their threshold value, but providing a "0" output if there is no input stimulus, or if the input stimulus falls below the threshold value. In essence, they provide at their outputs a

---

[1] A table of random numbers was used to determine which S-unit is connected to which A-units.

binary replica of the input pattern. As we have said before. the nature of
input stimuli can be acoustical, mechanical, electrical, etc. For illustrative
purposes, we shall assume here optical input patterns and a two-dimensional
arrangement of sensory units as shown in Fig. 9.52 a.

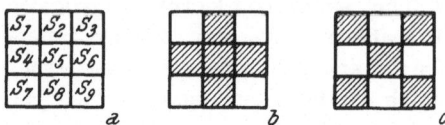

Fig. 9.52. Arrangement of S-Units and Two Sample Patterns

Fig. 9.52 b and 9.52 c show two examples of input patterns. For simplicity, we shall refer to them as the $+$ pattern and the $\times$ pattern. The
$+$ pattern stimulates the sensors S_2, S_4, S_5, S_6, and S_8 to provide a "1"
output; the $\times$ pattern stimulates S_1, S_3, S_5, S_7, and S_9.

The A-units are also threshold units. An A-unit provides a "1" output
only if the number of "1" inputs it receives exceeds a certain value. In the
above example, we find that the unit A_1 receives a single "1" input (from
S_5) if the $+$ pattern is applied, but it receives three "1" inputs if the $\times$
pattern is applied (S_1, S_3, S_5). Fig. 9.53 a summarizes the number of "1"
inputs to each A-unit for the $+$ and $\times$ patterns defined in Fig. 9.52.

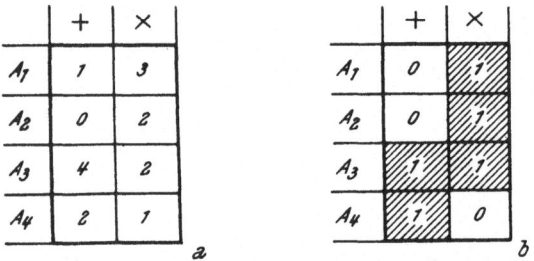

Fig. 9.53. Inputs and Outputs of A-Units for Sample Patterns

If we assume that the threshold of all A-units is such that a unit produces
a "1" output only if it receives two or more "1" inputs, we obtain the
responses summarized in Fig. 9.53 b. The two binary patterns at the output
of S-units shown in Fig. 9.52 b and c have become translated into the two
binary patterns shown in the left and right-hand columns of Fig. 9.53 b.

We note that a certain degree of classification is incidental to this
process: the infinite number of input patterns[1] produces one of 2^9 possible
patterns at the output of S-units, and one of 2^4 possible patterns at the

---

[1] Infinite because of sizes, shapes, and gradations.

output of A-units. In effect, the combination of S- and A-units classifies all possible input patterns into $2^4 = 16$ categories[1].

The R-*unit* is also a threshold unit. It receives its inputs from A-units. Here, however, each connection has a weight, W, associated with it. The R-unit sums the weighted inputs from all A-units, and responds with a "1" output only if the weighted sum exceeds the threshold value. Individual weights may have positive, or negative values. A positive weight tends to stimulate the R-unit to provide a "1" output if the associated A-unit has a "1" output. A negative weight tends to inhibit the "1" output if the associated A-unit has a "1" output. If the A-unit has a "0" output, the value of the weight does not influence the response of the R-unit. For our present purpose, we shall assume that the threshold of the R-unit is such that the R-unit responds with a "1" output to a weighted sum $\geqslant 0$, and with a "0" output to a weighted sum < 0.

The weights are adjusted during a learning period depending upon, whether or not the R-unit provides the desired response. A number of rules for this adjustment has been proposed. Let us assume here that the "teacher" requests an adjustment of weights only if the R-unit provides the undesired response to an input pattern. The capability to accept such a request is indicated in Fig. 9.51 by a dashed line. The request causes a change of weights in only those input lines which are connected to A-units having a "1" output. The weights are increased by a numerical value of one if the R-unit provides a "0" output instead of a desired "1" output. The weights are decreased by the numerical value of one if the R-unit provides a "1" output instead of a desired "0" output.

Let us show this action in a specific example. Suppose we want to "train" the circuit to respond to the $+$ pattern of Fig. 9.52b with a "1" output, but to respond to the $\times$ pattern with a "0" output. We assume that all weights have the initial value zero. We now "show" the machine a series of $+$ and $\times$ patterns.

Suppose the first pattern is a $+$ pattern. The units A_3 and A_4 provide a "1" output; the units A_1 and A_2 provide a "0" output. Since all weights have the value zero, the weighted sum is zero, and the R-unit provides a "1" output. This is the desired response. We, therefore, do not request an adjustment of weights. If we now show the $\times$ pattern, the machine still responds with a "1" output. This response is "wrong". We request the machine to adjust weights. This adjustment is made only in weights W_1, W_2, and W_3 since only the units A_1, A_2, and A_3 provide a "1" output in case of the $\times$ pattern (See Fig. 9.53b). The weights are decreased since the machine provided a wrong "1" output. After the adjustment, the weights have the

---

[1] As we shall see later in more detail, the size or "content" of each category is determined by the number of units, their thresholds, and their interconnections.

values: $W_1=-1$; $W_2=-1$; $W_3=-1$; $W_4=0$; $W_5=0$. We now continue to show $+$ and $\times$ patterns until the weights are adjusted so that the machine responds correctly, that is, until it has learned the correct response. The resulting sequence of weight changes is reflected in Fig. 9.54. Shaded entries indicate that a weight is associated with an A-unit providing a "1" output. The shaded areas indicate, therefore, which weights are summed in the R-unit, but they also indicate which weights are subject to change in case of a wrong response. Arrows indicate that a weight is actually changed.

Pattern	$+$	$\times$	$+$	$\times$	$+$
W_1	0	0 ← -1	-1	-1	-1
W_2	0	0 ← -1	-1	-1	-1
W_3	0	0 ← -1	0	0	0
W_4	0	0	0 ← +1	+1	+1
Weighted sum	0	0	-1	-2	+1
Response	1	1	0	0	1
Desired response	1	0	1	0	1
Adjustment	none	de-crease	in-crease	none	none

Fig. 9.54. Teaching of "1" Response to $+$ Pattern

The weights assume their final values after only three trials. The machine has learned the desired response, and no further adjustment or teaching is required.

So far, the behavior of the circuit may still seem rather trivial. However, the arrangement has several remarkable properties:

The circuit has not been designed to recognize a specific pattern, it has been *trained* to do so after it was constructed. If it amuses us, we can re-train the machine to recognize other patterns. Fig. 9.55 shows, for example, a sequence which teaches the machine to respond with a "1" output to the $\times$ pattern, and with a "0" to the $+$ pattern. The sequence starts with weights having the final values shown in Fig. 9.54. In effect, we make the machine forget what it has learned previously, and teach it the most contrasting new response.

Of course, we may also train the machine to recognize other patterns

Problem 21: Teach the sample machine to distinguish the "letter" ☐ from the letter └. Assume that all initial weights are zero. Show the consecutive adjustment of weights in a table similar to the one in Fig. 9.54.

Pattern	$\times$	$+$	$\times$	$+$	$\times$	$+$	$\times$
W_1	$-1 \to 0$	0	0	0	$0 \to +1$	$+1$	
W_2	$-1 \to 0$	0	0	0	$0 \to +1$	$+1$	
W_3	$0 \to +1 \to 0$		$0 \to -1 \to 0$		0		
W_4	$+1$	$+1 \to 0$	$0 \to -1$		-1		-1
Weighted sum	-2	$+2$	0	0	-1	-1	$+2$
Response	0	1	1	1	0	0	1
Desired response	1	0	1	0	1	0	1
Adjustment	increase	decrease	none	decrease	increase	none	none

Fig. 9.55. Training of "1" Response to $\times$ Pattern

Remarkable also is the insensitivity of the machine against internal failures. The sample circuit will, for instance, be able to learn distinguishing between the $+$ and $\times$ pattern, even if any one of the following conditions exist:

a) a single S-unit fails, providing a constant "0" or "1" output;

b) a single connection between S and A-units is removed or added;

c) a single A-unit fails, providing a constant "0" or "1" output;

d) the threshold of any or all A-units increases or decreases by one;

e) a single "W" unit fails to change weights;

f) the threshold of the R-unit changes.

The circuit will not require re-training for several of these conditions, and it can be re-trained for a large number of specific combinations of indicated failures.

The machine is also able to recognize correctly a number of deformed or "noisy" input patterns. For example, the circuit with weights adjusted to the final values shown in Fig. 9.54 happens to classify correctly the input patterns given in Fig. 9.52 b and 9.52 c, even if any single input field is altered from bright to dark, or vice versa. In general however, training with noisy input patterns will be required for their correct classification.

To a limited extent, our primitive machine is also able to recognize patterns which are "off-registration". For instance, the circuit is able to distinguish the vertical bars shown in Fig. 9.56 a and b from the horizontal bars shown in Fig. 9.56 c and d.

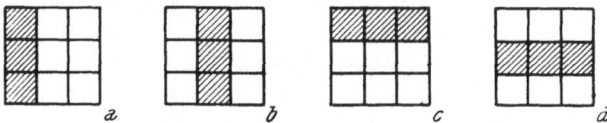

Fig. 9.56. Samples of Off-Registration Patterns

However, the circuit will not be able to distinguish without errors between vertical and horizontal bars in all three possible registrations.

Problem 22 (Voluntary): a) Teach the sample machine to distinguish the patterns in Fig. 9.56 a and b from the patterns shown in Fig. 9.56 c and d. Show the consecutive weights W_1 through W_4 while applying a sequence of all four input patterns.

b) Why is the sample machine unable to learn distinguishing between vertical and horizontal bars in all three possible registrations?

Problem 23 (Voluntary): What is the minimum error rate which the sample perceptron could achieve in distinguishing vertical and horizontal bars in all three registrations?

a) Assume that all registrations occur equally frequent.

b) Assume that in 80% of all cases, the bars are in central registration, and that the two possible off-registrations of a bar are equally likely.

9.9.2. Other Learning Structures

At this point it is well to remember that, up to now, we have discussed only one specific example of a perceptron, although the indicated example should have suggested several variations in design, operation, or application. In fact, the greatest value of the perceptron may well prove to be the wealth of new ideas which it has stimulated. Realizing that the detailed study of many of these ideas continues, and also that the discussion of learning structures is only incidental to the prime purpose of this book, let us omit the detailed description of other known specific structures. Instead, we shall simply indicate a number of possible directions for departure from the previously discussed sample perceptron, including those directions leading to structures already well established in their own right.

The indicated circuit may, obviously, be equipped with different numbers of S, A, or R-units. Having more S, or A-units, the circuit will be able to make finer distinctions between various input patterns. Having more than one R-unit, the circuit can learn to recognize more than two classes of input patterns. Each single R-unit may then respond to a specific class, or all the R-units together may provide a "coded" response to each class. The addition of a delay line or of some other information storage allows the circuit to accept sequential rather than spatial input patterns.

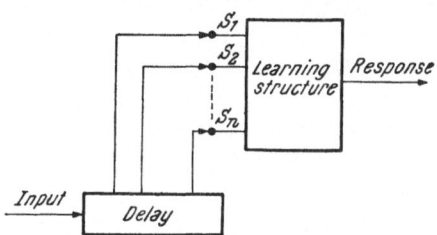

Fig. 9.57. Accomodation of Sequential Input Patterns

Of course, patterns consisting of a mixture of sequential and spatial signals can also be accommodated.

The connections between S and A-units do not have to be random as they are in the sample circuit. In fact, non-random connections may be preferable when the exact nature of input patterns is known, before a circuit is designed. On the other hand, there are several different possible approaches to "randomness": Each A-unit may be connected to the same number of S-units (although not the same S-units) in a random manner; the wiring may include multiple connections between single S and A-units, giving an S-unit more weight as far as the A-unit is concerned; and the randomness may include connections with reversed polarities, so that a certain S-unit may tend to inhibit an A-unit rather than stimulate it.

A-units may be cross-coupled, that is the output of one A-unit may serve as one of the inputs to another A-unit. In models postulating "instantaneous" responses, this may serve to increase the "contrast" of patterns; in models postulating finite propagation times, cross-coupling may be used to correlate a pattern with a previously shown pattern, or the same pattern in a previous position.

The threshold of R-units may be different from zero. In fact, most "neuron models" have a non-zero threshold. Some arrangements even contain adjustable thresholds. Their adjustment is governed by rules similar to the ones for the adjustment of weights. Fig. 9.58 shows a possible implementation.

The R-unit is not only connected to A-units, but also to a constant "1" source. The latter connection has the adjustable weight W_T. All weights are adjusted identically. An increase in the weight W_T biases the R-unit such that it provides a "1" output for a smaller weighted sum of the remaining

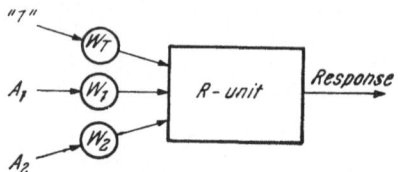

Fig. 9.58. R-Unit With Adjustable Threshold

inputs. It, therefore, decreases the effective threshold. Conversely, a decrease in the weight W_T increases the effective threshold.

Many different rules for the adjusting of weights during the learning period are conceivable. Frequently proposed or implemented is a scheme in which the weights are increased or decreased during a single trial until the desired response is achieved .The adjustment may be just large enough to stimulate the correct response, or it may be such that the weighted sum has the same magnitude as before the trial, but the opposite polarity. In another scheme, the change in a weight is proportional to the weight itself. The purpose of this approach is to change the weight of that input by the largest amount which contributes most to the erroneous response.

Many proposals are concerned with the use of analog rather than digital weights. Although this approach should not result in any basically different characteristic, it may be much easier to implement. One particular implementation employs RC-circuits to store the weights. Their value decays exponentially with time, so that the machine gradually forgets what it has learned unless some learning takes place continuously. This feature may be desired in an application where the input pattern varies with time in "shape" or "amplitude", or both[1].

Finally, there exist concepts in which an adjustment of weights takes place not only after a wrong response, but—to a limited extent—also after correct responses. One might compare this to a learning process in which the pupil is reinforced in his understanding by favorable reactions of the teacher (or rewards), and not only motivated to change his approach by adverse comments (or punishment).

Some learning structures have adjustable weights not between A and R-units, but between S and A-units. Others employ weighted connections

---

[1] Such a circuit might, for instance, have learned to recognize the Sonar "signature" of a vessel. It continues to recognize the signature, even if it varies with the distance and the maneuvering of the ship.

between all units. Under such circumstances, it is difficult to make generally applicable distinctions between sensory, associative, and response units, and it becomes preferable to consider the single threshold unit as the basic building block, and the overall structure as a network of such blocks.

The characteristics of individual threshold units may vary widely. Fig. 9.59 is meant to represent the most general case.

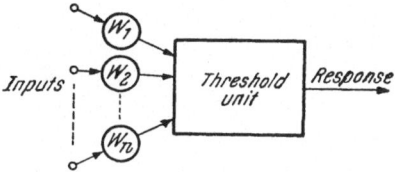

Fig. 9.59. Representative Diagram of a Threshold Unit

In some designs, the weights are fixed. In others, they are variable. Weights may be represented by digital or by analog values. There may be digital, analog, or fixed inputs. Outputs may be digital, or analog in nature. Finally, input values may be summed linearly, or combined non-linearly.

A feeling for the capabilities and the limitations of the various designs is conveyed by "transfer functions" which describe the output as a function of the input variables. For this purpose, the n inputs of a threshold unit are usually treated as a single n-dimensional vector, while a plot or graph is used to indicate the output state for the different regions of the n-dimensional "input space". Let us demonstrate this approach by a few, relatively simple examples.

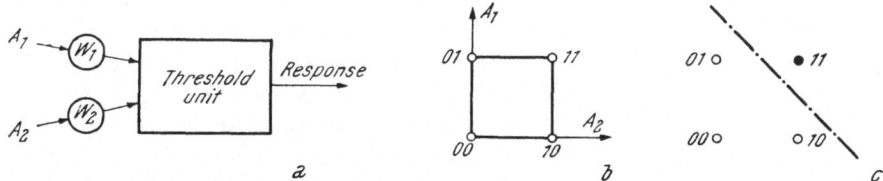

Fig. 9.60. A Threshold Unit, Its Input Space, and Transfer Function

Fig. 9.60a is meant to represent a threshold unit with two binary inputs, linear summation, adjustable threshold, adjustable weights, and one binary output. The two-dimensional vertex frame in Fig. 9.60b shows the input space represented by the four possible combinations of input states for the two binary input variables. Fig. 9.60c indicates the output function for one possible adjustment of threshold and weights. In this particular diagram, a "1" output is generated only for the input state "11". In general,

the input space is separated into two regions, one with "1" output, and another with "0" output. For units with linear summation, the regions are separated by a straight line as in Fig. 9.60 c. The line can be moved and turned into any desired position by the adjustment of threshold and weights. Fig. 9.61 a shows an example of another possible positioning. In contrast, Fig. 9.61 b shows a separation which is impossible to achieve with a linear threshold unit.

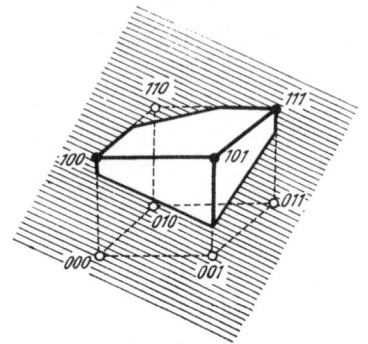

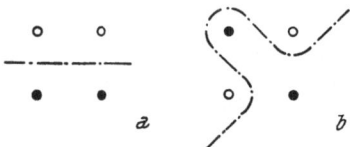

Fig. 9.61. Realizable and Non-Realizable Separation of the Input Space

Fig. 9.62. Separation of a Three-Dimensional Input Space by a Plane

In order to accomplish such a separation, it is necessary to combine inputs non-linearly, as for instance in a logic circuit, or to employ a network of several linear threshold units.

Problem 24 (Voluntary): Show the simplest network of linear, two-input threshold units which is able to learn any logic function of two binary input variables.

The transfer function of linear threshold units with more than two inputs can be shown similarly. The transfer function of a circuit with three inputs requires, for example, a three-dimensional vertex frame for its representation. The input space is divided by an adjustable plane into the two regions with "0" and "1" outputs respectively as shown in Fig. 9.62.

Transfer functions of units with more than three inputs can be represented by hypercubes[1]. The input space is separated by "hyperplanes". Although such representations have only a limited practical value, they convey an intuitive understanding of the capabilities and limitations of the circuits which they represent.

Transfer functions of threshold circuits with analog inputs are similar to those of units with binary inputs. The only essential difference is that the input space is continuous rather than discrete. Fig. 9.63 shows, as an example, the transfer function of a circuit with two inputs.

---

[1] See paragraph 3.4.1.

The input space is divided by a straight line into two regions with "0" and "1" output respectively. Again, the separation line can be positioned as desired. Transfer functions of circuits with three or more inputs require for their representation three or more-dimensional "graphs". The input space is then divided by planes or hyperplanes.

Let us now turn to networks containing several threshold units. The network, no matter how complex it is, performs a "compartmentation" of the input space. Individual compartments are bound by the lines, planes, or more generally, hyperplanes the position of which is determined by the adjustment of the individual threshold units. Specific outputs are assigned to each compartment. An adjustment of weights during the learning period corresponds to a movement of hyperplanes so that an acceptable compartmentation of the input space results. The number of threshold units in the net, their design, and also the manner in which they are interconnected, places certain restrictions on the number of compartments and on the "mobility" of compartments. With a sufficiently large number of threshold units and proper interconnection, the input space can be separated in as many compartments as desired. In general however, the number of compartments will be limited. Compartments, will be adjusted so that the typical patterns shown during the learning period are properly classified.

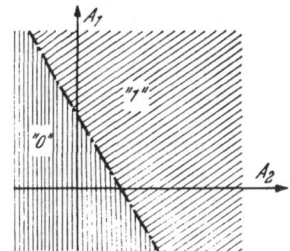

Fig. 9.63. Transfer Function of a Linear Threshold Circuit with Two Analog Inputs, A_1 and A_2

Many non-typical patterns may fall arbitrarily in one or the other compartment. The more similar an input pattern is to a typical pattern, the better is the chance that it will be classified identically with the typical pattern.

The foregoing remarks touched only lightly on the existing, and in many respects massive, theoretical foundation for the concept of learning. In spite of this foundation and the success of many existing adaptive special purpose devices, there is not yet a clear answer to the problem of applying the concept of learning to computer organizations. Intuitively, of course, the approach is promising: the human brain, which served as inspiration for many learning structures, is, after all, an excellent computer in many respects.

Problem 25: A speculative approach to the implementation of computers by learning structures might be to have threshold units "learn" the functions normally performed by individual logic circuits. Try to find a simple network of threshold units which is able to learn both, the function of a full binary adder, and of a full subtracter.

References

1. CANADAY R. H.: Two-Dimensional Iterative Logic, AFIPS Conference Proceedings, vol. 27, part 1, pp. 343—353. Fall Joint Computer Conference, 1965.
2. MINNIK R. C.: Cutpoint Cellular Logic, Transactions IEEE, vol. EC-13, No. 6, pp. 685—698. Dec. 1963.
3. MINNIK R. C.: Cobweb Cellular Arrays, AFIPS Conference Proceedings, vol. 27, part 1, pp. 327—341. Fall Joint Computer Conference, 1965.
4. SHORT R. A.: Two-Rail Cellular Cascades, AFIPS Conference Proceedings, vol. 27, part 1, pp. 355—369. Fall Joint Computer Conference, 1965.
5. HOLLAND J. H.: An Iterative Circuit Computer, etc., Proceedings Western Joint Computer Conference, pp. 259—265. San Francisco. May 1960.
6. COMFORT W. T.: A Modified Holland Machine, Proceedings Fall Joint Computer Conference, 1963.
7. GARNER, and SQUIRE: Iterative Circuit Computers, Workshop on Computer Organization. Washington: Spartan Books. 1963.
8. GONZALEZ R.: A Multilayer Iterative Computer, Transactions IEEE, vol. EC-12, No. 5, pp. 781—790. Dec. 1963.
9. SLOTNICK, BORK, and McREYNOLDS: The SOLOMON Computer, Proceedings Fall Joint Computer Conference, Philadelphia. Dec. 1962.
10. KNAPP, and TURNER: A Centrally Controlled Iterative Array Computer, Technical Documentary Report, Rome Air Development Center, RADC-TDR-64-251. July 1964.
11. SENZIG, and SMITH: Computer Organization for Array Processing, AFIPS Conference Proceedings, vol. 27, part 1, pp. 117—128. Fall Joint Computer Conference, 1965.
12. HAWKINS, and MUNSEY: A Parallel Computer Organization and Mechanization, Transactions IEEE, vol. EC-12, No. 3, pp. 251—262. June 1963.
13. McCORMICK: The Illinois Pattern Recognition Computer, Transactions IEEE, vol. EC-12, No. 5, pp. 791—813. Dec. 1963.

Selected Bibliography

Polymorphic Computers

PORTER R. E.: The RW 400 — A New Polymorphic Data System, Datamation, pp. 8—14. Jan./Feb. 1960.
ESTRIN, BUSSEL, TURN, and BIBB: Parallel Processing in a Restructurable Computer System, Transactions IEEE, vol. EC-12, No. 5, pp. 747—755. Dec. 1963.

Hybrid Computation

BURNS M. C.: High-Speed Hybrid Computer, National Symposium on Telemetering, San Francisco. 1959.
CONNELLY M. E.: Real-Time Analog Digital Computation, Transactions IRE, vol. EC-11, No. 1, p. 31. Feb. 1962.
HAGAN T. G.: Ambilog Computers: Hybrid Machines for Measurement System Calculation Tasks, 17th Annual Conference, Instrument Society of America, New York. Oct. 1962.
SCHMID H.: An Operational Hybrid Computing System Provides Analog-Type Computations with Digital Elements, Transactions IEEE, vol. EC-12, No. 5, pp. 715—732. Dec. 1963.
TRUITT T. D.: Hybrid Computation, IEEE Spectrum, pp. 132—146. June 1964.

RIORDAN, and MORTON: The Use of Analog Techniques in Binary Arithmetic Units, Transactions IRE, vol. EC-14, No. 1, pp. 29—35. Feb. 1965.

Digital Differential Analyzers

FORBES G. F.: Digital Differential Analyzers, 4th Edition, Private Print, 1957 (13745 Eldridge Ave, Sylmar, Cal.).

GSCHWIND H. W.: Digital Differential Analyzers, in Electronic Computers, ed. by P. VON HANDEL. Vienna: Springer. Englewood Cliffs: Prentice Hall. 1961.

SHILEIKO A. U.: Digital Differential Analyzers. New York: McMillan. 1964.

Micro-Programmed Computers

WILKES, and STRINGER: Microprogramming and the Design of Control Circuits in Electronic Digital Computers, Proceedings Cambridge Phil. Soc., vol. 49, part 2, pp. 230—238. April 1953.

MERCER R. J.: Micro-Programming, Journal ACM, vol. 4, pp. 157—171. Apr. 1957.

BLANKENBAKER J. V.: Logically Micro-Programmed Computers, Transactions IRE, vol. EC-7, pp. 103—109. June 1958.

WILKES M. V.: Micro-Programming, Proceedings Eastern Joint Computer Conference, pp. 18—20. Philadelphia. Dec. 1958.

KAMPE T. W.: The Design of a General-Purpose Microprogram Controlled Computer with Elementary Structure, Transactions IRE, vol. EC-9, pp. 208—213. June 1960.

GRASSELLI A.: The Design of Program-Modifiable Micro-Programmed Control Units, Transactions IRE, vol. EC-11, No. 3, pp. 336—339. June 1962.

GERACE G. B.: Microprogrammed Control for Computing Systems, Transactions IEEE, vol. EC-12, No. 5, pp. 733—747. Dec. 1963.

BRILEY R. E.: Pico-Programming: A New Approach to Computer Control, AFIPS Conference Proceedings, vol. 27, part 1, pp. 93—98. Fall Joint Computer Conference, 1965.

Machines with Cellular Organization

UNGER S. H.: A Computer Oriented Toward Spatial Problems, Proceedings IRE, vol. 46, No. 4, p. 1744. Oct. 1958.

HOLLAND J. H.: A Universal Computer Capable of Executing an Arbitrary Number of Subprograms Simultaneously. Proceedings Eastern Joint Computer Conference, Boston. Dec. 1959.

STEWART R. M.: Notes on the Structure of Logic Nets, Proceedings IRE, vol. 49, No. 8, pp. 1322—1323. Aug. 1961.

COMFORT W. T.: Highly Parallel Machines, Workshop on Computer Organization. Washington: Spartan Books. 1963.

GREGORY, and REYNOLDS: The SOLOMON Computer, Transactions IEEE, vol. EC-12, No. 5, pp. 774—775. Dec. 1963.

Multiple Processing Techniques, Technical Documentary Report, Rome Air Development Center, RADC-TDR-64-186. June 1964.

List Processors

NEWELL, SHAW, and SIMON: Empirical Explorations of the Logic Theory Machine; A Case History in Heuristics, Proceedings. Western Joint Computer Conference, pp. 218—230. Feb. 1957.

GREEN B. F.: Computer Languages for Symbol Manipulation, Transactions IRE, vol. EC-10, No. 4, pp. 729—735. Dec. 1961.

MUTH, and SCIDMORE: A Memory Organization for an Elementary List Processing Computer, Transactions IEEE, vol. EC-12, No. 3, pp. 262—265. June 1963.

PRYWES, and LITWIN: The Multi-List Central Processor, Proceedings of the 1962 Workshop on Computer Organization, pp. 218—230. Washington: Spartan Books. 1963.

WIGINGTON R. L.: A Machine Organization for a General Purpose List Processor, Transactions IEEE, vol. EC-12, No. 5, pp. 707—714. Dec. 1963.

Associative Computers

ROSIN R. F.: An Organization of an Associative Cryogenic Computer, Proceedings Spring Joint Computer Conference, pp. 203—212. May 1962.

DAVIES P. M.: Design for an Associative Computer, Proceedings Pacific Computer Conference, pp. 109—117. March 1963.

ESTRIN, and FULER: Algorithms for Content Addressable Memories, Proceedings Pacific Computer Conference, pp. 118—130. March 1963.

EWING, and DAVIES: An Associative Processor, Proceedings Fall Joint Computer Conference, pp. 147—158. Nov. 1964.

FULLER, and BIRD: An Associative Parallel Processor with Application to Picture Processing, Proceedings Fall Joint Computer Conference, pp. 105—116. 1965.

Learning Structures

ROSENBLATT F.: Perceptron Simulation Experiments, Proceedings IRE, vol. 48, No. 3, pp. 301—309. March 1960.

MINSKY M.: Steps Toward Artificial Intelligence, Proceedings IRE, vol. 49, No. 1, pp. 8—30. Jan. 1961.

HAWKINS J. K.: Self-Organizing Systems — A Review and Commentary, proceedings IRE, vol. 49, No. 1, pp. 31—48. Jan. 1961.

COLEMAN P. D.: On Self-Organizing Systems, Proceedings IRE, vol. 49, No. 8, pp. 1317—1318. Aug. 1961.

HIGHLEYMAN W. H.: Linear Decision Functions, with Application to Pattern Recognition, Proceedings IRE, vol. 50, No. 6, pp. 1501—1514. June 1962.

SIMMONS, and SIMMONS: The Simulation of Cognitive Processes, An Annotated Bibliography, Transactions IRE, vol. EC-11, No. 4, pp. 535—552. Aug. 1962.

CRANE H. D.: Neuristor — A Novel Device and System Concept, Proceedings IRE, vol. 50, No. 10, pp. 2048—2060. Oct. 1962.

ROSENBLATT F.: Principles of Neurodynamics, Perceptrons, and the Theory of Brain Mechanisms. Washington: Spartan Books. 1962.

MATTSON, FIRSCHEIN, and FISCHLER: An Experimental Investigation of a Class of Pattern Recognition Synthesis Algorithms, Transactions IEEE, vol. EC-12, No. 3, pp. 300—306. June 1963.

KAZMIERCZAK, and STEINBUCH: Adaptive Systems in Pattern Recognition, Transactions IEEE, vol. EC-12, No. 5, pp. 822—835. Dec. 1963.

FUKUNAGA, and ITO: A Design Theory of Recognition Functions in Self-Organizing Systems, Transactions IEEE, vol. EC-14, No. 1, pp. 44—52. Feb. 1965.

10. Miscellaneous Engineering and Design Considerations

Both, the overall and the detailed design of digital computers are to a large extent trial and error processes. One frequently starts with a tentative design, fills in details and critically evaluates the result. Normally, it is then necessary to go back in order to modify and improve the original concept, or to remove inconsistencies. This iterative process is typical for any systems design and can become rather complex since many economic, operational, and engineering considerations enter into it. Since it is quite impossible to provide generally applicable rules or guidelines, we shall restrict ourselves to indications of the complexity of the problem, indications of the range of the design parameters, and a discussion of the interdependence of the main characteristics. Furthermore, we shall pose questions to which no generally applicable answers exist, but to which answers consistent with given specific circumstances have to be found. In many instances, by necessity, only an over-all trend is indicated, and it is left to the reader to see the implications for a particular case. In other instances, it becomes more valuable to consider specific circumstances. Here then, the reader has to generalize to cover other situations. In any event, it is intended that the discussion is applicable not only to the problems of the computer designer, but also to the closely related problems of the computer user who has to select a computer with the appropriate characteristics for his application.

There are mainly four general characteristics by which a computer system can be described: the capabilities to perform certain functions or operations; the speed with which these functions are performed; the cost of the system; and its reliability. The computer designer has some control over these characteristics. He may, therefore, consider them as design parameters. On the other hand, his control is limited in several respects. First of all, the parameters are not independent. For example, he may have to pay for an increase in reliability by a decrease in speed. This is usually referred to as "trade-off". There are regions which are of no practical interest (for example, no capabilities at no cost, or infinite capabilities at infinite cost), and there may be specific requirements for, say, a minimum of 100,000 multiplications per second. We might compare the design of a computer with a multi-variable mathematical problem in which we are searching for an optimum within certain boundaries or constraints. Unfortunately, an optimum in one respect is very seldom an optimum in another respect, and it will be necessary to find reasonable compromises. Mathematically speaking, we put certain "weights" on the various characteristics and search for a weighted optimum. The individual weights depend upon the particular situation (the maximum weight might be given to reliability in one situation, but to speed in another)[1]. In order not to confuse the picture unnecessarily, let us first consider the interdependence of only two characteristics assuming that all other parameters remain constant. In this manner, it is possible to avoid the consideration of weights which, in any case, have to depend upon the specific circumstances.

10.1. Capability versus Cost

To begin with, let us make the assumption that the implementation of each capability to perform an individual operation such as multiply, punch cards, read tapes, etc. has a certain fixed cost. Consequently, the more capabilities a design has, the higher will be the cost. The selection-problem for the computer user is then basically simple. He should select that computer which gives him exactly the required capabilities, and no more, since the incorporation of features which are not needed only makes a computer more expensive, without making it more useful. The problem of the computer designer is also basically simple: He should design the most economic computer, that is one which exactly fits the customers requirements.

In reality, the problems are, of course, more complex. The computer user may, to a certain extent, substitute one requirement for another (e.g. he may be satisfied with having no paper tape capability if he has adequate card handling capabilities). Alternately, features which are not really required may still have some value.

The computer designer faces an even more serious problem. If he provides a computer taylor-made to a specific customer's requirement, he, in a sense, has to design a special-purpose computer which meets the requirements of only one customer. He may be left with a potential market for only one machine. The development of this machine has to be paid by the single customer, and the actual cost will be unreasonably high. If he provides a design with universal or maximum capabilities in order to meet the possible requirements of all users, his computer will be so expensive that he has no potential market at all.

---

[1] The problems of the computer user are similar. He should select a computer according to criteria, weighted so as to reflect the requirement of his particular application.

The two indicated extremes are, obviously, not practical. The designer or manufacturer has to settle for a computer with enough capabilities to "cover" a reasonably large market area so that the cost of an individual machine will also be reasonable. In addition, he has two approaches to make the situation more desirable from his point of view. He can "cover the market" not with a single design, but with a number of individual designs. He can also make his designs "modular", that is, he can make provisions so that individual units like memory banks, tape units, etc. can be added or removed from the system, or so that high-speed or low-speed arithmetic and control units can be used optionally in an otherwise identical system. In this manner, it is possible to provide a relatively good match to a wide variety of user- requirements at a reasonable cost. The flexibility of the modular design has the additional advantage that a computer system can be expanded or reduced so that the user can retain an economic system even with changing requirements. This property alone should normally have a considerable value to the user.

There is one further aspect of the problem which should be mentioned. Starting with a hypothetical design, the cost for the addition of a certain capability is in many instances comparatively small when a part of the existing hardware can be used. In any event, the comparison of an increase in capabilities versus the corresponding increase in cost (or the decrease in cost which can be achieved by a decrease in capabilities) is frequently very indicative of the correctness or desirability of a specific solution. We may compare this to the use of differentials in a mathematical problem in order to find a minimum or maximum.

So far, we have considered the problem solely from the hardware-standpoint. But we observe that the stored program computer, in itself, offers a large degree of flexibility. The program adapts the computer to a particular requirement. It covers, by itself, a number of individual applications. From a systems-standpoint, the "software" capabilities are as important as the hardware capabilities. The two extremes, i.e. to provide no software or universal software, are impractical. On the other hand, the computer manufacturer should avoid providing a software package with his computer which is oriented too much towards a single customers requirement for the same reasons for which he should not design a special-purpose computer for a single customer.

Problem 1: How important or unimportant could the capability to handle paper tape become to a computer user ?

Problem 2: Should a computer designer incorporate a square root instruction in his design, (hardware), or should the computer manufacturer deliver a square root routine (software) with the system? Upon what factors should the decision depend? Does the size of the contemplated computer system have any influence upon the decision?

10.2. Speed versus Cost

The speed of a computer is to a large extent determined by the speed of its components. Of course, one can build faster and slower computers with the same components simply by changing the organization of the computer (e.g. parallel or serial computers). But, assuming identical organizations and the state of the art at a given time, there is a dependence of cost and speed which is qualitatively (not quantitatively) reflected in Fig. 10.1.

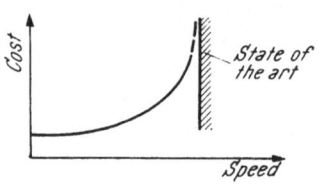

Fig. 10.1. Cost versus Speed

The graph is indicative of a situation where one tries to increase the speed of a computer by using faster and faster components. The limiting value in speed is then the state of the art. When we approach this limit, the cost of the computer becomes unreasonably high.

The general characteristic of the curve can be verified by a number of specific, but diversified examples. For instance, Fig. 10.2 may be representative of the cost per bit of computer storages using various components.

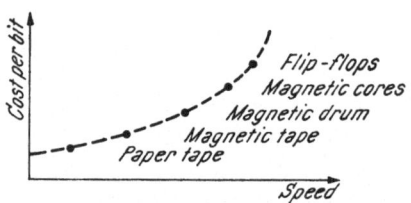

Fig. 10.2. Cost per Bit for Various Storage Media

Fig. 10.3 indicates a particular approach to high switching speeds. Suppose a transistor inverter or logic element can drive the inputs of seven other elements at slow speeds, say 1 Mc. However, the element can drive only four other elements at 7 Mc, three elements at 8 Mc, two elements at 9 Mc and only one other element at 10 Mc. If a total of seven "loads" has to be driven from a particular output, a cascading of elements is required as it is indicated in Fig. 10.3a. The total number of required elements (and the cost) increases then with speed as indicated in Fig. 10.3b. It is to be noted that it is not possible to build a computer using these elements which has a speed of 10 Mc, no matter how many elements we are willing to use.[1]

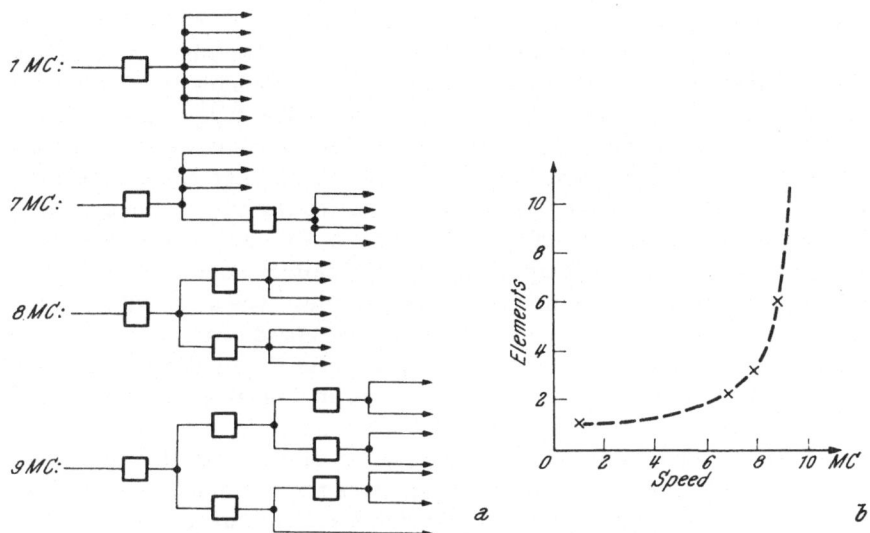

Fig. 10.3. Cost for High Speeds by Cascading

Let us assume for the moment that the curve in Fig. 10.1 is typical of the cost of a computer versus speed. Let us further assume that computer users are willing to pay an amount which is proportional to the speed. We have then the situation shown in Fig. 10.4.

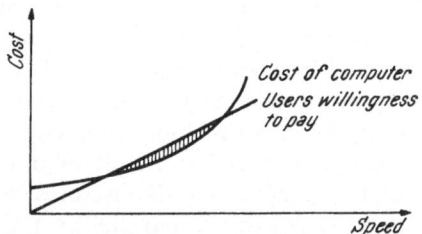

Fig. 10.4. Designer's and User's Aspects

We see that there is only a certain area in which the user is willing to pay the cost or more than the cost of the computer. The designer has to make sure that his machine falls within this range.

---

[1] Moreover, delays through the additional stages further complicate the design problem.

Again, the real situation is more complicated: The curves change with time. Advances in the state of the art will shift the curve representing the cost of the computer to the right, but competition between computer manufacturers will lower the line representing the users willingness to pay. But even if we disregard variations with time, the picture is over-simplified. The customer may not be willing to pay twice as much for twice the speed (or 100 times as much for 100 times the speed). The following simple consideration will show this. If a customer has the choice between a computer with a certain speed and two computers which each have half of the speed, but cost only half as much, he is likely to select the two slower machines. He has then the possibility of installing them in different or in the same location, he obtains some back-up in the event of failures, and the flexibility of independent operations at no extra cost. In effect, the users willingness to pay may increase less than linearly with the speed. In this respect, the situation for the computer manufacturer is worse than indicated. The area in which he can make a profit (shaded area in Fig. 10.4)is narrowed. In extreme cases, it may shrink to a point or be non-existent. However, there are other considerations which are in favor of the manufacturer. Even though he has very little or no influence over the customers willingness to pay, he can manipulate the cost curve for his design in several ways.

As we have mentioned before, the manufacturer may advance the state of the art. He then pursues research which may enable him eventually to lower the cost curve. This approach is, of course, a long-range enterprise, but may be necessary in order to remain competitive.

A more immediate control over the situation can be exercised by variations in the organization of computers. Let us remember that the previously discussed cost curve is only applicable to an increase in speed by the use of faster components, but otherwise identical organizations. Fortunately, an increase in speed can frequently be achieved much more economically by a different organization than by faster components. Let us take an example. Suppose we compare a serial and a parallel machine with equally fast components, each having the same memory capacity and the same word-length. The serial machine operates on one bit of information at a time, while the parallel machine operates on all bits of a word, say 36, simultaneously. The increase in speed of the parallel over the serial machine is then roughly in the order of 36 times. The increase in cost should be much less. True, certain components in the computer will be required 36 times rather than only once (e.g. the transmission lines, or the number of flip-flops in a working register, or the number of individual binary adders), but a very large part of the machine does not have to be implemented 36 times. For instance, still only one control unit is required. The control unit may even be simpler since it is no longer necessary to schedule operations on single bits, but only operations on full words. Also, the number of individual storage elements in the memory remains the same and so does probably the number of circuits for addressing. The argument could be extended to most any parallel operation in a computer system. Where an increase in speed by faster components requires a more than linear increase in cost, an increase in speed by parallel operations usually can be achieved more economically.

A third and rather intrigueing aspect of the problem offers itself when one considers intermixing high-speed and low-speed components or techniques in a single design. If it is possible to use high-speed components in certain critical areas, but more standard components in others, one may achieve high "effective speeds" rather economically. This approach has a far wider application than one would expect at the first glance. A fairly common example is the storage "hierachy" of a computer. Fast and expensive flip-flops are used as storage elements only in the working registers of the machine. Slower and certainly less expensive magnetic cores are used for the moderate sized internal memory. Larger amounts of information are stored in still slower and less expensive drum, disk, tape, or card storages. Here, the considerations are rather obvious. A computer which has only flip-flops as storage elements could be extremely fast as far as the access to information under all possible circumstances is concerned, but it would be prohibitively expensive. On the other hand, a computer having only punched cards as a storage medium might be comparatively inexpensive, but too slow to be attractive (assuming that it is at all possible to design such a machine). In effect, the memory hierarchy, constitutes an "optimum" configuration under the assumption that in the average problem frequent accesses are required to only a fraction of the total information, but accesses to other parts are (or can be scheduled) infrequently. An increase in access speed for the small portion is of course expensive, but increases the "effective speed" considerably. A decrease in access time for the remainder of the stored information would only increase the cost without much affecting the overall effective speed.

Analogous considerations can be applied to many other details of the computer organization. Let us take the arithmetic unit as an example. As we have seen previously, there exist certain schemes which allow a high speed execution of arithmetic instructions. The implementation of these schemes is fairly expensive, but they may be worth the cost, if they increase the effective speed of a computer sufficiently.

Problem 3: Consider the following hypothetical case. A particular computer spends 20% of the total running time with the execution of add (or subtract) instructions during the execution of a "typical program". It spends 4% of the time executing multiply instructions and 1% of the time in executing divide instructions. Suppose that it is possible to employ high-speed techniques which cut the execution time of add, multiply and divide instructions in half. The price for the implementation of each of these high-speed techniques amounts to approximately 1% of the computer cost.

a) Is it advantageous to incorporate high-speed techniques for add instructions, multiply instructions, divide instructions? Substantiate your opinion.

b) What additional information should you have in order to make a sensible decision?

Problem 4: How valuable can it become to a specific customer if the card handling speed in an otherwise identical computer system is doubled? How unimportant can this be to another customer?

10.3. Reliability versus Cost

Reliability is not simply another aspect of computers, but an important design parameter which can and has to be closely controlled. In some instances, the reliability of a computer may be the paramount design criterion (as in some military installations), but even for more standard applications the achievement of sufficient reliability poses actual technical problems. Let us take a hypothetical illustrative example. A computer contains 3000 vacuum tubes with an average useful life expectancy of 3000 hours for a single tube. We see that, on the average, one tube fails per hour. If it takes anywhere near an hour to find and replace the faulty tube, the computer is practically useless. There have been computers with as many as 18,000 vacuum tubes, and we may conclude that it took a great deal of courage and to build such machines.

The overall reliability of a system, such as a computer, decreases quickly with an increase in the number of individual units or components. If we assume that a computer system contains 20 units which each have a reliability of, say, 97% in a specified time interval, i.e. a 3% chance of failing during this interval, the overall reliability of the system in this time interval is only approximately 54%.

Problem 5: What overall reliability has a computer with 10^3 elements, each having 99,99% reliability? With 10^4 elements? With 10^5 elements?

Now that we realize that the reliability of computers can be a problem, the question of how we can control the reliability arises.

The most obvious approach is, perhaps, the selection and use of more reliable components for the construction of the computer. Here then, we attempt to control the overall reliability by controlling the reliability of the individual components. There is, however, a penalty. More reliable components are apt to be more expensive. Fig. 10.5 may be representative of the cost of components versus their reliability.

The cost increases very rapidly when high reliabilities are approached. A component with 99,999% reliability may cost a multiple of the amount for an equivalent component with 99% reliability, while, a component with 80% reliability may not be much more expensive than a component with 10% reliability (if this range of component reliabilities would be of practical interest at all). The cost approaches an infinite value for a 100% reliable component. In other words, a 100% reliable component cannot be built. It is interesting to note that the curve in Fig. 10.5 is very similar to the one in Fig. 10.1. Similar considerations apply, therefore, to the achievement

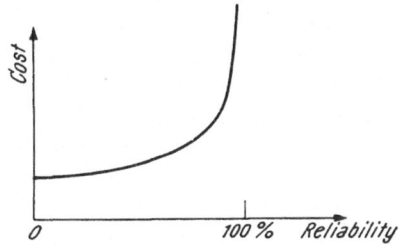

Fig. 10.5. Component Cost versus Reliability

of high reliabilities by extremely reliable components as to the achievement of high speeds by extreme fast components. Unless reliability is of utmost importance, as for the computer in a space probe which must be designed for long trouble-free service, the designer is better off using more conventional components, and achieving the required reliability by other approaches. In this connection it is worthwhile to note that the reliability of components may depend very much upon the manner in which they are used. As an example, the useful life of a vacuum tube, and therefore its reliability, may be extended considerably if the tube is operated well below maximum or nominal values. Also, circuits may be designed in such a manner that they still work satisfactorily even if component characteristics degrade. In this manner, circuit design can be used to increase the component reliability. Furthermore, the component reliability may depend on environmental conditions like temperature, vibration, etc. By controlling these conditions one can also influence the component reliability.

All measures discussed so far, influence the intrinsic reliability of a design. However, computer users are usually interested in what we might call the effective reliability. This is already reflected in the manner in which computer users refer to reliability. Rather than expressing the reliability in percent for a certain time, computer users speak of good time, up-time, down-time or mean error free time. These terms imply that a computer failure does not render the computer useless, but that the trouble can be fixed. If we have two machines with the same intrinsic reliability, but troubles can be diagnosed and corrected faster in one of them, the computer user has an effectively more reliable machine (the ratio of up-time to down-time or of up-time to total time is larger). Undoubtedly, the speed with which troubles are diagnosed and corrected can depend to a large extent upon the computer maintenance technician, but the computer designer can aid this process in a number of ways and can thereby exercise some control over the effective reliability.

For one, the designer can organize the computer in a clear and orderly manner, so that troubleshooting will be simplified. He can also provide an adequate number of indicators and test points to shorten this process. Preventive maintenance procedures with the aim of correcting faulty situations before they affect the actual operation can be established. For instance, components subject to ageing in a computer can be periodically checked and components with degraded characteristics can be replaced before they cause the associated circuits to fail. Mechanical parts can be periodically cleaned and lubricated before they cause failures. All such preventive maintenance procedures require, of course, time in which the computer is not available for actual use. In a sense, they decrease, therefore, the computer up-time. However, in many applications such maintenance periods can be easily scheduled and do not diminish the usefulness of the computer. By detecting and correcting potential causes for failures, there will be less unexpected failures and, therefore, a more reliable operation during the time which is scheduled for actual operation. Although the ratio of usable time to total time is not improved, preventive maintenance procedures can definitely improve the sometimes more important ratio of usable time to scheduled time and, correspondingly, what one might call the "mean error free time between unexpected failures".

A further, and very important measure to decrease unexpected down time is marginal checking, that is the testing of the computer under worse conditions than those encountered during actual operation.

For this purpose, usually, the supply voltages are varied within limits. In some instances also, the shape of clock pulses or the pulse repetition frequency is varied. Weak components will then cause failures, even though they may still work satisfactorily under normal conditions. By replacing these components or, in general, by correcting "marginal" situations one obtains a rather good assurance that components will not drift out of their operational range during the period for which the computer is scheduled for operation. It is to be noted that even though marginal checks are very effective against failures caused by gradual changes in components they are no guard against sudden failures, such as shorts in a transistor or diode, or against breakage of the filament in a vacuum tube.

Problem 6: What marginal checks can you propose for the read-amplifiers in a magnetic tape unit (which are essentially analog devices) ?

Problem 7: Which of the following measures of reliability would you consider to be more meaningful for a comparison study:

1. mean error free time,

2. $\dfrac{\text{down time}}{\text{time during which system is turned on}}$,

3. $\dfrac{\text{up time}}{\text{scheduled time}}$.

Assume that you
   a) operate a leased computer in 8 hour shifts;
   b) own a computer and would like to operate it around the clock;
   c) are evaluating missile guidance computers.

Last not least, one can employ error detection and correction techniques to increase the effective reliability of computers. However, since one cannot very well separate their trade-off versus cost from their trade-off versus speed, it may be better to consider them in a separate paragraph.

10.4. Error Detection and Correction Techniques

10.4.1. Error Detection

The nature of many computer programs is such that errors caused by either, the computer hardware or by faulty programs may show up only after hours of operation or days of diligent analysis. Of course, considerable computer time can be wasted in this manner. Automatic error detection can decrease this time considerably. Detection methods with various degrees of sophistication are in existence, encompassing the range from very simple to extremely complicated procedures.

In general, there are two approaches to automatic error detection: one can detect errors by check circuitry, that is by additional hardware, or by checks in the computer program, that is by additional computer time. In any case, automatic error detection has a certain cost. A detection of 100% of all possible errors is hard to achieve or, at least, is possible only at prohibitive costs. The evaluation of the pay-off is not simple and, as a matter of fact, it cannot be determined without reference to the specific circumstances. Instead of discussing the solutions applicable to particular instances, let us show the spectrum of techniques from which a solution can be selected.

Software Checks: The idea of all software checks is basically simple. Normally, a computer has sufficient capabilities to check its results. The computer may, for instance, extract a square root following a relatively complicated procedure, subsequently square the result by a simple multiplication, and compare the product with the original operand. Of course, such tests have to be programmed just like the calculation itself.

An extreme in this respect might be the comparison of results derived by completely independent and different mathematical methods. Such checks could give an almost complete assurance of the fault-free operation of the computer and of the computer programs. However, the computer time required to check the result will probably be of the same order of magnitude as the computer time required to derive the result. The "effectiveness" of the computer would be decreased by a factor of approximately two. Moreover, the chance for an error (even though detected) would be increased simply because of increased running time. Even then, the procedure is not foolproof. For instance, errors in the computer output devices, say, the printer, are sometimes beyond program-checks.

One category of software checks deserves specific mentioning here. These are the computer test programs. They are written with the purpose of not only detecting errors, but also of pin-pointing the trouble. Frequently we refer to them as diagnostic routines. The aim of such routines is the check of every individual circuit in the computer under every possible operational condition. Errors are printed according to type and location. A memory test, for instance, will check every storage location. Failures may cause the computer to print the address of the associated storage location; the bit which failed; and whether a "1" was changed to a "0" or a "0" to a "1". A complete test, and especially the complete pin-pointing of errors, is a goal which one can approach, but never fully achieve in practice. Even so, test programs are essential for the computer maintenance. The maintenance technician is usually thoroughly familiar with the test programs, while he usually has no knowledge whatsoever about production programs. A failure of a production program tells him nothing, but the fact that there is a fault somewhere in the machine. (Even this he may doubt, as it is well possible that the production program itself contains an error). A failure in a test program gives the maintenance man at least an idea where to look for the trouble. Furthermore, test programs are usually written in such a manner that a particular test can be repeated over and over again, so that there is a chance to examine indicators or waveforms, and to find the error. For these reasons, a maintenance man will almost always use test programs rather than production programs for both, preventive and corrective maintenance procedures.

Hardware Checks: While the programmed checks can, at least theoretically, be used equally well to verify the soundness of the computer hardware and of computer programs, hardware checks lend themselves more naturally to the confirmation of the proper operation of the computer hardware. Where program checks require a redundancy in time, i.e., time which is not spent for actual computations, hardware checks require redundant circuitry, i.e., circuits which are not used for useful computations.

Again, there is a wide range of possible error detection circuits and techniques. Examples of rather simple error alarms are indicators for blown fuses, high or low voltage sensors, temperature alarms, etc. Fig. 10.6 shows as an illustration the schematic of a rather interesting circuit which at one time, has been used to detect the breakage of the filament in vacuum tubes.

Filament
Supply
Transformer

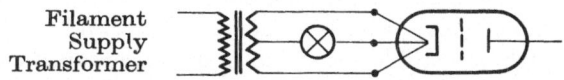

Fig. 10.6. Test Circuit and Indicator for Broken Filaments

When one side of the center-tapped filament breaks, the bridge-like circuit becomes unbalanced and the indicator light comes on. With one glance down the array of tube mounts, one can immediately detect a burned out tube.

All such circuits can be very helpful in the quick diagnosis of certain kinds of troubles, but none of them checks the correctness of the internal operation of the computer. A few examples of simple circuits of the latter kind might be alarm circuits which detect illegal instruction codes, illegal addresses, illegal number representations, arithmetic overflows, or wrong responses of peripheral equipments. Such circuits can be very straightforward and uncomplicated, and it is probably worthwhile to incorporate them into almost any design.

These alarms, together with the alertness of an experienced computer operator who can interpret indicators, movements of tape reels, noises of mechanical units like card readers, printers, etc., may give a very good indication of whether or not the computer goes through the right motions for the solution of a problem[1].

The techniques which we have discussed so far can, undoubtedly, detect a great number of computer malfunctions. Perhaps, one could even say that almost any malfunction will sooner or later show up in one of the detectors (e.g. a malfunction of the arithmetic adder may eventually result in an illegal address). However, if one desires a more complete and especially a more immediate assurance of the proper operation of a computer, one has to employ more elaborate techniques. Again, we have to state that it is impossible to achieve 100% assurance at finite costs (for instance the check-circuitry might fail, or the check-circuitry for the check-circuitry might fail).

---

[1] Some computer systems have an amplifier and loudspeaker connected to some strategic point of the circuitry (e.g. the memory access control, some high-order bits in the accumulator or the instruction decoder). Skilled operators are able to monitor the progress of a computation very closely, just by listening to this loudspeaker.

All we can really do, is to achieve a high degree of assurance or, in other words, state with a certain high probability that the computer is working properly. Unfortunately, the simple green light on the computer which indicates that everything is well, can be implemented only at an infinitely high cost.

Let us now discuss some extremes in error detection by hardware. An, at least theoretically feasible, approach is the duplication of equipment. Suppose we implement each logic function of the machine twice. When we now compare the outputs of both implementations and find agreement, we have a very high probability that both implementations work properly since it is extremely unlikely that both fail in exactly the same manner and at exactly the same time. Let us show this approach in the following diagram.

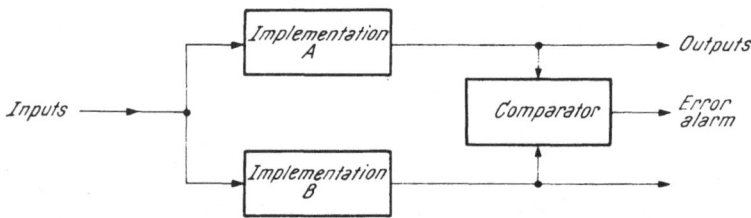

Fig. 10.7. Error Detection by Duplication of Circuitry and Comparison

The comparison might be performed for individual computer elements like flip-flops and AND- or OR-circuits, or for computer units like adders, counters, etc., or for entire computers. Each of these approaches has its relative advantages and disadvantages. The comparison of the outputs of two identical computers requires a relative small number of comparator circuits, but the comparison of all individual circuits can give an immediate indication of the location of the trouble.

It is interesting and revealing to compare the hardware duplication scheme with the previously discussed program-checks. The most complete assurance can be obtained in both cases by running a problem twice and by comparing the results. In one case we spend twice the necessary time plus the time for checking, in the other case, we spend twice the necessary hardware plus the additional hardware for the comparator. In their effectiveness, the methods can be considered equivalent.

Problem 8: Try to estimate the degree of assurance of fault-free operation one can achieve with the hardware duplication scheme. What are some of the parameters upon which the degree of assurance depends?

Incidentally, one may take the diagram shown in Fig. 10.7 as the starting point of some interesting theoretical speculations. The duplication of circuits may not necessarily require a doubling of the cost. Both circuits normally provide valid outputs so that the output load can be split and less powerful drivers can be used. One can also modify the scheme so that one circuit provides complementary outputs to the other. This is illustrated with a specific example in Fig. 10.8.

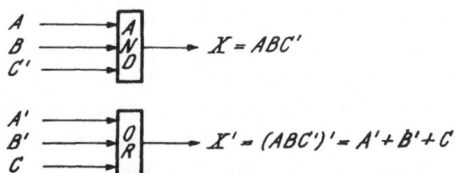

Fig. 10.8. Complementary Implementation

Simultaneously with each required logic function, we develop its logic complement. A failure in any one of the implementations will cause an equality of the outputs (rather than an inequality as indicated in Fig. 10.7). The scheme may have the advantage that the total implementation requires almost no logic inverters since the prime of any function is readily available.

All the error detection schemes which require a full duplication (or more than a duplication) of hardware are undoubtedly not very practical due to their high cost. They have been seriously considered only for computer installations for which the assurance of proper operation is the paramount design criterion, as for some isolated military applications. In all other cases, the computer designer has to look for compromises which give a reasonable assurance at a reasonable cost.

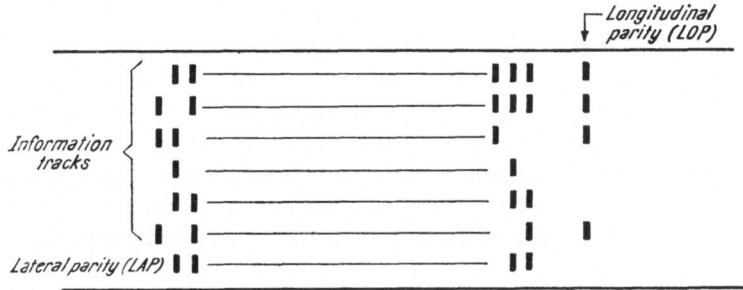

Fig. 10.9. Lateral and Longitudinal Parity Check

Two thoughts may help the designer to find such a compromise. First of all, an error detection is necessary only for those parts or units of a computer whose inherent reliability does not by itself give sufficient assurance of the proper operation. In other words, if one considers partial error detection, one should begin with the more unreliable parts of a machine. Secondly, it is not necessary to have an immediate detection of every error as long as there is sufficient probability that any faulty component will produce a detectable error in a sufficiently short time.

A frequently used technique which follows this line of thinking is the parity checking of magnetic tape recordings. Tape units are normally less reliable than other parts of a computer. This is caused partly by imperfections in the magnetic tape itself and partly by dust which interfers with the writing or reading process. In order to overcome this situation, redundant information is recorded which is subsequently used to verify the recording. Fig. 10.9 shows a commonly used recording format.

Six regular information tracks are recorded on the tape. The seventh track contains so-called lateral parity bits (LAP). The state of these parity bits is such that always an odd (in some cases always an even) number of "1's" is recorded in a line across the tape. When the tape is read back, the parity is checked, that is, a circuit determines whether or not an odd number of "1's" has been read in a line. An error alarm is given if an even number of "1's" is found. It is apparent that the misinterpretation of any single bit in a line (due to either a wrong recording or wrong reading) will cause such a lateral parity error. The lateral parity check constitutes a simple method to detect single errors in any line with 100% probability. We have to note, however, that there still is a finite probability for undetected multiple errors in a line. For instance, if two bits in a line become reversed, the parity check is unable to detect the error. Such an event is, of course, much more unlikely than a single error. Furthermore, if there are double errors, it is very likely that a recorded block of information contains also single errors which can be detected, so that the absence of any parity errors within a block of information assures the correctness of the information with a very high degree of probability.

To increase this probability further, sometimes a longitudinal parity check (LOP) as indicated in Fig. 10.9 is added. An inspection of the figure will convince us that both checks together will detect any double error within a block of information, that is, it will be impossible to change two bits of information within a block of information in such a manner that neither the lateral nor the longitudinal check detects a discrepancy.

Problem 9 (Voluntary): Assume that experience shows that in the average one out of 100,000 bits is mis-recorded and that one out of 100,000 bits is mis-read during the actual operation of a magnetic tape storage.

a) What are the chances for a single error in one line across the tape?

b) What are the chances for a double error in one line across the tape?

c) What are the chances for an error in a block of information containing 720 lines?

d) What are the chances for an undetected error in such a block of information, if only a lateral parity check is performed?

e) What are the chances for an undetected error in such a block of information if both, lateral and longitudinal parity checks are performed?

f) How many errors would you expect in a 1,000 ft reel of magnetic tape, recorded at a density of 200 lines per inch?

It is interesting to compare the parity checking technique with the possible scheme of recording all information twice. Let us show this in a diagram.

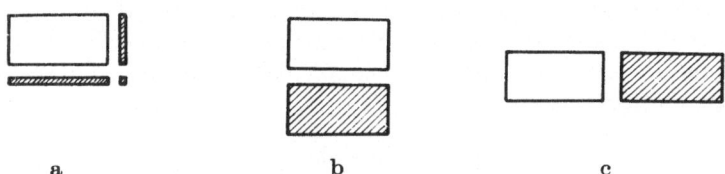

a   b   c

Fig. 10.10. Checks by Redundant Information
a) Parity Check
b) Parallel Recording
c) Sequential Recording

The shaded areas represent redundant information. Comparatively little redundant information is recorded with the parity checking technique shown in Fig. 10.10a. Both, the parallel and sequential recording, Figs. 10.10b and c, require an amount of redundant information equal to the useful information. In a sense, the scheme shown in Fig. 10.10b is equivalent to the previously discussed check by hardware duplication. Twice the number of read and write amplifier plus some check circuitry is required. The scheme shown in Fig. 10.10c is analogous to the previously discussed software checks. No increase in hardware is required, but an equal amount of time is spent working on useful and on redundant information, plus some time for the checking.

The parity check requires some increase in hardware, and some increase in time. We may, therefore, consider the technique as a compromise between the other two schemes. The compromise is a very good one. Neither the increase in hardware, nor the increase in required time is anywhere near the factor of two as in other schemes. Still the check gives a very good assurance. As a matter of fact, using both lateral and longitudinal checks, any single and any double error in a block of information can be detected, whereas neither parallel nor sequential recording detects double errors with a probability of 100%. It is noteworthy that the complete failure of a read or write amplifier is detected with a probability approaching 100% when one uses the parity check, while the sequential recording gives no such assurance.

Problem 10: Suppose a magnetic tape storage has no provisions for parity checks. Can you make the computer record and check a small amount of redundant information in such a manner that the validity of the information can be established ? What would be a simple way to derive and check such redundant information by computer program ?

The parity check is a very good example of how one can find a sound technical solution to a problem by avoiding extremes. It is entirely conceivable to apply this philosophy to the error detection in parts of a computer other than magnetic tape storages. The parity check itself is used in some existing computers to monitor the internal transmission of information[1] and the functioning of the computer memory. But there exist also some investigations of the use of redundant information for the checking of arithmetic circuits.

Problem 11 (Voluntary): Compare the probability for an undetected parity checking of information consisting of 6×720 bits when you use error in a block versus duplicated recording.

Problem 12 (Voluntary): Assume that a weak component causes an error rate as high as 1 bit in 100 bits in one channel of a magnetic tape storage. What are the chances of detecting such a malfunction within one block of information without error detection techniques; with longitudinal and lateral parity checks; with the duplicated recording of information ?

10.4.2. Error Correction

Where automatic error detection provides basically only an increased assurance of proper computer operation and, therefore, influences the reliability only indirectly, automatic error correction has a direct bearing on the reliability of a computer operation. Obviously, if failures are detected and automatically corrected, the effect is that of a more reliable machine.

Most approaches to error correction can be considered as extensions of error detection techniques. For one, error detection is a prerequisite for error correction (if errors can be detected only with a certain probability, errors can be corrected only with a limited probability). Furthermore, automatic error correction, similar to error detection, requires alternate paths or alternate operations to establish correct functions, in spite of failures in the hardware. The techniques are usually quite costly and similar cost considerations apply as for error detection schemes.

---

[1] Incidentally, the use of redundant information for validity checks is not restricted to computers but can and is applied to the transmission of information over communication links. This application has been studied in great detail.

Fig. 10.11 shows what one can consider as a brute force approach to error correction.

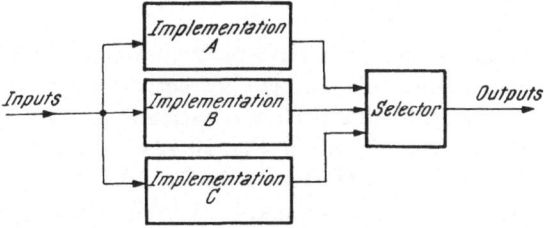

Fig. 10.11. Error Correction by Triplication of Hardware

Three separate circuits or components implement equivalent functions. A selector connected to their outputs compares the outputs, and selects one of them for transmission. Normally, all three outputs are identical. However, if there should be a failure in one of the implementations, the outputs disagree. One of the two agreeing outputs is selected, while the third (disagreeing) output is disregarded. The probability for a correct output is very high since it is extremely unlikely that two implementations fail in exactly the same manner at the same time.

The scheme is clearly an extension of the error detection technique indicated in Fig. 10.7. As such, it can be applied to individual components or to entire computers. Again, the first approach has the advantage of pinpointing faulty components, and the latter approach the advantage of a lower number of required selectors. In any case, the scheme is very costly and has a rather low efficiency (more than three times the hardware

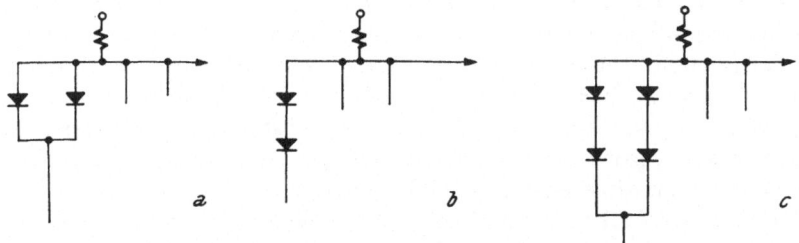

Fig. 10.12. Circuits which are Insensitive to Single Failures

in order to increase the intrinsic reliability by a limited amount, still not achieving 100% reliability). The scheme is, therefore, more valuable as a theoretical extreme, rather than as a practically feasible approach.

Again, one can think of variations in this scheme. Instead of performing three equivalent functions simultaneously in three different implementations, one could perform equivalent functions in a sequential fashion by one implementation. Here, however, the chance for a consistent error (caused by a faulty component) is relatively large.

A basically different approach is illustrated in Fig. 10.12.

Each of the circuits is insensitive to a particular type of single component-failure, that is, the circuits will perform correctly even if one of the diodes fails. The circuit in Fig. 10.12a performs satisfactorily even if one of the two diodes is "open", the circuit in Fig. 10.12b still works if one of the diodes becomes "shorted", and the circuit in Fig. 10.12c permits satisfactory operation with one bad diode, no matter whether it is open or shorted. Although we have illustrated the principle with a diode AND- (or OR-) circuit, it can be applied to other components like transistors, relay contacts, etc. If desired, one could also find circuits which permit double or triple errors.

It is questionable whether the shown technique can be considered as an error correction scheme. Particularly, no error detection in the true sense of the word is performed. On the other hand, the effect of the technique is equivalent to a true error correcting scheme. A correct operation in spite of component failures is possible and, therefore, the reliability of a design is increased. The high cost, probably, restricts the practical use of the scheme to relatively unreliable components in critical applications.

It is also possible to use redundant information for error correction. If we refer to the parity check indicated in Fig. 10.9, we see that a single error within a block of information causes both, a lateral and a longitudinal parity error. These parity errors determine the row and the column of the incorrectly recorded or reproduced bit. It is, therefore, possible to determine which bit in a block of information is incorrect. Since there are only two possible conditions for this bit, we have only to complement the bit in order to obtain the correct information. The scheme, as is, will not work for double errors, that is, not all double errors can be corrected, although they can be detected. But again, it is at least theoretically possible to construct and record redundant information in such a manner that a correction of multiple errors is made possible.

Undoubtedly, all the error correction techniques which we have discussed so far, are not used very extensively at present, probably because they require a significant amount of additional hardware, and because sufficient reliability can be achieved without them. On the other hand, practically all computer installations rely to some extent on error corrections. Mostly, these will be manual corrections after an error has been found by one means or another. Individual areas in which corrections can be performed automatically are iteration processes where errors during a computation may mean a few more iterations before the correct solution is found, and, perhaps, the correction of obviously wrong input data (e.g. detecting a punched hole in a card where there obviously should not be one). There are also some instances where the entire computer operation can be considered as an error correction process. For instance, statistical smoothing procedures may be applied to input data in order to correct errors in a measurement or to eliminate noise components of a signal.

In addition there is at least one area where errors are frequently corrected by program. This is the correction of errors in magnetic tape storages. Errors are here so frequent that one simply cannot afford to repeat, say, a two-hour computer run because an error occurred during the last minute of the run. Moreover, such an error should not be considered as an equipment failure (and charged as down-time), since its cause may be entirely beyond the control of the computer manufacturer or the maintenance technician. The method which is here applied is fairly simple. A read-head checks the recorded information during the write operation. If an error is found, the tape is backed up and the information is re-recorded. If there is no success during a few tries, the recording is erased, the tape is advanced a few inches and the information is recorded again on a presumably better spot on the tape. Alternately, if an error is found during a read operation, the tape is backed up and a new attempt is made.

Problem 13 (Voluntary): Estimate the increase in effective reliability for a magnetic tape storage which can be achieved by the above technique.

10.5. Computer Evaluations

So far, we have mainly discussed the trade-offs between capabilities. speed, reliability, and cost. The problems encountered are difficult enough, But, undoubtedly, there are many more design parameters which have to be considered, and certainly, there are instances, where one can trade not only one against another, but against several other parameters. Moreover, the design goals are usually only insufficiently defined, and there are no good measures for parameters such as capabilities or speed. On the other hand, the failure to find a reasonable compromise in a single trade-off may mean a design which is not competitive. There is, however, a consolation for the computer designer facing this dilemma: the computer user is in a not much better position to evaluate the merits of particular designs for his application. A mathematician would list all parameters of interest, such as size, weight, cooling requirements, ease of programming, software capabilities, etc. He would define measures for all of them. He would give each parameter a certain weight reflecting its value for the particular application. For a particular design he would then compute a figure of merit. Repeating the process for other designs, he could select the best design for his application. He might also investigate the influence of changes in each design parameter upon the figure of merit. He could then find the optimum values for all design parameters. However, these approaches are not, or at least not at present, feasible for several reasons:

First of all, it is almost impossible to define meaningful measures for many parameters such as capabilities or, perhaps, convenience. Even speed is difficult to define. Certainly, one can determine the speed for individual operations, as for instance an addition, but how should we rate the respective speed of two computers when the first adds faster, but the second needs less time for a division? If we measure the speed according to the significance in our application (e.g. the time required to solve a specific problem containing a particular "mix" of additions and divisions), we bias our measure and cannot apply it to other applications of the computer. We could define an "average" problem (and this has been attempted for arithmetic speeds, e.g. 40% additions and subtractions, 5% multiplications, 2% divisions and the remainder for shifts, transfers, tests, logic operations, etc.), but then there is the danger that this mix is not representative for a great many individual applications, although we may have a superficial measure to compare the arithmetic speeds of different computers in general. The remaining approach, i.e. to consider all individual speeds as independent parameters, very much complicates all evaluations and trade-offs.

A second consideration which makes an "exact" approach unfeasible is the difficulty of finding a formula which combines different parameters in a meaningful figure of merit. Let us take an oversimplified example. We want to select a computer. The only arithmetic operations which have to be performed are additions. Under these circumstances, our figure of merit certainly should be proportional to the speed of additions, or the "speed" for short. On the other hand, the wordlength has a bearing on the figure of merit. The longer the word, the more accuracy we get with one addition. For the moment, let us assume that twice the accuracy is twice as valuable to us, in other words, the figure of merit shall be proportional to the wordlength. The reliability of the design influences our evaluation. With a low reliability, we get usable results only during a fraction of the total time. So let us assume that our figure of merit is proportional also to the ratio of the usable to the total time. Last, but not least, the cost of the computer is to be considered. The more expensive the computer, the less attractive it is. In other words, the figure of merit is inversely proportional to the cost. The formula for our figure of merit stands now as follows:

$$\text{Figure of Merit} \sim \frac{\text{Speed} \times \text{Wordlength}}{\text{Cost}} \times \left(\frac{\text{Usable Time}}{\text{Total Time}} \right) \quad (10.1)$$

Certainly, we can use this formula to rate different designs. But is this really a meaningful measure? Let us find out what is wrong with it.

For one, it is an oversimplified assumption that the value of a computer depends linearly upon the shown parameters. For instance: the operands in our hypothetical problem may have only a limited number of bits, say 24. A wordlength beyond 24 bits is then practically wasted. If we would show the value of the wordlength in a graph, we might obtain the curve shown in Fig. 10.13.

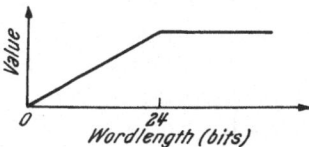

Still this representation is simplified. There might be a small additional benefit in a wordlength beyond 24 bits. Also, the value of a wordlength between zero and 24 bits is probably not a linear function (a wordlength of 12 bits may require in effect three, instead of two individual additions to achieve a 24-bit precision). Similar considerations can be applied to other parameters (e.g. the actual workload may require only a certain number of additions, and any speed beyond that value may be practically useless). In general then, the merit of individual parameters should be reflected by more complex expressions, such as polynominals or, perhaps, by graphs.

Fig. 10.13. Hypothetical Value of the Wordlength

The second fallacy is the assumption that individual merits always combine in the given manner to represent the total figure of merit. It can well be that, for a particular application, a 10% increase in speed is as valuable as a 10% decrease in cost as it is reflected in equation (10.1) But what about the case where cost is really a secondary consideration compared to speed or reliability? What we here say is: It should be possible to put more or less emphasis (more or less relative weight) on individual parameters. This is very easily accomplished in a formula which adds, rather than multiplies individual merits:

$$\text{Figure of Merit} = W_1 \times \text{Speed} + W_2 \times \text{Wordlength} + W_3 \times \text{Reliability}$$

$$+ W_4 \div \text{Cost} \tag{10.2}$$

We simply select appropriate weights (W's) to reflect the varying emphasis. Although an equivalent formula is sometimes applied to computer evaluations, the derived figure of merit may be even less representative of the actual value of a computer than the previous measure. For instance, a wordlength of zero bits, that is a computer with no practical value whatsoever, still has a non-zero figure of merit.

There is a third reason why Equation (10.1) does not give a generally applicable figure of merit. The value of individual parameters may depend upon the manner in which the computer is operated. For instance, cost cannot be represented by a fixed value, but there are one-time costs (at the purchase price or the installation cost) and daily costs (like costs for maintenance, spare parts, supplies). Furthermore, the costs may vary with the use (e.g. the second shift rental is usually less expensive than the prime shift rental). The reliability may depend upon environmental conditions, like the ambient temperature, the noise environment, or the regulation of the line voltage, etc.

Finally, Equation (10.1) does not take into consideration many and very important aspects. In our particular example for instance, the input/output speed has not been taken into account, although it may be an important consideration. As a matter of fact, it could be the only important consideration if only trivial arithmetic operations are to be performed on a large amount of data. Other aspects which are not reflected in Equation (10.1), but really should be, are differences in the capabilities of the system layout (e.g. is the input/output independent from arithmetic operations, how many channels are there etc.), the ease of programming, the power consumption, cooling requirements, weight, size, etc.

Now that we have contemplated the problem in some detail, we must come to the conclusion that the evaluation or the comparison of computers cannot be based purely upon objective, mathematical procedures. A practical solution of the problem would be the programming of problems typical for the intended application and test runs on different computers. Theoretical computer evaluations will have to rely to a large extent upon intuition supplemented by as much knowledge as possible. Such knowledge is required mainly in three areas: an understanding of the kind of parameters which are involved, their possible and practical ranges, and the possible trade-offs.

Problem 14: How would you rate two computers which are for all practical purposes identical, except that one costs 10% less than the other, but requires 10% more cooling capacity? Could there be circumstances which would cause you to reverse your decision?

Problem 15: Suppose you have two computers A and B. Computer A has exactly twice the speed of Computer B and costs twice as much. Each has a mean time between failures of one hour. All other characteristics are identical. Would you consider computer A or computer B a better buy. Could there be circumstances which would cause you to select the "worse" buy?

Problem 16: Can you think of ways in which you as computer user can take advantage of an unnecessarily high computer speed to compensate for a relatively low reliability.

Problem 17: Suppose a square root command is optional with the computer that you are going to buy. Its incorporation would increase the price of the computer by 1%. Should you buy it?

Problem 18 (Voluntary): Can you imagine a computer application or a computer system in which the attachment of a second printer would make the system twice as valuable to you? Can you imagine a case where the second printer has no value to you?

Go through the same exercise at your leisure with twice the memory capacity, the memory speed, the arithmetic speed, the number of I/O channels and/or all other computer characteristics you can think of.

10.6. Engineering Check List

As we have seen, it is not so much the brilliant solution of a particularly difficult design problem which makes a "good" or an "excellent" computer, but rather a reasonable compromise of its over-all aspects. Even the consideration or neglection of trivial engineering details can make a significant difference in this respect. Let us, therefore, attempt to compile a short check list of engineering aspects. Even if the list is not complete, it should suggest additions applicable under the specific circumstances. We should keep in mind that, even for a single aspect, there usually exists a whole range of possible solutions. As an example, our list may contain the entry "degree of minimization". If we take the role of a computer manufacturer who mass-produces computers, the saving of even a single component becomes important. However, if we imagine ourselves to be designers who build a single circuit for a specific application, the saving of a component is rather unimportant. As a matter of fact, if we spend any time at all with minimization efforts, the cost of the engineering hours may well exceed the cost of the saved components. In general, it is not always best to achieve a design which is ideal as far as a single aspect is concerned. Instead, the designer has the responsibility to make a reasonable choice within the possible range. This choice has to be consistent with other aspects and the overall design objective.

Basic Approach and Implementation: Do I use the appropriate principle under the given circumstances? (e.g. digital, analog, or combinations thereof). Did I give proper consideration to the possible use of magnetic, mechanical, pneumatic, optical etc., instead of electronic components in the system?

Do I make reasonable use of simultaneous or decentralized operations? (e.g. serial versus parallel arithmetic, simultaneous execution of instructions, several smaller or slower units, or even computers, instead of one large or fast one).

Is the basic implementation the most advantageous one, under the circumstances? (e.g. diode-transister or pure transister circuits, pulse or DC circuitry, NOR, NAND or majority logic, electronic, magnetic, cryogenic components, etc.)

How do the capabilities match the requirements? (e.g. do I want a special purpose computer, i.e. an exact fit, or a general purpose computer, i.e. a less perfect fit, but a larger potential market?) Does the system contain luxury features which really are not worth the cost to implement them? Are there additional features which could be provided at very low or no cost?)

Is the system balanced in its capabilities? (e.g. an input/output speed which is compatible with the internal speed for the given application, compatible speeds and sizes of the different levels of the memory hierarchy, the execution time of individual instructions, the number and type of different instructions).

Are there adequate provisions for future expansion in the system? (particularly modular design with the possibility of attaching additional units, or replacement of units by others with the purpose of increasing or decreasing input/output capabilities, arithmetic speed, memory capacity etc.)

Do I really provide a systems concept? (e.g. not only a hardware system with system control, but interrupt and/or status lines and monitoring by "executive" program.

Reliability: Are there sufficient safety margins? (e.g. in the timing of all internal operations, in the load to be driven by each component, in the rating of components?) Did I consider all possible cases and especially the worst case?

Does the life of the components allow adequate reliability to be achieved? Do I have to employ error detection and correction techniques?

Operational Characteristics: Do I provide a clear and functional layout of all operating controls? Does the operator get some assurance of the working condition? (some indication, movement, or even noise should be apparent).

Does he get a warning of improper working? Are there too many, or too few indications? Does the system force the operator to deviate unnecessarily from etablished operating procedures?

Are there fault indicators or interlocks guarding against operator errors? How foolproof is the system?

How much control over the executive program or the "system" do I want the operator to have?

Environment: What approach should I take to each single environmental condition? Shall I design for a wide range (e.g. 0° C to 60° C), shall I have the computer automatically shut off if normal limits are exceeded (e.g. supply voltage drops 10% under nominal value) or shall I do nothing (e.g. no provisions guarding against somebody who would perhaps submerge the computer in sea water). For what unusual conditions do I have to provide (e.g. the doors are open for maintenance, but the equipment still has to be cooled sufficiently.)

Maintenance: Does the system have a logical arrangement of components? Are the diagrams and manuals clear and easy to read? How many levels of diagrams shall I provide (e.g. block diagrams, logic diagrams, circuit diagrams). Is the number of different modules and their size reasonable? Could I reduce the spare parts requirement by changes in the design?

Are the components easily accessible? Is the number of test points and indicators sufficient? How good and helpful are the diagnostic test routines? Are the test programs written in such a fashion that they can be run (and give meaningful indications) even if components in the system fail?

Can maintenance procedures harm the system? (e.g. the removal of a bias voltage, the grounding of a single line, or the removal of a circuit card may overload certain components).

How much built-in test circuitry shall I provide? Marginal checks? Slow clock? Stepping of operations or clock cycles? Test and disconnect switches? Are the circuits sufficiently separated or isolated (e.g. the accidental application of a high voltage to a single logic terminal should not damage a large number of transistors, a short in one peripheral unit should not put the entire system out of commission, manipulations in a turned off piece of peripheral equipment, or turning it on and off should not interfere with the operation of the remainder).

Detailed Design: Should I use familiar components, or existing ones, or even develop new components? Shall I use precision components or allow wide enough design margins to accommodate components with normal tolerances? How much effort should I put into minimization? Did I consider all functional circuits (not logic circuits) in enough detail (e.g. slicers, clippers, amplifiers, line-drivers, transmission lines, emitter followers)? Did I consider all auxiliary circuits in detail (e.g. interlocks, fault circuits, power sequencing circuits). Did I consider abnormal conditions (e.g. power disconnected to a specific unit, some of the components removed, common types of failures)? Do I synchronize all external inputs where necessary?

Bibliography

PETERSON W. W.: On Checking an Adder, IBM Journal of Research and Development, vol. 2, pp. 166—168. Apr. 1958.

BROWN D. T.: Error Detecting and Correcting Codes for Arithmetic Operations, Transactions IRE, vol. EC-9, No. 3, pp. 333—337. Sept. 1960.

PETERSEN W. W.: Error Correcting Codes. New York: The MIT Press and John Wiley. 1961.

BROWN, TIERNEY, and WASSERMAN: Improvement of Electronic Computer

Reliability through the Use of Redundancy, Transactions IRE, vol. EC-10, No. 3, pp. 407—415. Sept. 1961.

LLOYD, and LIPOW: Reliability, Management, Methods, and Mathematics. Englewood Cliffs: Prentice-Hall. 1962.

WILCOX, and MANN: Redundancy Techniques for Computing Systems. Washington: Spartan Books. 1962.

PIERCE W. H.: Adaptive Decision Elements to Improve the Reliability of Redundant Systems, IRE International Convention Record, part 4, pp. 124—131. March 1962.

MESYATSEV P. P.: Reliability of the Manufacture of Electronic Computing Machines, Moscow: Mashgiz, 1963. Translation available through Joint Publication Research Service, US Department of Commerce, JPRS: 26,687.

MORGAN et al.: Human Engineering Guide to Equipment Design. New York: McGraw-Hill. 1963.

KEMP J. C.: Optimizing Reliability in Digital Systems, Computer Design, vol. 2, No. 1, pp. 26—30. Jan. 1963.

EINHORN S. J.: Reliability Prediction for Repairable Redundant Systems, Proceedings IEEE, vol. 51, No. 2, pp. 312–317. Feb. 1963.

DAHER P. R.: Automatic Correction of Multiple Errors Originating in a Computer Memory, IBM Journal of Research and Development, pp. 317–324. Oct. 1963.

MALING, and ALLEN: A Computer Organization and Programming System for Automated Maintenance, Transactions IEEE, vol. EC-12, No. 5, pp. 887—895. Dec. 1963.

ROSENTHAL S.: Analytical Technique for Automatic Data Processing Equipment Acquisition, AFIPS Conference Proceedings, vol. 25, pp. 359—366. Spring Joint Computer Conference, 1964.

JOSLIN, and MULLIN: Cost-Value Technique for Evaluation of Computer System Proposals, AFIPS Conference Proceedings, vol. 25, pp. 367—381. Spring Joint Computer Conference, 1964.

HERMAN, and IHRER: The Use of a Computer to Evaluate Computers, AFIPS Conference Proceedings, vol. 25, pp. 383—395. Spring Joint Computer Conference, 1964.

GOLDMAN H. D.: Protecting Core Memory Circuits with Error Correcting Cyclic Codes, Transactions IEEE, vol. EC-19, No. 3, pp. 303—304. June 1964.

BUCHMAN A. S.: The Digital Computer in Real-Time Control Systems, part 5 — Redundancy Techniques, Computer Design, pp. 12—15. Nov. 1964.

PIERCE W. H.: Failure Tolerant Computer Design. New York: Academic Press. 1965.

ATKINS J. B.: Worst Case Circuit Design, IEEE Spectrum, pp. 152—161. March 1965.

GURZI K. J.: Estimates for Best Placement of Voters in a Triplicated Logic Network, Transactions IEEE, vol. EC-14, No. 5, pp. 711—717. Oct. 1965.

HOWE W. H.: High-Speed Logic Circuit Considerations, AFIPS Conference Proceedings, vol. 27, part 1, pp. 505—510. Fall Joint Computer Conference, 1965.

DAVIES R. A.: A Checking Arithmetic Unit, AFIPS Conference Proceedings, vol. 27, part 1, pp. 705—713. Fall Joint Computer Conference, 1965.

FORBES, RUTHERFORD, STIEGLITZ, and TUNG: A Self-Diagnosable Computer, AFIPS Conference Proceedings, vol. 27, part 1, pp. 1073—1086. Fall Joint Computer Conference, 1965.

POFOLO J.: Computer Specification Guideline, Computer Design, vol. 4, No. 12, pp. 42—48. Dec. 1965.

11. Unusual Computer Components

In this chapter we shall discuss unusual or unconventional computer components, that is, such components which presently have no wide-spread application, but which offer, at least in some respects, potential advantages over conventional components. In addition, some components will be discussed which almost certainly have no practical value. These are included when their operating principle is theoretically interesting, and when the latter might conceivably be used with novel components.

11.1. Magnetic Components

Magnetic components are rather attractive to the computer designer for several reasons: they possess an inherent high reliability; require in most applications no power other than the power to switch their state; and are potentially able to perform all required operations, i.e., logic, storage, and amplification. It is, for instance, conceivable to build an all-magnetic computer, let us say for a space probe, which is extremely reliable, and which may be in a "half-dormant" state for long periods of time, in which it requires almost no power, but which is immediately ready for operation when it becomes desired.

11.1.1. Magnetic Cores as Logic Elements

The basic properties of magnetic cores and the use of cores in computer memories have been discussed in chapters 5 and 8. Cores which are used as logic elements may have slightly different magnetic properties or construction[1]. This, however, is of little significance for our present purpose.

Fig. 11.1 shows the basic arrangement for employing a core as logic element.

Suppose the core has been saturated by a set signal, let us say in clockwise direction. If we now apply a drive pulse, the flux in the magnetic core is reversed, and a signal will appear at the output terminals. However, if the core had not been set previously, the flux in the core will not be changed by the drive input and, consequently, no output signal will be produced.

---

[1] Some switching cores are, for instance, laminated.

This behavior is similar to that of a gate: inputs produce outputs or no outputs depending upon whether the core has been set or not. Of course, there is also an essential difference between a core and a gate. The core

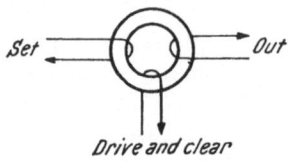

Set   *Out*

Drive and clear

Fig. 11.1. The Magnetic Core as Logic Element

does not stay set, but each drive pulse resets the core and no further outputs are produced, unless the core is set again. Core logic circuits will, therefore, be different from conventional switching circuits. However, before we investigate these characteristics in detail, let us consider another fundamental property of the basic circuit.

The core can be considered as a magnetic amplifier. It can produce more energy on the output than there is energy required on the input. This energy is ultimately delivered by the drive pulse, but, considering only the set input and the output, we can control a larger amount of energy by a smaller amount of energy and have, therefore the property of an amplifier. Referring to Fig. 11.2, we may see this more clearly.

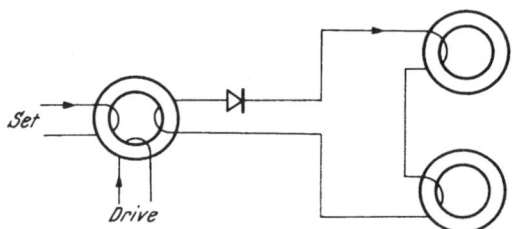

Set

Drive

Fig. 11.2. The Magnetic Core as Amplifier

The core acts similar to a transformer. While the core is being set, no load appears on the output (any current induced into the output winding would have to pass the diode in reverse direction). A relatively small amount of energy is, therefore, sufficient to set the core. However, when the drive pulse resets the core, the full load appears on the output (the current induced into the output winding passes through the diode in forward direction). A large drive current is required to overcome the flux produced by the induced current and a relatively large amount of energy, supplied by the drive pulse, is available at the output. This energy is sufficient to set several other cores, if so desired.

So far we have found three capabilities of the magnetic core. It may act as a storage device, as a simple gate or switch, and as an amplifier. If we should be able to show that we can also implement more complicated logic functions, we have found a universal computer component. The implementation of logical functions is not complicated. Fig. 11.3 shows the basic approach.

The core has two set windings. It can be set by a current through either one, or the other (or both) windings. The drive pulse will produce an output in either case, and the arrangement implements a logical OR-circuit.

If small currents were used for setting the core, one could obtain a condition where one set current would not be sufficient to switch the flux, whereas simultaneous currents through both windings would set the core.

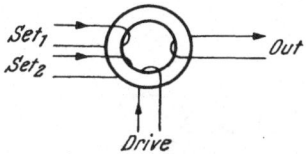

Fig. 11.3. Representation of an OR-Circuit

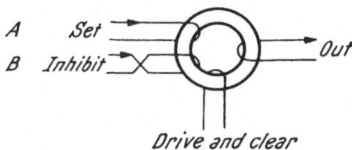

Fig. 11.4. Representation of an Inhibit-Circuit

Since both set inputs must be present to produce an output, a logical AND would be implemented. Even though a similar principle is widely used for the selection of cores in a memory, switching circuits normally do not contain AND-circuits of this type since they have a relatively narrow margin of operation as far as set currents are concerned[1].

The logic operations of AND and NOT are usually implemented indirectly by inhibit circuits as shown in Fig. 11.4.

One of the input windings has its leads reversed. If no inhibit input is present, the core is set in the usual manner. If an inhibit is present simultaneously with a set input, the two excitations counteract each other and prohibit a setting of the core. If only the inhibit input is present, the core is cleared (or stays cleared). Denoting the set input by A, and the inhibit input by B, the output produced by the drive pulse may be expressed as: AB'.

Connecting the set input to a logic "1", i.e., connecting the input to a source which always delivers a set current at the time when we may have an inhibit input, we obtain a NOT-circuit. We obtain an output if the logic proposition connected to the inhibit winding is not present, and no output if it is present. Denoting the latter input by B, the output follows B'.

Problem 1: As we have seen in chapter 4, the logic functions of OR and NOT are sufficient to implement any logic function. Since we have just shown how to implement the NOT by an inhibit circuit, OR and INHIBIT are also sufficient to implement any logic function. Design a core switching circuit which implements the logic AND. (Note: It is helpful

---

[1] In addition, the arrangement would unnecessarily restrict the operating speed.

to study the remainder of paragraph 11.1.1 before attempting to solve
the problem.)

Since cores cannot be set and read out[1], simultaneously the transfer of
information through each stage of a core switching system is necessarily
subject to delay. As far as a single core is concerned, an operation cycle
consists of two phases: the setting and the read-out. When the output of
one core is used as input to another core, the operation cycles of the two
cores must be staggered. This phasing is demonstrated by the magnetic
core shift register shown in Fig. 11.5.

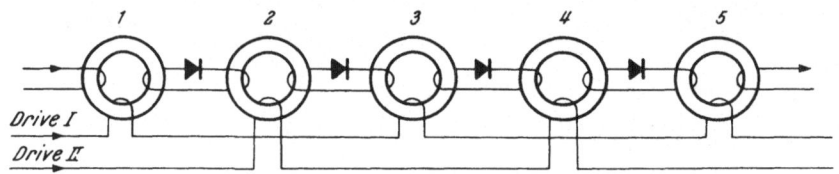

Fig. 11.5. Magnetic Core Shift Register with Two Cores per Bit

The "Drive I" signal reads all odd numbered cores and sets the even
numbered cores. In this scheme, half of the cores are cleared at any one
time. Therefore, 2 n cores are required to store n bits of information and
we speak of a "two-core-per-bit logic." One complete cycle of a Drive I plus
a Drive II signal is required to shift the information by one "place". A
similar phasing of DI's and DII's must be employed for any logic design,
that is, the output of a core driven by DI can be used as input to only those
cores which are driven by DII, and vice versa. In some instances we may
find it necessary to insert cores into logic circuits which serve no other
purpose than to match the phasing of two logic propositions.

Problem 2: Design a magnetic core half adder. Label clearly which cores
are driven by DI and which by DII.

We may consider only the odd numbered cores as information storages
and the even numbered cores as auxiliary devices to transfer information
between odd numbered cores. We may then call the even numbered cores
"interstages". A single interstage is able to store one bit of information
temporarily. It delivers received information after a certain elapsed time.
Using elements, other than cores, we are able to construct such delays and,
therefore, construct interstages. Fig. 11.6 shows a commonly used interstage
employing a capacitor as temporary storage element.

The drive pulse clears and reads out all cores. The output signals then

---

[1] The flux in the core must be reversed for each operation.

charge the capacitor of the interstage. The capacitor delivers a current to the next core in the chain which sets the core (after the drive pulse has ceased.) It is to be noted that the diode in the interstage prevents the energy from flowing from the capacitor back to the input core. In this scheme, each core can store one bit of information. We speak, therefore, of a "one-core-per-bit" logic. The obvious advantage of the scheme lies in the economy as far as

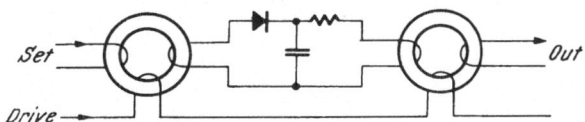

Fig. 11.6. Interstage for One-Core-per-Bit Logic

the number of cores required is concerned. Disadvantages include the use of the diode, the capacitor, and possibly the resistor, which are less reliable than cores. Furthermore, the drive pulse timing is rather critical. A pulse which is too short will not deliver sufficient energy to the interstage. A drive pulse which is too long will override the set pulse delivered by the interstage. This latter property may be used to advantage. A long drive pulse clears the shift register or, for this matter, any core in the system.

Problem 3: Design the following core circuits with capacitive interstages:
a) a half adder;
b) the equivalent of a conventional flip-flop with set and clear inputs.

11.1.2. The Transfluxor

Like regular magnetic cores, transfluxors[1] are magnetic devices made of a magnetic material with an almost rectangular hysteresis loop. But, where the magnetic core has a single hole and is magnetically entirely saturated in one of two directions, the transfluxor has several apertures and may be saturated with several distinct flux patterns by various currents through these apertures.

In order to understand the operation of a transfluxor, let us first investigate in general what happens when a current passes through a hole in a magnetic material. Suppose a current passes through a single aperture in a large sheet of magnetic material as shown in Fig. 11.7a. Due to the current, i, a magnetic field intensity, H, is produced around the aperture. The field is strongest in the immediate vicinity of the aperture and diminishes in strength, with increasing distance.

Suppose that the material has an ideal rectangular hysteresis loop as shown

---

[1] Transfluxors are also referred to as: fluxor, magnistor, logicor or multi-apertured core.

in Fig. 11.7 b and that the current, i, is of such magnitude that the field intensity, H, in the vicinity of the hole, is larger than the coercive strength H_c, but smaller than H_c outside of a certain diameter. In the vicinity

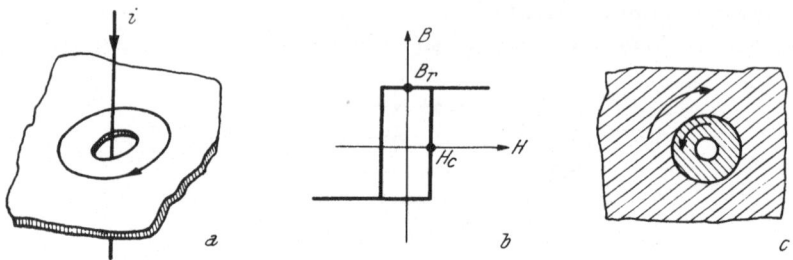

Fig. 11.7. Flux Patterns Produced by a Current Through an Aperture in a Magnetic Material

of the hole then, any remanent magnetism is overcome and the resulting magnetic flux has the direction determined by the direction of the current. Outside this area, an already existing magnetic flux is not affected. The diameter of the affected area varies with the magnitude of the current. By first applying a large current which magnetizes in clockwise direction and then a smaller current which magnetizes in counterclockwise direction, we may for instance create a flux pattern similar to that shown in Fig. 11.7 c.

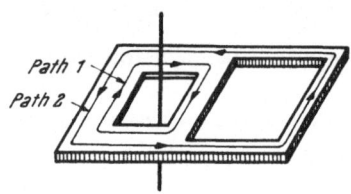

Fig. 11.8. Flux Pattern in a Piece of Magnetic Material with Two Apertures

Flux patterns of this type are not necessarily circular and concentric with the hole, but follow the shape of the material. Fig. 11.8 shows a possible flux pattern for a rectangular piece of magnetic material with two rectangular apertures.

The flux along path 1 can be changed by a relatively small current. Due to the much greater length of path 2, a much higher current is required to affect the flux along this path, in spite of the fact that parts of this path are fairly close to the current carrying aperture.

Let us now look at the detailed operation of a relatively simple transfluxor with two apertures as shown in Fig. 11.9 a. By a large current through the control winding, we can produce the flux pattern shown in Fig. 11.9 b. By a subsequent application of a smaller current in reverse direction, the flux pattern of Fig. 11.9 c is generated.

Suppose such a transfluxor is in the "unblocked" condition (Fig. 11.9 c) and an AC current is applied to the input winding. This current is of sufficient magnitude to affect the area around the small aperture but does not affect the flux along the much longer path around both apertures. The current produces an alternating magnetic field around the small aperture. The corresponding magnetic flux changes its direction periodically and induces an AC voltage into the output winding.

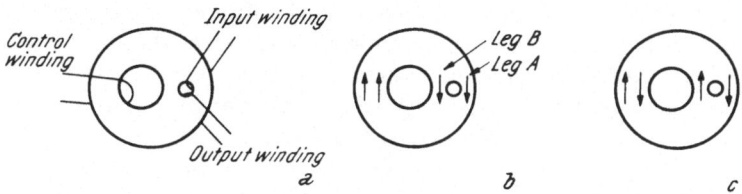

Fig. 11.9. Basic Operation of a Two-Apertured Transfluxor

If the transfluxor is in the "blocked" state as shown in Fig. 11.9b, the current through the input winding can not change the magnetic flux pattern. We notice that an increase of the flux in a clockwise direction around the hole would require an increase of the flux in leg A. This is not possible since leg A is already saturated. But if no increase of flux in "leg" A is possible, no change of flux can take place in leg B, since the magnetic flux is continuous. In a very similar manner, the already saturated leg B prohibits the establishing of a counter-clockwise flux around the small aperture. Since the magnetic flux pattern is not changed by the input current, no voltage will be induced into the output winding.

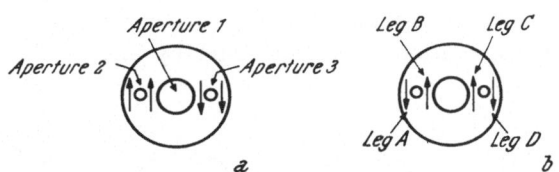

Fig. 11.10. Three-Apertured Transfluxor

The simple arrangement of Fig. 11.9 proves to be equivalent to a gate. If the transfluxor is blocked, the "gate" is closed and no outputs are produced. The unblocked transfluxor corresponds to an open gate where outputs are produced by inputs. It is worthwhile to note that the transfluxor gate has a built in "memory". Once it is opened or closed it remains in this state until it is switched by another control signal.

The operation of the two-hole transfluxor as shown in Fig. 11.9 is somewhat critical with respect to the unblock or set pulse. A too large pulse will "overset" the transfluxor, i.e. the flux in part of leg B (Fig. 11.9b) will be reversed; a too small pulse leaves parts of leg A unset. This difficulty can be overcome by using transfluxors with more than two holes. Fig. 11.10 shows a possible arrangement with three apertures.

The block or clear pulse through aperture 1 saturates the core as shown in Fig. 11.10 a. The set pulse through aperture 2 switches then the flux to the state shown in Fig. 11.10 b. Since only leg A provides a return path for any flux change to the right of aperture 2, the flux change is limited to the amount which one leg can accommodate. Therefore, the flux will change only in one leg to the right of aperture 2. Since leg C is the closest leg which can change, leg C, and only leg C will change, independent of how large the set current is, provided the current has the minimum required intensity. Similar arrangements, where the amplitude of block and set currents are not critical, can be made with transfluxors which have more than three apertures.

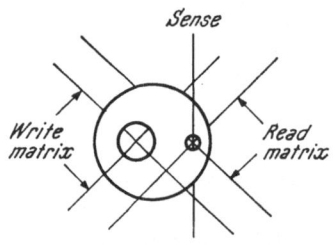

Fig. 11.11. Non-Destructive Memory with Transfluxors

Let us now look at some applications of transfluxors.

Non-Destructive Memories: The read-out of conventional core memories destroys the stored information, as we have seen in chapter 8. This normally necessitates a "restore" operation. Since it is possible to "read" the state of a transfluxor repeatedly without destroying the stored information, it is possible to build non-destructive memories. An arrangement employing coincident current selection and two-apertured transfluxors is indicated in Fig. 11.11. Information is written into the storage in very much the same manner as into normal core memories, i.e., by a write matrix. The read operation is accomplished by means of a separate read matrix.

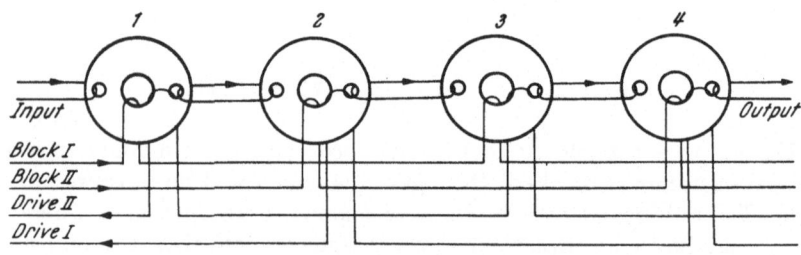

Fig. 11.12. Transfluxor Shift Register

The matrix currents are limited in amplitude so that only that transfluxor which is located at the intersection of the two energized matrix wires is affected. The presence or absence of a signal on the sense winding during a read operation is determined by the stored information. It is interesting to note that the arrangement allows independent read and write operations to be performed simultaneously at different addresses.

Shift Registers: Let us consider the arrangement of the three-hole transfluxor shift register shown in Fig. 11.12.

The information is shifted in a 4-phase cycle. The Block I pulse clears all odd numbered transfluxors. The Drive I pulse sets then all odd numbered stages, that is, it transfers information from the even numbered to the odd numbered transfluxors. Subsequently, the Block II pulse clears all even numbered transfluxors, and the Drive II pulse transfers information from the odd to the even numbered transfluxors. This cycle is repeated for a repeated shift.

The flux patterns of a single odd numbered transfluxor are shown in more detail in Fig. 11.13.

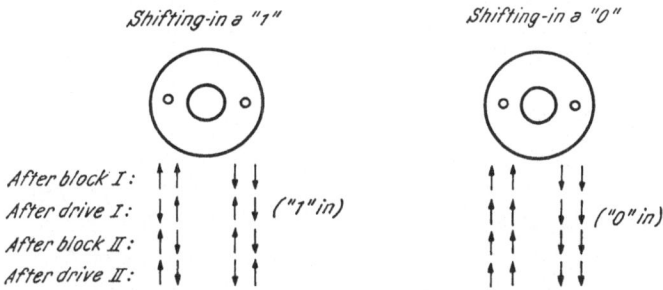

Fig. 11.13. Flux Patterns of an Individual Transfluxor

The flux patterns for even numbered transfluxors are the same, except that I and II pulses are interchanged. The given scheme requires two transfluxors per bit. Its advantage over magnetic core shift registers[1] lies

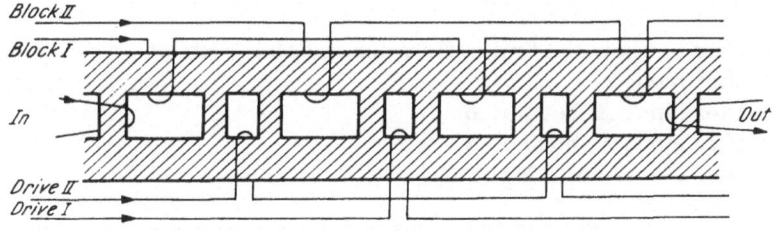

Fig. 11.14. Transfluxor Lattice Shift Register

in the fact that the margin of operation is relatively wide and that no diodes (susceptible to failures) are required.

---

[1] See paragraph 11.1.1.

30*

It is interesting to note that one can build a transfluxor shift register with a single strip of magnetic material, as shown in Fig. 11.14.

Control pulses are applied in the following sequence: Block I, Drive I, Block II, Drive II.

Problem 4 (Voluntary): Show the consecutive flux patterns for the above register and for shifting a single "1" through the system. Assume that all stages are initially blocked (i.e. in the "0" state). Input a single "1" and shift this "1" through the register by a repeated sequence of DII, BI, DI, BII signals.

Techniques, similar to those used with cores, can be used to implement the logic functions OR and INHIBIT. A transfluxor with two separate set windings, each driven by sufficient current, represents an OR-circuit. It is set by either one input, or the other, or both. Inverting the leads on one set winding, we can inhibit the setting by the other and implement an inhibit circuit. The circuit in Fig. 11.15 represents the "exclusive OR".

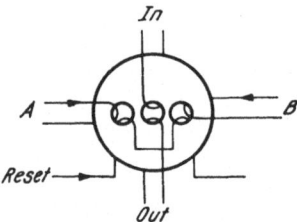

Fig. 11.15. Exclusive OR-Circuit

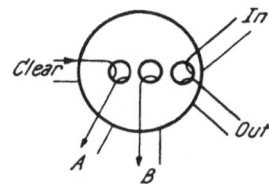

Fig. 11.16. Transfluxor
Sequencing Circuit

In order to obtain an output, the flux around the center hole must be continous, which is possible only for the conditions $AB' + A'B$.

Problem 5: Show the flux pattern for the above circuit:
a) after reset,
b) after input A, but not B,
c) after input B, but not A,
d) after input A and input B.

More complicated logic functions can be implemented by transfluxors with more holes and some rather ingenious arrangements have been built.

It is interesting to note, that one can also build transfluxor circuits which require a proper time sequencing of inputs, before they deliver an output. Fig. 11.16 shows as an example a circuit which requires inputs in the sequence A, B before an output is delivered.

Problem 6: a) Show the resulting flux patterns for the following sequence of inputs: reset, *A* input, *B* input.

b) Show the flux pattern after the following sequence: reset, *B* and then *A*.

11.1.3. The Parametron

The basic circuit of a parametron is shown in Fig. 11.17a. The circuit elements L_1, L_2, and C constitute a resonant circuit which is tuned to a frequency f. Due to the ferro-magnetic cores of L_1 and L_2, the inductance of the circuit is not constant, but varies with the magnetic saturation of the cores 1 and 2. The flux density and, hence, the inductance is varied periodically by an exciting current i_1 (which a frequency $2f$) in such a manner that the original oscillation in the resonant circuit is sustained.

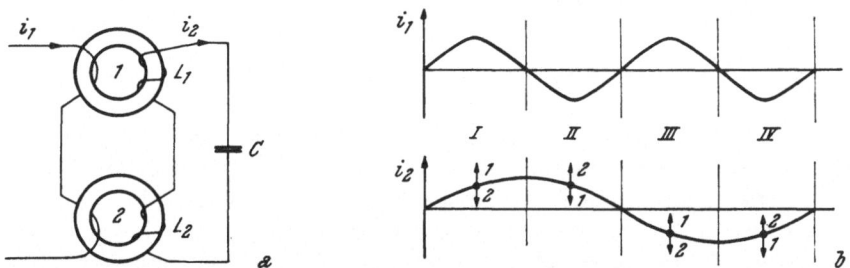

Fig. 11.17. Basic Circuit of a Parametron and Relationship of Oscillating Current i_2 and Exciting Current i_1

For a better understanding of this effect, let us consider the diagram given in Fig. 11.17b. During the time period I, the exciting current i_1 is attempting to increase the current i_2 by induction via core 1 and attempting to decrease the current i_2 by induction via core 2. If both cores had identical magnetic properties during this time, the effects would cancel and the exciting current i_1 would not influence the oscillation current i_2. However, during time I, the currents i_1 and i_2 pass core 1 in such a way that the magnetic flux generated by one current opposes the magnetic flux generated by the other, whereas both currents produce a magnetic flux in the same direction in core 2. Consequently, core 2 is magnetically more saturated than core 1, L_1 is greater than L_2, and core 1 induces more energy into the resonant circuit than core 2. The net effect is an increase in the current i_2. During time period II, the exciting current i_1 is attempting to decrease the current i_2 by induction via core 1, and attempting to increase the current i_2 by induction via core 2. Since now core 1 is more saturated than core 2, the latter couples more energy into the system and the net effect is a sustaining of current i_2. In time period III, both currents have changed their direction. Core 2 is less saturated and delivers more energy than core 1. Current i_2 is increased. During period IV, core 1 is less saturated, and current i_2 is sustained.

A parametron not only sustains a once existing oscillation, but increases its amplitude up to a maximum value which is determined by circuit parameters and the connected load. When the parametron is no longer excited, the oscillation deteriorates. The parametron has, therefore, at least in some respects the properties of an amplifier. Contrary to an amplifier, it forces the oscillation into one of the two well-defined phase-relationships with the exciting current shown in Fig. 11..18

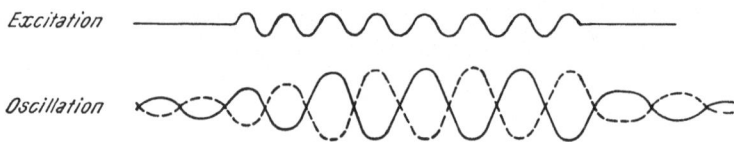

Fig. 11.18. The Two Modes of Oscillation

Since there are two possible modes of oscillation which differ in phase by 180°, the parametron has the properties of a binary element. The "1" or "0" states are defined with respect to a phase standard.

In order to "set" a parametron to the "0" or "1" state, a low amplitude oscillation of the proper phase is introduced into the resonant circuit via a coupling transformer (see Fig. 11.19). The exciting current is then applied. The oscillation is amplified and sustained as far as amplitude and phase are concerned, even after the input signal is no longer present.

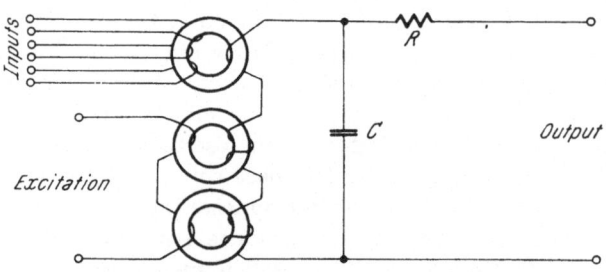

Fig. 11.19. Complete Parametron Circuit

In order to facilitate the transfer of information throughout a parametron-system, excitations are applied in a cyclic manner. Fig. 11.20 illustrates the usual "three-phase" excitation and shows the arrangement of a parametron delay line or shift register.

Suppose that in a certain instant, the first parametron in the chain is excited by the wave form I (frequency $2f$) and contains a "1" (i.e. oscillates with a frequency f and a defined phase with respect to a standard). This causes a low amplitude oscillation with the phase "1" in the second parametron. When now the excitation by wave form II

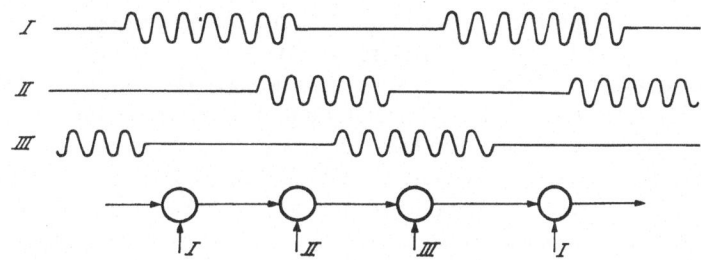

Fig. 11.20. Transfer of Information in a Parametron Shift Register

starts, the second parametron will amplify this oscillation with the proper phase and sustain it even after excitation I ceases to exist. The information has been transferred from the first to the second parametron. While the second parametron oscillates with full amplitude, a low amplitude oscillation of proper phase is coupled into the third parametron. As soon as excitation III begins, the oscillation in the third parametron builds up to the full amplitude, and the information has been shifted one more place.

In an identical manner, i.e. by a series connection of parametrons excited by the wave forms I, II, III, the transfer of information is facilitated throughout any logical network of parametrons.

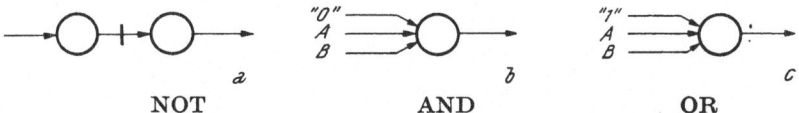

NOT   AND   OR

Fig. 11.21. Implementation of NOT, AND, OR with Parametrons

If it is desired to transmit the logic complement of the contents of one parametron to the next, the output of the first parametron is connected to the input of the second parametron with reversed leads. If, for instance, the first parametron oscillates with the phase "0", an oscillation of the phase "1" is stimulated in the second parametron, and vice versa. Reversed leads are usually indicated by a bar across the transmission path as shown in Fig. 11.21 a. Such a transmission is, in effect, equivalent to a logic NOT operation.

In order to perform logic operations other than NOT, the parametron is provided with more than one input. The state of a parametron is then determined by a majority principle: it will assume the state which the majority of the inputs have. In order to avoid any ambiguity, there is always an odd number of inputs connected to a parametron, each carrying an oscillation of one of the two possible phases.

Fig. 11.21 b shows the implementation of an AND-circuit. One of the three inputs to the parametron is permanently connected to an oscillation of the phase "0". An oscillation with the phase "1" can therefore be stimulated in the parametron only if both other inputs oscillate with the phase "1" and override the zero input. Consequently, the state of the parametron is determined by the logic product of A and B.

Fig. 11.21c similarly, represents an OR-circuit. One input is permanently connected to a logic "1". If at least one of the other inputs represents a "1", the parametron assumes the state "1" and implements, therefore, the logic sum $A + B$.

Problem 7: What logic functions are implemented by parametrons which have
  a) Three inputs labelled A, B, C
  b) Five inputs labelled "1", "1", A, B, C
  c) Five inputs labelled "0", "0", A, B, C
  d) Five inputs labelled "1", A, A, B, C
  e) Five inputs labelled "0", A, A, B, C

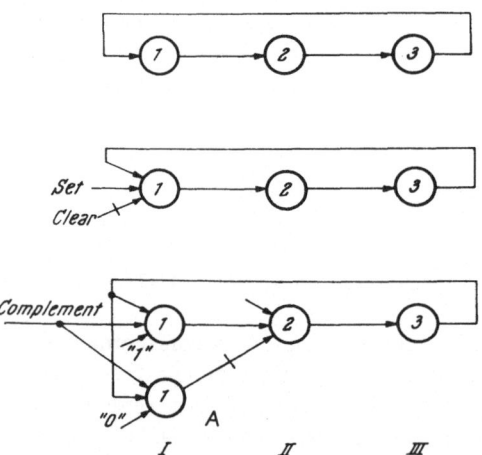

Fig. 11.22. Parametron Circuits Equivalent to a Conventional Flip-Flop

So far, we have seen that the parametron can, in principle, implement all the basic operations of a universal computer element, i.e. amplification, storage, delay and logic operations. Let us now consider circuits which illustrate the particularities of a parametron system.

Suppose we want to build a circuit equivalent to a normal flip-flop, which can store one bit of information for an arbitrary length of time and which can be set, cleared and perhaps complemented by signals applied to specific inputs. Assuming a 3-beat excitation we use three parametrons to store the information during the three cycles. (See Fig. 11.22a) The information re-cycles through the parametrons I, II, III, I etc. Depending upon the time at which the information is to be read out, output connections are made to parametrons 1, 2, or 3. Fig. 11.22b shows inputs only to the first of the three parametrons. All inputs have therefore to come from parametrons excited by phase III. Set inputs (to be exact: set inputs with phase "1" *and* clear inputs with phase "0") stimulate oscillations of phase "1" in the first parametron, regardless of the previous state of the "flip-flop". Clear inputs (i.e. oscillations with phase "1" on the clear input *and* oscillations with phase "0" on the set input) stimulate oscillations of phase "0" in the first parametron.

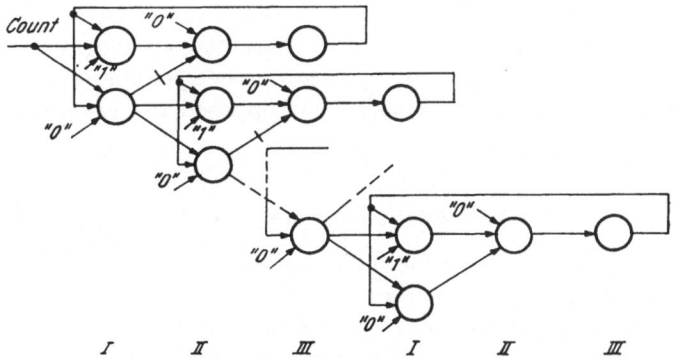

Fig. 11.23. *n*-Stage Binary Counter

Fig. 11.22c shows a parametron "flip-flop" with a complement input. Parametrons 1, 2, 3 recycle the state of the flip-flop in the usual manner. *A* is an auxiliary parametron which assumes the "1"-state only if the state of the flip-flop is a "1" during the excitation cycle III *and* there is a complement input. If these conditions are met, parametron 2 will assume the "0"-state so that, in effect, the flip-flop changes its state from "1" to "0". If the flip-flop contains a ' 0" when the complement signal arrives, parametron 1 will be set to the "1" state so that again the flip-flop is complemented.

Several of these complement flip-flops can be combined to produce a binary counter as indicated in Fig. 11.23.

The complement input to a counter stage can be derived from the output of the auxiliary parametron in the next lower stage since this output, as we have seen, assumes the "1"-state only if the state of the stage is a "1" and a complement signal arrives.

Although these circuits may seem rather complicated compared to circuits with conventional elements, the overall circuitry of a logic system may become simpler due to the logic flexibility of an individual parametron.

Problem 8: Design a binary full adder consisting entirely of parametrons. Use as few parametrons as possible and develop the sum and the carry with as little time delay as possible.

The parametron is one of a group of devices which may be called parametric phase-locked oscillators (PLO). Common to this class of devices is the stimulation of oscillations by the variation of one of the parameters in a resonant circuit (parametric excitation). The energy for this oscillation is delivered by a "pump". In case of the parametron the pump changes the inductance of a resonant circuit periodically. It is, however, equally possible to build a parametric oscillator which is excited by the variation of the capacitance in the circuit (e.g. by using diodes as variable capacitances). The latter approach is more feasible for operation at high frequencies (in the microwave region) since it is then possible to build resonant circuits in the "distributed constant" (as opposed to "lumped constant") techniques.

11.1.4. Magnetic Film Memory Devices

The use of magnetic films for the construction of storage devices, rather than the use of sizeable amounts of magnetic material as in a core is attractive for several reasons: Less energy is required to drive an element, due to the decreased volume of material which has to be switched; the switching time is usually shorter, due to the decrease in eddy currents; and the mass fabrication of the device may be simpler. The magnetic properties of films are normally identical to those of larger amounts of material, unless the film becomes extremely thin (say 10^{-5} centimeters). When those properties are employed, the device is expressly identified as "thin film" device.

The *Ferrit-Sheet* memory bears a rather close resemblance to a normal core memory. A relatively large sheet of magnetic material is deposited on an insulating board. A number of holes through this arrangement accomodate drive, inhibit and sense wires as in a core memory. The magnetic material in the neighborhood of a hole acts as the equivalent of a magnetic core.

The Rod-Memory: The basic arrangement of a magnetic rod cell is shown in Fig. 11.24:

A glass rod carries a deposited conductor and, on top of it, an electroplated magnetic film. The film can be magnetized by coils around the rod. The usual arrangement consists of four coils per cell wound on top of each other: the X- and Y-drive windings; an inhibit winding; and a sense winding. The function of these is equivalent to the windings in a coincident current magnetic core memory[1]. The part of the magnetic film covered by

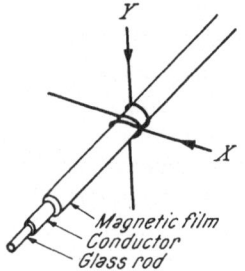

Fig. 11.24. Arrangement of a Rod-Cell

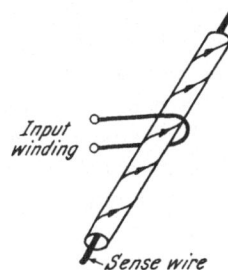

Fig. 11.25. Transformer Action of Helical Devices

the coils may be magnetized in one of two directions and can, therefore, act as binary storage element. The switching speeds which can be attained with this device are higher than those of magnetic cores.

Twistors, Tensors, Bit-Wires, and Bit-Tubes are similar to the rod in appearance, but use a helical magnetization of the material. One of the advantages of these devices lies in the transformer action depicted in Fig. 11.25.

The magnetic material is fabricated so that it forces any magnetic flux into a helical path. The direction of the flux can be reversed, for example, by a current through the input winding as shown in Fig. 11.25. When the flux switches, the change in the circular component of the magnetic field induces a voltage into the sense wire. This voltage is larger than the voltage which would be induced into one turn of a conventional output winding around the arrangement.

Let us now show the detailed operation for one of these devices, say the twistor. (See Fig. 11.26.)

A thin foil of magnetic material is wrapped around a conventional wire in helical fashion. The arrangement thus prefers a magnetic flux in direction of the helix. The wire itself serves as drive and sense line. One additional drive line is wound around the twistor, frequently in form of a flat metal strip.

In order to write, i.e. to set up a flux in one of the two possible directions, the twistor wire is energized by current i_1 and the drive winding by a current i_2. The currents are of such magnitude that neither of them in

---

[1] See paragraph 8.3.2.

itself is sufficient to change the magnetic flux in the foil. The current through the twistor produces a circular magnetic field, the current through the drive winding produces an axial field in the foil. At the intersection of the twistor and the drive winding, both fields superimpose and produce a helical field strong enough to affect the magnetic flux. If both currents have the direction indicated in Fig. 11.26, the resulting flux has the shown direction. For opposite currents, a flux in the opposite direction is produced. Defining one direction as the logic "1" and the other as the logic "0", we have a binary storage element.

In order to read the stored information, the flux is set to the "0" direction by a strong current through the drive winding. If there has been a "1" stored previously, the twistor wire will produce an output signal. For a "0", there will be no signal. The given arrangement is fairly simple, but can be used only for memories with linear selection[1]. For such a memory, the drive winding is associated with an address (word drive) and the sense wire acts as information conductor. The word drive is common to all bits of a "storage location" and an information conductor is common to a specific bit in all words.

A true coincidence selection can be achieved with the arrangement shown in Fig. 11.27.

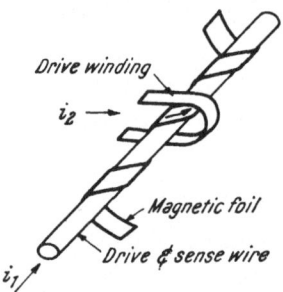

Fig. 11.26. Basic Twistor
Cell

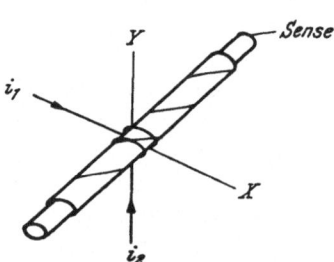

Fig. 11.27. Twistor Cell with
Coincidence Selection

Each cell carries two drive windings. For both, the read and the write operation, a single cell at the intersection of both drive lines is driven by the matrix. The twistor wire serves now only as a sense line. We notice that the drive produces a strictly axial field in the cell, and the satisfactory operation depends upon the twistors ability to force the flux into a helical path.

The *Tensor* and the *Bit-Wire* have the basic properties of the twistor, and differ only with respect to their manufacturing processes. The tensor

---

[1] Compare paragraph 8.3.1.

is an electrically and magnetically conducting wire. Originally, the wire preferred an axial magnetic flux, but by twisting and keeping the wire under tension, a helical flux is now preferred[1]. The bit-wire consists of a normal wire which is electro-plated with a magnetic material, while a helical magnetic field is applied. Due to this process, a helical flux is preferred by the material. Devices without a preference for a helical flux can be operated in a helical manner. Fig. 11.28 shows the principle.

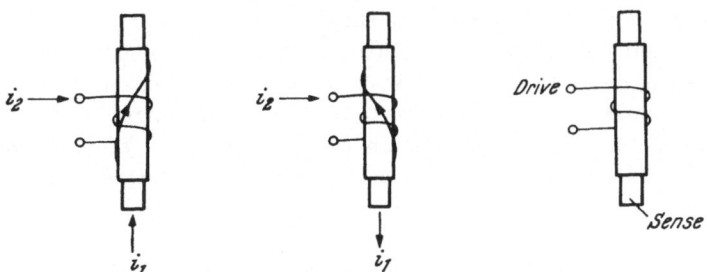

Fig. 11.28. Helical Operation of Non-Helical Devices

A conventional wire carries a coat of uniform magnetic material. For writing "1's" or "0's", a helical flux of one or the other direction is produced by the coincidence of an axial and a circular magnetic field. In order to

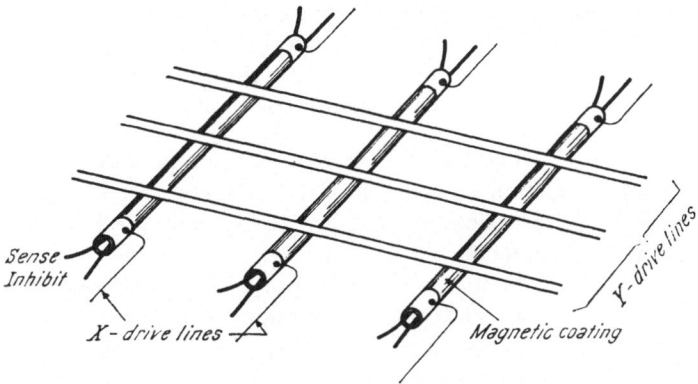

Fig. 11.29. Bit-Tube Memory

read the written information, the flux is switched to an axial direction by a current through the drive winding. Depending upon its previous direction, a signal of one or the other polarity is induced into the sense wire.

---

[1] The tensor is no magnetic-film device and only of historical interest.

The *Bit-Tube* is more complicated to manufacture than other helical devices, but its greater flexibility allows a more convenient operation. A metallic or metal-coated tube carries a coating of magnetic material with preference for a helical magnetic flux. The tube itself serves usually as drive line, but can also accommodate one or more additional wires. Fig. 11.29 shows the arrangement of a bit-tube memory with coincidence selection.

Individual cells of the memory are selected by the matrix-like arrangement of X- and Y-Drive lines. At the intersection of two energized drive lines, a helical field is produced with sufficient magnitude to switch the flux. The magnetic flux may be switched to one of the two possible directions depending upon the direction of the drive currents. For a write operation, both drive lines are energized regardless of whether a "0" or a "1" should be written. A current through the inhibit winding, counteracting the X-drive, suppresses a change in flux if a "0" is to be written.

In order to read, both drive lines are again energized, but now with currents in reverse directions. The flux in cells containing "1's" reverses, and an output signal is induced into the sense wire. Cells containing "0's" produce no output signal. Since the X-drive line and the sense line are routed in parallel, the X-drive will induce a signal into the sense winding regardless of whether or not a "1" was contained in the selected cell. To avoid any ambiguity, the X-line is energized first, then after the undesired transient has died out, the Y-line is energized and the sense is observed.

Compared to other helical devices, the bit-tube has the advantage of lending itself easily to true coincident selection, while still retaining the advantages of helical flux devices for both the read and the write operation.

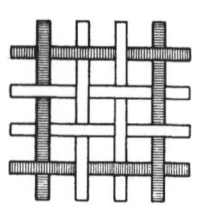

Fig. 11.30.
Woven-Screen
Memory Cell

Problem 9: Draw a simple (preferably perspective) sketch showing how the address register and the information register are to be connected to the memory elements

a) in a twistor memory with linear selection

b) in a bit tube memory with coincidence selection.

Woven-Screen Memory: Fig. 11.30 shows a single cell of a woven screen memory.

The memory is fabricated by weeving bare, and insulated wires on a loom. Originally bare wires are then electroplated with a film of magnetic material, while the insulated wires are not affected by the plating process. The cells formed by the plated wires (shaded wires in Fig. 11.30) act as binary storage elements: they can be magnetized in clockwise or counterclockwise direction. In effect, they are the equivalent of square shaped magnetic cores. In order to control the operation of the memory, each cell is threaded by four control wires. The function of these corresponds to the functions of the control wires in a core memory with coincident-current selection. Individual "active" cells are usually separated by "buffer" cells which minimize the crosstalk.

Problem 10: Assume that a clockwise magnetization of the cell shown in Fig. 11.30 represents a stored "1", while a counter-clockwise magnetization represents a "0". Show the direction of the currents in the drive lines and the inhibit line

    a) for writing a "1",
    b) for writing a "0".

Thin Film Cells: One of the advantages of thin film storage devices is the short switching time which can be achieved because of the small amount of magnetic material involved. Cells of this type may be manufactured by evaporation, masking and printed circuit techniques. Moreover, the unique properties of extremely thin films make film memories even more attractive. A disadvantage of film cells is the relative low output signal. This not only makes noise a serious problem, but also requires high-gain amplification. The inherent delay of amplifiers offsets to some extent the extremely fast switching times of thin film devices.

Properties of Thin Films: Thin films of proper composition and fabrication show a decided preference for a magnetic flux in an "easy" direction[1].

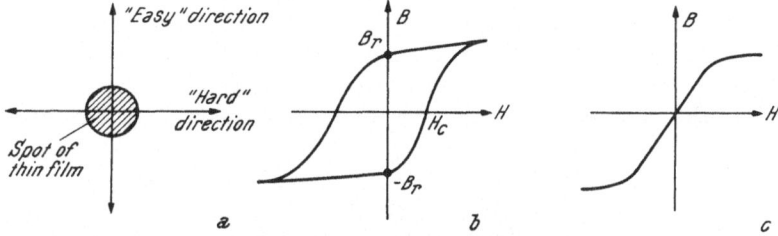

Fig. 11.31. Hysteresis of Thin Films

This preference is indicated in the hysteresis loops shown in Fig. 11.31. If the magnetization by an external magnetic field is parallel to the easy direction, the material exhibits a hysteresis loop as shown in Fig. 11.31b, which indicates a relatively high remanence.

Flux reversals perpendicular to the easy direction (sometimes referred to as the "hard" direction) are characterized by the single hysteresis curve

---

[1] They may be fabricated by vacuum deposition of the magnetic material on glass substrates in the presence of an external magnetic field.

of Fig. 11.31c. This latter curve indicates that there is no remanent flux
in the hard direction. Once the exciting field H is removed, the induction B
and, therefore, the magnetic flux disappears. Consequently, any remanent
magnetism must be in the easy direction. In other words, if a spot of thin
film is left alone, it will show a magnetization in the easy direction if it
shows any magnetization at all. Corresponding to the two points labelled
B_r (in Fig. 11.31b), there are two stable states in which the remanent flux
can remain. (Both are in the easy direction, but have opposite signs).

Suppose now we apply an external field in the hard direction to an
already magnetized spot. We can consider the resulting flux to be composed of two components, one in the easy, and one in the hard direction.
A relatively small component (corresponding to a weak external field) in

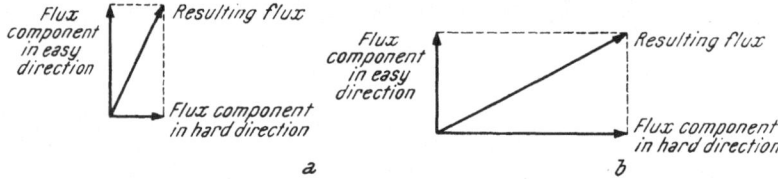

Fig. 11.32. Rotation of the Flux in a Thin-Film Spot

the hard direction will result in a total flux which has almost the easy
direction (Fig. 11.32a). The stronger the external field in the hard direction
is, the more will the resulting flux deviate from the easy direction
(Fig. 11.32b). In other words, by increasing the field in the hard direction,
we can rotate the resulting flux away from the easy direction. The component of flux in the hard direction, and hence the rotation of the resultant
field, persists only so long as the external field is applied. As soon as it is
removed, the flux in the hard direction disappears and the resulting

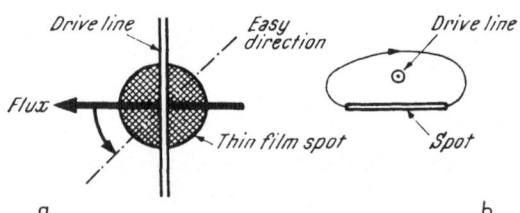

Fig. 11.33. Write Operation

field rotates back to the easy direction. This property, i.e., that the flux
pattern may be rotated by an external field and that it will return to its
original configuration when the external field is removed, is the basis for
the operation of thin film memory cells.

Let us assume we have a spot of thin magnetic film and a drive line arranged as shown in Fig. 11.33.

A current i with the shown direction magnetizes the spot. If the current is sufficiently strong, the flux in the material will rotate into the indicated direction regardless of any previous magnetization. At the termination of the drive current, the flux will rotate to the nearest easy direction (to the lower left in Fig. 11.33a). A drive current in opposite direction would produce the opposite flux. Terminating the drive, the flux would rotate to the opposite easy direction (to the upper right in Fig. 11.33a).

Defining the two possible stable states as "1" and "0" respectively, we have a binary storage element and the possibility of writing into it.

In order to read the stored information, a renewed drive is applied so that the flux rotates to the direction of the sense line (Fig. 11.34).

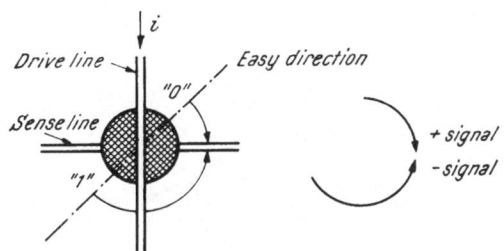

Fig. 11.34. Read Operation

Depending upon whether a "0" or "1" was stored previously, the flux will rotate in clockwise or counterclockwise direction and induce a positive or negative signal into the sense line. After completion of the read operation, the flux will rotate to the "0" direction.

In principle, it should be possible to build a thin film memory with true coincident selection operated like a core memory or the bit-tube memory. The thin film memories reported so far use a linear selection with inhibit-like control. A possible write mechanism is shown in Fig. 11.35.

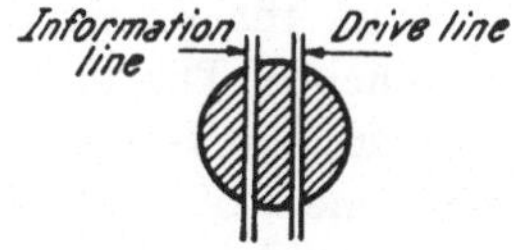

The single drive line of Fig. 11.33 is replaced by two wires: a drive line and an information line. In order to write a "1", the drive and information lines carry a current in the same direction. The resulting field is strong enough to switch the flux in the spot to the "1" direction. When a "0" has to be entered, the flux change is suppressed (by a current in opposite direction through the information line) although the drive line is still

Fig. 11.35. Linear Selection with Inhibit-Like Control

energized. Currents in the information line are somewhat smaller in magnitude than those through the drive line. This avoids disturbing other spots covered by the same information line. The disturbance of other spots by the drive line is not critical since all those spots are written into at the same time (linear selection). Of course, one can use a single line instead of separate drive and information lines. The current in this line has then to be controlled so that its magnitude is equivalent to the combined currents of the drive and information line in the previously discussed scheme.

The thin film spots may be round or rectangular. Experiments have also been performed with continuous thin film sheets. With the latter arrangement, only the material in the neighborhood of the intersection of drive lines is affected by the drive currents and acts as "spot".

Fig. 11.36 shows the design of a woven thin film memory in which the magnetic material is deposited on a cylindrical surface. The coating exhibits a circumferential easy direction, and circular flux patterns represent stored "0's" and "1's".

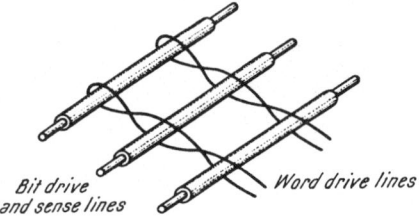

Fig. 11.36. Woven Thin-Film Memory

In order to write, a relatively strong current is applied to the word drive line which rotates the flux to an almost axial direction. A smaller current through the bit drive line is then used to overcome the circumferential remanence. The direction of the bit current depends upon the bit to be written, and serves so to establish a clockwise or counterclockwise circular component of the flux. When now the word current, and subsequently the bit current, is removed, the flux "slips" into the desired circumferential direction. In order to read, a current is applied to the word line, and the output of the bit line (now serving as sense line) is observed. Depending upon the information stored, the flux rotates in different directions, and produces output signals of opposite polarities.

Compared with other thin film designs, the arrangement has the advantage that the flux representing the stored information closes entirely through magnetic material. Consequently, neighboring cells disturb each other less.

11.1.5. Nondestructive Read for Magnetic Devices

As we have previously seen, magnetic cores are usually reset to the "0" state in order to read the stored information. Any previously stored information is, consequently, lost and we speak of a "destructive" read. If the information is to be retained, either a "restore" operation is required or a nondestructive read method has to be devised. This latter approach may increase the operating speed of normal memories, since no time for the restoring of information is required, but it is even more attractive for use in "associative" memories, where the read-out may be by bit, rather than by word, and where restore operations are most inconvenient to handle.

There are several possible approaches to a nondestructive read. One of them has been shown already, the use of transfluxors instead of regular cores[1]. Following is a short description of other principles or techniques. Most of these nondestructive read methods apply enough excitation to the core to disturb the flux temporarily but not enough to reverse the flux permanently. Practically all these methods are characterized by producing output signals of relatively low power so that they can be used essentially only for memories, but not for core switching circuits where the output of one core has to drive one or more cores without the benefit of amplification.

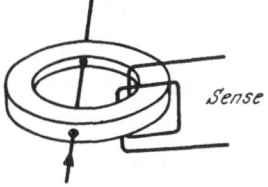

Disturbing Read: Suppose we provide a normal core with an additional aperture as shown in Fig. 11.37.

An interrogating current is applied to a wire through this aperture in order to read, i.e. to detect the magnetic state of the core. Fig. 11.38

Fig. 11.37. Core with Additional Aperture for Disturbing Read

shows the resulting fields (stretched into a plane). Assumed here is a magnetic saturation of the material from the left to the right (corresponding to a counterclockwise magnetization in Fig. 11.37) and an interrogate current of the direction indicated in Fig. 11.37.

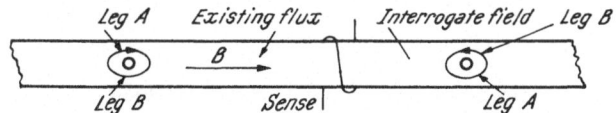

Fig. 11.38. Interrogating Fields in the Core of Fig. 11.38

A part of the interrogating field aids the existing flux, another part counteracts it. Particularly, if the core was previously saturated as in-

---

[1] See paragraph 11.1.2.

dicated in Fig. 11.38, no increase of flux in leg A can take place, only a reduction of flux in leg B. The net effect is, therefore, a decrease in the circular flux component. This decrease induces a signal of specific polarity into the sense winding. If the core was originally saturated in a direction opposite to the one shown in Fig. 11.38, the interrogation also decreases the flux but now in opposite direction, and a signal of opposite polarity is induced into the sense winding.

An extremely powerful interrogation signal may reduce the net flux in the core to zero as shown in Fig. 11.39, but it will never generate a net flux opposite to the one previously established.

Fig. 11.39. Flux Produced by a Rather Strong Interrogating Current

Even in this case, the magnetic "elasticity" of the core will re-establish a small net flux in the original direction, after the interrogate pulse has ceased. Any subsequent interrogation decreases the flux to zero, and the (small) net change in flux can be repeatedly detected by the sense winding without re-writing the original information.

Problem 11: Show the threading of drive, sense and interrogate lines in a core memory with linear selection which uses the disturbing read.

A different approach to a disturbing read is indicated in Fig. 11.40.

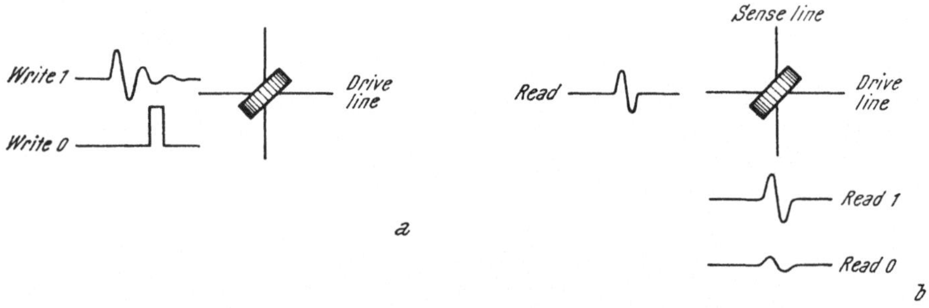

Fig. 11.40. Alternate Disturbing Read

The approach uses the fact that disturbing a core produces a relatively small signal when the core is magnetically saturated, but a relatively large signal when the core is de-magnetized. The writing of information is indicated in Fig. 11.40a. A relatively strong DC current saturates the core for the writing of a "0". For the writing of a "1", the core is demagnetized by an

AC signal of decreasing amplitude. In order to read, a bi-polar signal of
low amplitude is applied to the core as indicated in Fig. 11.40b. The signal
disturbs the flux-state, but is not sufficient to alter it permanently. If the
core is saturated, (i.e. a "0" is stored) only a small output signal is produced.
However, if the core is de-magnetized (i.e. storing a "1"), a much stronger
output signal is generated.

Readout by Radio Frequencies: Let us assume a normal core memory
with coincidence selection. The writing is performed as usually by energizing matrix lines with current pulses. In order to select a core for reading,
RF frequencies are applied to the X and Y drive lines which intersect the
desired core (see Fig. 11.41).

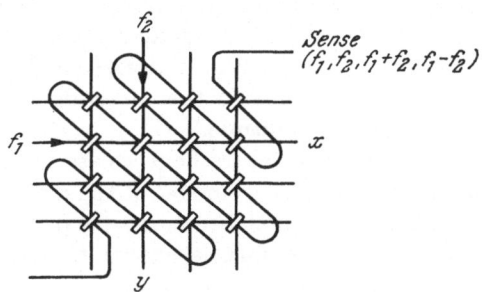

Fig. 11.41. Readout by Radio Frequencies

A frequency f_1 is applied to the proper X drive line, and a frequency f_2
is applied to the proper Y drive line. The amplitudes of the RF signals are
relatively small so that the magnetic state of the cores is not disturbed. Un-
selected cores and RF cross talk induce either the fequency f_1 or f_2 into
the sense winding. The selected core, however, acts as an RF mixer and
induces the beat frequencies of f_1 and f_2, as well as the fundamentals. To
understand this effect we have to consider the selected core as a nonlinear
circuit element which has two frequencies on its input and produces, as
any other nonlinear device would do, also the sum and difference frequencies, $f_1 + f_2$ and $f_1 - f_2$ (in addition to some higher order terms) on its output. Analyzing the output of the sense winding, we find a mixture of induced
frequencies. The frequencies f_1 and f_2 are due to various sources, but the
sum and difference frequencies are due only to the selected core. The
phase of these frequencies depends upon the magnetic state of the selected
core, that is to say, the beat frequencies produced by a core which stores
a "1" will have the opposite phase than the beat frequencies produced by a
core which stores a "0". By filtering the beat frequencies from the output
of the sense winding and subsequent phase detection, the state of a selected
core can be read without destroying the stored information.

Problem 12 (Voluntary): Explain in simple terms why the magnetic state of the selected core influences the phase of the beat frequency.

Readout by Quadrature Fields: There is a number of devices in existence which employ quadrature fields to detect the magnetic state of a storage core. By a quadrature field, we mean a magnetic field which is externally applied so that it is oriented at right angles to the flux representing the

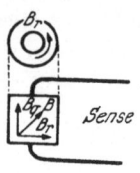

Fig. 11.42.
Storage Core
with Applied
Quadrature
Field

stored information. Fig. 11.42 shows the arrangement in principle. A core is magnetized by a flux B_r representing the stored information. An externally generated quadrature flux B_q is oriented at right angles to B_r. The total flux B, resulting from B_r and B_q, has a direction intermediate to the directions of B_r and B_q, but has a magnitude which is practically equal to B_r. (B_r is assumed to saturate the material so that the quadrature field can only change the direction of the flux but not increase its magnitude.)

By the application of the quadrature field we essentially can rotate the direction of the flux in the core. The removal of the quadrature field allows the flux to rotate back into its original direction with its original magnitude[1].

The rotation of the magnetic flux is shown in some more detail in Fig. 11.43.

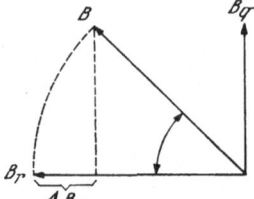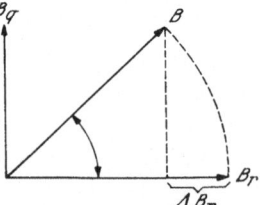

Fig. 11.43. Rotation of the Flux by the Quadrature Field

Fig. 11.42b represents the previously discussed case. Fig. 11.42a shows the diagram applicable to a core in which the flux B_r has the opposite direction corresponding to the storage of the complementary binary state. We see from the diagrams that the quadrature field rotates the flux clockwise or counterclockwise, depending upon the stored information. The circular component of the flux in the core which links the sense winding is in both cases decreased by an amount ΔB_r. A small signal is, therefore,

---

[1] This description is somewhat over-simplified. The first disturbance of the core by a quadrature field will reduce the flux to some extent but subsequent disturbances have very little effect on the magnitude of the flux.

induced into the sense winding in both cases, but, since the decrease of flux has different directions, the induced signal for one case is of opposite polarity than that for the other case. In order to detect the state of a core (non-destructively), it is then only necessary to apply momentarily a quadrature field and to observe the polarity of the signal induced into the sense winding.

There are many possible arrangements for producing the quadrature field. Fig. 11.44 shows a few of them.

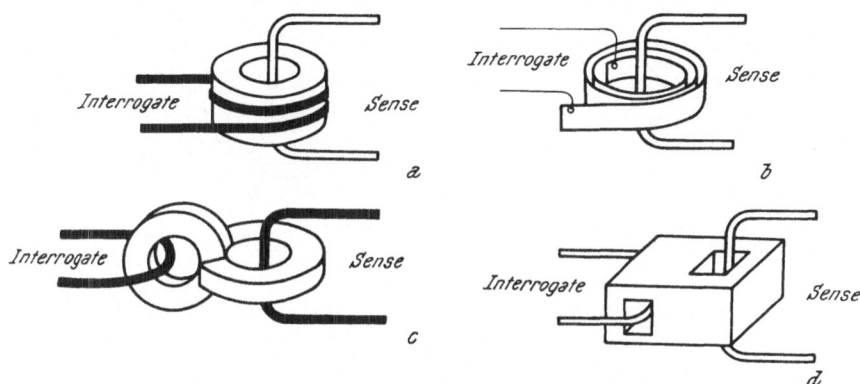

Fig. 11.44. Various Devices with Readout by Quadrature Fields

The device shown in Fig. 11.44a consists of a normal, sintered core with an interrogate winding around the outer edge. A current through the interrogate winding produces a quadrature field in the core. The flux lines of the quadrature field, by necessity, have to close through the surrounding air so that a relatively large interrogate current is required to produce a modest quadrature flux density. Fig. 11.44b shows an arrangement in which the laminated core material is used to carry the interrogating current. The devices in Fig. 11.44b and c allow a closure of the quadrature flux through magnetic material. The flux density of the quadrature field is, therefore, higher than in the two other devices, however, the area or the volume in which the quadrature field interacts with the storage flux is smaller.

Problem 13: Estimate how much voltage is induced into the sense wire by application of the quadrature field in comparison to a flux reversal. Assume the device shown in Fig. 11.44d and substantiate your estimate.

Readout with Pulses of Short Duration: Magnetic cores exhibit a particular behavior which offers the potential capability of a non-destructive read: they are rather insensitive to pulses of short duration, that is, the permanent flux in a core is little affected by very short current pulses or the corresponding pulses of the magnetic field strength. The mechanics of this phenomenon are perhaps not quite understood, but it seems that at least part of the effect can be explained in the following manner: When the magnetic flux in a core is switched by a current through the aperture, the complete flux reversal takes a finite time. A flux reversal takes place first on the inside of the core (where the driving field strength is the largest) and then propagates to the outside of the core. At a certain time during the process we may have a flux pattern as indicated in Fig. 11.45.

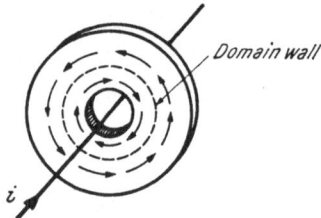

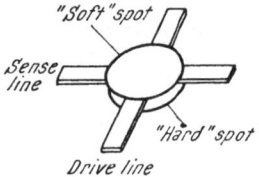

Fig. 11.45. Partially Switched Core   Fig. 11.46. Coupled Film Element

The current i has switched the flux on the inside but not yet on the outside of the core.

If the driving current is cut off before a complete flux reversal has taken place, the core is left in a partially switched state. If enough material is left in the original state, the core will restore itself to the original flux direction throughout its volume. With gross simplification, we might say that the magneto-motive force (mmf) of the unswitched volume overcomes the mmf of the already switched volume. The restoration process can be aided by the application of a DC bias current of proper direction, or by the proper selection of magnetically different materials for the inside and outside of the core.

The detection of the stored information is accomplished in the usual manner: by observing the output of a sense winding. The partial switching of the flux in the core produces an output signal only if the original magnetization of the core (say storage of a "1") and of the driving current oppose each other, but it does not take place if they are in the same direction (storage of a "0"). The amplitude of the output signal (and also the signal to noise ratio) can be increased if the short driving pulse is preceeded by a weaker one in opposite direction. The driving pulse itself may have a fairly large amplitude, provided the time during which it is applied is short.

Nondestructive read methods have been devised not only for magnetic cores, but also for thin film devices. One such device, the woven thin film memory has been shown previously[1]. The read current applies a quadrature field, the storage flux rotates away from the circular easy direction, and a signal is induced into the sense line. When the drive current is removed, the flux slips back into the easy direction. A similar approach can be devised for flat thin films. Here, however, the storage flux closes through air, and the rotation causes an appreciable weakening of the storage flux. This difficulty can be overcome by using two layers of films as indicated in Fig. 11.46.

The flux in one spot returns through the other spot so that the gap in the magnetic path has been considerably reduced when compared to single spots. Moreover, the two spots may have different composition. One of them is relatively hard to switch, and its flux is little affected by the small read current. The flux in the other is easily rotated. The output signal during a read operation is almost entirely due to the rotation of the flux in the "soft" spot. The magnetization of the "hard" spot helps to restore the original magnetization of the soft spot after the read current has terminated.

A basically different approach is indicated in Fig. 11.47. The approach

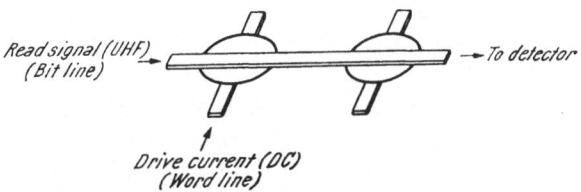

Read signal (UHF)
(Bit line)   → To detector

Drive current (DC)
(Word line)

Fig. 11.47. Readout by Resonant Absorbtion

uses the magnetic resonance of thin films at UHF- frequencies.With a given geometry, the resonant frequency is determined by the magnetic orientation of the film. The geometry and the frequency of the read signal are selected such that the resonance of an individual spot is excited only if it stores a "1" *and* if its flux is rotated appropriately by a DC drive current. The resonance of the spot causes an absorbtion of UHF energy and, consequently, a detectable attenuation of the UHF signal on the bit line.

As a final possibility let us mention the optical readout. The method could use the effect that the light reflected by a thin film spot has a direction of polarization which is dependent upon the direction of magnetization.

Fixed Storages: A nondestructive read for magnetic storage devices can be achieved with rather simple techniques if one is satisfied with a "read only" storage. The contents of such a storage cannot be altered electrically, rather it is necessary to perform mechanical alterations in order to change the "stored" information. Obviously such devices are of use only for the permanent or semi-permanent storage of information. However, in this

---

[1] See Fig. 11.36.

application the devices have their merits, expecially since their construction can be relatively simple and inexpensive compared to conventional memories.

Fig. 11.48 shows the fundamental arrangement of a *wired core* memory[1]. The stored information is determined by the manner in which the device is wired.

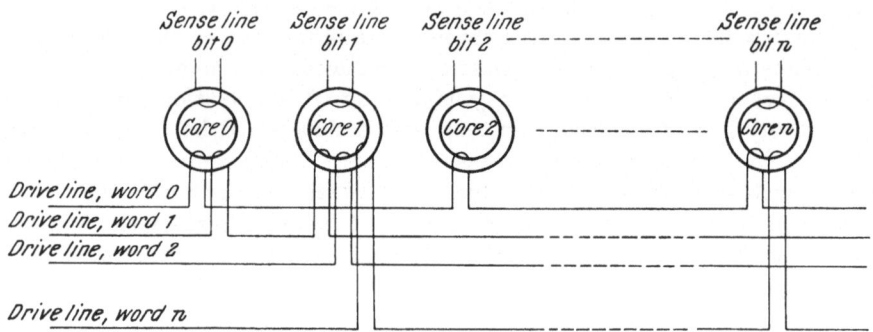

Fig. 11.48. Wired Core Memory with One Core per Bit

The memory contains one core per bit. If the word-length is, say, 36 bits, the memory contains only a total at 36 cores, regardless of the number of words which are stored. The word drive lines thread the cores or by-passe them according to the stored information. In this manner, each word drive may or may not produce an output on a specific bit sense line. For instance, the energizing of the drive line for word 0 in the arrangement shown in Fig. 11.48 will produce an output from cores, 0, 2, n since it is threaded through these cores, but it will not produce an output from core 1 since it by-passes this core. In effect, the word 0 "contains" the information 1 0 1 .... 1. An alteration of the contents would require the re-threading of the word drive line according to the new information.

The actual design of a wired core memory may deviate somewhat from the arrangement shown in Fig. 11.48. For one, the memory needs a clear or reset line (threaded through all cores) so that the drive lines can repeatedly reverse the flux in the selected cores. Also, the word selection (omitted in Fig. 11.48) may be simplified by threading "address lines", rather than word drive lines through the core, and operating them with inhibit-like controls. The storage capacity (i.e. the number of words which can be accomodated) is limited by the number of wires which can be threaded through a core. Fig. 11.49 gives an alternate arrangement which practically has no re-

---

[1] This device is also referred to as wire rope memory, Dimond Switch, or wire-in memory.

strictions as far as the number of words is concerned, but is limited in the number of bits per word.

One core per word is provided. A word drive line switches a single core and the bit sense lines will pick up "0's" or "1's" dependent upon whether it threads or by-passes the core. For example, in the arrangement of

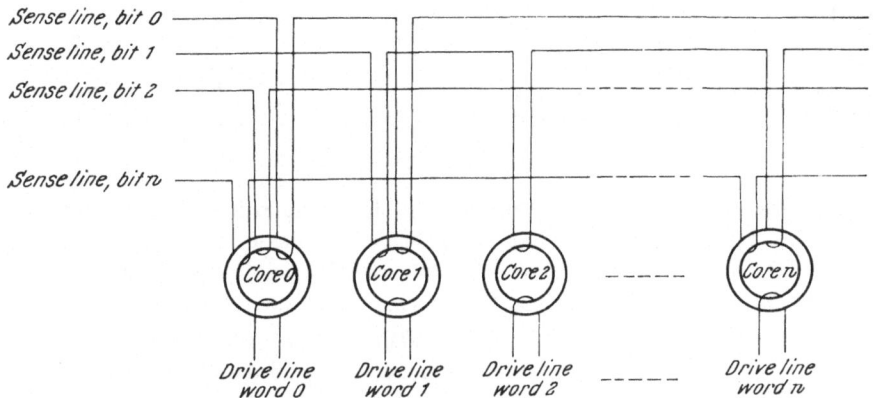

Fig. 11.49. Wired Core Memory with One Core per Word

Fig. 11.49, the sense lines for bits 0, 2, n thread the core associated with word 0 whereas the sense line for bit 1 bypasses this core. The "contents" of word 0 is, therefore, 101 ... 1 (The same as in the arrangement of Fig. 11.48). The arrangement of Fig. 11.49 requires more cores than that of Fig. 11.48 (if the number of words in the memory exceeds the number of bits in a word), but the storage capacity in words can be larger and the construction (i.e. the wiring of larger memories) is simplified. Here, one may have, for example, a true coincidence selection by X- and Y-drive lines as in normal core memories. The design of such a memory may be identical to a normal memory except that it will contain only one "core plane" and a peculiar threading of all sense lines through this single plane[1].

Problem 14: Show the threading of drive and sense lines for a fixed core memory

a) with one core per bit

b) with one core per word

[1] The term "nondestructive read" is perhaps not quite correctly applied to "read only" memories. As far as cores are concerned, a restore operation (i.e. a clear or reset) is required before the device can be interrogated again, but this operation is independent of the stored information (all cores are cleared or reset), and very simple compared to the restore operation in conventional core memories.

The memory shall contain the following information:

	Word 0	Word 1	Word 2 –	000	Word 62	Word 63
Bit 0	0	1	0		0	1
Bit 1	1	0	1		1	1
Bit 2	0	1	0		1	0
–						
Bit 34	1	1	1		1	0
Bit 35	1	0	1		1	0

Several read-only storages employing the same principle as wired core memories, but using a somewhat different construction have been proposed. Fig. 11.50 shows a design which in its organization is equivalent to the memory of Fig. 11.48, but uses *transformers* instead of cores.

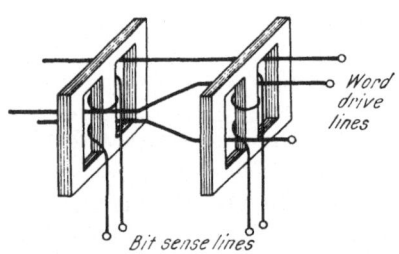

Fig. 11.50. Transformer Storage

When one of the word lines is driven, it will produce an output signal in each bit position. However, the polarity of this signal depends upon which of the two sides of the transformer core is threaded by the word line. One polarity is used to represent "0's", and the other to represent "1's". Compared with wired core memories, the arrangement has the advantages that no reset cycle is required, that relatively large output signals are produced, and that crosstalk is reduced. Moreover, the transformer core can be designed so that it can be easily "opened". This simplifies the alteration of the information content of an already existing memory.

The construction of a *magnetic rod* store is shown in Fig. 11.51.

The information content of the store is determined by the manner in which the information wires thread the rods. In one mode of operation, one of the solenoids is driven and the output of all information wires is observed. In an alternate mode, one of the information wires is driven, and the output of all solenoids is observed. The two modes correspond to those of a wired core memory with one core per word and one with one core per bit. The main advantage of the device lies in the easy changeability of the information content: it has been proposed to print the information wires on cards, and to stack these cards tightly onto the rods. Moreover, cards are fabricated identically so that they carry both, left and right by-passes for each rod.

Undesired paths are removed by punching holes into the cards at the appropriate positions.

The principle of a rather interesting magnetic read-only device is shown in Fig. 11.52.

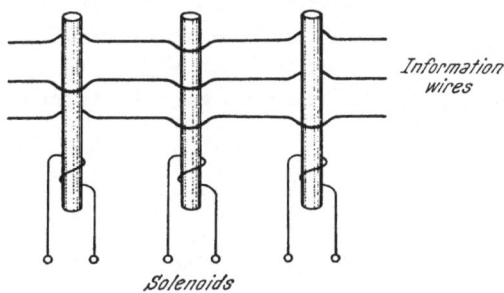

Fig. 11.51. Magnetic Rod Store

Word drive lines are arranged at right angles to the bit sense lines. The bit sense lines can be coupled to word drive lines by ferrite *slugs* or *pegs*. Word drives produce sense outputs or no outputs (ones or zeros) depending upon whether slugs are inserted or omitted at the appropriate intersection of drive and sense lines. The advantage of this arrangement is that the information content can be changed without changing the wiring of the device, simply by mechanically inserting or removing slugs.

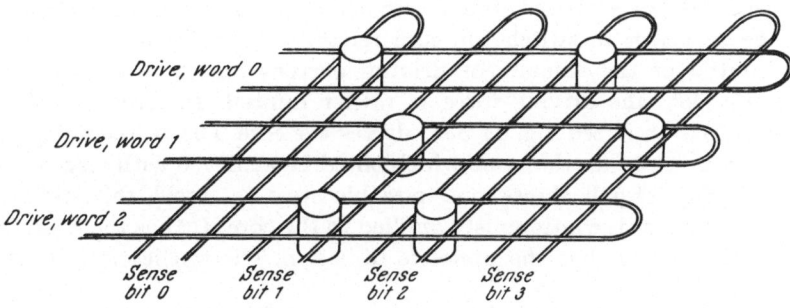

Fig. 11.52. Ferrite Peg or Slug Memory

There is at least one other possible approach to constructing fixed memories with magnetic devices. A small permanent magnet or a bias winding carrying a DC current can be used to restore the magnetic state of selected bit locations in an otherwise conventional memory. Working models of both twistor and core memories have been built using this approach[1].

11.1.6. Miscellaneous Magnetic Devices or Techniques

The Biased Switch: In many instances it is desired to have a coincidence selection of magnetic cores[2]. One of the encountered difficulties is that only a relatively small driving current can be allowed which unnecessarily restricts the switching speed of the core. Fig. 11.53a indicates some details of the problem.

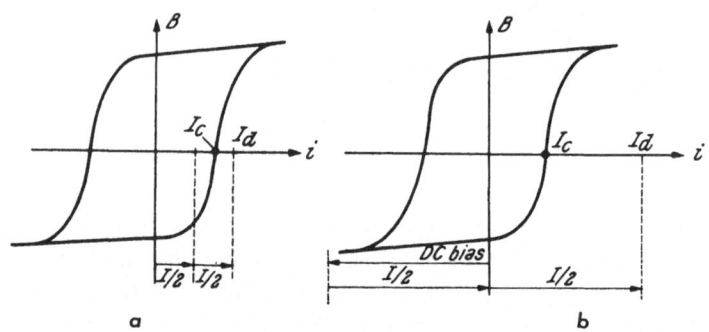

Fig. 11.53. Coincidence Selection of a Magnetic Core
a) Conventionally, b) With DC Bias

The current $I/2$ must be of small enough magnitude so that the flux in the core is not switched, but a current of twice this magnitude (i. e. I) must be sufficient to switch the core reliably. In addition to requiring a close control of the amount of driving current and a relatively square hysteresis loop, the driving force is rather limited. In essence, only the excess of I_d over I_c (see Fig. 11.53a) drives the core. The switching speed of the core is much lower than one which could be achieved with large driving currents. Fig. 11.53b shows one possible way to avoid this difficulty. A DC bias current is continuously applied to the core. $I/2$ has the magnitude of the bias current, but the opposite direction. The application of $I/2$ re-

---

[1] In addition, a number of non-magnetic read-only memories have been proposed or constructed. These include capacitive, resistive, and inductive devices. The information content is usually card-changeable. Capacitive devices use holes in a metallic shield to facilitate the coupling between selected matrix wires. Resistive and inductive devices use punched holes to disable undesired current paths.

[2] For example for the implementation of the logic AND in core switching circuits (see paragraph 11.1.1) or for the selection of cores in a core matrix as in core memories (see paragraph 8.3).

sults, therefore, in the removal of any external driving force. The flux in the core is with certainty not switched. The application of two times $I/2$ results in a large driving force (note the excess of I_d over I_c in Fig. 11.53b), and the core switches very fast. Using the indicated principle, we are able to build a fast and reliable core AND-circuit.

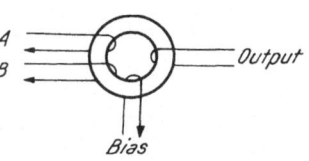

The DC bias normally keeps the core in the reset state. The application of both, A and B inputs is required, before the core is switched and an output is produced. The bias returns the core to its normal state after the inputs are de-energized.

Fig. 11.54. Biased Switch

The device shown is referred to as "biased switch" and is frequently used for the address selection in memories. In this application, the output of the biased switch is connected to the drive lines of a core or twistor array. The output current of the switch is directly used as driving current for these devices.

Problem 15: What DC bias and what amplitudes of driving currents would you use for a biased switch with triple coincidence selection (a core AND-circuit with three inputs)? Show your choice in a diagram similar to Fig. 11.53b.

The Multipath Core: Biasing can be applied very well to core switching. However, it cannot be applied without modification to the coincidence selection of memory cores. As we see from Fig. 11.53b, the DC bias resets the core and does, therefore, not allow the retention of two static states. This difficulty can be solved by the use of a multipath core similar to the one shown in Fig. 11.55.

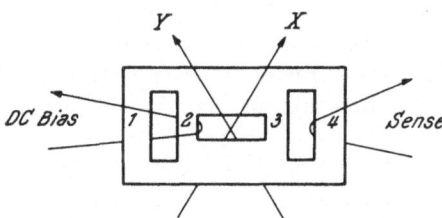

Fig. 11.55. Multipath Core

A DC bias current establishes a counterclockwise flux around the leftmost aperture. Due to the manner of winding and due to the geometry of the device, the bias current does not affect the flux pattern in the right half of the device.

When we apply a driving current through the center aperture, the bias must be overcome before any flux in the right half of the device is changed. In particular, if the driving current attempts to magnetize the device in

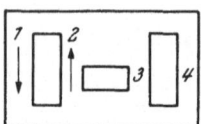

Fig. 11.56. Flux Pattern Produced by the Bias Current

clockwise direction, the flux in leg 1 must be reversed against the bias, before any of the flux in legs 3 or 4 is changed. If the driving current magnetizes counterclockwise, the flux in leg 2 must be reversed against the bias. Essentially then, no matter in what direction we drive, a bias must be overcome. The principle of the biased switch can now be applied to the coincidence selection of storage devices.

The state of the multipath core is not affected if only one of the X- or Y-drive lines is energized. (The DC bias prohibits any change of flux). In order to write a "1", both the X- and Y-drives of the selected core are

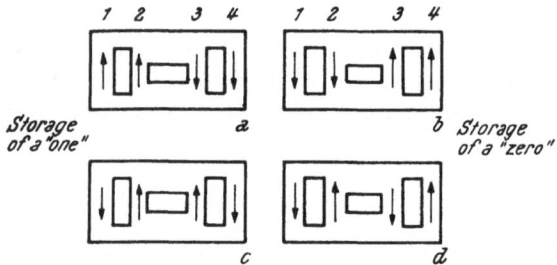

Fig. 11.57. Storage of Ones and Zeros in a Multipath Core

energized in the same direction, the bias is overcome, and a continuous flux around the center aperture, say in clockwise direction, is established (see Fig. 11.57a). When the driving currents have been removed, the bias re-establishes a counter-clockwise magnetization around the left-most aperture, whereby the direction of the flux in leg 1 and 3 is reversed (Fig. 11.57c).

In order to write a "0" (or in order to read), both X- and Y-drive lines carry a current in opposite direction. The flux pattern of Fig. 11.57b is established. When the drive has ceased to exist, the bias establishes the pattern shown in Fig. 11.57d. As in a conventional memory, the sense winding around leg 4 will pick up a signal for reading a "1", but no signal for reading a "0". We notice that for reading a "0" (i.e. a transition from pattern 11.57d to pattern 11.57b), the flux in leg 4 is not reversed, whereas for reading a "1" (i.e. a transition from pattern 11.57c to pattern 11.57b), the flux is reversed.

The advantage of a memory with multipath cores, compared to a conventional core memory with coincidence selection, is the relatively large driving current which can be applied and the resulting comparatively high speeds.

Two-core-per-bit memories: The principle of a memory using two cores per bit is indicated in Fig. 11.58.

One of the cores is magnetized in clockwise direction for the storage of a "1" and counterclockwise for the storage of a "0". The other core has always the opposite magnetization to the first core, so that for both, the storage of a "0" and the storage of a "1" the two cores have opposing magnetizations. The write operation is, of course, more complicated than that of one-core-per-bit memories, but presents no insurmountable difficulties.

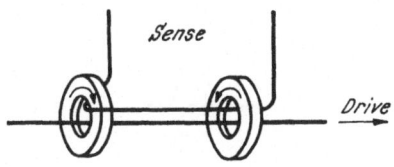

Fig. 11.58. Two-Core-per-Bit-Memory

During the read operation, a current is applied in the same direction to both cores (see Fig. 11.58). By necessity then one of the cores switches and produces an output signal. The signal has opposite polarities for reading "1's" and "0's", so that the state of a bit can be detected by polarity discrimination rather than by amplitude discrimination as in conventional core memories.

The Magnetic Core as Multi-Level Device: Storage cores are normally employed as binary devices i.e. they are magnetically saturated in one of two directions depending upon the binary information they store. If we would control the amount of flux in a core, in addition to its direction, we could design a multi-level storage device. For instance, one might operate a core as a 5-level device with the following five hypothetical states:

1 Full clockwise saturation

2 Some clockwise flux

3 Zero net flux

4 Some counterclockwise flux

5 Full counterclockwise saturation.

A potentially feasible technique for controlling the amount of flux in a core is indicated in Fig. 11.59.

The amount of flux which is changed in a core is determined by the amount of voltage which is applied to the drive winding and the length of time during which it is applied. Ideally, that is if the drive winding has zero resistance and if there are no fields external to the core, the amount of flux which is changed in the core is given by the voltage time integral as:

$$\Delta \Phi = \int e \, \mathrm{d}t \, ^1$$

A drive pulse of specific amplitude and duration would, for instance, change the core from state A to state B (see Fig. 11.59). The core traverses a "minor loop" (in contrast to the major loop with which we were concerned

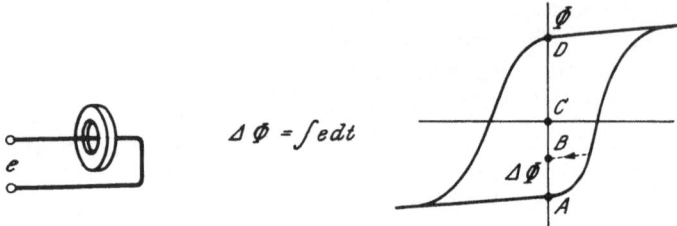

Fig. 11.59. Integrating Action of a Magnetic Core

so far.) By unit pulses of appropriate amplitude and duration, we could achieve flux changes $\Delta \Phi$ so that, for example, two unit pulses would drive the core from state A to state C, and four unit pulses would drive the core from state A to state D (see Fig. 11.59). Essentially we would have implemented the 5-level storage device described above. Such a device could be very attractive for the construction of counters (counting unit pulses.)

Although magnetic cores have been successfully operated as multi-level devices in the laboratory and in equipment prototypes, the unavoidably small margin of operation makes a wide-spread future application questionable. Driving current amplitudes and durations have to be kept to close tolerances. The device is rather sensitive to changes in ambient temperatures and the reading operation requires an unusually fine degree of discrimination.

Problem 16: Show the consecutive states of an integrating core during a sequence of two positive and three negative driving pulses. Assume that the core is originally in a magnetically neutral state. Draw a hysteresis loop similar to the one in Fig. 11.59 and show all transitions by arrows.

11.2. Superconductive Components

A number of low-temperature digital devices[2] uses the effect of superconductivity. The resistance of certain materials not only becomes very

---

[1] This equation can be obtained from the better known formula $e = -\mathrm{d}\Phi/\mathrm{d}t$ which describes the amount of voltage induced by a change in flux.

[2] Frequently (and not quite correctly) referred to as cryogenic devices.

small at low temperatures, say 0° K, to 17° K but disappears completely. Experiments set up to demonstrate this effect, show for instance that a current once induced into a superconducting lead ring remains for years without detectable deterioration.

The exact temperature at which a conductor becomes superconducting, depends upon the material, the strength of any ambient magnetic field, and to some extent, upon the mechanical stress to which the sample is subjected. It seems that, in general, it is easier to achieve superconductivity in materials which have a fairly low conductivity at normal temperatures, whereas it is relatively hard (or so far impossible) to produce superconductivity in normally good conductors.

Fig. 11.60 shows the approximate regions of conductivity and superconductivity for the two elements tantalum and niobium.

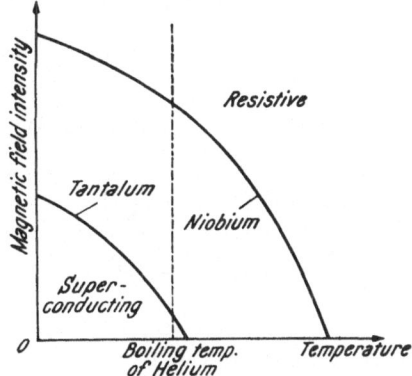

Fig. 11.60. Regions of Conductivity and Superconductivity Tantalum and Niobium

The curves shown are representative also for other materials exhibiting superconductivity. We see that the transition from the conducting state to the superconducting state takes place at higher temperatures when no, or when only a weak magnetic field is present, but requires lower temperatures when the magnetic field is stronger.

11.2.1. The Cryotron

Let us consider the arrangement shown in Fig. 11.61.

The device is immersed in a liquid helium bath maintained at its boiling temperature. Tantalum — at this temperature and in the absence of any magnetic field — is superconducting. However, the application of a relatively small magnetic field will cause the tantalum wire to loose its superconductivity (see Fig. 11.60). Such a field can be conveniently applied by a current through the superconducting niobium control wire. By the application or the removal of currents through the control winding, the tantalum wire can be easily switched from the superconducting to the conducting state, and vice versa.

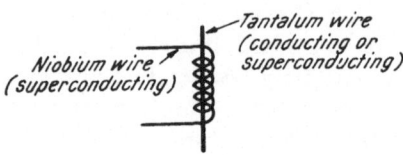

Fig. 11.61. Basic Cryotron

The cryotron can operate as an amplifier and perform logic operations. In order to understand these properties, let us first consider the circuit shown in Fig. 11.62.

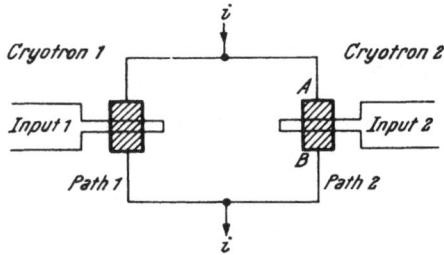

Fig. 11.62. Sample Circuit with Cryotrons

The two cryotrons in this circuit are film cryotrons. They are fabricated by depositing layers of appropriate materials on glass substrates. Their principle of operation is identical to that of the arrangement given in Fig. 11.62 although better operating characteristics at a lower cost can be achieved.

It is assumed that not more than one cryotron receives a control input at any one time, so that the current always finds a superconducting path from the top to the bottom of Fig. 11.62. Suppose for the moment that a control current is applied to cryotron 1. The cryotron becomes resistive and, consequently, the total current is switched to the superconducting path 2. If input 1 is deenergized and input 2 is energized, cryotron 2 becomes resistive and the current i switches to path 1.

Relatively large values of the current can be switched with fairly small control currents; thus the device acts as an amplifier. One might be tempted to think that the amount of current which is controlled depends entirely upon the (external) current supply and that the amplification factor is therefore essentially infinite. However, the load current through the cryotron produces its own magnetic field so that there is a practical limit to the amount of current which a superconducting cryotron can accommodate. The magnetic fields of currents larger than this limit make the cryotron resistive. The current amplification which can be achieved is therefore finite.

Although a superconducting cryotron has zero resistance, its inductance is finite. Consequently, a finite voltage must be momentarily applied before a superconducting current can be established. If we again consider the circuit in Fig. 11.62 and assume that the current i propagates at the moment through path 1, we see that we can remove the control current on input 2 (i.e. make the cryotron number 2 superconducting) without switching current from path 1 to path 2. A finite voltage would have to be applied between the points A and B in order to establish current through path 2. This, however, cannot be done without making cryotron number 1 resistive.

If one wants to make certain that always only one of two possible paths carries a superconducting current, one can make one path resistive with the current in the other path (see Fig. 11.63).

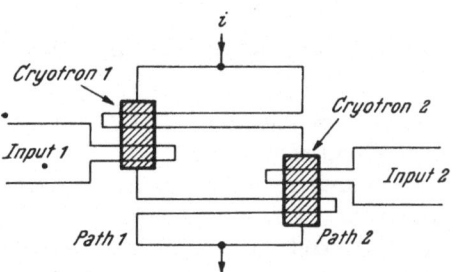

Fig. 11.63. Basic Cryotron Flip-Flop

A current through path 1 makes cryotron 2 resistive and current through path 2 makes cryotron 1 resistive. By necessity, one path is resistive if the other path is superconductive. The current can be switched from path 1 to path 2 and vice versa by control currents applied to the inputs. The device is equivalent to a conventional flip-flop with set and clear inputs.

A single cryotron in Fig. 11.63 acts as NOR-circuit. (The cryotron is superconductive only if neither of its inputs carries a control current.) Since we know that the NOR-operation is sufficient to perform any logic function and that the cryotron is capable of performing amplification and storage we have shown that the cryotron is a universal computer element.

Problem 17: How would you arrange one or more cryotrons to represent:
a) a NOT-circuit
b) an OR-circuit
c) an AND-circuit

Problem 18: How can you make the state of the cryotron flip-flop in Fig. 11.63 available for other circuits? In other words, how would you use the flip-flop "output" as inputs to other cryotrons?

Cryotron storage cells frequently employ circulating persistent currents to represent binary information. Fig. 11.64 shows one of many possible arrangements.

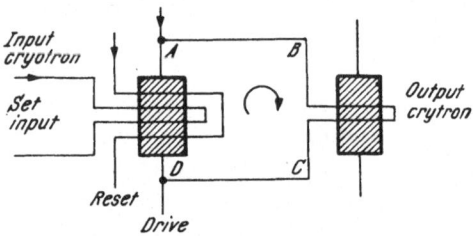

Fig. 11.64. Cryotron Storage Cell with Provisions for Input, Output and Reset

A recirculating current in the loop $ABCD$ represents the storage of a "1"; no current in the loop represents a "0". Once the current is set up, no external power is required to maintain it, since the loop is superconducting in the absence of any set or reset inputs. We speak of a persistent current. The contents of the cell can be read non-destructively by detecting whether the output cryotron is resistive or superconducting. The cell can be cleared rather easily by energizing the reset input. The input cryotron becomes then resistive, and the recirculating current ceases. The read and reset operations are fairly straightforward. In order to understand the write operation, it is necessary to realize that the inductance of a cryotron control winding is many times that of the main cryotron path[1]. Suppose now that a drive current is applied to the circuit of Fig. 11.64, while it is at rest. The current will select the path with the lowest inductance, that is, it will flow from A to B via the input cryotron[2]. If on the other hand, a set input is present while the drive current is applied, the input cryotron is

---

[1] One also speaks of the "grid" and the "gate" of a cryotron.
[2] This description is rather simplified. Since there is no resistance, the current divides inversely proportional to the inductances of the various paths. A comparitively small amount of current will flow in the path $BCDA$.

resistive and the total drive current is deflected to the path $ABCD$. Depending upon the state of the set input, the drive current flows either directly from A to B, or from A to B via B and C. This behavior is shown more clearly in the top part of Fig. 11.65. Current carrying conductors are shown in heavy lines.

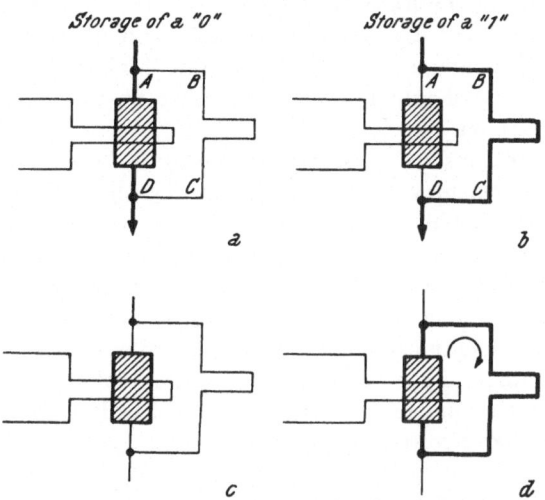

Fig. 11.65. Write Operation for the Storage Cell Shown in Fig. 11.65.

When the storage cell is in the state depicted in Fig. 11.65b and the drive is disconnected subsequent to the removal of the input current, a persistent current around the path $ABCD$ is set up. (Essentially the energy $\frac{LI^2}{2}$ which is stored in the inductance of the control winding produces a voltage between the points B and C which overcomes the rather small inductance of the path DA.) However, if the storage cell is in the state depicted in Fig. 11.65a when the drive current is removed, no persistent current is set up. (The inductance of the path AD is rather small so that the stored energy $\frac{LI^2}{2}$ is likewise rather small. The voltage produced between points A and D is not sufficient to set up an appreciable amount of current through the relatively large inductance of the path CB[1].)

Problem 19 (Voluntary): Design a cryotron shift register using the basic storage cell shown in Fig. 11.65. How many cryotrons per bit do you have

---

[1] If we assume an approximate ratio of 50 : 1 for the inductance of the path BC versus the inductance of path AD, the persistent current will also have an approximate ratio of 50 : 1 for the storage of a zero and a one. The small persistent current for the storage of a zero is not sufficient in magnitude to make the output cryotron resistive.

to use ? Show the sequence of control signals which has to be applied for a shift by one bit.

The cryotrons which we considered so far, are "crossed film" cryotrons. The grid is oriented at right angles to the gate. Consequently, the grid and the gate are magnetically de-coupled. There is no crosstalk from one to the other. "In-line" cryotrons have their grid in parallel to the gate. There is a certain amount of crosstalk, but higher impedances can be achieved, when this should be desired, simply by increasing the physical length of the cryotron.

Although cryotrons are presently used only in laboratory set-ups, their inherent advantages may lead to very interesting practical possibilities. Mass fabrication should make cryotrons very inexpensive compared to conventional components. Their speed may become comparable (switching times much shorter than one microsecond have already been achieved). The packing density of cryotron circuits may be extremely high. The power consumption of a computer may be in the order of one watt. The only serious disadvantage of cryotrons is the requirement for low temperatures. Present refrigerators are bulky and use several orders of magnitude more power than the cryotrons they have to cool.

11.2.2. Superconductive Storage Cells

In addition to cryotrons, a number of other superconductive devices are being investigated as potential memory elements. Most of these devices are switched from the superconducting to the conducting state by the current in the main path. No control grid is, therefore, required. The fabrication of these devices is simpler than that of cryotron cells. However, they require, as a rule, a destructive readout.

Fig. 11.66. Persistor Memory Cell

The Persistor: The basic circuit of a persistor is shown in Fig. 11.66.

An inductance parallels a conductor. Both are normally superconducting. However, the conductor will become resistive when it carries a current which is larger than a certain critical value. When both elements are in the superconducting state, the device can accommodate recirculating persistent currents. Clockwise and counterclockwise persistent currents represent the storage of binary zeros and ones.

In order to write (i.e. to set up a current of desired direction), a current of approximately twice the critical value is externally applied to the device (see Fig. 11.67).

At the first moment, practically the full amount of current will flow in the low inductance conductor. The conductor becomes resistive and more and more current will be deflected to the inductance. When the external drive is terminated, the energy stored in the inductance will set up a persistent current in the device. The magnitude of the persistent current is limited to the critical value. If we assume a direction of the drive current

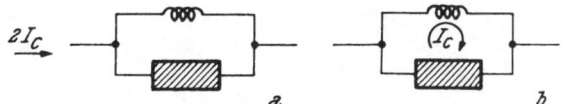

Fig. 11.67. Persistor Write Operation

as shown in Fig. 11.67a, the persistent current will have a clockwise direction (see Fig. 11.67b). A counterclockwise persistent current can be set up by a reversed drive current.

In order to read the content of the device, again an external drive current of approximate magnitude $2\,I_c$ is applied (see Fig. 11.68).

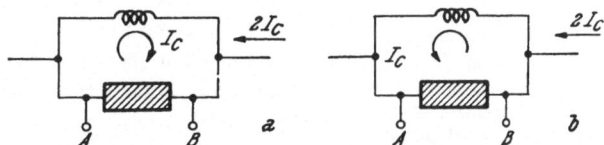

Fig. 11.68. Persistor Read Operation

If the persistent current and the drive current are additive in the conductor as shown in Fig. 11.68a, the conductor carries a current of magnitude $3\,I_c$. It becomes resistive, and a voltage appears between points A and B. If the currents are subtractive as in Fig. 11.68b, the conductor carries a current of magnitude I_c. It remains superconducting, and no voltage appears between the points A and B. The presence or absence of an output voltage signifies then the logic state of the device. It is to be noted that the read operation sets up a persistent current with a direction independent of the previously stored information. A restore operation is required, if the information is to be retained.

The Trapped Flux Memory Element: The physical appearance of an individual trapped flux memory element or Crowe cell can be seen in Fig. 11.69a.

An insulating base carries a film of superconducting material. A hole in the film is bridged by a crossbar. Binary zeros and ones are represented by superconducting persistent currents around the hole as shown in Fig. 11.70 b and c. The superconducting currents "trap" the magnetic flux linking the crossbar. The name of the device is derived from this property.

Persistent currents of one or the other direction can be set up by current pulses through the drive conductor. If the drive current exceeds the critical value, the crossbar becomes momentarily resistive. The persistent current

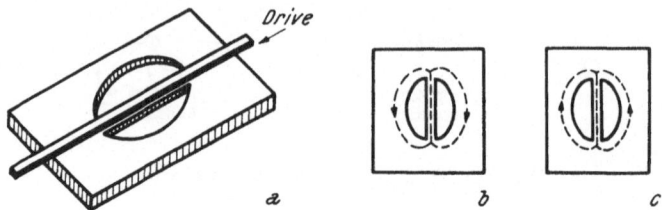

Fig. 11.69. Crowe Cell

deteriorates. When the drive, current is removed a persistent current is set up again. The direction of this persistent current depends only upon the direction of the drive current.

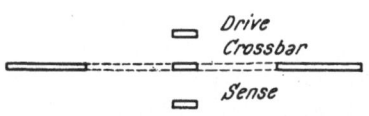

Fig. 11.70. Cross Section of a Crowe Cell with Drive and Sense Conductor

To permit the reading, the Crowe cell carries a sense conductor opposite to the drive conductor (see Fig. 11.70).

The sense conductor is normally perfectly shielded from the drive conductor by the superconducting film. A signal is induced only when the crossbar becomes resistive and the cell switches. The polarity of the signal depends upon the direction of the reversal.

Switch times in the order of 10 nanoseconds can be achieved. Unfortunately, there are thermic effects which restrict the actual operating speeds to lower values. It requires times in the order of 100 nanoseconds before a once conductive crossbar returns to the superconducting state.

The Continuous Sheet Memory: The continuous sheet memory has a principle of operation similar to that of the Crowe cell. Persistent currents in a figure "8" pattern are set up in a film of superconducting material. Here, however, no holes in the film are required.

X- and *Y*-drive lines carry the write currents. The magnetic field produced by an individual energized drive line is not strong enough to switch the superconductive sheet to the conductive state. Only in the vicinity of the intersection of the energized *X*-line with the energized *Y*-line, does the magnetic field strength exceed the critical value. The sheet becomes resistive. Any previously existing persistent currents deteriorate. When the drive terminates, persistent currents are set up as shown in Fig. 11.71. The direction of the persistent currents depends upon the direction of the drive currents.

A sense line, oriented at 45° with respect to the drive lines and located beneath the memory plane, facilitates the read operation. The sense line links all "storage cells" in a memory plane and is normally perfectly

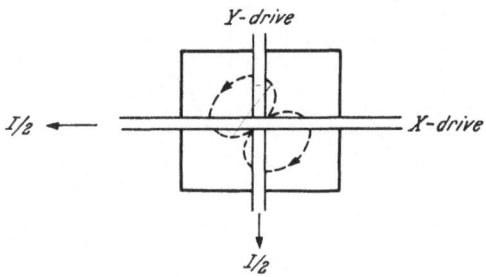

Fig. 11.71. Continuous Sheet Memory Cell

shielded from the drive lines. The shield is destroyed only at the specific selected location. The drive induces a signal into the sense line whose polarity depends upon the direction of the previously existing persistent current of the cell and the direction of the drive current.

11.3. The Tunnel Diode

The tunnel diode[1] is a semiconductor device exhibiting the current-voltage characteristic indicated in Fig. 11.72.

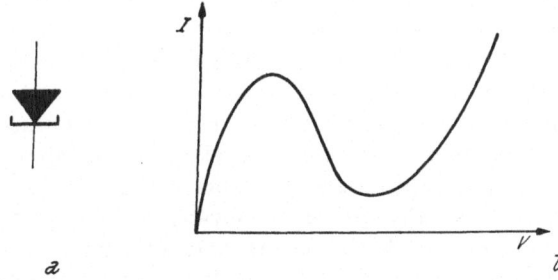

Fig. 11.72. Symbol and Current-Voltage Characteristic of a Tunnel Diode

It is convenient to consider this diagram as representing the normal behavior of a semiconductor diode as reflected in Fig. 11.73a, super-imposed by the tunneling effect as indicated in Fig. 11.73b.

---

[1] Also referred to as Esaki diode.

"Tunneling" is a quantum-mechanical phenomenon which allows electrons to penetrate a barrier of electric potential at the diode junction, even if, according to classical theory, they do not have sufficient energy to do so.

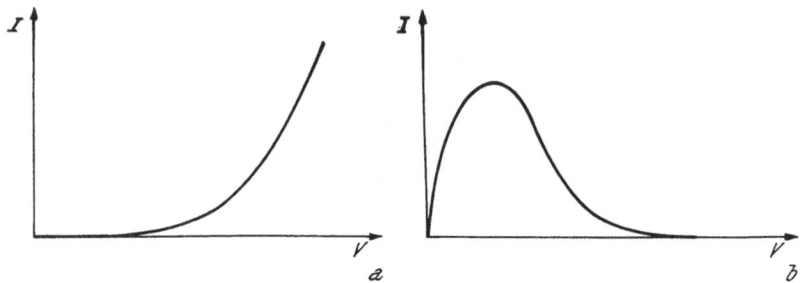

Fig. 11.73. Characteristic of a Normal Diode and of the Tunneling Effect

The effect may be compared to a tunnel in a mountain separating two bodies of water. In general, no water is exchanged, unless the water level in one of the bodies rises above the crest of the mountain. However, under certain conditions, the tunnel opens so that an exchange can take place. The "electron tunnel" opens only if the diode material has the proper composition, and only for a relatively narrow range of applied voltage (see Fig. 11.73b). The electrons travel through the tunnel with nearly the speed of light: they disappear at one side of the junction, and appear almost instantaneously at the other side.

We see from Fig. 11.73 that the resistance of the tunnel diode, i.e. the ratio of voltage versus current, is variable. Of particular interest is the region in which the current decreases when the voltage is increased: the resistance is negative. The negative resistance range provides the tunnel diode with the potential capability to act as an amplifier, just as the positive resistance of other devices enables them to act as attenuators.

The simplicity of the tunnel diode, its reliability, its speed of operation, and its low power consumption are desirable properties. Its disadvantages include the characteristically low signal levels, and the fact that two terminals must be shared by both, inputs and outputs.

The Tunnel Diode as Storage Element: The basic mode of operation of most tunnel diode circuits can be seen from Fig. 11.74.

A voltage, V_0, is applied to the tunnel diode through a resistor, R. Due to the current, I, through the combination, there will be a voltage drop accross the resistor, and the tunnel diode experiences a voltage, V, which, in general, is smaller than the voltage V_0. This relation between I and V is indicated by the straight "load line" in Fig. 11.74b. Alternately, the voltage accross the tunnel diode must be related to the current as reflected in the superimposed diode characteristic. We see, an equilibrium is possible only at the three points labelled A, B, and C. Point B represents an unstable condition[1], so that the circuit can practically be in only one of

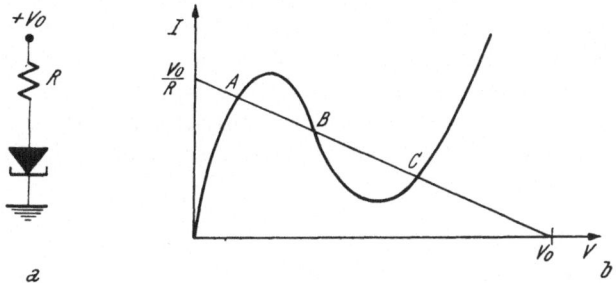

Fig. 11.74. Basic Tunnel Diode Storage Element

the two states A, or C. State A is characterized by a relatively low resistance of the tunnel diode, a large current, a large voltage drop accross the resistor, and a small voltage drop accross the diode. State C is characterized by a relatively high resistance, a small current, a small voltage drop accross the resistor, and a large voltage drop accross the diode. It is customary to associate the logic state "1" with the high resistance state, and the logic state "0" with the low resistance state.

For practical applications, the storage element is provided with input and output terminals as shown in Fig. 11.75a.

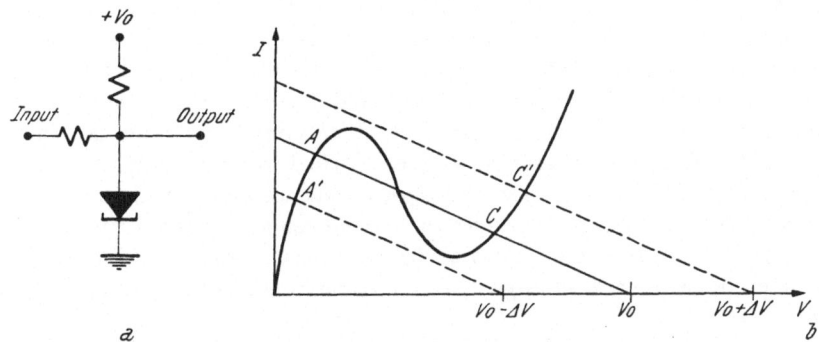

Fig. 11.75. Complete Tunnel Diode Storage Element

---

[1] Any increase in current causes a decrease in the resistance, which, in turn, increases the current. Conversely, a decrease in current causes an increase in resistance, and so on. The changes continue until stable operation either at point A, or C is achieved.

In order to switch the element to the "1" state, a positive going signal is applied to the input terminal. This has the same effect as increasing the supply voltage to the value $V_0 + \Delta V$, for which there is only one possible state (C' in Fig. 11.75b). When the input signal is removed, the circuit "slides" into state C. Conversely, a negative going input signal is applied to set the storage element to the "0" state. The supply voltage drops to a value $V_0 - \Delta V$, and the circuit is forced to assume the state A'. It slides into state A, when the signal is removed. The logic state of the element can be detected by sensing the voltage at the output terminal.

The basic circuit can be varied in a number of ways. Fig. 11.76 shows several possibilities.

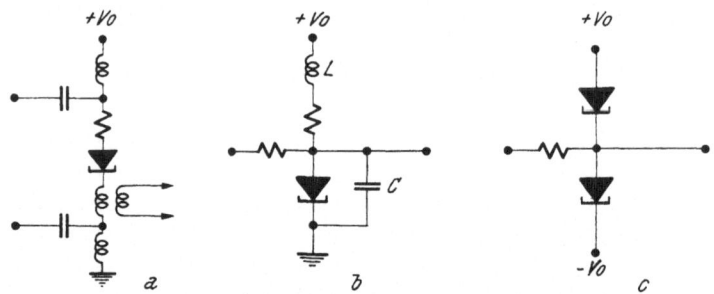

Fig. 11.76. Variation of the Basic Circuit

Fig. 11.76a represents a circuit with AC coupling. Input signals may be applied to one, the other, or both terminals simultaneously (with opposing polarities). An output signal is produced whenever the device changes states. Fig. 11.76b shows a device which, in the manner of a complement flip-flop, can be complemented by pulses of the same polarity. A prerequisite for the proper operation is that the input signal has a shape which stimulates a resonance-like response in the L and C components of the circuit. Fig. 11.76c, finally, represents a symmetrical circuit with two tunnel diodes[1]. One of the diodes is in the high-resistance state, while the other is in a low-resistance state. The circuit can be set and cleared by signals of the same magnitude, but of opposite polarities. Similarly, the two logic states produce output voltages of the same magnitude, but of different sign.

Tunnel Diode Memories: Several designs of tunnel diode memories have been investigated. Fig. 11.77 shows one of the simplest.

The operation of the memory is very similar to that of the magnetic core memory discussed in paragraph 8.3.1: it has a linear selection, and is

---

[1] Frequently referred to as Goto pair.

operated in the clear/write, read/restore mode. The individual tunnel diode receives its supply voltage from two sources, the bit line, and the word line. These lines are biased so that a set tunnel diode remains set, and a cleared diode remains cleared.

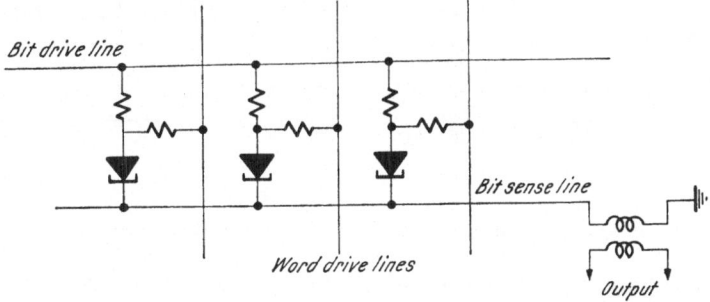

Fig. 11.77. Tunnel Diode Memory with Linear Selection

A storage location is *cleared* by a momentary drop of the voltage on the word line. All diodes associated with this line assume the "0" state. In order to *write*, the voltage on the word line is increased to a value which is slightly higher than normal, but not sufficient to set any diodes. Simultaneously, the voltage on the bit lines of those bit positions into which a "1" is to be written, is increased. Again, this increase, in itself, is not sufficient to set any tunnel diode. However, where the energized bit and word lines intersect, the voltage increase is sufficient to set the diode. In order to *read*, the storage location is cleared. The switching of the diodes which store a "1" causes an appreciable current change in the bit sense line, and induces a signal in the output transformer.

Problem 20 (Voluntary): Draw a diagram similar to the one in Fig. 11.75, and show the voltages and load lines for clearing a memory element, for writing a "0", and for writing a "1".

We have seen here the operation for only one design. Many variations have been proposed which reduce undesired signals, use voltage instead of current read-outs, connect tunnel diode cells serially rather than in parallel, and use additional linear or non-linear components to increase the margin of operation.

Tunnel Diode Shift Registers: Fig. 11.78 shows a tunnel diode shift register design which incorporates several storage elements of the type shown in Fig. 11.75.

The circuit, in itself, is bi-directional, but the phasing of supply voltages enforces a shift of information from left to right. Suppose for the moment that tunnel diode TD_1 stores a "1", while power phase I is applied. When phase II is applied, the "1" becomes transferred to TD_2. With phase III it is shifted to TD_3, and with the following phase I, to TD_4. When a tunnel diode stores a "0", the following diode is not set during the "on" phase of

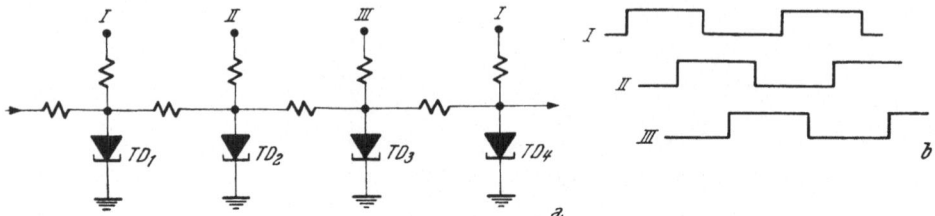

Fig. 11.78. Three-Phase Tunnel Diode Shift Register

its power supply. All diodes are cleared during the "off" cycle of their power supplies. In some respects, the "off" cycle is as important as the "on" cycle: it separates the individual bits stored in the register[1].

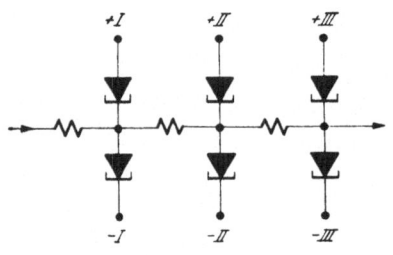

Fig. 11.79. Alternate Shift Register Design

With the given arrangement, a register requires three diodes for each bit of storage capacity. The number of storage elements per bit and the mode of operation is similar to those of the parametron[2]. In fact, the resemblance becomes striking when we consider the circuit shown in Fig. 11.79.

The design employs the storage elements of Fig. 11.76c. The arrangement has the advantage that the stages have to provide only relatively little signal power at their outputs: signals must only be strong enough to overcome the mis-balance of the two diodes in the next stage. Fig. 11.80 shows a third alternative for tunnel diode shift registers.

The interstage contains a conventional diode which prevents a propagation of signals from right to left. The supply voltages are applied in a two-phase cycle, and only two diodes per bit are required. The average voltage on the supply buses is such ·that the tunnel diodes are neither cleared nor set, but can retain their state. When supply I assumes a more positive voltage, all odd numbered diodes are set if their even numbered neighbors to the left contain "1's". All even numbered stages are then cleared by the negative swing of supply II. Next, the even numbered diodes are set if their

---

[1] Note that if power would be applied to all diodes simultaneously, they would all assume the same logical state.

[2] See paragraph 11.1.3.

neighbors to the left contain "1's", and, finally, the even numbered diodes are cleared by the negative swing of supply II. The information has become shifted by one bit-position.

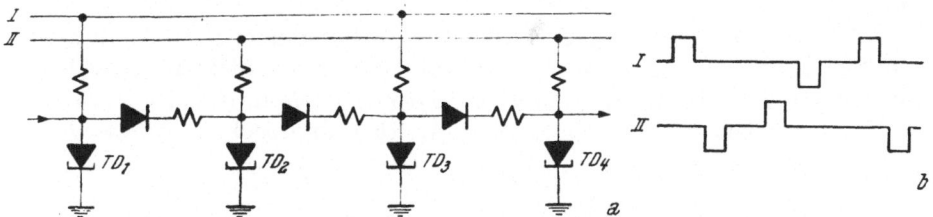

Fig. 11.80. Two-Phase Shift Register with Diode Interstage

Tunnel Diode Logic Elements: The diagram of a simple tunnel diode logic circuit is shown in Fig. 11.81.

We have encountered the basic circuit already in Figs. 11.75, 11.77, and 11.78, but perhaps did not realize that it acted as logic element. The circuit is set by positive signals on the input. The values of the resistors can be selected such that a positive signal on one of the inputs is sufficient to set the element. The circuit acts then as OR-circuit. However, with a different selection of resistors, the element acts as an AND-circuit: signals on both

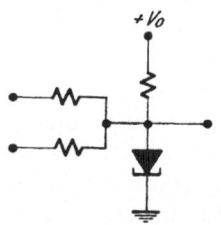

Fig. 11.81. Tunnel Diode AND or OR-Circuit

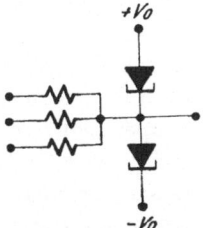

Fig. 11.82. Majority Element

inputs must be present to set the element. In order to clear the element, the supply voltage is removed. In fact, elements of this type are usually operated with the three-phase power cycle indicated in Fig. 11.78 b.

Problem 21: a) Draw a diagram of the current-voltage relations of the circuit in Fig. 11.81 when it is operated as an AND-circuit. Show the load line for zero, one, and two inputs.

b) What are the reasons that this circuit is normally only set by input signals, but cleared by the removal of supply voltages?

Fig. 11.82 shows the diagram of a majority element. It is operated in the manner of the shift register shown in Fig. 11.79. The circuit has an odd number of input terminals, and will assume the state of the majority of its inputs. The applications of this circuit are almost identical to that of the parametron[1].

A basic approach to implementing the logic function of NOT is indicated in Fig. 11.76a. A pulse transformer (with reversed leads) can provide a positive pulse, corresponding to a "1" output, when an element storing a "0" is switched by a probe signal. Another approach is indicated in Fig. 11.83a.

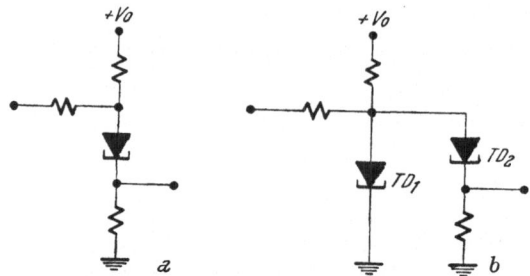

Fig. 11.83. Tunnel Diode NOT Circuit

A momentary positive signal on the input increases the voltage accross the tunnel diode and switches it to the high-resistance state. The current through the circuit decreases and produces a more negative level on the output. A negative going signal on the input switches the diode to the low-resistance state and produces a smaller voltage at the output. In order to achieve better operational margins, the circuit is usually employed in the form shown in Fig. 11.83b. Tunnel diode TD_2 corresponds to the diode in Fig. 11.83a. When it is in its low-resistance state (providing a high output voltage), tunnel diode TD_1 draws relatively little current. However, when TD_2 is in its high resistance state, TD_1 draws a relatively high current due to its non-linear characteristics. It diverts current from the path through TD_2 and, in effect lowers the already low output voltage.

Problem 22: Show that at least one of the tunnel diode circuits in paragraph 11.3 provides amplification.

11.4. Miscellaneous Devices

In addition to the circuits we have seen in this chapter, there is a number of experimental digital devices including multi-region semiconductor,

---

[1] See paragraph 11.1.3.

microwave, cryoelectric, ferroelectric, mechanical, optical, hydraulic, and even chemical devices. In their present state of development they do not seem to have the properties which would make their general application likely. Moreover, their discussion would require the treatment of related theories, which is well beyond the scope of this book. Let it suffice that the following bibliography contains several references for the interested reader.

Bibliography

Magnetic Cores as Logic Elements

GUTERMAN, KODIS, and RUHMAN: Logical and Control Functions Performed with Magnetic Cores, Proceedings IRE, vol. 43, No. 3, pp. 291—298. March 1955.

AUERBACH, and DISSON: Magnetic Elements in Arithmetic and Control Circuits, Elec. Eng., vol. 74, pp. 776—770. Sep. 1955.

LOEV, MIEHLE, PAIVINEN, and WYLEN: Magnetic Core Circuits for Digital Data Systems, Proceedings IRE, vol. 44, No. 2, pp. 154—162. Feb. 1956.

NEWHOUSE, and PRYWESS: High-Speed Shift Registers Using One Core per Bit, Transactions IRE, vol. EC-5, No. 3, pp. 114—120. Sep. 1956.

EINHORN S. N.: The Use of the Simplex Algorithm in the Mechanization of Boolean Functions by Means of Magnetic Cores, Transactions IRE, vol. EC-10, No. 4, pp. 615—622. Dec. 1961.

Transfluxors

RAJCHMAN, and LO: The Transfluxor, Proceedings IRE, vol. 44, No. 3, pp. 321—332. March 1956.

ABBOT, and SURAN: Multihole Ferrite Core Configurations and Applications, Proceedings IRE, vol. 45, No. 8, pp. 1081—1093. Aug. 1957.

PRYWES N. S.: Diodeless Magnetic Shift Registers Utilizing Transfluxors, Transactions IRE, vol. EC-7, No. 4, pp. 316—324. Dec. 1958.

CRANE H. D.: High-Speed Logic Systems Using Magnetic Elements, etc., Proceedings IRE, vol. 47, No. 1, pp. 63—73. Jan. 1959.

GIANDOLA, and CROWLEY: The Laddice — A Magnetic Device for Performing Logic, Bell System Tech. Journal, vol. 38, pp. 45—71. Jan. 1959.

CRANE, and DE RIET: Design of an All-Magnetic Computing System, Transactions IRE, vol. EC-10, No. 2, pp. 207—232. June 1961.

LEAYCRAFT, and MELAN: Characteristics of a High-Speed Multipath Core for a Coincident-Current Memory, Transactions IRE, vol. EC-11, No. 3, pp. 405—409. June 1962.

LUSSIER, and SCHNEIDER: All-Magnetic Content Addressed Memory, Electronic Industries, pp. 92—96. March 1963.

ANDERSON, and DIETMEYER: A Magnetic Ternary Device, Transactions IEEE, vol. EC-12, No. 5, pp. 911—914. Dec. 1963.

Phase-Locked Oscillators

MUROGA: Elementary Principle of Parametron and its Application to Digital Computers, Datamation, vol. 4, No. 5, pp. 31—34. Sept./Oct. 1958.

TERADA H.: The Parametron — An Amplifying Logical Element, Control Engineering. April 1959.

GOTO E.: The Parametron, a Digital Computing Element which Utilizes Parametric Oscillation, Proceedings IRE, vol. 47, No. 8, pp. 1304—1316. Aug. 1959.

ONYSHKEVYCH, KOSOMOCKY, and LO: Parametric Phase-Locked Oscillator —
Characteristics and Applications to Digital Systems, Transactions IRE,
vol. EC-8, No. 3, pp. 277—286. Sept. 1959.

HILIBRAND, MUELLER, STOCKER, and GOLD: Semiconductor Parametric Diodes
in Microwave Computer, Transactions IRE, vol. EC-8, No. 3, pp. 287—296.
Sept. 1959.

ABEYTA, BORGINI, and CROSBY: A Computer System Using Kilomegacycle
Subharmonic Oscillators, Proceedings IRE, vol. 49, No. 1, pp. 128—135.
Jan. 1961.

Magnetic Film Devices

BOBECK H. A.: A New Storage Element Suitable for Large-Sized Memory
Arrays — The Twistor, Bell System Technical Journal, vol. 36, pp. 1319 —
1340. Nov. 1957.

BOBECK, and FISCHER: A Reversible Diodeless Twistor Shift Register, Journal
of Applied Physics, vol. 30, pp. 39S—44S. Apr. 1959.

MEIER: A Millimicrosecond Switching and Storage Element, Journal of Applied
Physics, vol. 30, pp. 45S—46S. Apr. 1959.

BITTMAN E. E.: Thin-Film Memories, Transactions IRE, vol. EC-8, No. 2,
pp. 92—97. June 1959.

KOLK, and DOHERTY: Thin Magnetic Films for Computer Applications, Datamation, vol. 5, No. 5, pp. 8—12. Sept./Oct. 1959.

SCHWARTZ, and SALLO: Electro-Deposited Twistor and Bit Wire Components,
Transactions IRE, vol. EC-8, No. 4, pp. 465—469. Dec. 1959.

RAFFEL, CROWTHER, ANDERSON, and HERNDON: Magnetic Film Memory Design, Proceedings IRE, vol. 49, No. 1, pp. 155—164. Jan. 1961.

DAVIES, and WELLS: Investigation of a Woven Screen Mass Memory System.
AFIPS Conference Proceedings, vol. 24, pp. 311—326. Fall Joint Computer
Conference, 1963.

KOENIG M.: Magnetic Thin Films as Digital Storage Devices — part 2, Computer
Design, pp. 16—24. March 1964.

TOWNSEND, and FOX: Cylindrical Memory Device Characteristics, Transactions
IEEE, vol. EC-13, No. 3, pp. 261—268. June 1964.

CHONG, REID, and TURCZYN: Medium Speed Mass Random Access Memory,
Rome Air Development Center, Technical Documentary Report, RADC-
TR-64-571. March 1965.

MAEDA, TAKASHIMA, and KOLK: A High-Speed, Woven Read-Only Memory,
AFIPS Conference Proceedings, vol. 27, part 1, pp. 789—780. Fall Joint
Computer Conference, 1965.

Nondestructive Readout

WIDROW: A Radio-Frequency Nondestructive Readout for Magnetic Core
Memories, Transactions IRE, vol. EC-3, No. 4, pp. 12—15. Dec. 1954.

THORENSEN, and ARSENAULT: A Nondestructive Read for Magnetic Cores,
Proceedings of the Western Joint Computer Conference, pp. 111—116.
March 1955.

NEWHOUSE: The Use of Domain Wall Viscosity in Data Handling Devices,
Proceedings IRE, vol. 45, No. 11, pp. 1484—1492. Nov. 1957.

LOONEY D. H.: A Twistor Matrix Memory for Semipermanent Information.
Proceedings Western Joint Computer Conference, pp. 36—40. March 1959.

DeBUSKE, JANIK, and SIMONS: A Card Changeable Nondestructive Readout
Twistor Store, Proceedings Western Joint Computer Conference, pp. 41—46.
March 1959.

WANLASS, and WANLASS: BIAX High-Speed Magnetic Computer Element, IRE Wescon Conference Record, No. 5, pp. 46—48. Sept./Oct. 1959.

BOUTWELL, and CONN: The BIAX Magnetic Element, Datamation, vol. 5, No. 5, pp. 46—48. Sept./Oct. 1959.

LAMBERT: Nondestructive Readout of Metallic—Tape Computer Cores, Transactions IRE, vol. EC-8, No. 4, pp. 470—474. Dec. 1959.

KILBURN, and GRIMSDALE: A Digital Computer Store With Very Short Read Time, Proceedings IEEE, vol. 47, No. 11, pp. 567—572. Nov. 1960.

GRAY R. L.: An Electrically Alterable Nondestructive Twistor Memory, Transactions IRE, vol. EC-9, No. 4, pp. 451—455. Dec. 1960.

KUTTNER P.: The Rope Memory — A Permanent Storage Device, AFIPS Conference Proceedings, vol. 24, pp. 45—57. Fall Joint Computer Conference, 1963.

PICK, GRAY, and BRICK: The Solenoid Array — A New Computer Element, Transactions IEEE, vol. EC-13, No. 1, pp. 27—35. Feb. 1964.

WIESNER E. P.: Memory has Nondestructive Readout of Standard Ferrite Cores, Electronic Design News, pp. 44—45. March 1964.

BUTCHER I. R.: A Prewired Storage Unit, Transactions IEEE, vol. EC-13, No. 2, pp. 106—111. Apr. 1964.

WIESNER E. P.: Read-Only Memory, Electronic Design News, pp. 36—38. Apr. 1964.

CLEMSON, and KUTTNER: Applications of Rope Memory Devices, Computer Design, pp. 12—22. Aug. 1964.

BAKER W. A.: Memory Elements — A New Twist, Data Systems Design, pp. 9—15. Jan. 1965.

Study and Investigation of Techniques for Constructing Medium-Speed Random-Access Memory, Rome Air Development Center, Technical Report, RADC-TR-64-538. March 1965.

LEWIN M.: A Survey of Read-Only Memories, AFIPS Conference Proceedings, vol. 27, part 1, pp. 775—787. Fall Joint Computer Conference, 1965.

MAEDA, TAKASHIMA, and KOLK: A High-Speed, Woven Read-Only Memory, AFIPS Conference Proceedings, vol. 27, part 1, pp. 775—787. Fall Joint Computer Conference, 1965.

MAY, POWELL, and ARMSTRONG: A Thin Magnetic Film Computer Memory Using a Resonant Absorption Nondestructive Readout Technique, AFIPS Conference Proceedings, vol. 27, part 1, pp. 801—808. Fall Joint Computer Conference, 1965.

SIDMU P. S.: Development of an E-Core Read-Only Memory, AFIPS Conference Proceedings, vol. 27, part 1, pp. 809—818. Fall Joint Computer Conference, 1965.

Miscellaneous Magnetic Devices and Techniques

RHODES, RUSSEL, SAKALAY, and WHALEN: A 0.7-Microsecond Ferrite Core Memory, IBM Journal of Research and Development, vol. 5, No. 3, pp. 174—182. July 1961.

SMITH D. O.: Proposal for Magnetic Domain-Wall Storage and Logic, Transactions IRE, vol. EC-10, No. 4, pp. 708—711. Dec. 1961.

LEACRAFT, and MELAN: Characteristics of a High-Speed Multipath Core for a Coincident-Current Memory, Transactions IRE, vol. EC-11, No. 3, pp. 405—409. June 1962.

FREEMAN J. D.: New Idea in Counting — Incrementally Magnetized Cores, Electronics, pp. 40—43. June 15, 1962.

SHAWBENDER, WENTWORTH, LI, HOTCHKISS, and RAJCHMAN: Laminated
Ferrite Memory, AFIPS Conference Proceedings, vol. 24, pp. 77—90. Fall
Joint Computer Conference, 1963.

General References on Magnetic Devices

RAJCHMAN J. A.: Magnetics for Computers — A Survey of the State-of-the-
Art, RCA Review, vol. 20, pp. 92—95. March 1959.
LOONEY D. H.: Recent Advances in Magnetic Devices for Computers, Journal
of Applied Physics, vol. 30, No. 4, pp. 38S—42S. Apr. 1959.
MEYERHOFF A. J. ed.: Digital Applications of Magnetic Devices. New York:
John Wiley and Sons. 1960.
RAJCHMAN J. A.: Computer Memories — A Survey of the State-of-the-Art,
Proceedings IRE, vol. 49, No. 1, pp. 104—127. Jan. 1961.
HAYNES J. L.: Logic Circuits Using Square-Loop Magnetic Devices — A Survey,
Transactions IRE, vol. EC-10, No. 2, pp. 191—203. June 1961.
BENNION, CRANE, and ENGELBART: A Bibliographical Sketch of All-Magnetic
Logic Schemes, Transactions IRE, vol. EC-10, No. 2, pp. 203—206. June
1961.
LOONEY D. H.: Magnetic Devices for Digital Computers, Datamation, vol. 7,
No. 8, pp. 51—55. Aug. 1961.
YOVITS M. ed.: Large Capacity Memory Techniques for Computing Systems,
New York: Mcmillan. 1962.
RAJCHMAN J. A.: Memories in Present and Future Generations of Computers.
IEEE Spectrum, pp. 90—95. Nov. 1965.

Superconducting Elements

BUCK D. A.: The Cryotron — A Superconductive Computer Component, Pro-
ceedings IRE, vol. 44, No. 4, pp. 482—493. Apr. 1956.
CROWE J. W.: Trapped-Flux Superconducting Memory, IBM Journal of Re-
search and Development, vol. 1, No. 4, pp. 294—303. Oct. 1957.
CRITTENDEN, COOPER, and SCHMIDLIN: The Persistor — A Superconducting
Memory Element, Proceedings IRE, vol. 48, No. 7, pp. 1233—1246. July
1960.
NEWHOUSE, BREMER, and EDWARDS: An Improved Film Cryotron and Its
Applications to Digital Computers, Proceedings IRE, vol. 48, No. 8,
pp. 1395—1404. Aug. 1960.
SLADE A. E.: Cryotron Characteristics and Circuit Applications, Proceedings
IRE, vol. 48, No. 9, pp. 1569—1576. Sept. 1960.
COHEN M. L.: Thin Film Cryotrons, Proceedings IRE, vol. 48, No. 9, pp. 1576—
1582. Sept. 1960.
EDWARDS, NEWHOUSE, and BREMER: Analysis of a Crossed Film Cryotron Shift
Register Transactions IRE, vol. EC-10, No. 2, pp. 285—287. June 1961.
BURNS, ALPHONSE, and LECK: Coincident-Current Superconductive Memory,
Transactions IRE, vol. EC-10, No. 3, pp. 438—446. Sept. 1961.
STEWART, OWEN, LUCAS and VAIL: Persistent Current Memory Circuit, Pro-
ceedings IRE, vol. 49, No. 11, pp. 1681—1682. Nov. 1961.
SLADE A. E.: A Cryotron Memory Cell, Proceedings IRE, vol. 50, No. 1,
pp. 81—82. Jan. 1962.
BRENNEMAN A. E.: The In-Line Cryotron, Proceedings IEEE, vol. 51, No. 3,
pp. 442—449. March 1963.

HARMAN M. G.: A New Form of Cryotron Logic Circuitry, Transactions IEEE, vol. EC-12, No. 5, pp. 568—570. Oct. 1963.

FRUIN, and NEWHOUSE: A New Crossed-Film Cryotron Structure with Superimposed Controls, Proceedings IEEE, vol. 51, No. 12, pp. 1732—1736. Dec. 1963.

AHRONS, and BURNS: Superconductive Memories, Computer Design, pp. 12—19. Jan. 1964.

PORTER S. N.: A New Configuration for Faster Cryotron Circuits, Transactions IEEE, vol. EC-13, No. 1, pp. 56—57. Feb. 1964.

CASWELL H. L.: Thin-Film Superconducting Devices, IEEE Spectrum, pp. 84—99. May 1964.

BURNS L. L. et al.: Cryoelectric Random Access Memory, Rome Air Development Center, Technical Documentary Report, RADC-TDR-64-376. Nov. 1964.

Tunnel Diode

GOTO E., et al.: Esaki Diode High-Speed Logical Circuits, Transactions IRE, vol. EC-9, pp. 25—29. March 1960.

CHOW W. F.: Tunnel Diode Digital Circuits, Transactions IRE, vol. EC-9, pp. 295—301. Sept. 1960.

BERGMAN R. H.: Tunnel Diode Logic Circuits, Transactions IRE, vol. EC-9, pp. 430—438. Dec. 1960.

SIMS, BECK, and KAMM: A Survey of Tunnel-Diode Digital Techniques, Proceedings IRE, vol. 49, No. 1, pp. 136—146. Jan. 1961.

YOURKE, BUTLER, and STROHM: Esaki Diode NOT-OR Logic Circuits, Transactions IRE, vol. EC-10, No. 2, pp. 183—190. June 1961.

One-Tunnel-Diode Flip-Flop, Proceedings IRE, Correspondence in March, June, Sept., Nov. 1961, and Feb. 1962.

AXELROD, FARBER, and ROSENHEIM: Some New High-Speed Tunnel-Diode Logic Circuits, IBM Journal of Research and Development, vol. 6, No. 2, pp. 158—169. Apr. 1962.

RENTON, and RABINOVICI: Tunnel Diode Full Binary Adder, Transactions IRE, vol. EC-11, No. 2, pp. 213—217. Apr. 1962.

RABINOVICI B.: Tunnel Diode Shift Register, Proceedings IRE, vol. 50, No. 4, p. 473, Apr. 1962.

CARR, and MILNES: Bias Controlled Tunnel-Pair Logic Circuits, Transactions IRE, vol. EC-11, No. 6, pp. 773—779. Dec. 1962.

CRAWFORD, PRICER, and ZASIO: An Improved Tunnel Diode Memory System, IBM Journal of Research and Development, pp. 199—206. July 1963.

Series Coupled Tunnel Diode Memory, Computer Design, pp. 8—11. Nov. 1963.

COOPERMAN M.: 300 Mc Tunnel Diode Logic Circuits, Transactions IEEE, vol. EC-13, No. 1, pp. 18—26. Feb. 1964.

SEAR B. E.: Design of Modular 250 Mc Circuits, Computer Design, pp. 20—25. Apr. 1964.

CRAWFORD, MOORE, PARISI, PICCIANO, and PRICER: Design Considerations for a 25-Nanosecond Tunnel Diode Memory, AFIPS Conference Proceedings, vol. 27, part 1, pp. 627—636. Fall Joint Computer Conference, 1965.

Miscellaneous Devices

ANDERSON J. R.: Ferroelectric Elements for Digital Computers and Switching Systems, Elec. Engineering, vol. 71, pp. 916—922. Oct. 1952.

MOLL, TANNENBAUM, GOLDEY, and HOLONYAK: PNPN Transistor Switches, Proceedings IRE, vol. 44, No. 9, pp. 1174—1182. Sept. 1956.

ANDERSON J. R.: A New Type of Ferroelectric Shift Register, Transactions IRE, vol. EC-5, No. 4, pp. 184—191. Dec. 1956.

BRAY T. E.: An Electro-Optical Shift Register, Transactions IRE, vol. EC-8, No. 2, pp. 113—117. June 1959.

WORTHER, and REDIKER: The Cryosar — A New Low-Temperature Computer Element, Proceedings IRE, vol. 47, No. 7, pp. 1207—1213. July 1959.

Solid State Products Inc., A Survey of Some Basic Trigistor Circuits, Bulletin D 410—02. July 1959.

MELNGAILIS I.: The Cryosistor — A New Low-Temperature Three-Terminal Switch, Proceedings IRE, vol. 49, No. 1, pp. 352—354. Jan. 1961.

IZUMI H.: The Silicon Cryosar, Proceedings IRE, vol. 49, No. 8, pp. 1313—1314. Aug. 1961.

MELNGAILIS, and MILNES: The Cryosistor, Proceedings IRE, vol. 49, No. 11, pp. 1616—1622. Nov. 1961.

PETTUS, and YOUNG: Magnetoresistive Effect as a Possible Memory Device, Proceedings IRE, vol. 49, No. 12, pp. 1943—1944. Dec. 1961.

JOHNSTON R. C.: Cryosar Memory Design, Transactions IRE, vol. EC-10, No. 4, pp. 712—717. Dec. 1961.

MELNGAILIS, and REDIKER: The Madistor — A Magnetically Controlled Plasma Device, Proceedings IRE, vol. 50, No. 12, pp. 2428—2435. Dec. 1962.

GERRITSEN H. J.: Operation of a Memory Element Based on the Maser Principle Proceedings IEEE, vol. 51, No. 6, pp. 934—935. June 1963.

TESZNER, and GICQUEL: Gridistor — A New Field-Effect Device, Proceedings IEEE, vol. 52, No. 12, pp. 1502—1513. Dec. 1964.

BIARD, BONIN, MATZEN, and MERRYMAN: Optoelectronics as Applied to Functional Electronic Blocks, Proceedings IEEE, vol. 52, No. 12, pp. 1529—1536. Dec. 1964.

BHATIA M. K.: Electro-Optics Generate and Position Characters, Electrical Design News. Jan. 1965.

REIMAN, and KOSONOCKY: Progress in Optical Computer Research, IEEE Spectrum, pp. 181—195. March 1965.

CHANG, DILLON, and GIANOLA: Magneto-Optical Variable Memory, etc., Journal of Applied Physics, vol. 36, pp. 1110—1111. March 1965.

HAWKINS, and NAVE: Large Signal and Transient Characteristics of Electrochemical Amplifiers, Proceedings IEEE, vol. 53, No. 11, pp. 1707—1713. Nov. 1965.

Index

Access 171, 285, 311, 319—326, 363, 412, 437
— by addressing 171, 285
— by content 291, 326, 412, 416
— minimum 324, 325
— random 171, 311, 321
— sequential 171, 311, 321, 323
Accumulation of partial products 201
Accumulative multiplication 212, 301
Accumulator 137—141, 158—177, 191, 201—215, 220, 223—225, 238—240, 256—259, 264—273, 277, 287, 376, 398, 405
Accumulator field 412
Active cell or module 392, 395, 416
Adaptive Systems 416—427
Add instruction 276, 282, 286, 300
Adder 134—158, 195—199, 213, 267—273
— address, index 261, 309, 310
— binary 134—142
— carry by-pass 195
— carry look-ahead 196
— carry save 199, 219
— completion recognition 197
— decimal 142—155
— full 135
— half 134, 136
— ripple 142, 195
Addition 6, 7, 20, 179—199, 264, 387
— accumulative 136, 270
— algebraic 185—195
— floating-point 255—262
— high-speed 195—199
— logic 27
— of residue numbers 20
— of sign bit 191
— of the complement 179—184
— parallel 138—149
— serial 136—138, 149—155
— two step 139, 140
Address 171, 285, 310
— base 289
— data as 291
— effective 289
— recorded 325, 326
— return 173, 409, 410

Address adder 309, 310
Address channel 325
Address counter 322, 324
Address matrix 382
Address memory 386
Address modification 173, 290, 309, 310
Address register
— channel 343
— control 381, 384—386
— memory 174, 311, 316, 320, 322, 324
— program 176, 276, 308, 409
— real-time 353
— stack 407
— storage 308
Address string 386
Addressing 171, 285
— abbreviated 289
— block 287
— by content 291, 326, 412, 416
— by counting 322
— by links 401
— by recorded addresses 325
— by tags 288
— immediate 286
— implicit 287
— implied 287
— indirect 288, 398
— relative 288
— truncated 289
Algebraic addition and subtraction 185—195
Algebraic operations 23, 38
Algorithm 161
— divide 224, 227, 233
— perform 307, 387
— square root 237, 241, 242
Alignment of coefficients 255—257, 261, 263
Alphanumeric code 18
Amplification, amplifier 83, 101, 107, 314, 460, 470, 472, 501, 508
Analog components 371, 372
Analog computer 1, 2, 370—372, 375
Analog representation 1, 373, 424, 426
Analog to digital converter 371, 374

AND, logic function 27, 61, 248
AND-circuit 46, 49, 55, 65, 461, 472, 495, 513
Arithmetic module 415
Arithmetic operations 6—9, 179—273,
Arithmetic Registers 177, 202, 224, 245, 264—273
Arithmetic unit 164, 174, 177, 179—273, 352, 357, 369, 393, 399, 405, 437
— cellular 399
— distributed 369
— multiple 393, 398
— with stack structure 405
Array
— cellular 387, 392, 399
— cobweb 390
— cutpoint 389
— majority 388
— three-dimensional 393, 394
Array processor 387
Associative computer 411—416
Associative memory 293—295, 326—331, 411, 416
Associative unit 417, 418

B-box, B-register 289
Bank 359—364
Base 7—14, 19—21, 184
Base-1 complement 184, 186, 191—195
Bi-phase recording 95
Bi-polar recording 93
Biased switch 494, 495⌐
Binary addition 7, 134—142, 185—199
Binary coded decimal numbers 14—16
Binary counter 115—123, 473
Binary number system 5, 7
Binary subtraction 155—158, 185—195
Binary variable 24, 80
Biological System 416
Biquinary code 15, 16
Bit 79
Bit counter 320
Bit drive, sense 312, 317, 476, 491, 492, 493, 511
Bit plane 316
Bit tube 475, 478
Bit wire 476, 477
Block access 287, 326
Block, variable length 345
Boolean Algebra 23—43, 129
Borrow 156, 245, 377
Bound register 310
Branching 168, 171, 381, 383, 404
Buffer 315, 332, 351, 405
Bus 347, 389, 512
Byte 252, 344, 358

Calculator 1
Capability 432—434, 452, 454
Carry 7, 123, 134—155, 206
— completion recognition 197, 277
— end 182, 191
— end around 194, 198
Carry by-pass 195, 199
Carry flip-flop 137, 151, 209, 220
Carry look-ahead 196
Carry propagation 138, 140, 142, 195—199
Carry save adder 199, 271
Cathode follower 71
Cell
— active 392, 395, 416
— cobweb 390
— Crowe 505
— cutpoint 389
— superconductive 502—507
— thin film 479—482
— two-rail 391
Cellular organizations 387—400
Channel
— address 325
— clock 322
— input/output 342, 356—364
— marker 322
Channel register 343—346
Characteristic 19, 252
Check
— hardware 442—448
— marginal 440
— parity 446, 447
— software 441, 442, 447
— status 336, 338
Checklist 455—457
Circuit
— AND 46, 49, 55, 65, 461, 472, 495, 513
— core 459—463, 494—498
— cryotron 499—504
— diode 46—56
— logic 45—76, 459—474, 494, 495, 499—501, 507—515
— NAND 61, 81
— NOR 61, 81, 501
— NOT 56, 70, 461, 471, 514
— OR 46, 55, 71, 461, 468, 471, 513
— parametron 469—474
— pulse 66, 67, 74, 75, 459—469, 494—498, 510
— transfluxor 463—468
— transistor 56—69
— tube 70—75
— tunnel diode 507—514
— with inverted levels 55, 69, 75, 76
Circular shift 113, 244, 269
Classification of patterns 417, 425—427
Clear/write cycle 315, 318, 462, 467, 511

Clock 94, 101, 107, 109, 152, 279, 322, 323
Cobweb cell 390
Code
— alphanumeric 18
— binary coded decimal 14—19
— error detection and correction 19, 447, 450
— function 332
— Gray or inverted binary 16—18, 31, 126—129
— minor function 296
— operation 173, 274, 282—284, 295
— residue 19—21
— status 336, 339
Coded numbers 14—21
Coded response 423
Coefficient 19, 252—262
Coincidence detection 320, 323
Coincidence selection 315, 316, 466, 475, 478, 481, 485, 491, 494, 497
Comfort organization 393
Command 164, 274
Command matrix 382, 386
Communication 166, 331, 347—365
Comparison 245, 253, 258, 320, 444, 449
Compartmentation of input space 427
Complement 25, 179
— 1's 184
— 2's 184
— 9's 15, 180, 184
— B-1 184, 186, 193
— logic 25, 246
— true 180, 186, 190
Complementation of flip-flop 82, 473
Completion recognition adder 197
Compute-independent input/output 343, 352—358
Computer
— analog 1, 2, 370—372, 375
— associative 411—416
— digital 1, 2, 160—178
— digital differential analyzer 2, 369, 375—379
— externally programmed 167
— highly parallel 369, 392—400, 411
— hybrid 2, 370—375
— iterative circuit 387, 392
— list processing 400—411
— micro-programmed 379—387
— modular 359, 433
— plugboard programmed 169, 170
— polymorphic 367, 368
— restructurable 367
— serial versus parallel 114, 436
— special purpose 325, 432
— stored logic 381
— stored program 170—178, 393, 433
— variable instruction 381

Computer variable structure 367
Computer circuits 104—158
Computing plane 394
Concentrated units 370
Concurrent access 363, 412
Concurrent operations 392
Condition
— jump 169, 244, 290, 303—305, 381, 409
— marginal 440
— status 177, 336, 339, 344
— unexpected 336, 337
Conditional jump 169, 244, 290, 303—305, 381, 385, 404, 409
Conjunction 27
Conjunctive normal 34
Constraint 432
Content addressable memory 293—295, 326—331, 411, 416
Continuous sheet memory 506, 507
Control
— geometric, mode 397, 398, 400
— input/output 331—347
— internal 164, 174—177, 273—310
Control address register 381—386
Control memory 380—385
Control plane 394
Control unit 164, 174—177, 273—310
Controlled substitute, transfer 251
Conversion
— of fractions 12—14
— of Gray code 17
— of information 114, 115
— of integers 9—12
Converter 371, 374
Core
— magnetic 89—91, 459—463, 483—491, 494—499
— multi-apertured 463
— multipath 495—497
Core logic 460—462
Core memory 312—319, 483—488, 490, 491
Core plane 316
Cost 431—441, 452—454
Counter 115—134
— address 322, 324
— assembly/disassembly 344
— binary 115—123
— bit 320
— decade 123—126
— flexible 129—134, 280, 281, 283
— flip-flop 115—123
— Gray code 126—129
— mileage 179
— parametron 473
— recirculating 123
— up-down 117
— word 320, 343

Counting 5
Crowe cell 505
Cryoelectric 515
Cryogenic 498
Cryotron 499—504
Cutpoint cell 389, 390
Cycle time 319, 364

Darlington compound 58
Data as addresses 291
Data channel 356
Data synchronizer 355
De Morgan's theorem 34, 37
Dead space 326
Decimal adder 142—155
Decimal codes 14—19
Decimal counters 123—126
Decision circuits 45
Decoding matrix 53, 280, 282, 312, 315
Decrement 290, 376
Degradation 368
Delay 85, 96—98, 199, 275—278, 462, 472, 479
Delay line 99—101, 320, 321
Descriptor 326, 414
Design 431—457
Design, modular, 359, 433
Design parameter 430, 451—454
Design philosophy 1, 160—164
Destructive read 79, 90, 314, 315, 318, 476, 478, 481, 511
Detection of errors, faults 340, 441—448
Detector
— coincidence 321—323
— equality 245, 250, 321
— match 327, 330
Diagnostic program, routine 442
Digital differential analyzer 2, 369, 375—379
Digital to analog converter 371, 374
Diode 46
— Esaki 507
— steering 88
— tunnel 507—514
Diode circuits 46—56, 507—515
Diode matrix 53—55
Disjunction 28
Disjunctive normal 34
Disk memory 92, 321
Distributed hardware 369, 370
Disturbing read 483—485
Divide fault 222, 225, 226
Divide instruction 301
Dividend 221, 225, 226, 237
Division 220—235, 263, 264
— binary 220—228
— decimal 228—231
— floating point 263, 264
— high-speed 231—235

Division non-restoring 226, 229, 232
— of fractions 225, 226
— of integers 223—225
— restoring 226, 229
— with test feature 223, 228
Divisor 221
Dont care 41, 132, 154, 328
Double-length, double-precision 201, 225, 228, 262, 273, 300, 301
Down time 439
Drive line 312—317
Drum memory 92, 321
Dual-rank registers 111, 120, 141, 268—273
Dumping 173
Duodecimal numbers 5, 8
Duplication of hardware 444, 447
Dynamic flip-flop 101—102, 106, 121—123
Dynamic storage elements 96—102, 469—474

Easy direction 479—482
Eccles-Jordan circuit 80
Effective reliability 439
Effective threshold 424
Effectiveness 442
Element
— cryogenic 498
— logic 45—76, 459—474, 494, 495, 499—501, 507—515
— magnetic 89—96, 459—498
— majority 388, 472, 514
— storage 78—102, 474—511
— superconductive 498—507
— universal 79, 460, 472, 501
Emitter follower 68
End carry 182, 191
End of list symbol 402
End of record 336
Engineering 431—457
Equality detector 245, 250, 321
Erase head 109
Error detection and correction 19, 340, 441—451
Esaki diode 507
Evaluation of computers 451—455
Exchange 359
Exchange register 308, 347, 350
Exclusive OR 28, 250, 468
Execute instruction 307, 409
Execution 275, 276, 333, 379, 383
Executive processor 363
Expansion of system 359, 362, 433
Exponent 19, 252, 258, 259, 261, 310
External function, operation 177, 305, 332, 333, 335, 343
Externally programmed computer 167
Extract operation, 246, 251

Failure insensitivity 421
Fault 339, 340, 441—448
Ferrite core 89—91, 312—319, 459—
    463, 483—491, 494, 499
Ferrite peg memory 493
Ferrite sheet memory 474 ·
Ferroelectric 515
Figure of merit 451—453
Film, magnetic 474—482, 488, 489
Fixed storage 489—494
Flag 295
Flip-flop 80—89, 101, 102, 473, 501
Flip-flop counter 116—123
Flip-flop register 104—107
Floating-point arithmetic operations
    252—262
Floating-point representation 18, 19,
    252, 253
Flux 90, 314, 459, 463, 475—478
Flux rotation 480—482, 486
Fraction 12—14, 211, 225, 226
Function code 332
Function translation 175, 273, 282 − 285

Gate
— inhibit 74, 75
— pentode 73
— pulse 66, 67, 75
— transfluxor 465
— transistor 62, 63
— triode 72
General theorem 34, 36
Geometric control 397
Gonzalez organization 394
Goto pair 510
Gray code 16—18, 31, 126—129
Growth capability 359, 362, 433

Half-select 316
Hardware checks 442—448
Hardware duplication, triplication 444,
    447, 449
Helical flux 475—478
Heuristic programming 416
Hexadecimal numbers 8
Hierarchy of storages 437, 456
High-speed techniques 195—199,
    214—220, 231—235, 437
Highly-parallel machines 369, 392—
    400, 411
Holland organization 392, 393, 416
Hybrid computer 2, 370—375
Hypercube 33, 426
Hyperplane 426
Hysteresis 90, 463, 479, 498

Illegal codes 443
Increment 290, 376
Index 289, 305, 309, 310

Indicator 327, 421, 456
Information 79, 105, 108, 110, 114,
    173, 174, 243, 332, 341, 359, 361
— as macro-program 385
Information bus 347
Information register 174, 311, 313
Inhibit gate, 74, 75
Inhibit winding 316, 461, 468
Initialization of program 294
Input pattern 417, 427
Input plane 399
Input space 426, 427
Input/output 349—357
Input/output channel 342—347, 356—
    360
Input/output exchange 359
Input/output instructions 305, 332,
    344, 345
Input/output processor 356
Input/output unit 165, 174, 177, 331—
    347
Insensitivity to failure 421
Instruction 164, 172—176, 379, 392
Instruction format 286, 295—298, 380,
    384, 403, 408, 412
Instruction mix 452
Instruction register 175, 282, 286,
    308, 381, 408, 409
Instruction repertoire 276, 282, 299—
    308, 380, 393
Instruction split 284
Integer 9—12, 14, 211, 223
Integrand 375—377
Integrating core 498
Integrating one-shot 98
Integrator 372, 375—378
Intelligence 163
Interlace 321, 324
Interrupt 277, 309, 337, 339, 344, 346,
    350
Intersection 27
Interstage 462, 463
Inverted binary code 16—18, 31,
    126—129
Inverter 56, 70
Item 401, 402
Iterative circuit computer 387, 392

Join 28
Joint set 28
Jump 168, 172, 176, 303
— conditional 169, 244, 290, 303—305,
    381, 385, 404, 409
— return 173, 409

Karnaugh map 25, 31

Lattice 467
Learning 163, 417—427

Linear selection 312, 476, 481, 482, 490, 511
Link 401, 407
Linking of I/O operations 346
List 401, 404
List processor 400—411
Lockout 350, 352
Logic
— cellular 388—391
— core 459—463
— majority 471, 472
— mathematical 23—42
— NOR 61
— one core per bit 463
— symbolic 23—43
— transistor 63—66
— transistor-diode 59—62
Logic AND 27, 61, 246, 248, 461, 472, 495, 513
Logic circuits 45—76, 459—474, 494, 495, 499—501, 507—515
Logic comparison 245
Logic complementation 25, 246
Logic conjunction 27
Logic diagrams 31
Logic disjunction 28
Logic elements 45—76, 459—474, 494, 495, 499—501, 507—515
Logic function 24—37
Logic instructions 301
Logic intersection 27
Logic join 28
Logic minimization 37—43
Logic NAND 61
Logic NOR 61, 501
Logic NOT 25, 56, 70, 246, 461, 471, 514
Logic operations 243—252
Logic OR 27, 46, 55, 71, 246, 461, 468, 471, 513
Logic product 27, 248
Logic proposition 23
Logic set 30
Logic sum 27, 247
Logic symbols 45
Logic truth-table 32
Logic union 28
Logic universe 25
Logic variable 24
Look-ahead adder 196
Look-ahead register 310
Lookup, table 291

Macro-instruction 379, 381, 383
Macro-program 385
Magnetic components 89—96, 459—498
Magnetic core 89—91, 312—319
Magnetic film devices 474—482, 488, 489

Magnetic recording 92—96
Magnetostrictive delay line 100
Magnitude 186, 244
Maintenance 440, 442, 457
Majority array 388
Majority element 388, 472, 514
Manchester mode 95
Mantissa 19, 252
Marginal checks, conditions 440, 457
Marker 113, 322
Mask 246, 256, 339
Mass storage 92, 321
Master clear 133
Match 326, 327, 330
Mathematical logic 23—42
Matrix
— address 382, 383
— command 382, 386
— decoding 53, 280, 282, 312, 315
— diode 53—55
— of drive lines 312, 315, 466, 478, 485, 495, 507, 511
— shift 216, 271, 273
Memory 171, 174, 310—331
— address 386
— allocation of 171, 403
— associative 293—295, 326—331, 411, 416
— buffer 315, 405
— content-addressed 293—295, 326—331, 411, 416
— continuous sheet 506, 507
— control 380—385
— core 312—319, 483—488, 490, 491
— delay line 320, 321
— disk, drum 92, 321
— ferrite peg, slug 493
— ferrite sheet 474
— magnetic film 474—482
— read only 382, 384, 490—494
— rod 492, 493
— rope 490, 491
— thin film 479—482, 489
— transfluxor 466
— transformer 492
— woven screen 478, 479
Memory bank 359—364
Memory conflict 346, 352
Memory cycle 319, 364
Memory exchange 359
Memory plane 316
Memory search 292, 293
Memory types 311, 312
Micro-program 299, 307, 408
Micro-programmed computer 379—387
Microwave 370, 474
Mileage counter 179
Minimization 37—43, 455

Minimum access 324, 325
Minimum product 40
Minimum sum 40
Minimum system 357
Minor function 296
Mode control 397
Mode, recording 93—95
Model layout 174—178
Modification of program 173, 393
Modular design, system 359, 433
Module 61, 359
— active 392, 395, 416
Modulo, modulus 17, 130, 180
Monitoring of external operations
  335—341
Multi-level device 497, 498
Multiapertured core 463
Multipath core 495, 496
Multiple arithmetic units 375, 393, 398
Multiple generator 213, 229
Multiple match 330
Multiple selector 272
Multiplexer, multiplexing 331, 359,
  360, 364
Multiplicand 200
Multiplication 6, 200—220, 263, 264
— accumulative 212, 301
— by addition and subtraction 215, 216
— by uniform shifts 217, 219
— high-speed 214—220
— of floating-point numbers 263, 264
— of fractions 211, 212
— of integers 211, 212
— with residue numbers 20
Multiplier 200
Multiply instruction 300
Multiprocessor 363
Multivibrator 80, 96

NAND 61, 86
Neuron 417, 423
No-address instruction 287
No-carry signal 197
No-information fault 341
Noise 372, 421, 479
Non-destructive read 466, 483—494
Non-restoring algorithm 229—231,
  241—243
Non-return to zero 94
Non-volatile storage 79, 323
NOR 61, 81, 501
Normalization of floating-point re-
  sults 257, 263
Normalize instruction 307
NOT 25, 56, 70, 461, 471, 514
Number conversions 8—14
Number systems 5—14, 19—21
Numbers
— coded 14—19

Numbers floating-point 18, 19, 252, 253
— residue 19—21

Octal numbers 7, 9, 12
On-line communication 337, 354, 364,
  365
One core per bit logic 463
One-shot 96—98
Op-code 175, 273, 274, 282, 286,
  295—297, 381, 386, 403
Operand 165, 167, 171, 172, 285, 288,
  310, 403, 412
OR 27, 46, 55, 71, 246
Organization 160—178, 427, 436
— associative computer 411—417
— cellular 387—400
— Comfort 393
— Gonzales 394, 395
— Holland 392, 393
— list processor 400—411
— micro-programmed computer 379—
  387
— model 175
— multiprocessor 363
— planar 399
— polymorphic 367—369
— Solomon 396, 397
— Vamp 398
Output plane 399
Overflow 137, 138, 148, 182, 191, 193,
  261, 376, 443

Parallel addition 138—142, 195—199
Parallel information 114
Parameter 431, 451—454
Parametron 469—474
Parity 94, 446, 447
Patch 39
Pathbuilding 392, 393, 395
Pattern 417—423, 427, 464—467
Pattern articulation unit 399
Pentode 73
Perceptron 417—427
Perfect induction 30, 52
Perform algorithm instruction 307, 387
Peripheral devices, equipment 174,
  331—337, 344, 345, 357—363
Peripheral processor 363
Peripheral storage 321
Persistent current 502, 505, 507
Persistor 504
Phase-locked oscillator 474
Polish notation 406
Polymorphic computer 367—369
Priority interrupt 339
Processor 243, 355—363, see also
  "computer" and "organization"
Product 200
— logic 27, 248

Product minimum 40
— standard 36, 37
Program 164, 167—176, 276, 289, 293, 294, 333—337, 340, 356, 393, 408, 410
— heuristic 416
— macro 385
— micro 299, 307, 379, 383, 384, 403, 408
— recursive 411
— test 442
— with list structure 403, 404
Program address 176, 177, 273, 276, 286, 308, 409
Program branch 168, 171, 381, 383, 404
Program-independent input/output 343, 352—358
Program interrupt 277, 337—339, 344, 346, 350
Program memory 396
Program plane 394
Program recovery 335, 337
Program storage 164, 170, 171, 403
Propagation of carries 138, 140, 142, 195—199
Propagation of signals 370
Proposition 23
Pulse 66, 67, 74, 75, 83, 97, 459—469, 494—498, 510
Pump 474

Quadrature field 486, 487
Quinary numbers 7
Quotient 221

Radicand 238
Random access 171, 311, 321
Random bit 372
Random connection 417, 423
Read 79, 89, 90, 310, 314, 317, 320, 462
— by quadrature fields 486, 487
— by resonant absorption 489
— by RF frequencies 485
— destructive 79, 90, 314, 315, 318, 476, 478, 481, 511
— disturbing 483—485
— non-destructive 466, 483—494
Read fault 340
Read/restore 315, 318, 511
Real-time 337, 353
Recirculating register 107—109, 112—114, 123
Recording, magnetic 92—96
Recovery 98, 335—337
Recursive program 411
Redundancy 442, 443, 447
Register 104—105
— address monitor 310

Register arithmetic 177, 202, 223, 224, 245, 264—273
— available space 408
— bound 310
— buffer 332
— channel 343, 345
— control address 381—386
— destination 387
— dual-rank 111, 120, 141, 268—273
— exchange 308, 347, 350
— in control unit 175, 308—310
— in memory 174, 175, 311—315
— index 289, 305, 309, 310
— input/output 175, 177, 332, 342, 343, 345, 350
— instruction 175, 282, 286, 308, 381, 408, 409
— integrand 376, 377
— interrupt 310, 339
— look-ahead 310
— macro-instruction 380, 381
— memory address 174, 311
— memory information 174, 311, 313
— micro-instruction 380, 381
— program address 175, 176, 273, 282, 286, 308, 409
— quotient 223
— real-time address 353
— stack 407, 408
— status 310, 339
— storage address 308
Reliability 162, 431, 438—441, 450, 452, 459
Repeat instruction 290
Residue number system 19—21
Response 419, 423
Response unit 417, 419
Restore 315, 318
Restoring of operands 226, 229, 241, 245, 250, 259
Restructurable computer 367
Resume signal 277, 280
Return address 173, 409, 410
Return jump 173, 409
Return to zero 94
Rod memory 492, 493
Rope memory 490, 491
Rounding 203, 256, 257
Rotation of flux 480—482, 486

Satellite computer 361
Scalefactor 221, 222, 244
Search instruction 293, 306, 307
Sector 322
Selection 310, 311, 331
— coincident 315, 316, 466, 475, 478, 481, 485, 491, 494, 497
— linear 312, 476, 481, 482, 490, 511
Selector channel 360, 364

Self-organizing system 416, 427
Sense line 312, 315, 466, 474—478, 481, 484, 485, 487, 491, 493, 496, 507, 511
Sensory unit 417
Sequencer 167, 176, 274—282, 468
Sequential access 171, 311, 321, 323
Serial addition 136—138, 149—155
Serial information 114
Set 23, 25
Shift instruction 303
Shift matrix 216, 271, 273
Shift register 109—115, 462, 467, 470, 471, 511—513
Shifting 109—115, 243, 244, 267, 386
Sign bit, position 185, 191, 194
Sign extended shift 244
Software 433
Software checks 433, 441, 442, 447
Solomon computer 396—398
Space, available 408
Space, input 426, 427
Special purpose computer 325, 432, 456
Speed 162, 369, 431, 434, 452, 453
Split phase recording 95
Sprocket 94, 335
Square root 235—243, 441
Squareness ratio 91
Stack 404—411
Standard product 36, 37 39
Standard sum 35, 36, 39
State-of-the-Art 434
Status 177, 306, 336, 338, 344
Stealing of cycles 346, 347
Steering diodes 88
Steering signals 388
Storage — see memory
Storage elements 78—102, 474—511
Storage hierarchy 437, 456
Storage location 174, 285, 311, 320
Stored logic computer 381
Stored program computer 170—178
Subset 30
Substitute, controlled 251
Subtracter 155—158, 185
Subtraction 155—158, 264
— algebraic 185—195
— by the addition of the complement 179—184
— floating-point 255—262
Sum
— logic 27, 247
— minimum 40
— standard 35, 36, 39
— weighted 419
Superconductive elements 498—507
Switching center 367
Switching circuits 45—76, 459—474, 494, 495, 499—501, 507—515

Symbol Logic 23—43
Synchronization 107, 279, 280, 457
System
— adaptive 416—427
— biological 416
— modular 359, 433
— polymorphic 367, 368
— self-organizing 416
System executive 363
System expansion 359, 362, 433, 456

Table lookup 291
Table search 293, 306, 307
Tag 288, 295
Tensor 476, 477
Ternary numbers 7
Test 440—448
Test instruction — see conditional jump
Test program 442
Thin film 479—482, 488, 489
Threshold unit 417, 418, 425, 427
Trade-off 431, 451, 454
Training 419, 420
Transfer
— controlled 251
— highly parallel 415
— parallel 150
— serial 108
Transfer function 425
Transfer instruction 415
Transfer rate 334
Transfluxor 463, 469
Transformer storage 492
Transistor 56
Transistor circuits 56—69
Transistor gate 62, 63
Translation 330, 418
Translator, function 175, 273, 283—285
Transmission line 66, 99, 370
Transmission of information 105, 108, 249, 251, 348, 386, 387
Transmit instruction 300—303
Triode 71
Triode gate 72
Triplication of hardware 449
Truncation 256
Truth table 32
Tube circuits 70—75
Tunnel diode 507—514
Twistor 475
Two-core-per-bit memory 497
Two-rail cell 391

Underflow 261
Union 28
Universe 25
Unorthodox concepts 367—427
Unusual components 459—515

Up-down counter 117
Up time 439

Vacuum tube circuits 70—75
Variable, binary, logic 24
Variable instruction computer 379—387
Variable structure computer 381
Veitch diagram 25
Venn diagram 25, 31
Vertex frame 33, 425
Volatile storage 79, 321

Weight 419, 424
Weighted code 15
Wilke's scheme 382
Wiring of core planes 318
Word counter 320, 343
Word drive 312, 476, 482, 490, 492, 493, 511
Woven screen memory 478, 479
Woven thin film memory 482
Write fault 340
Write operation 310, 313, 316, 466, 475, 478, 481, 482, 496, 506, 511